Turbulent Shear Flows 2

Selected Papers from the Second International Symposium on Turbulent Shear Flows, Imperial College London, July 2–4, 1979

Editors:

L. J. S. Bradbury F. Durst B. E. Launder
F.W. Schmidt J. H. Whitelaw

With 310 Figures

Springer-Verlag Berlin Heidelberg New York 1980

Leslie J. S. Bradbury

Department of Mechanical Engineering, University of Surrey,
Guildford, Surrey GU2 5XH, England

Franz Durst

Sonderforschungsbereich 80 der Universität Karlsruhe, Kaiserstrasse 12,
D-7500 Karlsruhe 1, Fed. Rep. of Germany

Brian E. Launder

Department of Mechanical Engineering, University of Manchester, Institute of Science and Technology, PO Box 88, Manchester M60 1QD, England

Frank W. Schmidt

Mechanical Engineering Department, The Pennsylvania State University,
University Park, PA 16802, USA

James H. Whitelaw

Department of Mechanical Engineering, Imperial College of Science and Technology,
Exhibition Road, London SW7 2BX, England

ISBN 3-540-10067-9 Springer-Verlag Berlin Heidelberg New York
ISBN 0-387-10067-9 Springer-Verlag New York Heidelberg Berlin

Library of Congress Cataloging in Publication Data. International Symposium on Turbulent Shear Flows, 2d, London, 1979. Turbulent shear flows 2. Bibliography: p. Includes index. 1. Turbulence–Congresses. 2. Shear flow–Congresses. I. Bradbury, Leslie John Stanley, 1936–. II. Title. TA357.I59 1979 532′.0527 80-14198

Composed by Elsner & Behrens GmbH, Oftersheim
Offset printing and bookbinding: Konrad Triltsch, Graphischer Betrieb, Würzburg
2153/3130-543210

Preface

The articles appearing in this volume have been selected from contributions to the 2nd Turbulent Shear Flow Symposium held at the Imperial College London, July 2–4, 1979. The present collection, comprising a little over one quarter of the papers presented at the Symposium, reflects some of the best current research in five of the areas addressed by the meeting: mathematical modelling of turbulence; two-dimensional thin shear flows near walls; coherent structures; environmental flows; and complex flows involving recirculation and/or three-dimensional straining. A chapter of the book is devoted to each of these areas. Each chapter begins with an introductory article written especially for this volume which provides both a guide to the technical papers and an impression of some of the other important reseach in the area not represented in the papers themselves.

Our aim has been to provide a book which captures the flavour of the Symposium while remaining of lasting scientific interest. In this connexion, a large proportion of the contributors have accepted the editors' invitation to rewrite and expand the version of their papers preprinted for the Symposium. The perspective added by six months' reflection (the conference papers being submitted in April 1979) coupled with a lively discussion at the Symposium has in several cases led to the introduction of major new material. In one case, authors felt that their papers so complemented each other that the results of two conference pre-prints appear in this volume as a completely rewritten joint paper.

Without the Symposium, there could have been no volume of proceedings and it is, therefore, particularly appropriate to acknowledge the financial support kindly provided by the National Science Foundation, the US Air Force European Office of Aerospace Research and Development, the US Army Research Office, London, and the US Office of Naval Research, London. As the host institution, the Imperial College provided extensive assistance while the Pennsylvania State University (which had hosted the 1st Symposium in 1977) provided valuable liaison on many aspects of organisation. We record also our thanks to the Fluids Engineering and Heat Transfer Divisions of the American Society of Mechanical Engineers for their assistance in publicity.

The technical content of the Symposium was shaped in large measure by a scientific advisory committee who undertook the task of reviewing nearly two hundred 1,000-word abstracts of papers offered for presentation and who later acted as chairmen of the technical sessions. The committee consisted of:

J.C. André, EERM/GMD, 92100 Boulogne, France
L.H. Back, Jet Propulsion Laboratory, Pasadena, Calif., USA
H.A. Becker, Queen's University, Kingston, Ont., Canada
P. Bradshaw, Imperial College, London, U.K.
S. Corrsin, The Johns Hopkins University, Baltimore, Md., USA
J.J. Domingos, University of Lisbon, Lisbon, Portugal
C. du P. Donaldson, ARAP, Princeton, N.J., USA
H. Fiedler, Technische Universität Berlin, Berlin, Fed. Rep. of Germany
I. Gartshore, University of British Columbia, Vancouver, B.C., Canada
V.W. Goldschmidt, Purdue University, W. Lafayette, Ind., USA
A.D. Gosman, Imperial College, London, U.K.
R. Günther, Universität Karlsruhe, Karlsruhe, Fed. Rep. of Germany
T.J. Hanratty, University of Illinois, Urbana-Champaign, Ill., USA
R.J. Herring, NCAR, Boulder, Colo, USA
W.P. Jones, Imperial College, London, U.K.
P.N. Joubert, University of Melbourne, Melbourne, Australia
P.A. Libby, University of California, La Jolla, Calif., USA
J.L. Lumley, Cornell University, Ithaca, N.Y., USA
O. Martynenko, Heat and Mass Transfer Institute. Minsk, USSR
J. Mathieu, Ecole Centrale de Lyon, Ecully, France
H. McDonald, Scientific Research Associates Inc., Glastonbury, Conn., USA
Y. Mori, Tokyo Institute of Technology, Tokyo, Japan
K. Owen, Owen International, Palo Alto, Calif., USA
W.C. Reynolds, Stanford University, Stanford, Calif., USA
W. Rodi, Universität Karlsruhe, Karlsruhe, Fed. Rep. of Germany
J.C. Rotta, DFVLR-AVA, Göttingen, Fed. Rep. of Germany
A.K. Runchal, Dames & Moore, Los Angeles, Calif., USA
I.Wygnanski, Tel-Aviv University, Tel-Aviv, Israel
J.C. Wyngaard, CIRES, Boulder, Colo., USA

Finally, it is a pleasure to record our thanks to the authors for meeting (or *nearly* meeting) the various deadlines we set and to Springer Verlag for their cooperation and flexibility in producing a book in harmony with our aims.

Karlsruhe, June 1980 *The Editors*

Contents

Part I Turbulence Models

Part II Wall Flows

Part III Complex Flows

Part IV Coherent Structures

Part V Environmental Flows

Part I

Turbulence Models

Introductory Remarks

B. E. Launder[1]

Sonderforschungsbereich 80 an der Universität Karlsruhe
D-7500 Karlsruhe 1, Fed. Rep. of Germany

The seven technical papers in the present chapter are all concerned, in one way or another, with providing a mathematical model for certain types of problem in turbulence. A few years ago one heard talk about devising a "universal" model of turbulence. Today, in a more sanguine vein, the proponents of a particular approach to turbulence see their methods as coping with a more limited range of problems and phenomena. Yet it is intrinsically at devising calculation procedures with *width* of applicability that the present papers are aimed. Here we remark that even the simplest of the closure levels dealt with in this chapter is more elaborate than any of the models used in part 3: *Complex Flows*. While there are several reasons for this a major one is that a serious numerical study of a two- or three-dimensional elliptic shear flow presses very hard upon one's computer core and/or computer budget allocation. At the practical level there is thus a very strong incentive to keep the closure as simple as possible.

The competing demands of generality and simplicity mean that any closure can be regarded as a compromise. The fact that the compromise is struck, by different workers, in such different terms, is testament to the sheer diversity of turbulent flow phenomena and to the difficulty of providing a general yet tractable mathematical description. In a nutshell the papers included in this first section fall into two groups. In the first are those concerned with providing closures for use in general numerical schemes; three deal with different aspects of second-moment closures and one with predicting the probability density function of scalars. The second group discusses treatments which, although not suitable for widespread use in predicting inhomogeneous turbulent flow, are nevertheless helpful in that they can provide direct tests of closure hypotheses used in simpler models or generate statistical data of a test flow more accurate and complete than could an experiment of the physical kind. The following paragraphs will consider in a little more detail the individual contributions and their connexion with related research.

For the relative newcomer to second moment closures the present group of three papers does not provide a broad enough view to serve as a satisfactory starting point. The chapter on *Developments in Reynolds Stress Closures* in **Turbulent Shear Flows – I** provides an easier point of entry with several papers having a direct bearing on the present contributions. Indeed the paper by *Rotta* can be regarded as a direct extension of one by the same author [1] in that volume. The flow of interest is the threedimensional boundary layer and the details of its skewing in approaching separation. In [1] *Rotta* has shown that the approximation of the mean-strain part[2] of the pressure-strain correlation is crucial to an adequate prediction. The present

---

[1] On leave from the University of California, Davis, CA 95616, USA.

[2] Sometimes called the "linear" part or the "rapid" part.

work documents the results of calculations based first on his T-model and subsequently on an extension of what is sometimes called the "quasi-isotropic" model. The calculated behaviour is by no means in full accord with the measurements – though the results are not less valuable for that. Indeed they underline the point made in the introductory article to Turbulent Shear Flows – I, namely that the quasi-isotropic model is not general enough to deal with complex strain fields.

While, with second moment closures, the approximation of pressure interactions is an ever present question mark, the representation of the dissipation rate of the stresses, via local isotropy, in terms of the dissipation rate of turbulent energy seemed a closed issue . . . until recently that is. However, the paper by *Wyngaard* and *Sundararajan* [2] in **Turbulent Shear Flows – I** reported departures from local isotropy in the skewness of the temperature fluctuations while here the paper by *Mjolsness* launches not exactly an onslaught on local isotropy but at least a persuasive case for keeping an open mind on the matter. His argument is that available experimental data that are *consistent* with local isotropy are insufficient to establish incontrovertibly that the fine scale motion *is* isotropic.

The paper by *Hanjalić* et al. addresses a separate question – that of the number of independent turbulent time scale that are needed. Other current closures use one; *Hanjalić* et al. suggest that, at least for certain flows, this is not enough. Accordingly they cut the energy containing part of the spectrum into two parts and develop distinct equations for the energy transfer rate from one region to another and for the energy dissipation rate (in single-scale closures the two quantities are assumed identical). As a postscript to that paper, we note that the equations for the partitioned energies [Eqs. (2) and (3) in the paper] may be obtained formally by decomposing the fluctuating velocity field u_i into contiguous but non-overlapping low- and high-wave-number parts, v_i and w_i. Multiplication of the x_i momentum equation by v_i and w_i and averaging produces the requisite equations which for a homogeneous incompressible flow at high Reynolds numbers may be written:

$$\frac{D\overline{v_i^2}/2}{Dt} = -\overline{v_i v_j}\frac{\partial U_i}{\partial x_j} \underbrace{- \overline{v_i \frac{\partial w_j w_i}{\partial x_j}} + \overline{w_i \frac{\partial v_i v_j}{\partial x_j}}}_{\epsilon_p}$$

$$\frac{D\overline{w_i^2}/2}{Dt} = -\overline{w_i w_j}\frac{\partial U_i}{\partial x_j} \overbrace{+ \frac{\overline{v_i \partial w_j w_i}}{\partial x_j} - \overline{w_i \frac{\partial v_i v_j}{\partial x_j}}} - \epsilon$$

where $\overline{v_i^2}/2 \equiv k_\mathrm{p}$ and $\overline{w_i^2}/2 \equiv k_\mathrm{T}$. If, as *Hanjalić* et al. assume, the spectral division is at sufficiently high wave number for $\overline{w_i w_j}$ to be isotropic the first term on the right of the second equation is zero. The advantage of the formal decomposition is that it identifies of what correlations the energy transfer rate, ϵ_p, is composed. Just as in the studies of *Hussain* and *Reynolds* [3] and *Liu* and his colleagues [4, 5] the energy transfer rate from one fluctuating field to another is seen to arise from triple correlations among components from both fields.

Because of the gross non-linearities of the reaction processes, calculations of turbulent flames have been based mainly on assumed forms of the probability density function (pdf) from information about their first and second moments. *Pope*'s work shows that the first *three* moments are needed to reproduce the variations of the pdf for the cases considered. The improvement over a two-moment description is striking. The paper also lends support to the argument that the pdf should adopt what *Pope* refers to as "its most likely value". In addition to the journals cited by *Pope,* the recent proceedings of the international symposia on combustion published by the Combustion Institute, provide several papers on the calculation of turbulent flames employing the pdf approach.

The paper by *Larcheveque* et al., which has been fused from two quite separate contributions presented at the Symposium, presents two different approaches to two-point closure: the test-field model and the eddy-damped Markovian procedure. Closure at the two-point level is seen by the authors as providing a means of studying turbulence decay processes and of refining the empirical coefficients in the second-moment (*one*-point) schemes discussed in preceding paragraphs. Unfortunately the earlier approximations of this type proved to be, in the authors' words "somewhat intractable computational tools". The two approaches presented in this paper represent different routes to simplification. Both approaches acquit themselves well for the decay of a scalar in isotropic turbulence. The study throws a good deal of computational light on why the decay exponent for the scalar field does not exhibit any universal connexion with that of the velocity field. The authors point out that from their computations the decay exponent is not really constant but that its variation is sufficiently slow that experiments covering only a decade or so of scalar variance could be interpreted as showing a power law decay.

Purists might feel that the final two papers are interlopers to this chapter for both resolve the main features of the turbulent motion by direct numerical simulation rather than by closure modelling. Their presence is essential, however, to underline the very rapid progress now being made in this field. The two studies address problems as different as one could wish for. That by *Leonard* considers the evolution of a turbulent spot arising from a local disturbance in a laminar boundary layer while that of *Riley* and *Metcalfe* studies the development of the axisymmetric wake far enough downstream for non-linearities in the mean convection to be unimportant; the spatial development can then be replaced by the numerically simpler time evolution. The *Leonard* paper employs an extension to 3 dimensions of the Lagrangian discrete vortex method of *Chorin* [6]. In **Turbulent Shear Flows – I** *Ashurst* [7] has presented a striking simulation of the evolution of the plane mixing layer with this method – a flow that is known to be largely dominated by nearly two-dimensional large scale structures.

The extension of the method by *Leonard* to three dimensions greatly enlarges the range of problems that may in principle be studied. Although the problem examined here employed up to 12,000 points in nearly 600 vortex filaments this proved insufficient to resolve the very finest scales of motion that are observed in experiment. A model is thus needed to prevent unphysical behaviour at high wavenumbers. The scheme adopted is to maintain a constant core size which, *Leonard* reports, is equivalent to using a subgrid viscosity dependent on the local rate of vortex stretching. The exploration of *Riley* and *Metcalfe* does not employ even a subgrid-scale model, the calculations being performed at sufficiently low Reynolds for the fine scales to be adequately resolved. One should note that their choice of flow is crucial in this regard for the turbulent Reynolds number of the asymptotic axisymmetric wake decays with time (or distance) while for most other free shear flows it grows. Despite the scatter in the statistical quantities and the short-cuts (valid in this study but not generally applicable) the impression conveyed is that by 1984, at any rate, such schemes may be finding application in directly resolving problems in industrially important flows. To a limited extent that situation has perhaps already arrived: *Grötzbach*'s [8] recent computations of buoyant channel flow have been published in a journal that at least professes to be concerned with engineering and design. A comprehensive review of this study and indeed of the whole subject of direct numerical simulation of turbulence is contained in the extensive survey by *Schumann* et al. [9] to be published shortly by Hemisphere.

Finally we note that the papers in this group are not the only ones in the volume concerned with the calculation of turbulent flows. The papers in the section on Complex Flows have already been mentioned; of these that by *Durst* and *Rastogi* is perhaps the most relevant to the concerns of the present group exploring as it does the effects of including streamline curvature corrections in the dissipation rate equation. All the papers in the Environmental Flows group consider fundamental closure questions, those by *Wyngaard* and *Lewellen* et al. aimed at second-(or third-) moment closure and that of *Chatwin* and *Sullivan* developing statistical results for the dispersal of a turbulent patch.

References

1. J. C. Rotta: "A Family of Turbulence Models for Three-Dimensional Boundary Layers", in *Turbulent Shear Flows I,* ed. by F. Durst, B. E. Launder, F. W. Schmidt, J. H. Whitelaw (Springer, Berlin, Heidelberg, New York 1979) p. 267
2. J. C. Wyngaard, A. Sundararajan: "The Temperature Skewness Budget in the Lower Atmosphere and Its Implications for Turbulence Modeling", in *Turbulent Shear Flows I,* ed. by F. Durst, B. E. Launder, F. W. Schmidt, J. H. Whitelaw (Springer, Berlin, Heidelberg, New York 1979) p. 319
3. W. C. Reynolds, A. K. M. F. Hussain: "The mechanics of an organized wave in turbulent shear flow: 3. Theoretical models and comparisons with experiment". J. Fluid Mech. *54,* 263 (1972)
4. J. T. C. Liu, L. Merkine: On the interactions between large-scale structure and fine-grained turbulence in a free shear flow: I – The development of temporal interactions in the mean. Proc. R. Soc. London Ser. A: **352**, 213 (1976)
5. A. Alper, J. T. C. Liu: On the interactions between large-scale structure and fine grained turbulence in a free shear flow: II – The development of spatial interactions in the mean. Proc. R. Soc. London Ser. A: **359**, 497 (1978)
6. A. J. Chorin: Numerical study of slightly viscous flow. J. Fluid Mech. **57**, 785 (1973)
7. W. T. Ashurst: "Numerical Simulation of Turbulent Mixing Layers via Vortex Dynamics", in *Turbulent Shear Flows I,* ed. by F. Durst, B. E. Launder, F. W. Schmidt, J. H. Whitelaw (Springer, Berlin, Heidelberg, New York 1979) p. 402
8. G. Grötzbach: Numerical investigations of radial mixing capabilities in strongly buoyancy-influenced vertical channel flows. Nucl. Eng. Des. **54**, 49 (1979)
9. U. Schumann, G. Grötzbach, L. Kleiser: "Direct Numerical Simulation of Turbulence". Lecture Notes for the course 'Prediction Methods for Turbulent Flows', von Kármán Institute, 1979 (to be published by Hemisphere Publishing Corp.)

Probability Distributions of Scalars in Turbulent Shear Flow

S. B. Pope

Massachusetts Institute of Technology, Department of Mechanical Engineering
Cambridge, MA 02139, USA

Abstract

A method of determining the statistically-mostlikely probability distribution of scalar quantities in turbulent flows has been described in previous works [2, 5]. The method is applicable to the joint probability density function (pdf) of any number of reactive or non-reactive scalars, and any number of moments can be incorporated. This paper reports, first, a comparison of most-likely pdf's with measured pdf's of passive scalars in turbulent shear flows. Secondly, the paper reports calculations of the evolution of the most-likely pdf of a reactive scalar in homogeneous turbulence. It is found that (with a single exception) the most-likely pdf based on the first three moments is in excellent quantitative agreement with the pdf's measured in shear flows. With only two moments, the most-likely pdf's are unable to adopt the sometimes-observed bimodal shapes. The calculations of reaction in homogeneous turbulence show that the pdf's are dominated by spikes at the two bounds. In addition it is found that, as the instantaneous reaction rate tends to infinity, the mean reaction rate tends to a universal function; this function is well represented by a simple empirical formula, Eq. (18).

Nomenclature

A_n	Coefficient in Eq. (1)
D	Jet diameter
H	Entropy (Eq. 7)
$p(\psi)$	Probability density function of ϕ
$\tilde{p}(\psi)$	Density-weighted pdf of ϕ
$q(\psi)$	*A priori* probability of ϕ
$Q(\psi)$	Arbitrary function
r	Radial distance
$S(\phi), S^*(\phi)$	Creation rate of ϕ, normalized creation rate
t, t^*	Time, normalized time
T, T_∞	Temperature, ambient temperature
$\underline{x}$	Position

Greek Symbols

α	Turbulent time scale/reaction time scale, Eq. (17)
Γ	Molecular diffusivity
λ	Number of moments
μ_n	nth central moment, $\mu_1 = \langle\phi\rangle$
ρ	Density
τ	Time scale Eq. (12)
$\phi, \langle\phi\rangle$	Scalar, mean of ϕ
ϕ'	Fluctuating or rms value of ϕ
ψ	Independent scalar variable

Introduction

Since the work of *Hawthorne* et al. [1], probability density functions have proved useful in the theoretical treatment of turbulent reacting flows. The representation of the pdf of a passive scalar as a function of its first and second moments has attracted much attention. Various authors have suggested a Gaussian, a beta-function distribution, a "clipped-Gaussian" and a double-delta function distribution (see [2] for references). In the model of *Bray* and *Moss* [3] a pdf for a single reactive scalar is prescribed and *Donaldson* and *Varma* [4] implicitly assume the joint pdf of reactive scalars to be comprised of delta functions. All these distributions are determined as functions of their first and second moments.

In previous work [2, 5] the author proposed a general method of determining single or joint pdf's of passive or reactive scalars in turbulent flows. The method is based on the assumption that the pdf is the statistically-most-likely distribution. An alternative view of this assumption can be based on the fact that the statistically-most-likely distribution contains a minimum of information. Consequently any other distribution contains more information, and the extra information is spurious. Thus, the assumption that the pdf is the statistically-most-likely distribution is the only assumption that excludes spurious information.

The method, which is described in the next section, is applicable to any number of scalars and any number of moments can be prescribed.

In the third section of the paper, the most-likely pdf's are compared with measurements made in turbulent shear flows. The data of *LaRue* and *Libby* [6] are of pdf's of temperature in the wake of a heated cylinder: the data of *Birch* et al. [7] are of pdf's of concentration in a turbulent methane jet in an atmosphere of air. Many of these pdf's are bimodal and, with a single exception, they are well represented by the most-likely distribution based on the first three moments. (Two moments are insufficient to describe bimodal distributions.)

Calculations of reaction in homogeneous turbulence are reported in the fourth section. These calculations are based on the solution of the differential equations for the mean and variance of a scalar which corresponds to the concentration of reaction products. An Arrhenius reaction rate is used and the mean reaction rate and rate correlations are determined from the assumption that the pdf adopts its most-likely value. The results show the evolution of the pdf which is dominated by spikes as the two bounds. For large reaction rates it is found that the mean reaction rate is a universal function for which an empirical formula is provided, Eq. (18).

Theory

In this section the results of previous work [2, 5] relevant to the present study are reported. If the first λ moments of the pdf $p(\psi)$ are known, then the most-likely value of p is

$$p_\lambda(\psi) = q(\psi) \exp\left(\sum_{n=0}^{\lambda} A_n \psi^n \right). \tag{1}$$

$q(\psi)$ is the *a priori* probability which, for nonreactive scalars, is a constant. $q(\psi)$ for a reactive scalar is given below, Eq. (8). The $\lambda + 1$ coefficients A in Eq. (1) are uniquely determined from the λ moments and from the condition that $p_\lambda(\psi)$ integrates to unity.

Let $\phi(\underline{x}, t)$ denote a scalar that obeys the conservation equation

$$\rho \frac{D\phi}{Dt} = \Gamma \nabla^2 \phi + \rho S(\phi), \tag{2}$$

where Γ is the molecular diffusivity and $S(\psi)$ is the rate of creation of ϕ. In a turbulent flow, the probability density function $p(\psi;\underline{x}, t)$ is defined so that the probability of ϕ being in the range $\psi < \phi < \psi + d\psi$ is $p(\psi)d\psi$. From this definition it follows that, if $Q(\phi)$ is any function, then its expected value (denoted by angled brackets) is

$$\langle Q[\phi(\underline{x}, t)]\rangle = \int p(\psi;\underline{x}, t)Q(\psi)d\psi. \tag{3}$$

By substituting, ψ and $(\psi - \langle\phi\rangle)^n$ for $Q(\psi)$ in Eq. (3), we obtain

$$\int p(\psi)d\psi = 1, \tag{4}$$

$$\int \psi p(\psi)d\psi = \langle\phi\rangle = \mu_1 \tag{5}$$

and

$$\int (\psi - \langle\phi\rangle)^n p(\psi)d\psi = \langle\phi'^n\rangle = \mu_n. \tag{6}$$

Thus, μ_1 is the mean of ϕ and $\mu_n (n > 1)$ is the nth central moment.

The most-likely pdf maximizes the entropy H subject to available information:

$$H \equiv -\int \ln [p(\psi)/q(\psi)]q(\psi)d\psi. \tag{7}$$

For a non-reactive scalar $[S(\phi) = 0]$, the transport Eq. (2) is linear in ϕ: from this it follows that the *a priori* probability $q(\psi)$ is a constant (unity, say). For a single reactive scalar, the expression given by *Pope* [2] reduces to

$$q(\psi) = [1 + S(\psi)\tau/\phi'], \tag{8}$$

where ϕ' is the rms fluctuation ($\phi' = \mu_2^{1/2}$) and τ is the time scale of dissipation of ϕ' [see Eq. (12) below].

If the available information is the first λ moments of $p(\psi)$, then the calculus of variations can be used to show that the distribution that maximizes H is

$$p_\lambda(\psi) = q(\psi) \exp\left(\sum_{n=0}^{\lambda} A_n \psi^n\right). \tag{9}$$

The $(\lambda + 1)$ coefficients A are uniquely determined from the λ moments and from the condition that $p_\lambda(\psi)$ integrates to unity. The subscript λ indicates that $p_\lambda(\psi)$ is the most-likely distribution based on a knowledge of the first λ moments.

In the next section $p_\lambda(\psi)$ $(2 \leqslant \lambda \leqslant 5)$ is compared with experimental data of pdf's of a non-reactive scalar in shear flows. The moments μ_n were either taken directly from the data or were determined by numerical integration of the pdf. Then Eqs. (4–6) and (9) were solved using Newton's method in order to determine the coefficients A and hence $p_\lambda(\psi)$.

The penultimate section reports calculations of the evolution of $p_2(\psi, t)$ for a reaction in homogeneous turbulence. In those calculations the first two moments were determined from their conservation equations. For constant-density flow in homogeneous turbulence these are:

$$\frac{d\langle\phi\rangle}{dt} = \langle S(\phi)\rangle \tag{10}$$

and

$$\frac{1}{2}\frac{d\langle\phi'^2\rangle}{dt} = \langle\phi' S(\phi)\rangle - \left\langle \Gamma \frac{\partial\phi}{\partial x_i}\frac{\partial\phi}{\partial x_i}\right\rangle. \tag{11}$$

The last term in Eq. (11) accounts for the destruction of fluctuations of ϕ and it can be used to define the time scale of decay τ:

$$\tau = \langle\phi'^2\rangle \Big/ \left\langle \Gamma \frac{\partial\phi}{\partial x_i}\frac{\partial\phi}{\partial x_i}\right\rangle. \tag{12}$$

The time scale τ, which is presumed to be constant, is used to normalize t and $S(\phi)$:

$$t^* \equiv t/\tau \quad \text{and} \quad S^*(\phi) \equiv S(\phi)\tau. \tag{13}$$

Then, Eqs. (10) and (11) can be written,

$$\frac{d\langle\phi\rangle}{dt^*} = \langle S^*(\phi)\rangle \tag{14}$$

and

$$\frac{1}{2}\frac{d\langle\phi'^2\rangle}{dt^*} = \langle\phi' S^*(\phi)\rangle - \langle\phi'^2\rangle. \tag{15}$$

With the assumption that the pdf of ϕ is $p_2(\psi)$, $\langle S^*(\phi)\rangle$ and $\langle\phi' S^*(\phi)\rangle$ can be determined from, for example

$$\langle\phi' S^*(\phi)\rangle = \int (\psi - \langle\phi\rangle) S^*(\psi) p(\psi) d\psi. \tag{16}$$

Thus, with given initial conditions for $\langle\phi\rangle$ and $\langle\phi'^2\rangle$ and a given normalized rate of creation $S^*(\phi)$, Eqs. (14) and (15) can be integrated in normalized time. This integration was performed numerically to produce the results reported in the final section.

Comparison with Experiment

In this section the most-likely pdf's are compared with measurements of passive scalars in turbulent shear flows. The measurements of pdf's of temperature reported by *LaRue* and *Libby* [6] were obtained in the far wake of a heated cylinder; and the measurements of *Birch* et al. [7] are of pdf's of concentration in a methane jet.

Plane Wake

LaRue and *Libby*'s data were obtained at an axial distance of 400 diameters downstream of the cylinder. At this location, the maximum observed temperatures are well below the temperature of the cylinder. Consequently the only effective bound on the temperature is the lower bound of the ambient temperature T_∞. The passive scalar is chosen to be $\phi \equiv T - T_\infty$, and hence $\phi \geqslant 0$.

Figures 1–5 show plots of $p(\psi)$ at increasing distances from the centre-line of the wake. The plots are normalized with the standard deviation of ϕ, ϕ'. The measured values of the moments were used to determine the most-likely pdf's based on the first two and three moments, $p_2(\psi)$ and $p_3(\psi)$. These are shown as dashed and full lines, respectively, and the experimental data are shown as circles.

Close to the centre-line (Figs. 1 and 2) the pdf's have a Gaussian shape except at the lower bound $\psi = 0$. However, the most-likely pdf based on the first two moments $p_2(\psi)$ does not follow the data; $p_3(\psi)$ does. Further away from the wake centre-line (Figs. 3–5) the pdf's become bimodal. Since $p_2(\psi)$ does not have sufficient degrees of freedom to adopt such a shape, it provides a poor representation of the measured pdf. On the other hand, the most-likely distribution based on three moments still follows the data closely. Indeed, at all five locations, discrepancies between $p_3(\psi)$ and the data are most likely less than experimental error.

From this comparison it can be concluded that, with a knowledge of only two moments, the most-likely distribution has a different shape from that measured. With a knowledge of three moments, not only is the shape correct, but also the quantitative agreement is excellent – at least for this flow.

Axisymmetric Jet

Birch et al. used laser Raman spectroscopy to measure the pdf of concentration ten nozzle diameters downstream of the exit of a methane jet. In this case the passive scalar ϕ is taken simply as the methane concentration: $\psi = 1$ corresponds to pure methane and $\psi = 0$ to pure air.

For variable-density flows ($\rho_{\text{methane}} \approx 0.55\ \rho_{\text{air}}$) it has been argued [8, 2] that it is appropriate to consider density-weighted quantities. The density-weighted pdf $\tilde{p}(\psi)$ is given by the relation

$$\langle \rho \rangle \tilde{p}(\psi) = \rho(\psi) p(\psi).$$

This relation was employed to determine $\tilde{p}(\psi)$ from *Birch* et al.'s measurements of $p(\psi)$, and the density-weighted moments were obtained by numerical integration of $\tilde{p}(\psi)$.

Figures 6–9 show $\tilde{p}_2(\psi)$ and $\tilde{p}_3(\psi)$ compared with the measurements at increasing distances from the jet axis. As before, near the axis (Figs. 6 and 7) the pdf's are Gaussian shaped but $\tilde{p}_2(\psi)$ is evidently discrepant. The curve of $\tilde{p}_3(\psi)$ passes through the data points. At $r/D = 1.49$ (Fig. 8) the measured pdf is bimodal and the shape appears to be similar to that of *LaRue* and *Libby*'s data. However, since neither $\tilde{p}_2(\psi)$ nor $\tilde{p}_3(\psi)$ are bimodal they provide poor representations of the measured pdf. Why this is so is discussed further below, but a clear answer is not evident. At the furthest radius measured (Fig. 9), the distribution ceases to be bimodal and is well represented by $\tilde{p}_3(\psi)$. As before, there is a noticeable difference between $\tilde{p}_2(\psi)$ and the data.

The distribution at $r/D = 1.49$ (Fig. 8) was investigated further. Figure 10 shows the data compared with the most-likely distributions based on four and five moments, $\tilde{p}_4(\psi)$ and $\tilde{p}_5(\psi)$. It may be seen that $\tilde{p}_4(\psi)$ is bimodal but a quantitative discrepancy remains. On the other hand, the curve of $\tilde{p}_5(\psi)$ passes through the data points. This confirms that with sufficient moments the most-likely distribution fits the data, but a five-parameter fit to fifteen data points can be expected to succeed.

Figs. 1–5. Comparison of most-likely distributions with data of *LaRue* and *Libby* [6]. Axial distance $x = 400\,d$: radial distance y is normalized with $l_c = (x - x_0)^{1/2} d^{1/2}$, where the virtual origin $x_0 = -40\,d$.

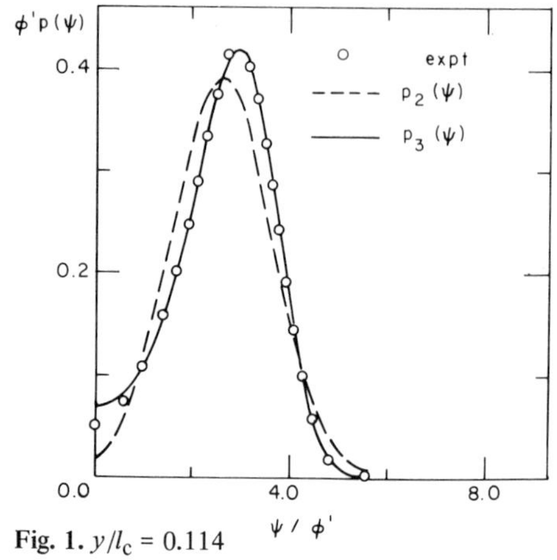

Fig. 1. $y/l_c = 0.114$

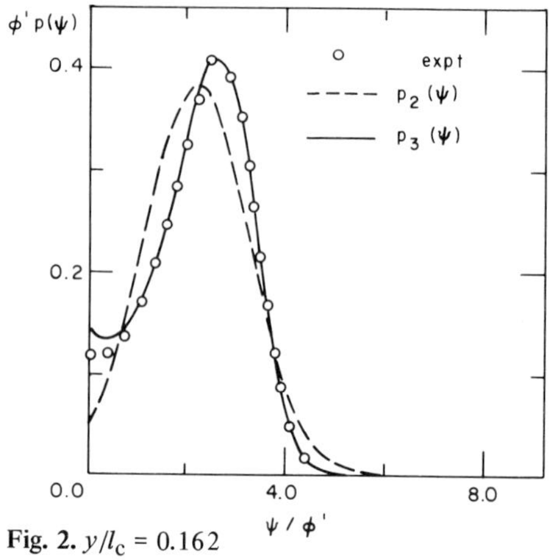

Fig. 2. $y/l_c = 0.162$

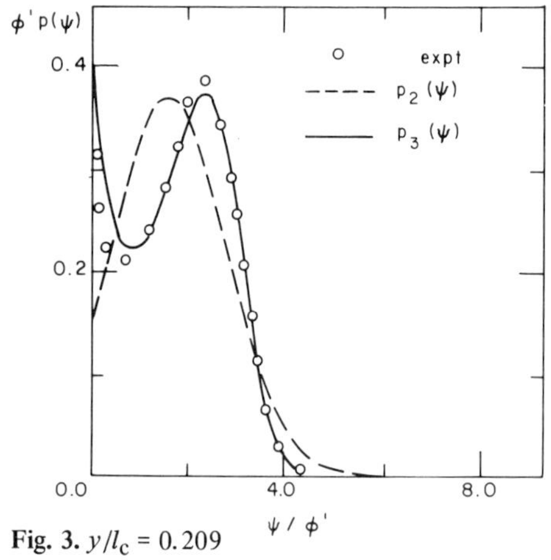

Fig. 3. $y/l_c = 0.209$

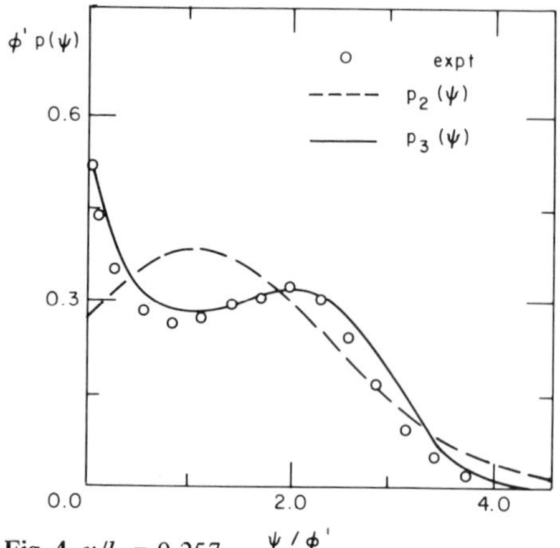

Fig. 4. $y/l_c = 0.257$

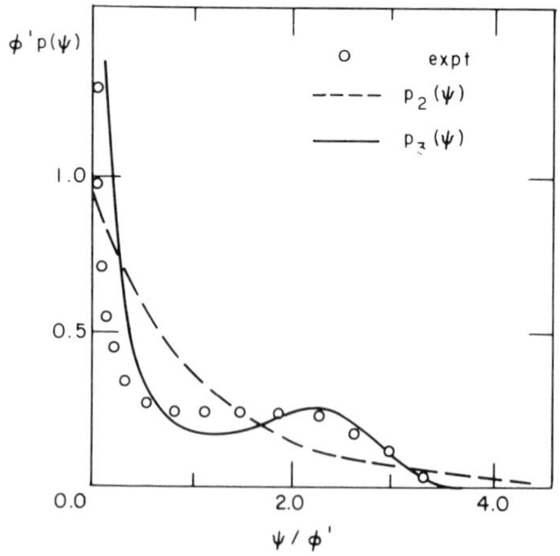

Fig. 5. $y/l_c = 0.294$

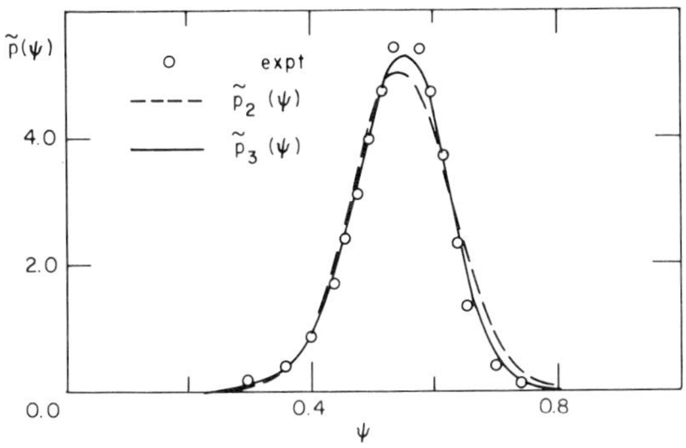

Fig. 6. $r/d = 0.0$

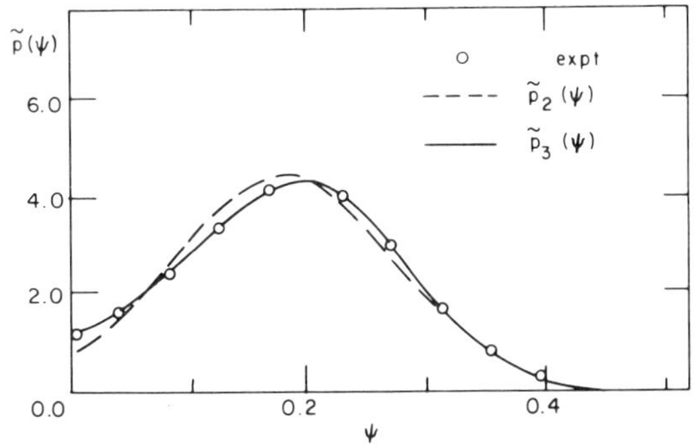

Fig. 7. $r/d = 1.3$

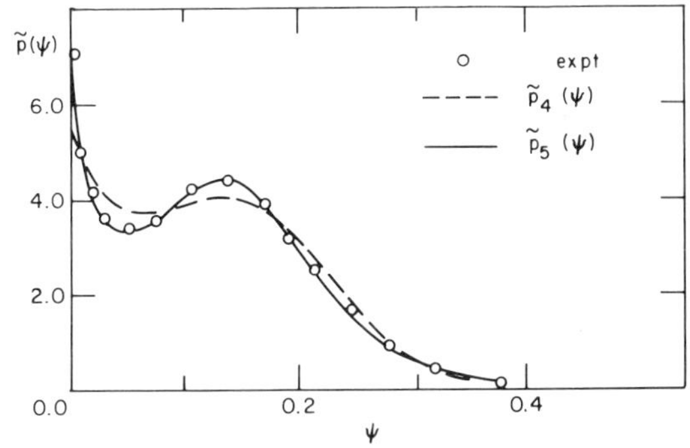

Fig. 8. $r/d = 1.49$

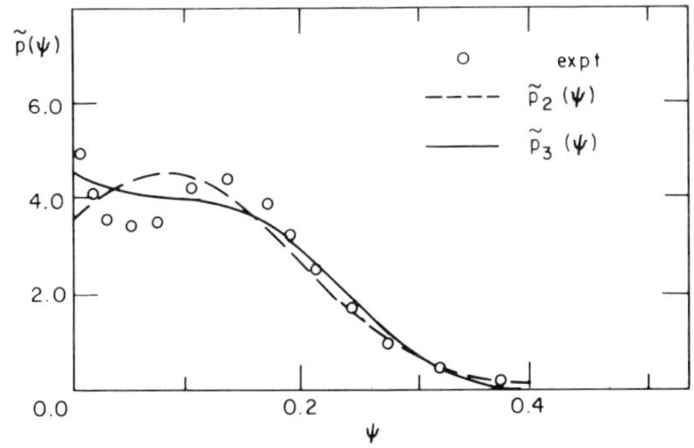

Fig. 9. $r/d = 1.8$

Figs. 6–10. Comparison of most-likely distributions with the data of *Birch* et al. [7]. Axial distance $x = 10\,d$.

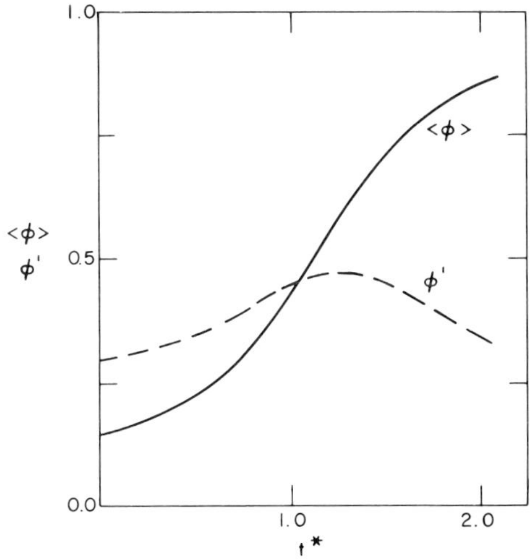

Fig. 10. $\langle\phi\rangle$ and ϕ' against t^* for $\alpha = 5$

Table 1. Skewness of distributions, data of [7]

	Skewness at $r/D = 1.3$	
	Weighted	Unweighted
Experimental	0.0572	0.0865
Most-likely distribution	0.2330	0.2620
Difference	−0.1758	−0.1755
	Skewness at $r/D = 1.49$	
	Weighted	Unweighted
Experimental	0.468	0.508
Most-likely distribution	0.690	0.741
Difference	0.222	0.233

Perhaps, contrary to the opinion expressed above, the unweighted pdf $p(\psi)$ should be considered rather than the density-weighted pdf $\tilde{p}(\psi)$. In order to investigate this question, the skewness of $p_2(\psi)$ and $\tilde{p}_2(\psi)$ were compared with the experimental values for $r/D = 1.3$ and $r/D = 1.49$. The results are shown in Table 1.

The difference between the experimental skewness and that given by $p_2(\psi)$ and $\tilde{p}_2(\psi)$ is a measure of the disagreement. It may be seen that, at both locations, the disagreement is nearly the same for weighted and unweighted pdf's. Consequently, the use of density-weighted pdf's cannot be deemed the cause of the observed discrepancy at $r/D = 1.49$.

In summary, for both the plane wake and the axisymmetric jet, in general there is excellent agreement between experimental data and the most-likely distribution based on three moments. The significant disagreement in one case remains unexplained. With only two moments prescribed, the most-likely distribution does not adopt observed bimodal shapes. Even where the pdf is not bimodal, there are discernible quantitative differences between $p_2(\psi)$ and the data.

Reaction in Homogeneous Turbulence

The evolution of the pdf of a reactive scalar in homogeneous turbulence was calculated by numerical integration of Eqs. (14) and (15) for $\langle\phi\rangle$ and ϕ' with the assumption that $p(\psi)$ is the most-likely distribution $p_2(\psi)$. The scalar ϕ is taken to represent the mass fraction of combustion products – $\psi = 0$ and $\psi = 1$ corresponding, respectively, to unburnt and fully burnt mixtures. Reaction proceeds according to the Arrhenius expression

$$S^*(\phi) = \alpha\, 6.11 \times 10^7\, \phi(1 - \phi) \exp[-30{,}000/(300 + 1{,}800\,\phi)]. \qquad (17)$$

The parameter α can be regarded as the ratio of the turbulent time scale to the reaction time scale. The multiplier 6.11×10^7 is chosen so that the maximum value of $S^*(\phi)$ is α: this maximum value occurs at $\phi = 0.933$.

For all the calculations reported, the initial conditions correspond, approximately, to the pdf $p_2(\psi)$ being a double-delta function distribution. The initial value of $\langle\phi\rangle$ is chosen so that reaction starts and continues at a reasonable rate.

Figure 10 shows the mean $\langle \phi \rangle$ and standard deviation ϕ' as a function of the normalized time t^*, for $\alpha = 5$. The mean increases monotonically (as it must) whereas ϕ' increases to a maximum of 0.47 before decaying. Since diffusion tends to decrease ϕ', the observed rise in ϕ' is due to reaction.

Figure 11 shows three pdf's taken from the same calculation (i.e. $\alpha = 5$). The arrows on the figure indicate the values of $\langle \phi \rangle$ for each pdf: these are $\langle \phi \rangle = 0.3$, 0.5 and 0.7. It may be seen that for each pdf there are spikes at zero and one with a relatively low value of $p(\psi)$ in between. This shape is precisely the one assumed in the models of *Bray* and *Moss* [3] and *Pope* [8]. The width of the spike at zero decreases as time (and hence $\langle \phi \rangle$) increases. The behaviour of the two spikes is more clearly seen on Fig. 12 where $p(0)$ and $p(1)$ are plotted against $\langle \phi \rangle$. The value at the upper bound $p(1)$ increases rapidly from its initial value of 15 to well over 100. The value at the lower bound $p(0)$ first falls, then rises to a maximum, and then falls again, ultimately to zero. In fact it can be shown from the pdf transport equation [10] that $p(0)$ must decrease monotonically. Ironically, the slight increase shown on Fig. 12 represents a violation of the second law.

It is generally held that as the instantaneous reaction rate $S(\phi)$ becomes large (compared with the turbulent frequency $1/\tau$) the mean reaction rate $\langle S(\phi) \rangle$ becomes independent of $S(\phi)$. In the present notation this implies that $\langle S^*(\phi) \rangle$ is independent of α for large α. This tenent

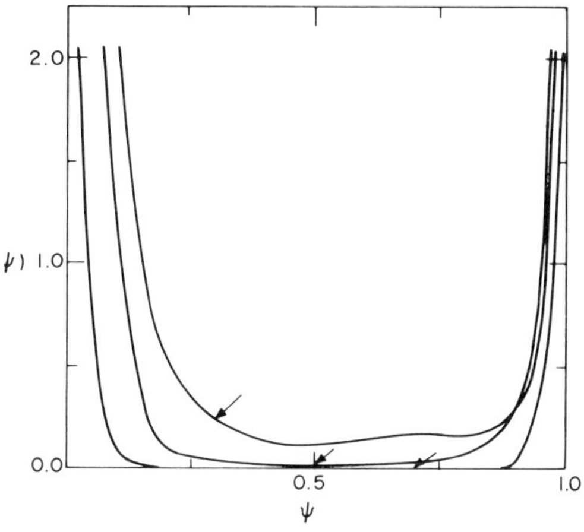

Fig. 11. $p(\psi)$ against ψ for $\alpha = 5$: $\langle \phi \rangle = 0.3$, 0.5 and 0.7

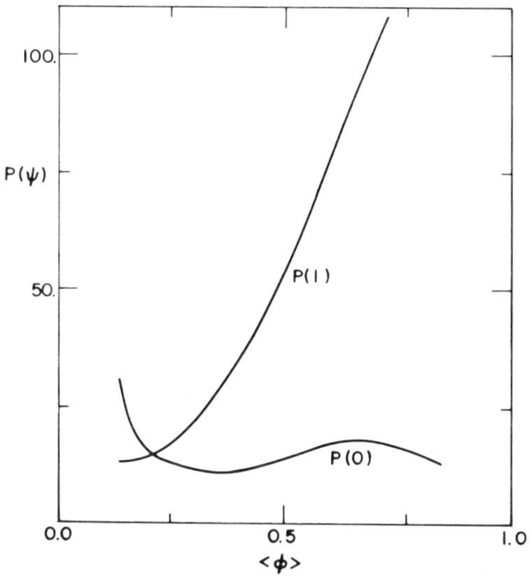

Fig. 12. $p(0)$ and $p(1)$ against $\langle \phi \rangle$ for $\alpha = 5$

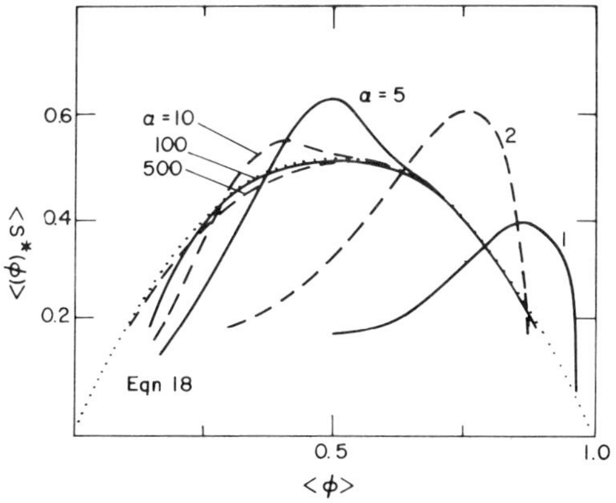

Fig. 13. $\langle S^*(\phi) \rangle$ against $\langle \phi \rangle$ for $\alpha = 1$, 2, 5, 10, 100 and 500. The dotted line is Eq. (18)

was investigated by integrating equation (14) and (15) for various values of α: the computed values of $\langle S^*(\phi)\rangle$ are plotted against $\langle\phi\rangle$ on Fig. 13.

For the smallest instantaneous reaction rate ($\alpha = 1$), the reaction cannot be sustained unless the initial value of $\langle\phi\rangle$ is greater than 0.5. Then the mean reaction rate starts at a value of 0.2, rises to 0.4 at $\langle\phi\rangle = 0.85$ and then falls rapidly. For the next two values of α (2 and 5) the same trend is observed but the higher maximum of $\langle S^*(\phi)\rangle = 0.6$ is reached sooner (at $\langle\phi\rangle = 0.7$ and 0.5 respectively). For these three cases $\langle S^*(\phi)\rangle$ is clearly a strong function of α. The independence of $\langle S^*(\phi)\rangle$ from α becomes evident for the larger values, $\alpha = 10$, 100 and 500. Although the three curves differ slightly where $\langle\phi\rangle$ is less than 0.5, over the other half of the range the three curves coincide. The curve for $\alpha = 5$ also joins this apparently universal curve at $\langle\phi\rangle = 0.7$.

The existence of a universal curve of $\langle S^*(\phi)\rangle$ against $\langle\phi\rangle$ supports the tenent that the mean reaction rate becomes independent of $S(\phi)$ for large $S(\phi)$. In addition, the universal relationship between $\langle S^*(\phi)\rangle$ and $\langle\phi\rangle$ can be used as a model of premixed turbulent combustion. It is fortunate then that the simple empirical formula

$$\langle S^*(\phi)\rangle = 2.2\,\langle\phi\rangle\,(1 - \langle\phi\rangle) \tag{18}$$

provides an excellent representation of the curve (see the dotted line on Fig. 13).

To conclude: for reaction in homogeneous turbulence, the differential equations for $\langle\phi\rangle$ and $\langle\phi'^2\rangle$ are closed by the assumption that $p(\psi)$ adopts its most-likely value. These equations have been solved numerically to produce the results reported in this section. The results confirm the expectations that the pdf is approximately a double-delta function distribution, and that the mean reaction rate becomes independent of the instantaneous rate. Equation (18) provides an empirical formula for the mean reaction rate $\langle S^*(\phi)\rangle$ which is in excellent agreement with the calculated dependence of $\langle S^*(\phi)\rangle$ upon $\langle\phi\rangle$.

References

1. W. R. Hawthorne, D. S. Wedell, H. C. Hottel: Third Symposium (International) on Combustion (1949) pp. 266–288
2. S. B. Pope: The statistical theory of turbulent flames. Philos. Trans. R. Soc. London Ser. A **291**, 529 (1979)
3. K. N. C. Bray, J. B. Moss: AASU Rpt. 335, University of Southampton (1974)
4. C. du P. Donaldson, K. A. Varma: Combust. Sci. Tech. **13**, 55–73 (1976)
5. S. B. Pope: A rational method of determining probability distributions in turbulent reacting flows. J. Non-Equilib. Thermodyn. **4**, 309 (1979)
6. J. C. LaRue, P. A. Libby: Phys. Fluids **17**, 1956 (1974)
7. A. D. Birch, D. R. Brown, M. G. Dodson, J. R. Thomas; "Studies of Flammability in Turbulent Flows Using Laser Raman Spectroscopy". 17th Symposium (International) on Combustion, The Combustion Institute (1979)
8. R. W. Bilger: Combust. Sci. Tech. **11**, 215 (1975)
9. S. B. Pope: Combust. Flame **29**, 235 (1977)
10. S. B. Pope: Combust. Flame **27**, 299 (1976)

On the Effect of the Pressure Strain Correlations on Three-Dimensional Turbulent Boundary Layers

J. C. Rotta

Deutsche Forschungs- und Versuchsanstalt für Luft- und Raumfahrt e. V.,
Aerodynamische Versuchsanstalt Göttingen, Bunsenstraße 10
D-3400 Göttingen, Fed. Rep. of Germany

Abstract

In order to test a previously proposed approximation for the pressure strain terms, comparative calculations of turbulent boundary layers on infinite swept wings are carried out, integrating the partial differential equations for the mean velocities and the transport equations for the shear stresses and kinetic energy. With a suitable choice of the coefficient appearing in the model, results are found to be in better agreement with experiments than with the simpler assumption that the vector of the part of the pressure strain terms depending on the mean rate of strain, is parallel to the mean rate of strain vector. Further consideration of the problem leads to another relationship for the pressure strain terms, which avoids the disadvantage of the first model of being related to the local mean flow direction. The latter model is shown to be less effective in practice, however.

Nomenclature

H Shape parameter, Eq. (2)

N Ratio of eddy viscosities, Eq. (3)

L Length scale of turbulence

$q^2/2 = \overline{u_i u_i}/2$ Kinetic energy of turbulence

p'_M Fluctuation of pressure, correlated with the mean rate strain

$R_{ij} = \overline{u_i(\boldsymbol{x}) u_j(\boldsymbol{x}+\boldsymbol{r})}$ Correlation function

$\boldsymbol{r}$ Vector with components r_i $(i = 1, 2, 3)$

T Empirical constant, Eq. (1)

U_i $(i = 1, 2, 3)$ Components of mean velocity

$U_r^2 = U_1^2 + U_2^2$

U_{re} Value of U_r at the outer edge of the boundary layer

U_s Component of mean velocity in direction of the outer streamline

u_i $(i = 1, 2, 3)$ Components of velocity fluctuation

$V(\boldsymbol{r})$ Volume of $\boldsymbol{r}$ space

$\boldsymbol{x}$ Vector

x_i $(i = 1, 2, 3)$ Cartesian coordinates, Fig. 1

x_{ig} Cartesian coordinates, x_{1g} in the direction of mean rate of strain vector, Fig. 6

x_m Coordinate in the direction of the undisturbed flow in the case of a swept wing

$y = x_3$ Coordinate normal to the wall

$\alpha_g = \tan^{-1}[(\partial U_2/\partial x_3)/(\partial U_1/\partial x_3)]$, Fig. 6

γ_p Angle between pressure strain vector and mean rate of strain vector, Fig. 6

δ Boundary layer thickness (for $U_r/U_{re} = 0.995$)

δ_{ij} Kronecker symbol ($\delta_{ij} = 1$ for $i = j$, $\delta_{ij} = 0$ for $i \neq j$)

κ_g Empirical constant, Eq. (5)

κ_{ss} Empirical constant, Eq. (1)

λ Empirical constant, Eq. (6)

ρ Density

Subscript

g Components in x_{ig} coordinates
Overbars denote time averages

Introduction

In a previous paper by the author [1], presented at the First International Symposium on Turbulent Shear Flows, the process of turbulent shear stress generation in three-dimensional boundary layers was studied by reference to the stress transport equations. That study proposed that the experimentally observed differences between the directions of the shear stress and the mean rate of strain are caused by the mean product of pressure and rate of strain fluctuations. Approximate relationships for these pressure strain terms were proposed, and turbulence models were derived which took account of these effects. Subsequently *Schneider* [2] applied the proposed mixing length model to boundary layer calculations and obtained good agreement with measurements. The present contribution provides some numerical results using the more complete model based on the stress transport equations. It also considers a different approach to approximating the pressure strain effects. A fuller account of the present work is provided (in German) in Ref. [3].

The T-Model

Consider a steady three-dimensional turbulent boundary layer on a plane wall in incompressible flow, Fig. 1. The proposed approximation to the part of the pressure strain terms, which is correlated to the mean rate of strain, is written as follows:

$$\overline{\frac{p'_M}{\rho}\left(\frac{\partial u_i}{\partial x_3}+\frac{\partial u_3}{\partial x_i}\right)} = \overline{u_3^2}\left[\kappa_{ss}\frac{\partial U_i}{\partial x_3}+(1-T)(1-\kappa_{ss})\left(\delta_{ij}-\frac{U_iU_j}{U_r^2}\right)\frac{\partial U_j}{\partial x_3}\right] \tag{1}$$

Equation (1) applies to $i = 1, 2$, and the last term has to be summed over $j = 1, 2$. This relation, referred to as the T-Model, rests on the simple assumption that the turbulence structure is symmetrical with respect to the local mean steamline [see Eqs. (15–19) of Ref. [1]]. The quantity T determines the deviation between the directions of the pressure strain vector and of the mean rate of strain vector[1]. For $T = 1$, both directions collapse.

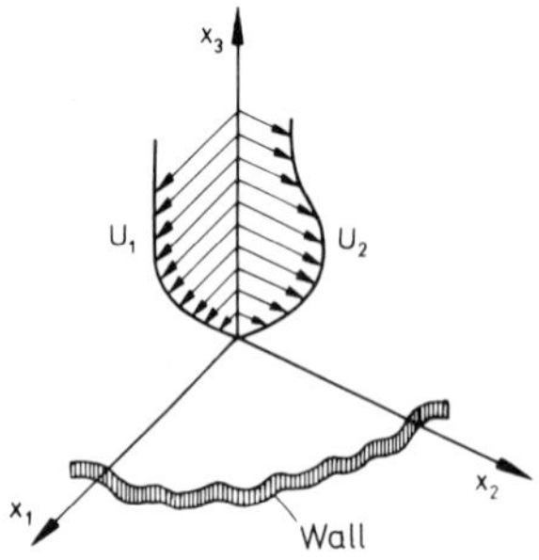

Fig. 1. Coordinate system

1 The resultants of the two tensor components of shear stress $-\rho\overline{u_iu_2}$, mean rate of strain, $\partial U_i/\partial x_3$, and pressure strain terms are denoted as vectors.

Numerical Results

Numerical calculations have been made in order to test the *T*-model and to study the effect of the parameter *T* on the results. The calculations are restricted to infinite swept wing cases. The partial differential equations for the mean velocity, shear stresses and kinetic energy of the velocity fluctuations have been solved by a finite difference procedure. The transport equations for the shear stresses are used in the form as given in Ref. [1].

Two cases have been treated, but only results of one case will be discussed here, namely the separating boundary layer on a 35 degree infinite swept wing, for which extensive measurements have been made by *van den Berg* et al. [4, 5]. The pressure distribution outside the boundary layer and the initial distributions of mean velocity, shear stress and turbulent energy are supplied in accordance with the experimental data.

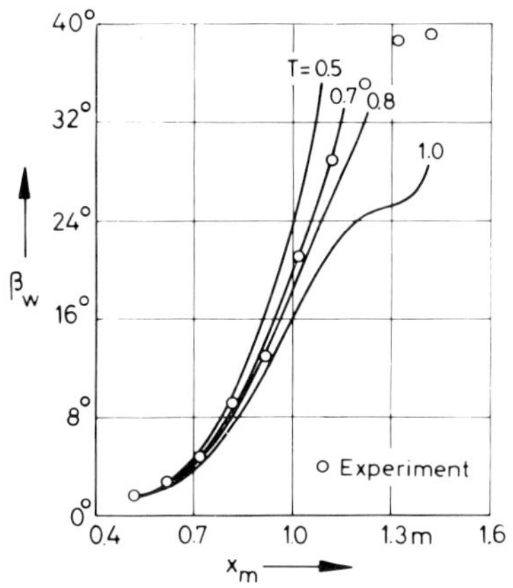

Fig. 2. Angle β_w between the limiting streamline at the wall and the outer streamline vs the coordinate in free stream direction. *T*-Model

A quantity, which is sensitive to the particular assumptions of the prediction method is the angle of the limiting wall streamline, β_w, measured with respect to the outer streamline. Fig. 2 shows the development of β_w vs the coordinate x_m in the direction of undisturbed flow. According to the experiment, β_w grows continously under the influence of the adverse pressure gradient; separation occurs at $x \approx 1.3$ m. Up to values of $x_m = 0.8$ m, the prediction shows only small differences for different values of *T*. At larger x_m, the calculated results for $T = 1$ (isotropic turbulence) are significantly below the experimental values, and no separation is indicated within the range of the calculation. The agreement improves for smaller values of *T*; for $T = 0.7$ the deviations are a minimum up to $x_m = 1.1$ m; separation is predicted at $x_m = 1.16$ m in this case. For still smaller values of *T* the agreement deteriorates.

A characteristic quantity of the mean velocity profiles is the shape parameter *H*, defined as the ratio of displacement thickness to momentum thickness,

$$H = \frac{\int_0^\delta \left(1 - \frac{U_s}{U_{re}}\right) dy}{\int_0^\delta \frac{U_s}{U_{re}} \left(1 - \frac{U_s}{U_{re}}\right) dy}, \tag{2}$$

the variation of which is shown in Fig. 3. As with two-dimensional boundary layers, the experimental values rise strongly under the influence of the adverse pressure gradient in the

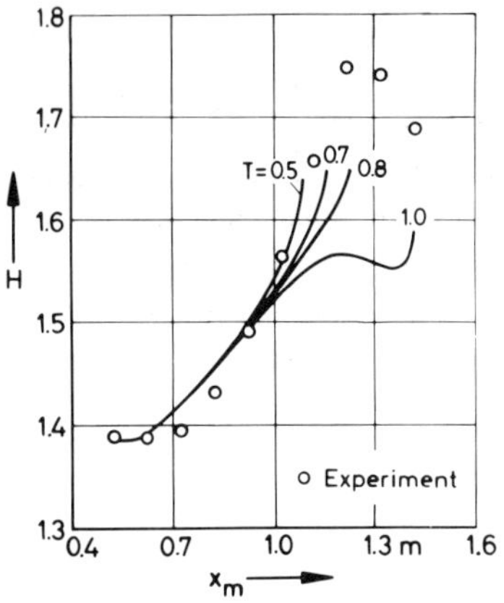

Fig. 3. Shape parameter H according to Eq. (2) vs the coordinate in free stream direction. *T*-model

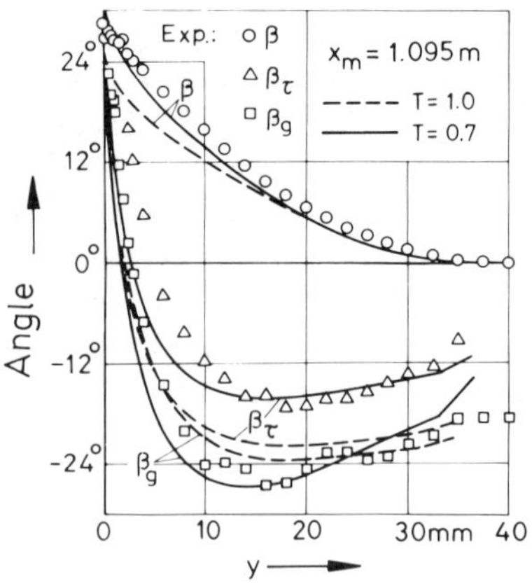

Fig. 4. Directions of mean flow, β, mean rate of strain, β_g, and shear stress, β_T, relative to the outer streamline vs the distance from the wall. x_m = 1.095 m. *T*-model

flow direction. The curves for $T = 1$ deviate markedly from the experimental values for $x_m > 0.9$ m. For smaller values of T, the curves come closer to the experiment.

The angle of local mean velocity, β, the angle of mean rate of strain, β_g, and the angle of Reynolds shear stress, β_T, are plotted versus the distance from the surface, y $(= x_3)$ in Fig. 4 for the station at x_m = 1.095 m. All angles are measured with respect to the direction of the mean streamline at the outer edge of the boundary layer. The calculations for $T = 1$ predict a difference between β_T and β_g which is due mainly to history effects. However, this difference is much smaller than that of the experiments. For a value of $T = 0.7$, the curves for β_T and β_g depart from each other and all three angles agree better with the experiment. This diagram provides clear evidence that the differences between β_T and β_g are largely caused by the pressure strain terms.

The effect of the assumed anisotropy, as expressed by the parameter T, becomes clear when the ratio

$$N = \frac{\tan(\beta_T - \beta)}{\tan(\beta_g - \beta)} \tag{3}$$

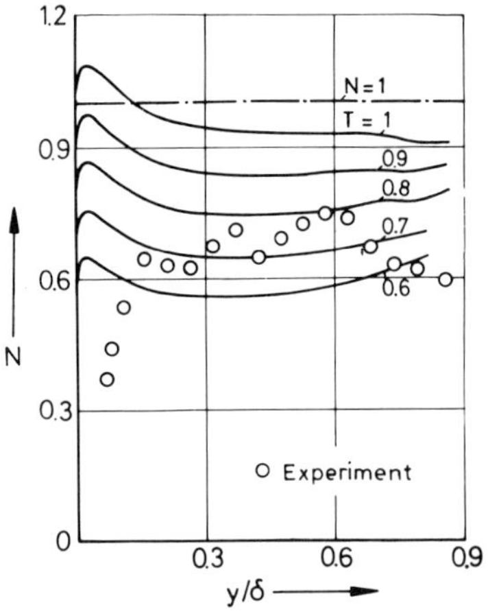

Fig. 5. Ratio N according to Eq. (3) vs the dimensionless distance from the wall. x_m = 1.095 m. *T*-Model

is considered. In terms of the eddy viscosity concept, N equals the ratio of the eddy viscosities transverse and parallel to the local mean velocity. The values of N determined from experiment and the curves resulting from the calculations assuming different values for T are plotted vs y/δ in Fig. 5 for the station at $x_m = 1.095$ m. $N = 1$ means that the shear stress and the mean rate of strain have the same directions. The calculations with $T = 1$ give $N \approx 0.9$ in the outer part of the boundary layer. Close to the surface a maximum value of N in excess of unity is caused by turbulent diffusion. The experimental results are considerably lower. While the scatter indicates that the determination of N from the experiments is rather inaccurate, for $T = 0.7$, the calculations agree roughly with the experiments.

Further Considerations

The above results have shown that, with $T = 0.7$, the model for the pressure strain terms clearly gives an improvement on the simpler isotropic assumption ($T = 1$). The arbitrariness of the basic assumption however, provokes further considerations. With most investigations, starting from the equation

$$\overline{\frac{p'_M}{\rho}\left(\frac{\partial u_i}{\partial x_3}+\frac{\partial u_3}{\partial x_i}\right)} = -\frac{1}{2\pi}\int_{V(r)}\frac{\partial U_j}{\partial r_3}\left(\frac{\partial^2}{\partial r_3\partial r_j}R_{i3}+\frac{\partial^2}{\partial r_i\partial r_j}R_{33}\right)\frac{dr_1dr_2dr_3}{|r|}, \quad i = 1, 2, \quad j = 1, 2, \tag{4}$$

which results from the solution of the Poisson equation for the pressure fluctuations, the derivatives $\partial U_j/\partial r_3$ to be taken at position $\boldsymbol{x} + \boldsymbol{r}$, are replaced by their values at $\boldsymbol{x}$. Then the mean rate of strain can be taken outside the integral sign. With some restrictions, the remaining integrals can be approximated by a linear combination of the Reynolds stresses. *Launder* et al. [6] have proposed such a relationship, which has in common with Eq. (1) that each component of the pressure strain terms depends on both, $\partial U_1/\partial x_3$ and $\partial U_2/\partial x_3$. But, putting the mean rate of strain outside the integral of Eq. (4), is an inadmissible simplification, because no information is provided about the three-dimensional character of the mean flow, as long as only the local values of the mean rate of strain are taken into account.

An alternative model, called the λ-model, uses Launder's relationship as the basis. A rectangular coordinate system x_{ig} is introduced, which is turned by the angle α_g with respect to the x_3-axis, such that the x_{1g}-axis is parallel to the vector of $\partial U_i/\partial x_3$ at the point under consideration, Fig. 6. Then locally $(\partial U_2/\partial x_3)_g \equiv 0$; however, the integrals for $j = 2$ of Eq. (4) will not vanish, and are assumed to be proportional to $(\partial^2 U_2/\partial x_3^2)_g$, the turbulence energy $q^2/2$ and the length scale L. On the other hand, $(\partial U_1/\partial x_3)_g$ has large values and the questionable assumption that the variation of $(\partial U_1/\partial r_3)_g$ is negligible within the range of integration, is retained. The analysis, details of which are given in Ref. [3], leads to relationships, which after simplification may be expressed:

$$\overline{\frac{p'_M}{\rho}\left(\frac{\partial u_1}{\partial x_3}+\frac{\partial u_3}{\partial x_1}\right)_g} = \kappa_g \frac{\overline{q^2}}{2}\left(\frac{\partial U_1}{\partial x_3}\right)_g, \tag{5}$$

$$\overline{\frac{p'_M}{\rho}\left(\frac{\partial u_2}{\partial x_3}+\frac{\partial u_3}{\partial x_2}\right)_g} = \lambda L \frac{\overline{q^2}}{2}\left(\frac{\partial^2 U_2}{\partial x_3^2}\right)_g. \tag{6}$$

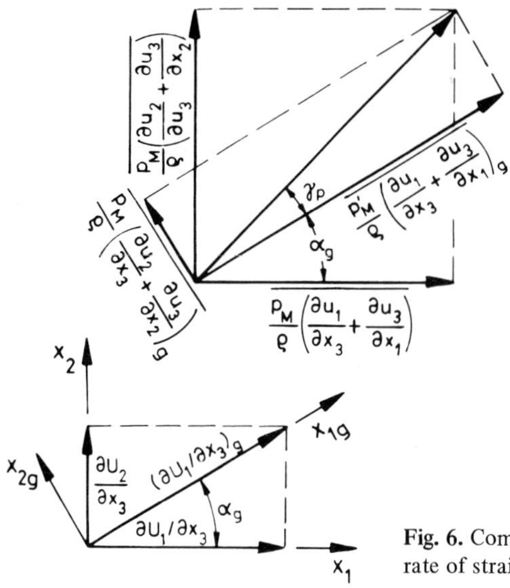

Fig. 6. Components in a coordinate system alined with the mean rate of strain vector

These forms, unlike those produced by the *T*-model are not related to the local mean flow direction.

Dividing Eq. (6) by Eq. (5) gives

$$\tan \gamma_p = \frac{\overline{\frac{p'_M}{\rho}\left(\frac{\partial u_2}{\partial x_3} + \frac{\partial u_3}{\partial x_2}\right)_g}}{\overline{\frac{p'_M}{\rho}\left(\frac{\partial u_1}{\partial x_3} + \frac{\partial u_3}{\partial x_1}\right)_g}} = \frac{\lambda}{\kappa_g} L \frac{\left(\frac{\partial^2 U_2}{\partial x_3^2}\right)_g}{\left(\frac{\partial U_1}{\partial x_3}\right)_g} . \tag{7}$$

Furthermore, since $\tan \alpha_g = (\partial U_2/\partial x_3)/(\partial U_1/\partial x_3)$ it follows that

$$\frac{\partial \alpha_g}{\partial x_3} = (\partial^2 U_2/\partial x_3^2)_g/(\partial U_1/\partial x_3)_g , \tag{8}$$

and we obtain from Eq. (7)

$$\tan \gamma_p = \frac{\lambda L}{\kappa_g} \frac{\partial \alpha_g}{\partial x_3} . \tag{9}$$

Thus, according to the λ-model, the angle, γ_p, between the vectors of the pressure strain terms and the mean rate of strain is proportional to the rate at which the direction of the mean rate of strain vector changes with the distance from the surface.

Calculations with the λ-model have been made with the same conditions as with the *T*-model. Because of the higher order derivative of the mean velocities in the former, the numerical solutions are more time consuming than those with the latter. The relevant results are reproduced in Figs. 7 to 10, and are to be compared with the diagrams of Figs. 2 through 5. The results with $\lambda = 0$ and those with $T = 1$ are practically identical. The second order derivative term improves the agreement with the experiments, if a positive value of λ is chosen, but the variation of λ has much less influence than the variation of *T*. For no value of λ do the

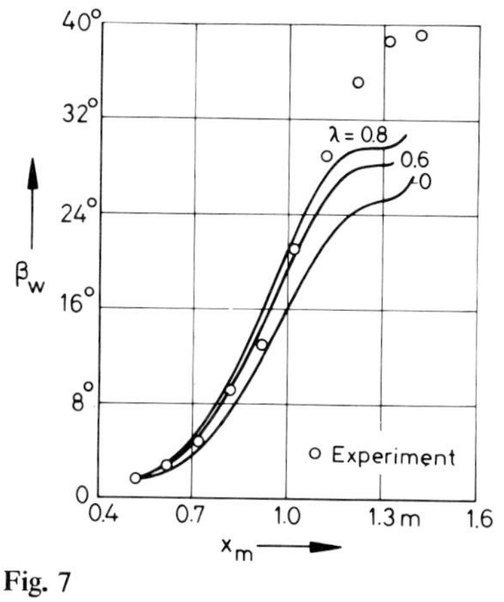

Fig. 7

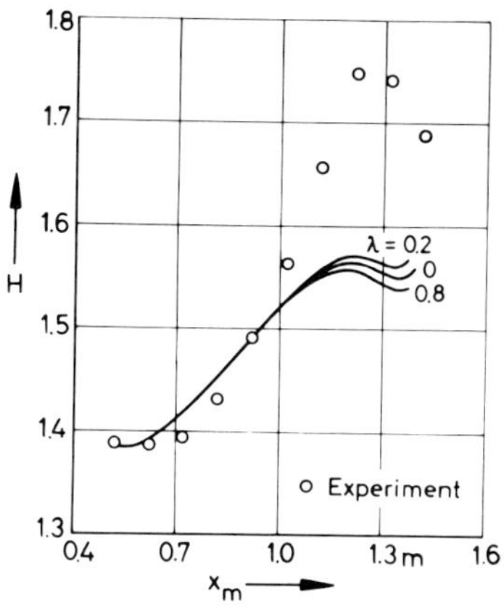

Fig. 8

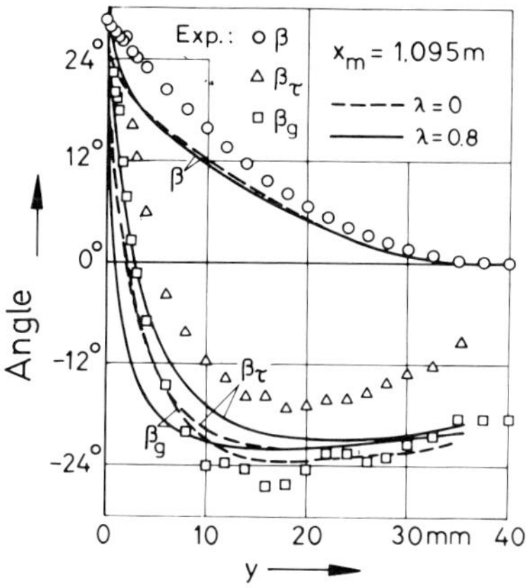

Fig. 9

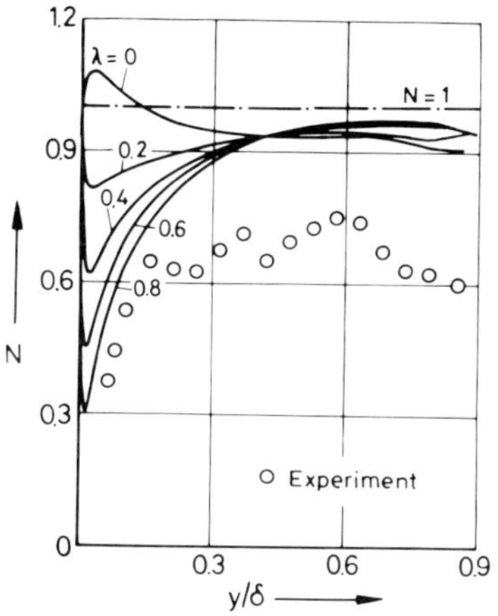

Fig. 10

Fig. 7. Angle β_w between the limiting streamline at the wall and the outer streamline vs the coordinate in free stream direction. λ-Model

Fig. 8. Shape parameter H according to Eq. (2) vs the coordinate in free stream direction. λ-Model

Fig. 9. Directions of mean flow, β, mean rate of strain, β_g, and shear stress, β_τ, relative to the outer streamline vs the distance from the wall. x_m = 1.095 m. λ-Model

Fig. 10. Ratio N according to Eq. (3) vs the dimensionless distance from the wall. x_m = 1.095 m. λ-Model.

calculations predict separation. This is seen from Fig. 7. Values of $\lambda > 0{,}8$ give no further improvement. Figure 8 shows that the shape parameter H is very little affected by the value λ. According to Fig. 9 the λ-model becomes effective mainly in the region closer to the surface. Here, with λ = 0.8 the angle of the mean rate of strain, β_g, and the angle of the Reynolds shear stress, β_τ, are further apart from each other, whereas the difference is small in the outer part of the layer. This behaviour is in accord with the variation of the ratio N [defined by Eq. (3)], which is plotted in Fig. 10. Strong dependence on λ occurs only at small y/δ: low values of N are observed here for λ = 0,8, growing up to nearly unity in the outer half of the layer. The tendency of the experimental values to drop towards the wall, if correct, may indicate that the ideas underlying the λ-model come near to reality in this part of the layer. Farther away from the surface, effects may come into action which are not included in the λ-model. The fact that N is rather independent of the value of λ in the outer region, may be the reason for the weak influence of λ on global boundary layer parameters like β_w and H.

The conclusion from these investigations is that the mechanism controlling the direction of the pressure strain vector is not yet understood. It appears to be very difficult to find an effective model, which has a better physical foundation than the T-model and, at the same time, meets the requirement of being numerically tractable without an excessive effort. At present, only the T-model appears to give useful results for practical purposes.

Acknowledgement. This research work is supported by the Deutsche Forschungsgemeinschaft.

References

1. J. C. Rotta: "A Family of Turbulence Models for Three-Dimensional Boundary Layers", in *Turbulent Shear Flows I,* ed. by F. Durst, B. E. Launder, F. W. Schmidt, J. H. Whitelaw (Springer, Berlin, Heidelberg, New York 1979) p. 267–278
2. G. R. Schneider: "Die Berechnung der turbulenten Grenzschicht am schiebenden Flügel unendlicher Spannweite mittels eines dreidimensionalen Mischungswegmodells". DLR-FB 77–73, Deutsche Forschungs- und Versuchsanstalt für Luft- und Raumfahrt e. V. (1978)
3. J. C. Rotta: "Eine theoretische Untersuchung über den Einfluß der Druckscherkorrelationen auf die Entwicklung dreidimensionaler turbulenter Grenzschichten". DLR-FB 79-05, Deutsche Forschungs- und Versuchsanstalt für Luft- und Raumfahrt e. V. (1979)
4. B. van den Berg, A. Elsenaar: "Measurements in a Three-Dimensional Incompressible Turbulent Boundary Layer in an Adverse Pressure Gradient under Infinite Swept Wing Conditions". NLR TR 72092 U, National Lucht- en Ruimtevaartlaboratorium (1972)
5. A. Elsenaar, S. H. Boelsma: "Measurements of the Reynolds Stress Tensor in a Three-Dimensional Turbulent Boundary Layer under Infinite Swept Wing Conditions". NLR TR 74095 U, National Lucht en Ruimtevaartlaboratorium (1974)
6. B. E. Launder, G. J. Reece, W. Rodi: Progress in the development of a Reynolds-stress turbulence closure. J. Fluid Mech. **68**, 537–566 (1975)

Possible Departures from Local Isotropy in the Small-Scale Structure of Turbulent Velocity Fields

R. C. Mjolsness

Theoretical Division, Group T-3, University of California,
Los Alamos Scientific Laboratory
Los Alamos, NM 87545, USA

Abstract

Two weakened similarity hypotheses are proposed as possible replacements for the two Kolmogorov similarity hypotheses as descriptions of laboratory turbulence. It is shown that these hypotheses yield consequences which are in full agreement with the whole range of laboratory, atmospheric, and oceanic measurements bearing on the question of the existence of local isotropy in turbulent flows. However, the weakened hypotheses are fully consistent with the non-existence of local isotropy. It is concluded that a new class of measurements – the measurement of cross-stream variations in the velocity correlations – would be required to establish the existence or non-existence of local isotropy. One consequence of the proposed weakened similarity hypotheses is that the Kolmogorov constant C is not a constant, but is, instead, a flow-dependent quantity, depending on velocity gradients. This is in at least qualitative agreement with the large scatter in values of C reported in the literature.

Nomenclature

$\hat{a}$, $\hat{a}_i$	Unit vector specified by velocity gradients in mean flow
A_0, A_1, A_2	Proportionality "constants" with weak functional dependences on velocity gradients in mean flow
$B_0(k^2/k_\eta^2)$	General positive function of wavenumber with weak dependence on velocity gradients
$B_{ij}(r)$	Two point velocity correlation
C	Kolmogorov constant, here taken to depend on velocity gradients
C_{ij}, C_1	Proportionality "constants" for the one-dimensional cross spectra and the spectrum, respectively, depending on velocity gradients in the mean flow
C_2, C_3	Proportionality "constants" in non-isotropic part of structure functions, dependent on velocity gradients
$D_{ij}(r)$, D^0_{ij}, D^1_{ij}	The tensor of structure functions, its isotropic component and its non-isotropic component, respectively
$D_{LL}(X)$, $D_{NN}(X)$	Longitudinal and lateral structure functions, respectively
$E(k)$	The spectrum
$F_{ij}(k)$, $F_{ij}(k_1)$	Three-dimensional and one-dimensional cross spectra, respectively
$f_{ij}\left(\frac{k_1}{k}, \frac{k_2}{k}, \frac{k_3}{k}\right)$	Dimensionless three-dimensional cross spectra of Hypothesis A
$G_{ij}\left(\frac{k_1}{k_\eta}, \frac{k_2}{k_\eta}, \frac{k_3}{k_\eta}\right)$	Dimensionless three dimensional cross spectra of Hypothesis B

$H(k_i/k)$	Function of wavenumber component
$I(\theta) = \{\theta: 0 \leqslant \theta \leqslant \pi/2$ if $k_1 > 0$, $\pi/2 \leqslant \theta \leqslant \pi$ if $k_1 < 0\}$	Integration domain for polar angle theta
$K(x)$	Function of proportionality in non-isotropic part of structure function
$\boldsymbol{k}, k_1$	Wavenumber vector
k	Magnitude of wavenumber vector
$k_\eta = (\epsilon/\nu^3)^{1/4}$	Kolmogorov wavenumber
l	Integration variable
$\boldsymbol{r}$	Position vector denoting separation of two points in the two-point functions
$\hat{\boldsymbol{r}}, \hat{\boldsymbol{r}}_i$	Unit vectors along $\boldsymbol{r}$
r	Magnitude of position vector
$\boldsymbol{x}$	General position vector
x_1	Local direction of mean flow
x	General argument of a function of one variable
$u_i(\boldsymbol{x}, t)$	Mean velocity vector
$v_\eta = (\epsilon\nu)^{1/4}$	Kolmogorov velocity
δ_{ij}	Kronecker delta function
$\delta(x)$	Dirac delta function
$\partial/\partial x^i$	Spatial gradient operator
ϵ	Local mean energy dissipation
$\Gamma(1/3)$	Gamma function of argument 1/3
η	Kolmogorov length scale
ϕ	Azimuthal angle of polar coordinates in wavenumber space
ν	Kinematic viscosity
θ	Polar angle relative to $\hat{\boldsymbol{r}}$ in wavenumber space

Introduction

It is often presumed that local isotropy of the small scale components of turbulent velocity fields has been demonstrated for several laboratory flows, for atmospheric boundary layers and for oceanic turbulence. There is even a large body of data which is consistent with local isotropy. Thus, the comparatively few and scattered observations which conflict with predictions based on local isotropy are regarded as insufficiently decisive to shake the consensus that local isotropy exists [1]. It is the purpose of the present paper to point out that the large number of observations consistent with local isotropy do not logically suffice to establish that local isotropy holds. The many observations all verify limited features of the flow field. In particular, the almost exclusive use of the time dependent signal of a single hot wire anemometer, together with the use of Taylor's "frozen turbulence" hypothesis to interpret the signal as the streamwise spatial dependence of a velocity component, means that the data systematically lacks the cross-stream derivatives of velocity components which are needed to fully establish local isotropy. The crossed wire data permit simultaneous measurement of two velocity components, but, again, do not yield cross-stream derivatives.

Even the numerous data establishing the $k^{-5/3}$ power law for velocity spectra in the atmospheric boundary layer are based only on streamwise spatial structure of the velocity correlations, and, thus, do not fully establish Kolmogorov's second similarity hypothesis [1, 2] even though the measurements are certainly consistent with this hypothesis. We will put forward a weakened form of similarity hypothesis, which permits much more general forms of "universal equilibria" than does the Kolmogorov hypothesis, violates local isotropy, and yet agrees with the experimental data that is used to support the local isotropy hypothesis. We put forward theoretical arguments which suggest that one should at least consider the possibility that local isotropy is violated. Naturally, the present hypothesis, permitting more complex equilibria, cannot have the esthetic appeal of the simpler Kolmogorov hypothesis. But since the more complex equilibria have consequences for turbulence modeling, and they cannot be ruled out by present evidence, it seems most reasonable to decide by conclusive measurements whether local isotropy holds in the inertial range. We mention several measurements that could provide additional evidence on this point.

The theoretical arguments suggesting that local isotropy holds in the dissipation range seem fairly strong, yet it isn't clear that they are necessarily valid. Again, we are able to display an example of a two-point velocity correlation tensor which agrees with the experimental data in the dissipation range supporting local isotropy, yet definitely violates local isotropy in the dissipation range. Thus, it is at least logically possible that local isotropy fails to hold to even in the dissipation range; present measurements do not rule out this possibility. Again, the question of whether local isotropy holds is best settled by conclusive measurements.

Weak Form of Similarity Hypotheses

We formulate two weakened similarity hypotheses for incompressible turbulent flows in terms of the mathematical properties of the two-point velocity correlation tensor

$$B_{ij}(\boldsymbol{r}) = \langle u_i(\boldsymbol{x}, t)u_j(\boldsymbol{x}+\boldsymbol{r}, t)\rangle = \int d^3k\, F_{ij}(\boldsymbol{k})e^{i\boldsymbol{k}\cdot\boldsymbol{r}}, \tag{1}$$

where the $\boldsymbol{x}$ and t dependences of B_{ij} have been ignored as usual for the large wavenumber (small $\boldsymbol{r}$) part of the spectrum.

Hypothesis A

For sufficiently large Reynolds numbers an inertial range exists in the $\boldsymbol{k}$ space of F_{ij}. When this occurs F_{ij} is determined by the mean energy dissipation rate ϵ, the mean velocity field and the geometry of the flow field according to the rule

$$F_{ij}(\boldsymbol{k}) = \epsilon^{2/3}\, k^{-11/3} f_{ij}\left(\frac{k_1}{k}, \frac{k_2}{k}, \frac{k_3}{k}\right). \tag{2}$$

Here f does not depend explicitly on ν but may depend locally, or, perhaps functionally on the mean velocity field and on the flow geometry. The $\boldsymbol{x}$ and t dependences of F_{ij} are contained implicitly in ϵ and f.

This is related to Kolmogorov's second similarity hypothesis [1, 2], but it assumes only what is necessary to achieve agreement with the data presently supporting the Kolmogorov hypothesis. According to the Kolmogorov hypothesis

$$f_{ij} = \frac{55\, C}{108\, \pi\Gamma(1/3)}\left(\delta_{ij} - \frac{k_i}{k}\frac{k_j}{k}\right), \tag{3}$$

where C, the constant of the longitudinal structure function, is a pure number independent of the particular high Reynolds number turbulent flow. This is one definite function in the infinite class of functions permitted by Hypothesis A.

The principal theoretical reason for thinking that the asymmetries permitted by Eq. (2) might be needed in the inertial range lies in the possibility that some significant fraction of the small scale structure of a turbulent velocity field may be produced by a few discrete steps of nonlinear evolution of instability, further convective distortion, and breakup, rather than via a nearly infinite cascading process involving adjacent Fourier components. The arguments that the production process necessarily leads to isotropy via multiple interactions of many degrees of freedom have less cogency in this case. However, even in the case of a cascade proceeding roughly along the lines originally envisioned, it is merely plausible, but not entirely certain, that all phase relations and anisotropic gradients would be lost during the cascade process.

The suggestion of several discrete steps for production of fine scale structure remains a theoretical possibility rather than a well validated scientific hypothesis, but some very impressive recent work [3] is not inconsistent with this suggestion. *Corrsin* and *Kollman* [3] give a numerical simulation of a homogeneous shear layer, to which *Taylor-Green* vortices [4] are added as a turbulence – producing finite perturbation. They find that a particular dynamic structure – a stagnation point with the "right" direction of energy transfer – becomes unstable and produces a fairly localized disturbance. One may further show [5] that the stagnation point will itself be a source of vorticity amplification via the localized and highly directional vortex stretching mechanism.

Apparently the first, and only, attempt to account for the fine scale features of a turbulent velocity field, including the inertial range, by means of a two-step process has been provided by *Saffman* [6], who has also questioned the existence of small scale isotropy in turbulent flows [7]. He is able to achieve agreement with most qualitative features of the fine scale structure by means of a "primary cascade", in which a concentration of vorticity is stretched out into a vortex sheet in the presence of the straining field of a general velocity field, and a "secondary cascade", in which the curved vortex sheet becomes unstable to the formation of *Taylor-Gortler* vortices [8, 9]. Of course, this sequence of ideas is speculative and heuristic until the whole theory is developed in much greater detail. Again, it may prove possible to develop a theory of formation of fine scale structure via several discrete steps along rather different lines. Nevertheless, the fact that there is a theoretical possibility of accounting for the fine scale structure of a turbulent velocity field via a production process of several steps seems to be an adequate motivation for formulating the similarity hypotheses from the point of view of clarifying what minimal assumptions need be made to account for the experimental evidence, as done in the present paper, rather than from the point of view of making the greatest simplifications which do not contradict experimental evidence, as done heretofore.

Next, we present a weakened form of Kolmogorov's first similarity hypothesis [1, 2] applicable to both the inertial and dissipation ranges. We introduce the Kolmogorov wave number k_η and velocity v_η and formulate

Hypothesis B

For sufficiently large Reynolds number the high wavenumber portion of the spectrum of F_{ij} is determined by ϵ, ν, the mean velocity field and the flow geometry according to the rule

$$F_{ij}(\boldsymbol{k}) = k_\eta^{-3} v_\eta^2 G_{ij} \left(\frac{k_1}{k_\eta}, \frac{k_2}{k_\eta}, \frac{k_3}{k_\eta} \right) . \tag{4}$$

Here G_{ij} may depend locally or, perhaps, functionally on the mean velocity field, and the flow geometry, but not explicitly on the gradient lengths of the mean flow field. The $\boldsymbol{x}$ and t dependence of F_{ij} are contained implicitly in ϵ and G_{ij}.

Of course, the violation of local isotropy seems less likely intuitively in the dissipation range than in the inertial range, but it cannot be excluded by the present evidence. The rough reduction of the data on one-dimensional energy spectra to a universal curve in the dissipation range suggests that the full dependence of F_{ij} on the flow field permitted by Eq. (4) may be excluded by experiment. However, it should be pointed out that the measured values of ϵ always *assume* local isotropy. Corrected values, allowing for measured rather than assumed cross-stream velocity derivatives, could lead to less scatter in the universal curve. Again, Hypothesis B is all that is necessary to achieve agreement with all the existing data, including dissipation range data, which is used to support the hypothesis of local isotropy.

Experimental Consequences of Hypothesis A

We discuss here the possible experimental consequences should Hypothesis A, rather than Kolmogorov's second similarity hypothesis, happen to be valid in the inertial range. We consider a point in a turbulent fluid and orient the coordinate system so that the mean flow is locally in the x_1 direction. We discuss the general implications of Eq. (2) as well as the specific implications of two model Fourier transformed spectral tensors

$$f_{ij}\left(\frac{k_1}{k}, \frac{k_2}{k}, \frac{k_3}{k}\right) = [A_0 + A_1 k^{-2}(\boldsymbol{k}\cdot\hat{\boldsymbol{a}})^2]\left(\delta_{ij} - \frac{k_i}{k}\frac{k_j}{k}\right), \tag{5}$$

where $\hat{\boldsymbol{a}}$ is a unit vector related to the flow field, A_0 and A_1 are constants determined by the mean velocity field, and secondly,

$$f_{ij}\left(\frac{k_1}{k}, \frac{k_2}{k}, \frac{k_3}{k}\right) = B_0\left(\delta_{ij} - \frac{k_i}{k}\frac{k_j}{k}\right) + \left[\delta\left(\frac{k_1}{k}\right)\delta\left(\frac{k_2}{k}\right)H\left(\frac{k_3}{k}\right)[\delta_{i1}\delta_{j2} + \delta_{j1}\delta_{i2}]\right.$$
$$\left. + \delta\left(\frac{k_1}{k}\right)\delta\left(\frac{k_3}{k}\right)H\left(\frac{k_2}{k}\right)[\delta_{i1}\delta_{j3} + \delta_{j1}\delta_{i3}]\right] \tag{6}$$

where H is a general smooth function decreasing at infinity. The tensors are chosen to illustrate the types of features that may be expected to result from additional measurements, should Hypothesis A be correct.

First, we note that all solutions of Eq. (1) satisfy the observed $k^{-5/3}$ law. Thus Hypothesis A obtains full agreement with the extensive sets of primarily atmospheric and oceanic hot wire measurements which establish the existence of the $k^{-5/3}$ law for one-dimensional longitudinal velocity spectra at high Reynolds number. These measurements are discussed in detail by *Monin* and *Yaglom* [1]. Here, we merely cite the measurements of *Grant* et al. [10], *Pond* et al. [11], and of *Sandborn* and *Marshall* [12] as instances of measured one-dimensional longitudinal velocity spectra which are in reasonable agreement with each other and with the existence of $k^{-5/3}$ law.

One may immediately verify that the general three-dimensional cross spectrum F_{ij} of Hypothesis A will yield the $k^{-5/3}$ law for the spectrum and for any of the measurable one-dimensional cross spectra. For an arbitrary one-dimensional cross spectrum, orient the axes so that the $\hat{\boldsymbol{k}}_1$ axis becomes the unintegrated direction and introduce spherical polar coordinates relative to this axis. Then each one-dimensional cross spectrum takes the form

$$F_{ij}(k_1) = \epsilon^{2/3}|k_1|^{-5/3}C_{ij}, \tag{7}$$

where

$$C_{ij} = \int_{I(\theta)} d\theta \sin\theta \int_0^{2\pi} d\phi |\cos\theta|^{2/3} f_{ij}(\cos\theta, \sin\theta, \cos\phi, \sin\theta \sin\phi). \tag{8}$$

Similarly, the spectrum takes the form

$$E(k) = \epsilon^{2/3} k^{-5/3} C_1 \tag{9}$$

where

$$C_1 = \frac{1}{2} \int_0^{\pi} d\theta \sin\theta \int_0^{2\pi} d\phi\, f_{ij}(\cos\theta, \sin\theta \cos\phi, \sin\theta \sin\phi). \tag{10}$$

Clearly C_{ij} depends on the particular function f_{ij} chosen in Hypothesis A and also on the direction chosen for the axis of the one-dimensional spectra, while C_1, which is related to the Kolmogorov "constant" C, depends only on the particular function chosen for f_{ij}.

Of course, the most general function f_{ij} of Hypothesis A would not satisfy the kinematic constraint

$$k_i F_{ij}(k) = 0, \tag{11}$$

which is demanded by the equation of continuity and is consistent with the (one-dimensional) spectra that have been measured in turbulent flows. Clearly, we must choose f_{ij} to be consistent with Eq. (11). We must also restrict f_{ij} to yield a C_1 which is consistent with the measured range of the Kolmogorov "constant" C, namely, within roughly 50% of 2.0, as discussed in Ref. [1].

The principal new content of Hypothesis A in the inertial range is that the Kolmogorov "constant" C of the longitudinal structure function, and the corresponding constants of the other measurable one-dimensional spectra, are not constants, but functions of the velocity gradients in the flow field. Doubts have occasionally been expressed [6, 7] about the universality of the Kolmogorov "constant". Here this variability follows as a direct consequence of our similarity hypothesis and is illustrated by reference to the model spectral tensors of Eqs. (5) and (6). We note that the longitudinal structure function

$$D_{LL}(r) = C \epsilon^{2/3} r^{2/3} \tag{12}$$

generated by these spectral functions gives the Kolmogorov "constant"

$$C = \frac{108\,\pi}{17 \cdot 11 \cdot 5} \Gamma(1/3) \{17 A_0 + [12 - 13 (\hat{\boldsymbol{r}} \cdot \hat{\boldsymbol{a}})^2] A_1\}, \tag{13}$$

and

$$C = \frac{108\,\pi}{11 \cdot 5} \Gamma(1/3) B_0, \tag{14}$$

for Eqs. (5) and (6), respectively. In these equations, A_0, A_1, B_0, and $\hat{\boldsymbol{a}}$ may be expected to be weak functions of local velocity gradients. In both Eqs. (13) and (14), the Kolmogorov "constant" is thus a weak function of velocity gradients, but in addition, this "constant" depends on the direction of $\boldsymbol{r}$ in Eq. (13).

Finally, both Eqs. (5) and (6) yield structure functions that contain deviations from isotropic form. Specifically, these tensors are given by

$$D_{ij}(r) = D_{ij}^0 + D_{ij}^1, \tag{15}$$

where D_{ij}^0 is the usual isotropic contribution generated by A_0 and B_0 terms in Eqs. (5) and (6) respectively,

$$D_{ij}^1 = C_2 \epsilon^{2/3} r^{2/3} \{-[6 + 11(\hat{r}\cdot\hat{a})^2]\hat{r}_i\hat{r}_j + [42 - 7(\hat{r}\cdot\hat{a})^2]\delta_{ij} - 9\,\hat{a}_i\hat{a}_j - 6(\hat{r}\cdot\hat{a})[\hat{r}_i\hat{a}_j + \hat{a}_i\hat{r}_j]\} \tag{16}$$

for Eq. (5), and

$$D_{ij}^1 = C_3\{\epsilon^{2/3} r_3^{2/3}[\delta_{i1}\delta_{j2} + \delta_{j1}\delta_{i2}] + \epsilon^{2/3} r_2^{2/3}[\delta_{i1}\delta_{j3} + \delta_{j1}\delta_{i3}]\} \tag{17}$$

for Eq. (6). In these latter two equations

$$C_2 = \frac{36\,\pi}{17\cdot 11\cdot 5}\,\Gamma(1/3)A_1, \tag{18}$$

and

$$C_3 = 3\,H(1)\Gamma(1/3). \tag{19}$$

The D_{ij}^1 of Eq. (16) contributes to the Kolmogorov "constant", while the D_{ij}^1 of Eq. (17) being traceless, does not. As noted earlier, it would be necessary to measure cross-stream components of D_{ij} to detect the presence of components such as the D_{ij}^1, and this has not been done.

Experimental Consequences of Hypothesis B

We discuss here some of the experimental consequences should Hypothesis B, rather than Kolmogorov's first similarity hypothesis, happen to be valid. We illustrate the effects to be expected by means of the model Fourier transformed spectral tensor

$$G_{ij}\left(\frac{k_1}{k_\eta}, \frac{k_2}{k_\eta}, \frac{k_3}{k_\eta}\right) = B_0\left(\frac{k^2}{k_\eta^2}\right)\left[\delta_{ij} - \left(\frac{k_\eta^2}{k^2}\right)\frac{k_i}{k_\eta}\frac{k_j}{k_\eta}\right] + \left[\delta\left(\frac{k_1}{k_\eta}\right)\delta\left(\frac{k_2}{k_\eta}\right) H\left(\frac{k_3}{k_\eta}\right)[\delta_{i1}\delta_{j2} + \delta_{j1}\delta_{i2}] + \delta\left(\frac{k_1}{k_\eta}\right)\delta\left(\frac{k_3}{k_\eta}\right) H\left(\frac{k_2}{k_\eta}\right)[\delta_{i1}\delta_{j3} + \delta_{j1}\delta_{i3}]\right] \tag{20}$$

where, again, the mean flow is locally in the x_1 direction, B_0 is a weak function of velocity gradients and a general function of $(k/k_\eta)^2$, and H is a general smooth function decreasing at infinity. The experimental consequences may be discussed by means of the spectral tensor of Eq. (20) and of the structure functions generated from it. When the two particular structure functions

$$D_{\mathrm{LL}}(rk_\eta) = 8\pi v_\eta^2 \int_0^\infty dl l^2 B_0(l^2) \left[\frac{2}{3} + \frac{\cos(lrk_\eta)}{(lrk_\eta)^2} - 2\frac{\sin(lrk_\eta)}{(lrk_\eta)^3} \right], \tag{21}$$

and

$$D_{\mathrm{NN}}(rk_\eta) = 8\pi v_\eta^2 \int_0^\infty dl l^2 B_0(l^2) \left[\frac{2}{3} - \frac{\sin(lrk_\eta)}{(lrk_\eta)} - \frac{\cos(lrk_\eta)}{(lrk_\eta)^2} + \frac{\sin(lrk_\eta)}{(lrk_\eta)^3} \right] \tag{22}$$

are introduced, the tensor of structure functions from Eq. (20) takes the form

$$\begin{aligned} D_{ij}(r) = {} & [D_{\mathrm{LL}}(rk_\eta) - D_{\mathrm{NN}}(rk_\eta)] k_\eta^2 r_i r_j + D_{\mathrm{NN}}(rk_\eta)\delta_{ij} \\ & + 2 v_\eta^2 [K(r_3 k_\eta)[\delta_{i1}\delta_{j2} + \delta_{j1}\delta_{i2}] + K(r_2 k_\eta)[\delta_{i1}\delta_{j3} + \delta_{j1}\delta_{i3}] \end{aligned} \tag{23}$$

where

$$K(x) = \int_0^\infty dl H(l)(1 - \cos lx). \tag{24}$$

There are thus non-isotropic corrections to the structure functions.

We note that the three-dimensional cross spectra of Eq. (20) and the structure functions of Eq. (23) satisfy all of the experimental relations used to support the assertion that local isotropy holds in turbulent flows, both in the inertial range and in the dissipation range, with the single exception of the $k^{-5/3}$ law in the inertial range, for which the more restrictive assumptions of Hypothesis *A* are needed. Specifically, the two tests that have been applied experimentally are, for velocity derivatives,

$$\left\langle \left(\frac{\partial u_1}{\partial x_1} \right)^2 \right\rangle = \frac{1}{2} \left\langle \left(\frac{\partial u_2}{\partial x_1} \right)^2 \right\rangle = \frac{1}{2} \left\langle \left(\frac{\partial u_3}{\partial x_1} \right)^2 \right\rangle, \tag{25}$$

where the mean flow is locally in the k_1 direction and for the one-dimensional cross spectrum,

$$F_{12}(k_1) = 0. \tag{26}$$

Now Eq. (23) is compatible with Eq. (25), while Eq. (20) is compatible with Eq. (26). Equation (25) has been verified by *Townsend* in grid turbulence [13] and also, and in much greater detail, in the wake behind a cylinder [14] and by *Champagne* et al. [15] in homogeneous shear flow. Equation (26) has also been verified experimentally at high wave numbers, or, equivalently, at high frequencies, by *Corrsin* in a round jet [16], by *Klebanoff* in a boundary layer [17], by *Laufer* in channel flow [18] and by *Champagne* et al. [15] in homogeneous shear flow. Additional experimental confirmations are cited by *Monin* and *Yaglom* [1], who also give references to the scattered literature of measurements which contradict the assumption of local isotropy.

The fact that the one-dimensional spectra appear to lie nearly on universal curves in the dissipation range [10–12] suggests that the full generality of Eq. (4) may not be needed in the dissipation range. Instead, the function $B_0(k^2/k_\eta^2)$ may approach a universal function, apart from a normalization factor, at very large wavenumbers. We note that the normalizing of all experimental results to the Kolmogorov scales involves computing the energy dissipation ϵ through the assumption of local isotropy. Using a full set of cross-stream velocity derivatives to evaluate ϵ

could modify most results and conclusions somewhat. In particular, it could happen that the universal curves referred to above are satisfied with greater accuracy (less scatter) than might appear from present measurements.

Finally, we observe that the deviations of Eq. (23) from the mathematical form of local isotropy, while substantial in size if B_0 and H are comparable, cannot be detected by the standard experimental criteria for the validity of local isotropy. Measurements of cross-stream variations in velocity correlations, which have not been carried out, would be required to disclose their presence.

Possible Two-Point Measurement Procedures

In principle, it would seem possible that any measuring technique for two-point velocity correlations, with or without time delay or spectral analysis, might serve as the basis for measurements that could test for the presence of non-isotropic components in the small scale structure of turbulent velocity fields. In particular, the techniques adapted by *Favre, Gaviglio,* and *Dumas* in their pioneering series of investigations, summarized in Ref. [19], on two-point, space-time, velocity correlations might be considered for such usage. However, the observations of *Tritton* [20] suggest that the possible interference effects of the supports of the hot wires may require further investigation before they can be safely dismissed. Possibly the technique suggested by *Bradshaw* [21], of seeing whether measurements are invariant to small changes in the size of the support structures for a given probe geometry, would suffice for this purpose. Even if the question of possible support interference is resolved satisfactorily, there may be considerable difficulty in making two-point measurements at sufficiently small spatial separations.

A method which might avoid some of the difficulties could be based on a two-point measurement scheme proposed by *Poreh* et al. [22]. In this scheme two hot wires, with independent electronics and linearized voltage outputs lie in the $x_2 - x_3$ plane and are oriented parallel to the x_3 axis, while the mean flow is parallel to the x_1 axis. The time average of the product of the output voltages is measured, noise terms drop out, and a spatial integral of the two-point velocity correlation is obtained. The time derivative of the corresponding average formed when one voltage is time delayed, when used with Taylor's frozen turbulence hypothesis, can be combined with the preceding measurement to obtain much of the information obtainable from measurements at a whole series of spatial separations of the two wires. The fact that the time-averaged product of the output voltages has a simple physical interpretation (proportional to a spatial integral of a two-point velocity correlation) in this scheme is dependent on the output voltage of each hot wire being strictly proportional to the fluid flow velocity transverse to the wire. For this reason it is probably worth taking unusual pains to linearize the anemometer output in this method.

The technique seems interesting because measurements have already been reported [22] at spatial separations of ~ 0.2 mm for the two wires, and this would seem to be sufficient for testing possible non-isotropy. Moreover, interference effects appear to be small in the reported measurements.

The lack of definite theoretical predictions for the mathematical form of the nonisotropic part of the two-point velocity correlations has the consequence that no definite predictions for the expected form of the output signal can be given, if non-isotropic effects are significant. However, it would be possible to use mathematical models of possible two-point velocity correlations to explore the possible consequences of non-isotropy on the spatially integrated two-point velocity correlation measured by experiment. For example, in cases where the wire separation and the wire lengths are small compared to the Taylor microscale, functions such as those of Eqs. (5) and (6) (more specifically, the Fourier transforms of such functions) might be used to affect the spatial integrations. When the physical lengths are large compared to the Taylor microscale, functions more adequately modeling the asymptotic form of the two-point velocity correlation at large spatial separation should be used.

Consequences for Turbulence Modeling

It has been pointed out [23] that the usual [24, 25] high Reynolds number modeling of the viscous terms in the Reynolds stress equations, namely, ignoring the viscous diffusion and replacing the energy dissipation effects by the factor

$$\frac{2}{3}\,\epsilon\delta_{ij} \tag{27}$$

on the right hand side of the Reynolds stress equation, has the potentiality for violating the realization conditions [26] of the Reynolds stress tensor. This is not to say that Reynolds stress tensors computed with the aid of Eq. (27) will often be unphysical, but only that this equation violates the principle of super-realizability formulated in Ref. [23] – that modelings of the physical groups of terms (the triple correlations, the pressure-velocity correlations, and the viscous terms) should separately satisfy the realizability criteria that the exact solutions of the Navier-Stokes equations do satisfy – and, thus, may on occasion give trouble.

It is usually felt that Eq. (27) is fully supported by experiment. Indeed, *Corrsin* [27] cites Refs. [13, 14, 16, 18] as evidence that local isotropy has been shown to hold in laboratory flows and that, *a fortiori,* Eq. (27) is valid. We have shown in previous sections of this paper that these, and other, works do not suffice to demonstrate that local isotropy holds in turbulent flows. In consequence, Eq. (27) need not be assumed, and no existing measurements suffice to specify the modeling of either the pressure-velocity correlations or the energy dissipation tensor. It should be regarded, at present, as an unsolved problem for Reynolds stress closure models of how to partition the terms of the model which are algebraic in the Reynolds stresses between the two physically distinct types of terms – the pressure-velocity correlations and the energy dissipation tensor.

The situation could be resolved by direct measurements of the pressure-velocity correlations or of all the components of the energy dissipation tensor. One could then see what terms of the turbulence model best represent the measured tensor. Measurements of the pressure-velocity correlations are thought to be extremely difficult. But measurements of the components of the energy dissipation tensor would involve nothing more intricate than measuring cross-stream variations of the velocity correlations – measurements which are needed in any case to establish the existence of local isotropy in turbulent flows

Acknowledgments. The author is indebted to Drs. F. H. Harlow and H. M. Ruppel for numerous helpful discussions and to Dr. H. A. Rose for a critical reading of an earlier version of this manuscript.

This work was supported by the U.S. Department of Energy.

References

1. A. S. Monin, A. M. Yaglom: *Statistical Fluid Mechanics: Mechanics of Turbulence,* Vol. 2 (The MIT Press, Cambridge 1975) Chap. 8
2. A. N. Kolmogorov: Dokl. Akad. Nauk **30,** 301–305 (1941)
3. S. Corrsin, W. Kollman: In *Turbulence in Internal Flows*, Turbomachinery and Other Engineering Applications, ed. by S. N. B. Murthy (Hemisphere, Washington 1977) pp. 11–28
4. G. I. Taylor, A. E. Green: Proc. R. Soc. Ser. A: *158,* 499–521 (1937)
5. R. C. Mjolsness, (To be submitted to Phys. Fluids)
6. P. G. Saffman: "Lectures on Homogeneous Turbulence", in *Topics in Nonlinear Physics,* Sec. 7, ed. by N. J. Zabusky (Springer, Berlin, Heidelberg, New York 1968) pp. 557–581
7. P. G. Saffman: "Problems and Progress in the Theory of Turbulence", *Structure and Mechanism of Turbulence,* Vol. II, ed. by H. Fiedler (Springer, Berlin, Heidelberg, New York 1978) pp. 273–305
8. G. I. Taylor: "Stability of a viscous liquid contained between two rotating cylinders". Philo. Trans. R. Soc. London, Ser. A: **223,** 289–343 (1923)
9. H. Gortler: "On the Three-Dimensional Instability of Laminar Boundary Layers on Concave Walls". NACA Technical Memorandum 1375 [Translation of "Mathematisch-Physikalische Klasse, Nachrichten von der Akademie der Wissenschaften in Göttingen, aus dem Jahre 1941"] pp. 1–26
10. H. L. Grant, R. W. Stewart, A. Moilliet: "Turbulence spectra from a tidal channel". J. Fluid Mech. **12,** 241–268 (1962)
11. S. Pond, R. W. Stewart, R. W. Burling: "Turbulence spectra in the wind over waves". J. Atmos. Sci. **20,** 319–324 (1963)
12. V. A. Sandborn, R. D. Marshall: "Local Isotropy in Wind Tunnel Turbulence". Technical Report of the Fluid Dynamics and Diffusion Laboratory, Colorado State University (1965)
13. A. A. Townsend: "Experimental evidence for the theory of local isotropy". Proc. Cambridge Philos. Soc. **44,** 560–565 (1948)
14. A. A. Townsend: "Local isotropy in the turbulent wake of a cylinder". J. Aus. Sci. Res. A1, 161–174 (1948)
15. F. H., Champagne, V. G. Harris, S. Corrsin: "Experiments on nearly homogeneous turbulent shear flow". J. Fluid Mech. **41,** 81–139 (1970)
16. S. Corrsin: "An experiment verification of local isotropy". J. Aeronaut. Sci. **16,** 757–758 (1949)
17. P. S. Klebanoff: "Characteristics of Turbulence in a Boundary Layer with Zero Pressure Gradient". NACA Report No. 1247 (1955)
18. J. Laufer: "Some recent measurements in a two-dimensional turbulent channel". J. Aeronaut. Sci. **17,** 277–287 (1950)
19. A. Favre, J. Gaviglio, R. Dumas: "Space-Time Correlations in Fluid Flows", in *The Mechanics of Turbulence* (Gordon and Breach, New York 1964) pp. 419–445
20. D. J. Tritton: "Note on the effect of a nearby obstacle on turbulence intensity in a boundary layer". J. Fluid Mech. **28,** 433–437 (1967)
21. P. Bradshaw: *An Introduction to Turbulence and Its Measurements* (Pergamon Press, Oxford 1971) pp. 184–185
22. M. Poreh, I. Landa, I. Kidron: "On the measurement of low level turbulence using a pair of sensors". Isr. J. Technol. **7,** 55–60 (1969)
23. R. C. Mjolsness: "Systematic Modeling Rules for Reynolds Stress Closures in Free-Shear Layers for Reynolds Stress Closures in Free-Shear Layers". Symposium on Turbulent Boundary Layers, American Society of Mechanical Engineers, Niagara Falls, NY (1979) pp. 139–144
24. B. E. Launder, G. J. Reece, W. Rodi: "Progress in the development of a Reynolds-stress turbulence closure". J. Fluid Mech. **68,** 537–566 (1975)
25. J. L. Lumley, B. Khajeh-Nouri: "Computational modeling of turbulence transport". Adv. Geophys. **18** A, 169–192 (1974)
26. U. Schumann: "Realizability of Reynolds-stress turbulence models". Phys. Fluids **20,** 721–725 (1977)
27. S. Corrsin: "Comments on 'Transport Equations in Turbulence' ". Phys. Fluids **16,** 157–158 (1973)

Multiple-Time-Scale Concepts in Turbulent Transport Modelling

K. Hanjalić[1], B. E. Launder, and R. Schiestel[2]

Mechanical Engineering Department, University of California,
Davis, CA 95616, USA

Abstract

The paper reports progress in developing a turbulence closure employing two or more independently calculated time scales with which to characterize the rates of progress of different turbulent interactions. The approach contrasts with that employed by earlier single-point models which adopt just a single time scale, proportional to the turbulence energy turnover time. The present treatment essentially divides the energy containing part of the spectrum into two regions which respond at different rates and in different ways to changes in the environment. The scheme may be regarded as providing an intermediate level of approximation between the relatively simple, but fallible, single-point closures and the vastly more elaborate two-point closures which have so far been applied only to simulating homogeneous flows.

The proposed approach requires only slightly more computational effort than single-scale schemes. Computational results are reported for several thin shear flows which show striking improvement in the level of agreement with experiment over that obtained with models employing only one time scale.

Nomenclature

- A — Anisotropy of turbulence $(\overline{u_i u_j} - 2/3\, \delta_{ij} k)^2/k^2$
- C_f — Skin friction coefficient $\tau_w/\frac{1}{2}\rho U_\infty^2$
- C_{P1}, C_{P2} ... etc. — Coefficients in turbulence model [defined following Eq. (15)]
- $\mathscr{D}_\phi$ — Net diffusion rate of φ [modelled by Eq. (16) or Eq. (20)]
- k — Turbulent kinetic energy
- k_P, k_T — Turbulent kinetic energies associated with production and transfer ranges
- l — Length scale associated with large-scale turbulent interactions
- M — Distance between bars of turbulence-generating grid
- P_k — Shear Production rate of turbulence energy [Eq. (2)]
- P — Static pressure
- U — Mean x-directed velocity
- U_i — Mean velocity in x_i direction
- U_0 — Reference x velocity at start of adverse pressure gradient
- $U_{℄}$ — Center-line velocity
- u_i — Fluctuating component of velocity in x_i direction

---

[1] Present address: Mašinski Fakultet, Sarajevo 71000, Jugoslavia.

[2] Present address: Inst. de Mécanique Statistique de la Turbulence, Marseille, France.

$\overline{u_i u_j}$	Kinematic Reynolds stress	δ_1	Boundary layer displacement thickness
$\overline{u_i u_{j\mathrm{P}}}$		ϵ, ϵ_1	Dissipation rate of turbulent energy
$\overline{u_i u_{j\mathrm{T}}}$	Reynolds stresses of "production" and "transfer" ranges	ϵ_P	Kinetic energy transfer rate out of production range
$\overline{uv}$	Kinematic turbulent shear stress in thin shear flow	ϵ_T	Kinetic energy transfer rate across inertial subrange
x	Streamwise distance coordinate	ν_t	Turbulent kinematic viscosity
y	Cross-stream coordinate	τ	Time scale of turbulent interactions
x_ref	Reference position	τ_w	Wall shear stress
β	Pressure-gradient parameter $(\delta_1/\tau_w)x(dp/dx)$		

1. Introduction

The fluid mechanics literature of the past sixty years abounds with proposals for calculating the correlations between fluctuating velocities that appear as unknowns in the Reynolds equations, i.e. the temporal- (or ensemble-) averaged forms of the Navier Stokes equations.

In devising approximations for these Reynolds stresses a choice must be made for the time scale or scales appropriate to the fluctuating velocity interactions. The Prandtl-Taylor mixing-length hypothesis [*Taylor* (1915); *Prandtl* (1925)] implicitly adopts the time scale provided by the mean rate of strain which, for a two-dimensional thin shear flow takes the form:

$$\tau \propto 1/\left|\frac{\partial U}{\partial y}\right| \tag{1}$$

where U is the mean velocity and y the direction of principal velocity gradient.

One limitation of Eq. (1) is readily apparent: as the mean strain vanishes the time scale tends to infinity, implying zero turbulent mixing. This objection is avoided in more recently proposed models by employing one or more (necessarily approximate) transport equations whose solution allows the local turbulent time scale to be computed. The earliest such scheme was that of *Kolmogorov* (1942) who suggested a heuristic equation for the rate of change, following a streamline, of a characteristic turbulence frequency, which we may interpret as τ^{-1}. Approaches intrinsically similar to Kolmogorov's have become increasingly popular over the last decade. The commonly adopted route is to solve approximate, closed equations for the transport of turbulent kinetic energy, k and its rate of dissipation ϵ; the time scale is then taken proportional to (k/ϵ).

The *way* that the time scale is employed depends, of course, on the nature of the closure hypotheses that are adopted. Closures like Kolmogorov's that assume the Reynolds stresses to be characterizable in terms of an effective viscosity take the kinematic turbulent transport coefficient, ν_t, proportional to l^2/τ which, in terms of the turbulent energy and dissipation rate, implies

$$\nu_\mathrm{t} \propto (k^{3/2}/\epsilon)^2/(k/\epsilon) = k^2/\epsilon$$

With the more elaborate and, potentially, more general second moment closures, the closed set of transport equations for the Reynolds stresses contain source, sink and transport processes which are held to proceed at rates proportional to ϵ/k.

It is recognized, however, that turbulence comprises fluctuating motions with a spectrum of sizes and time scales. Since different turbulent interactions are known to be associated with different parts of the spectrum the idea that one can mimic the response of the Reynolds stresses with a mathematical model containing just *one* time scale seems highly simplistic. The fact that, nevertheless, numerous successful applications of single-time-scale models have been reported in the literature reflects more that the flows considered have been fairly close to a spectral equilibrium than that the single-time-scale hypothesis is generally adequate.

In the present contribution we develop a simple multiple-time-scale model based on a rational extension of widely used single-scale equations and ideas. As with any single-point closure the model contains a number of coefficients whose magnitude must be fixed by reference to experiments. From the various shear flows considered to date, however, there seems reason to infer that the coefficients chosen within the framework of a multiple-time-scale closure can have a wider range of applicability than is the case when a conventional single-scale model is adopted.

2. The Physical Model

2.1. The Underlying Ideas

Figure 1 shows a sketch of a typical energy spectrum arising in a shear-flow at high Reynolds numbers. The quantity κ_1 denotes the wave number above which no significant mean-strain production occurs while κ_2 is the largest wave number at which viscous dissipation of turbulence energy is unimportant. Energy leaves the first region (the "production" region) at a rate ϵ_p and enters the high-wave-number or "dissipation" region at a rate ϵ. Between these two zones, occupying an intermediate range of wave numbers is the "transfer" region, across which we imagine a representative spectral energy transfer rate to be ϵ_T. It is from this rudimentary, highly simplified picture of the turbulent energy spectrum that the present closure is developed. Lest the spectral divisions appear to be too coarse to be useful, it should perhaps be remarked that published single-point closures take *no* cognizance of changes in the shape of the energy spectrum; indeed, there is implicit the assumption that $\epsilon_P = \epsilon_T = \epsilon$. While this state of affairs closely pertains in, say, a slowly developing flow such as a boundary layer on a flat plate it is not generally the case.

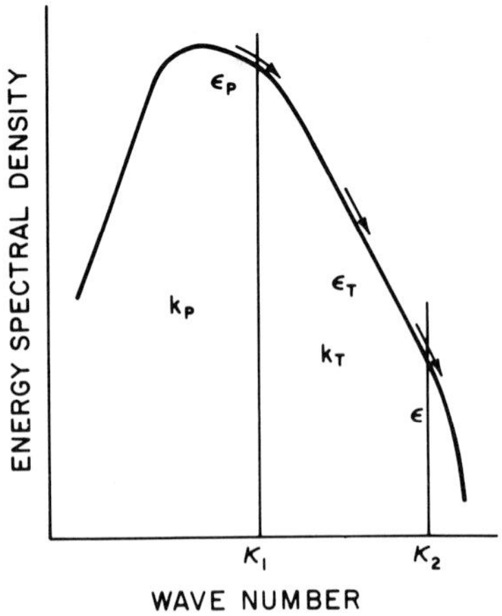

Fig. 1. The spectral division adopted in the model

The total turbulent energy, k, is assumed to be divided between the production range (k_P) and the transfer range (k_T). At high Reynolds numbers there is negligible kinetic energy associated with the dissipation range. In a homogeneous flow the level of these partitioned energies is controlled by the transport equations:

$$\frac{Dk_P}{Dt} = \underbrace{-\overline{u_i u_j}\frac{\partial U_i}{\partial x_j}}_{P_k} - \epsilon_P \tag{2}$$

$$\frac{Dk_T}{Dt} = \epsilon_P - \epsilon \tag{3}$$

where P_k denotes the production rate of turbulence energy by mean shear which, as remarked above, is assumed to be entirely contained in wave numbers less than κ_1.

2.2. The Energy Transfer Equations

The dynamic response of Eqs. (2) and (3) depends on the levels of the energy transfer rates ϵ_P and ϵ. The main task in our approach to providing a number of independent turbulent time scales is that of devising a pair of transport equations which adequately characterize the evolution of these transfer rates. We are guided in this principally by the available *single*-scale models that provide a diagnostic equation for the "energy dissipation rate". These may be expressed in the following form:

$$\frac{D\epsilon_1}{Dt} = C_{\epsilon 1} P_k \frac{\epsilon_1}{k} - \frac{\epsilon_1^2}{k}(C_{\epsilon 2} + C_{\epsilon 3} A) + \mathscr{D}_{\epsilon 1} \tag{4}$$

Here A denotes the turbulence anisotropy defined as

$$(\overline{u_i u_j} - \tfrac{2}{3}\delta_{ij}k)(\overline{u_i u_j} - \tfrac{2}{3}\delta_{ij}k)/k^2,$$

$\mathscr{D}_{\epsilon 1}$ denotes diffusive transport and the subscript 1 has been added to ϵ as a reminder that Eq. (4) is an equation for use in a single-scale scheme. The C's are coefficients to be determined experimentally. Table 1 gives the values recommended by two groups[3]. We note that the anisotropy A varies across a shear flow in a rather similar way to P_k/ϵ; the absence of any effect of anisotropy in the proposal of *Launder, Reece,* and *Rodi* (1975) (hereafter LRR) is

Table 1. Examples of coefficients in ϵ_1 equation

Authors	$c_{\epsilon 1}$	$c_{\epsilon 2}$	$c_{\epsilon 3}$
LRR (1975)	1.44	1.90	0
Zeman and *Lumley* (1979)	0.47	1.90	3.5

[3] In fact the Zeman-Lumley proposal cannot be exactly cast in the form of Eq. (4). The values given are however fairly representative of those implied by their model.

accordingly, accompanied by a larger value of the coefficient $C_{\epsilon 1}$ than that proposed by *Zeman* and *Lumley* (1979).

Now, despite its nominal role as a *dissipation* rate equation, Eq. (4) does not make sense as such; for it makes the local rate of change of ϵ_1 dependent on the local mean strain rate and the anisotropy of the stress field, neither of which, under conditions of local isotropy, can directly affect the dissipation rate. Several workers have remarked that the subject of Eq. (4) should correctly be regarded as a spectral energy transfer rate associated with the large-scale interactions; that is, in terms of our multiple-time-scale formulation, ϵ_P. Accordingly the initial form chosen for the ϵ_P equation was:

$$\frac{D\epsilon_P}{Dt} = C_{P1} P_k \frac{\epsilon_P}{k_P} - C_{P2} \frac{\epsilon_P^2}{k_P} + \mathscr{D}_{\epsilon P} \tag{5}$$

where the partitioned energy k_P replaces the total energy giving, as characteristic time scale, the energy turnover time of the large-scale motions.

In choosing the form that the corresponding equation for ϵ_T should take, certain basic requirements were evident. First, the equation should contain both source and sink terms on its right hand side since, in the decay of grid turbulence, the level of ϵ_T must decrease downstream while, if the level of the turbulence energy is raised, ϵ_T must, in due course, rise as a precursor to an increase in dissipation rate. It would, however, be contrary to established views of the spectrum to make ϵ_T respond (like ϵ_P) directly to an applied mean strain. Now, the factor which may be regarded as instrumental in raising ϵ_T is an increase in the energy flow rate from the production range into the transfer range. Accordingly, being guided by Eq. (5), the form adopted for the ϵ_T transport equation is:

$$\frac{D\epsilon_T}{Dt} = C_{T1} \frac{\epsilon_P \epsilon_T}{k_T} - C_{T2} \frac{\epsilon_T^2}{k_T} + \mathscr{D}_{\epsilon T} \tag{6}$$

We note that, in a flow where the turbulence energy generation is suddenly switched off, there is no necessity for ϵ_T (or ϵ) immediately to decrease since ϵ_P does not fall to zero. Thus, the energy dissipation rate now responds more slowly to the applied mean strain field. This characteristic makes the present form better able to represent rapidly changing turbulence fields than Eq. (4).

Initially it was planned to provide a transport equation for ϵ which, in structure, was similar to (but simpler than) Eqs. (5) and (6), *Launder* and *Schiestel* (1978). To keep the mathematical framework as simple as possible, however, we have here assumed spectral equilibrium between the transfer region and the dissipation regions:

$$\epsilon = \epsilon_T \tag{7}$$

We felt that, if there were significant improvements to be gained from using a multiple-time-scale approach, these advantages would be perceivable in a model that adopted just two independent spectral transfer rates. Later, if the results were promising but needed refinement, the spectrum could be divided in three, or thirty slices, according to taste and computer budget.

In making preliminary predictions of variously strained turbulent flow, it became evident that the generation term in Eq. (5) did not possess much width of applicability. For example, to reproduce the development of turbulence in an axisymmetric contraction a coefficient nearly twice as large was needed as for a simple shear flow. In the former flows energy generation is by *irrotational* straining. The implication seemed to be that this kind of deformation proved more effective in transferring energy across the spectrum than rotational shears. Such a preferential transfer may conveniently be introduced to the mathematical model by the addition to Eq. (5) of the following term containing the mean vorticity:

$$C'_{\mathrm{P}1} k_{\mathrm{P}} \frac{\partial U_l}{\partial x_m} \frac{\partial U_i}{\partial x_j} \epsilon_{lmk} \epsilon_{ijk} \tag{8}$$

where ϵ_{lmk} is the alternating third order tensor. The term vanishes in an irrotational flow and (apart from the sign of the coefficient) is positive in a rotational strain. The coefficient $C'_{\mathrm{P}1}$ thus needs to be *negative* to produce higher overall energy transfer rates in an irrotational strain.

It may be noted in passing that the term contained in Eq. (8) exhibits some similarity with an additional terms introduced recently by *Pope* (1978) to the (single-scale) dissipation equation:

$$\frac{k^2}{\epsilon} \left(\frac{\partial U_i}{\partial x_j} + \frac{\partial U_j}{\partial x_i} \right) \left(\frac{\partial U_i}{\partial x_k} - \frac{\partial U_k}{\partial x_i} \right) \left(\frac{\partial U_k}{\partial x_j} - \frac{\partial U_j}{\partial x_k} \right) \quad . \tag{9}$$

The term produces greater energy transfer rates when the mean flow vortex lines are stretched with passage downstream; it is of higher order than Eq. (8) and for this reason no attempt has yet been made to introduce its equivalent into Eq. (5).

It remains to assign the various coefficients in Eqs. (5), (6), and (8). The coefficients in Eq. (4), the single-scale counterpart of Eqs. (5) and (6), have usually been set to constant values and the initial impulse is to do likewise in the multiple-scale characterization. There are, however, two reasons for not doing so. Firstly, research on the much more elaborate spectral or "two-point" closures has shown the necessity of letting the middle range eddies "talk back" to the larger scales[4]. If both $C_{1\mathrm{P}}$ and $C_{2\mathrm{P}}$ are taken as constants the equation for ϵ_{P} is virtually independent of the medium scale motion since neither k_{T} nor ϵ_{T} appear in Eq. (5). Secondly, the adoption of constant coefficients in Eq. (4) was at least partly due to the absence of a suitable parameter of which they could become arguments. Now that the energy spectrum is divided into two parts, two parameters are available, $k_{\mathrm{P}}/k_{\mathrm{T}}$ and $\epsilon_{\mathrm{P}}/\epsilon_{\mathrm{T}}$. The first characterizes the shape of the energy spectrum, the second the degree of spectral imbalance. It is however a task beyond the scope of the present study to correlate the coefficients as functions of *two* independent parameters. The intuitive plumbing analogy illustrated in Fig. 2 suggests that the energy flow rate ϵ_{P} should be more dependent on the ratio of the partitioned energies than of their transfer rates. Correspondingly, in the transport equation for the energy transfer rate across the finer scale motion, the ratio of the transfer rates $\epsilon_{\mathrm{P}}/\epsilon_{\mathrm{T}}$ is clearly the preferred parameter since these finer-scale eddies should have no direct knowledge of the amount of energy contained in the largest motions. At present only one term in each of the transfer equations has

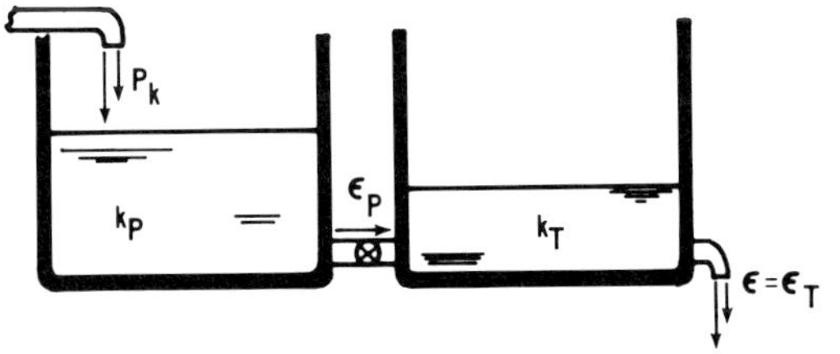

Fig. 2. Tank-and-tube analogy of spectral transfer of energy

[4] We are grateful to Dr. U. Frisch for this observation.

been allowed to depend on these energy or transfer-rate ratios; this limitation was again imposed to keep the task of optimisation within bounds.

A further basic point is where the division between the production and transfer region should be placed. If the partition is moved to too high wave numbers, such a small proportion of the total energy will be contained in k_T that the time for energy to cross the transfer region will be negligible. In this case we should find $\epsilon_T \doteq \epsilon_P$ and the calculated flow behavior would be negligibly different from that of a single scale model. If, on the other hand, the division is made at too low a wave number the assumption that there is zero energy production in the k_T equation becomes untenable. Provided due recognition is taken of these two limiting constraints, our experience is that the precise partitioning point does not significantly affect the predictions, provided coefficients have been appropriately tuned[5].

Turbulence decaying downstream of a grid quickly evolves to a form where energy generation is negligible. With the present model, this is described by the following system of equations.

$$\frac{Dk_P}{Dt} = -\epsilon_P \tag{10}$$

$$\frac{Dk_T}{Dt} = \epsilon_P - \epsilon_T \tag{11}$$

$$\frac{D\epsilon_P}{Dt} = -C_{P2}\frac{\epsilon_P^2}{k_P} \tag{12}$$

$$\frac{D\epsilon_T}{Dt} = C_{T1}\frac{\epsilon_P\epsilon_T}{k_T} - C_{T2}\frac{\epsilon_T^2}{k_T} \tag{13}$$

We assume that during the decay process the relative levels of k_P and k_T eventually reach a constant value. For this asymptotic state the solutions to Eqs. (10–13) imply $k_P \propto t^{-n}$. From Eqs. (10) and (12) it is readily deduced that

$$C_{P2} = 1 + 1/n \tag{14}$$

Numerous experiments on grid-turbulence have confirmed the power-law decay though there is a certain amount of variation in the values reported for n. *Lumley* and *Newman* (1977) have proposed an elaborate correlation in which the local exponent is dependent on the turbulence anisotropy and the Reynolds number. For the present we have preferred not to include such possible dependencies in our scheme. The choice of a fixed exponent $n = 1.25$ fits all the measurements within $\pm$ 0.2, a variation which must lie nearly within the uncertainty limits of the data considering that they rarely span more than one decade of energy level and that the effective time origin is unknown. Accordingly, from Eq.(14) we take C_{P2} = 1.8, where the ∞ subscript is a reminder that this value relates to an asymptotic state.

If we now proceed in the same way to consider the equations for k_T and ϵ_T the following relationship between the energy ratios and turbulence-model coefficients is obtained:

---

5 As we discuss later, however, it appears that it may be desirable to shift the spectral division to lower wave numbers and abandon the notion that the energy in the transfer range is isotropic.

$$\left(\frac{k_P}{k_T}\right) = \left(\frac{C_{P2} - C_{T2}}{C_{T2} - C_{T1}}\right)_\infty \tag{15}$$

There is thus a direct connection between the ratio of the partitioned energies and the source-term coefficients in the energy transfer equations. The coefficients in Eq. (15) have been chosen so that in this asymptotic state equal amounts of energy are present in the production and transfer ranges. In other types of flow, turbulence energy generation, which enters as a source in the k_P equation, will be positive; the ratio (k_P/k_T) will therefore always be higher than for asymptotic grid turbulence.

The following is the form for the coefficients, incorporating all the constraints discussed above, that led to best overall agreement with the test flows to be discussed later:

$$C_{P1} = 2.2; \quad C'_{P1} = -0.11; \quad C_{P2} = 1.8 - 0.3\left(\frac{k_P}{k_T} - 1\right)\Big/\left(\frac{k_P}{k_T} + 1\right)$$

$$C_{T1} = 1.08\, \epsilon_P/\epsilon_T; \quad C_{T2} = 1.15$$

The suggested dependence of C_{P2} on k_P/k_T ensures that the energy transfer rate from the large scale motion will be larger in a shear flow than in grid turbulence. The influence of (k_P/k_T) is more significant than may be supposed for while C_{P2} is rarely altered by more than 10% from its asymptotic value, this is of itself sufficient to change the rate of spread of a jet by some 40%.

The diffusion terms in both the partitioned energy and transfer-rate equations have been uniformly represented as:

$$\mathscr{D}_\phi = 0.22\,\frac{\partial}{\partial x_k}\left(\overline{u_k u_l}\,\frac{k_P}{\epsilon_P}\,\frac{\partial \phi}{\partial x_l}\right) \tag{16}$$

where ϕ stands for $k_P, k_T, \epsilon_P, \epsilon_T$. This formulation is essentially that proposed by *Daly* and *Harlow* (1970) save that here we use the time scale of the large scale motions rather than the turbulence-energy turnover time. In single-scale schemes it has generally not been possible to adopt the same diffusive coefficient for both the energy and the dissipation rate equation (see, for example, LRR) so the fact that the same coefficient now serves for all four dependent variables offers further peripheral support for a multiple-scale characterization.

Closure of the mean-flow equations for the calculations presented in Sect. 3 is achieved by applying, with minor adaptations, either the Reynolds-stress closure of LRR or the $k \sim \epsilon$ effective viscosity model [*Jones* and *Launder* (1972)]. In the former scheme the Reynolds stresses $\overline{u_i u_j}$ are obtained from the transport equations

$$\begin{aligned}\frac{D\overline{u_i u_j}}{Dt} &= -\left(\overline{u_j u_k}\,\frac{\partial U_j}{\partial x_k} + \overline{u_j u_k}\,\frac{\partial U_i}{\partial x_k}\right) - \tfrac{2}{3}\,\delta_{ij}\epsilon - 1.5\,\frac{\epsilon_P}{k_P}\,(\overline{u_i u_j} - \tfrac{2}{3}\,\delta_{ij}k) \\ &\quad + \phi_{ij2} + \phi_{ijw} \\ &\quad + 0.11\,\frac{\partial}{\partial x_k}\left[\frac{k_P}{\epsilon_P}\left(\overline{u_l u_k}\,\frac{\partial \overline{u_i u_j}}{\partial x_l} + \overline{u_l u_i}\,\frac{\partial \overline{u_k u_j}}{\partial x_l} + \overline{u_l u_j}\,\frac{\partial \overline{u_k u_i}}{\partial x_l}\right)\right]\end{aligned} \tag{17}$$

where ϕ_{ij2} and ϕ_{ijw} denote pressure interactions arising from mean strain and wall reflections. The detailed approximation for these processes is algebraically cumbersome and is not reproduced here; full details are provided in the original paper. The only difference between

Eq. (17) and that employed by LRR is that (ϵ_P/k_P) is used in place of (ϵ/k) as the time scale in representing return to isotorpy (the third term on the right) and diffusive transport (the last term).

The self-preserving jets have been calculated using the simpler turbulent viscosity hypothesis for the shear stress $\overline{uv}$

$$-\overline{uv} = 0.10\, k k_P/\epsilon_P \partial U/\partial y \tag{18}$$

and, for the *difference* between the normal stresses (which is what appears in the axial momentum equation):

$$(\overline{u^2} - \overline{v^2}) = k/3 \tag{19}$$

The reason for so doing was that, with the finite difference scheme adopted, the stress transport equations for the axisymmetric jet are rather awkward to solve [*Launder* and *Morse* (1979)]; moreover, our experience is that the simpler turbulent-viscosity characterization leads to nearly the same results as the stress-transport model for self-preserving free shear-flows. As individual normal stresses are not calculated in this scheme, Eq. (16) for the diffusion rates in the partitioned energy and transfer equations is replaced by

$$\mathscr{D}_\phi = 0.09 \frac{\partial}{\partial x_j} \left(\frac{k\, k_P}{\epsilon_P} \frac{\partial \phi}{\partial x_j} \right) \tag{20}$$

3. Presentation and Discussion of Results

We consider first the development of decaying, grid-turbulence passed through a sudden contraction in passage area. *Uberoi* (1957) has provided detailed experimental data for area contraction ratios from 4:1 to 16:1. As representative of this type of flow, Figs. 3 and 4 show features of the development in a 4:1 contraction. In the calculations Eqs. (2), (3), (5) (with (8)) and (6) have been solved for homogeneous conditions ($\mathscr{D}_\phi = 0$) by forward integration, supplying from experiment values of the mean velocity and the rate of turbulence energy generation rate. In this way the development of the turbulence energies could be calculated

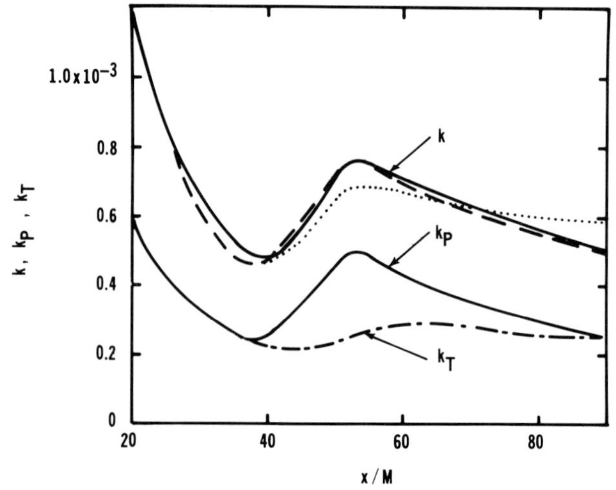

Fig. 3. Development of kinetic energies in 4:1 contraction: (---) experiment; (....) single-scale; (—) double scale

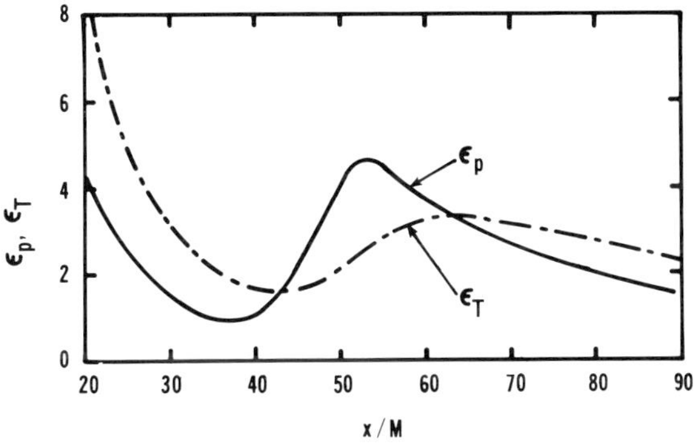

Fig. 4. Development of energy transfer rates through 4 : 1 contraction

without recourse to any closure relationship between stress and strain. The initial partitioning of energy between production and transfer ranges was taken as the equilibrium levels for decaying grid turbulence. Also shown in Figs. 3 and 4 is the behaviour calculated with the single-scale LRR scheme. The energy levels shown in Fig 3 show that much better agreement with measurements is obtained with the multiple-scale treatment. The single-scale model displays a too weak rise in energy through the contraction and a too slow decay downstream therefrom, in contrast to the virtually complete agreement shown with the present formulation. The reason for the differing behaviour may be inferred from the distributions of energy transfer rates shown in Fig. 4. The quantity ϵ_P exhibits a sharp rise on entering the contraction due to the action of the source term containing P_k. The dissipation rate of the single-scale scheme (ϵ_1) does likewise. The inertial transfer rate, ϵ_T, responds only sluggishly to the acceleration and, in fact, does not reach its maximum value until some distance *downstream* of the end of the acceleration. It is for this reason that, for the multiple-scale approximation, the calculated energy grows more rapidly through the acceleration yet falls off more steeply once the pressure gradient become zero.

The calculated mean velocity and turbulence energy profiles for the plane and axisymmetric turbulent jets in stagnant surroundings are presented in Figs. 5 and 6. The *Patankar-Spalding* (1970) numerical procedure was used to calculate these and all other inhomogeneous flows; approximately 25 cross stream modes were used in all cases. The calculated mean velocity profiles are nearly the same as those obtainable with a single-scale closure. There is an important difference, however, between the present calculations and those obtained with the standard $k \sim \epsilon$ model, as is brought out in Table 2. The spreading rate of the axisymmetric

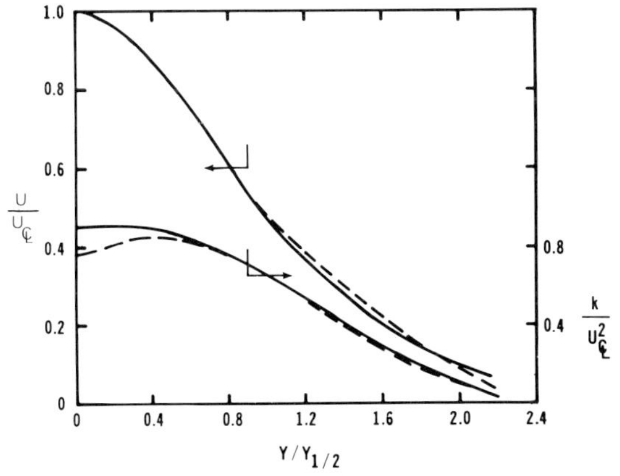

Fig. 5. Mean velocity and turbulence energy profiles across axisymmetric turbulent jet: (——) experiment [*Rodi* (1972)]; (---) present predictions

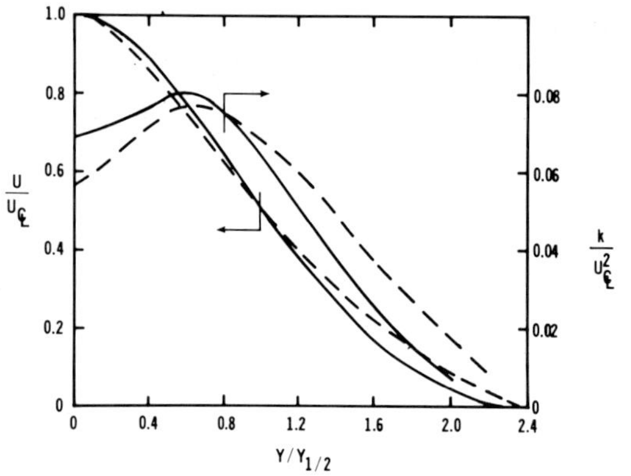

Fig. 6. Mean velocity and turbulence energy profiles in plane turbulent jet: (—) experiment [*Robins* (1972) (velocity); *Bradbury* (1965) (turbulence energy)]; (---) present predictions

Table 2. Rate of growth of half widths of round and plane jets

Flow	Expt.	Single scale	Present multi-scale
Plane jet	0.110	0.109	0.116
Axisymmetric jet	0.086–0.090	0.115	0.098

jet is observed to be about 20% less than that of the plane jet, whereas the standard $k \sim \epsilon$ model predicts a *faster* rate of spread. In contrast, the multiple-scale treatment produces the correct relative rates of spread for the two flows. It should be said that most of the improvement in this case arises from the addition of the term in Eq. (8) that amplifies the effect of irrotational strains. In the round jet, normal strains are several times more important than in the plane one; these lead (according to the present closure equations) to higher levels of ϵ_P and thus, from Eq. (18), to smaller effective viscosities which in turn cause the somewhat lower rate of spread.

Finally we consider two cases of boundary layers developing on smoth walls. In calculating these flows, the dependent variables are matched to the "law of the wall" and equivalent local-equilibrium expressions for the other dependent variables. In particular, the wall-adjacent energy transfer rates ϵ_P and ϵ_T are set equal to one another. In the case of the boundary layer developing in zero streamwise pressure gradient, calculations indicate that the ratio of these transfer rates does not differ by more than 10% from unity across the whole shear flow while, over the inner 80% of the boundary layer, more than 80% of the local turbulence energy is contained in k_P. The time scale (k_P/ϵ_P) thus varies in much the same way as (k/ϵ) and the behaviour predicted by the multiple-scale model does not differ in any important respect from those obtained with the single-scale scheme; both give acceptably close agreement with the experiments of *Klebanoff* (1955) (see LRR).

Figures 7–9 compare the measured and calculated development of a boundary layer subjected (after an initial region where the free stream velocity is uniform) to an adverse pressure gradient tailored so that the free stream velocity decreases downstream according to $(U_\infty/U_{\text{ref}}) \propto (x/x_{\text{ref}})^{-a}$. The experiments are those reported by *Bradshaw* (1969) for his "layer *C*". If the test section had been long enough the boundary layer would have reached an equilibrium state close to separation. The experiments, however, focus on the boundary layer *in transit* between different equilibrium states. The single-scale closure produces a too rapid transition and, beyond 30 in from the application of the adverse pressure gradient, the skin friction coefficient

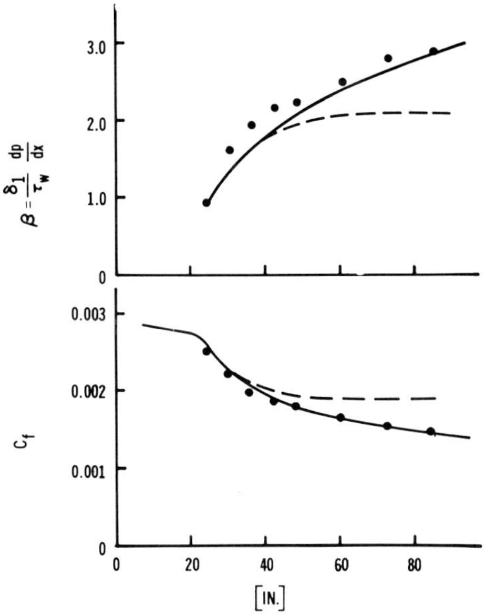

◀ **Fig. 7.** Development of boundary layer skin friction coefficient and pressure gradient parameter in an adverse pressure gradient: ● experiment [*Bradshaw* (1969)]; (---) single-scale model, [*Launder, Reece, Rodi* (1975)]; (——) present predictions

Fig. 8. Mean velocity profiles in boundary layer in adverse pressure gradient: (——) experiment [*Bradshaw* (1969)]; (---) present predictions

Fig. 9. Shear stress profiles in a boundary layer in an adverse pressure gradient: (○, ◇, ▼, □, ●) experiment [*Bradshaw* (1969)]; (—) present predictions

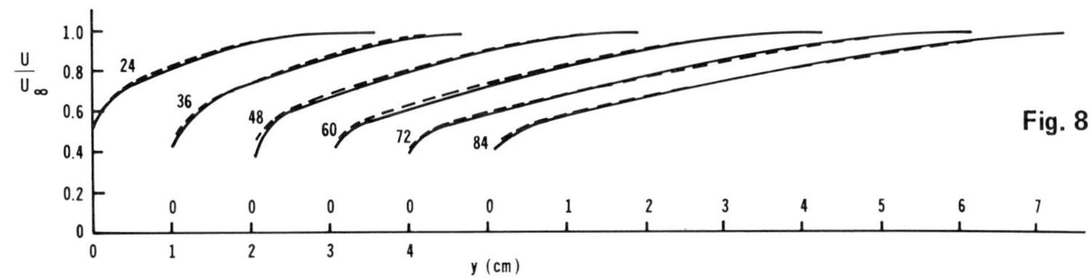

Fig. 8

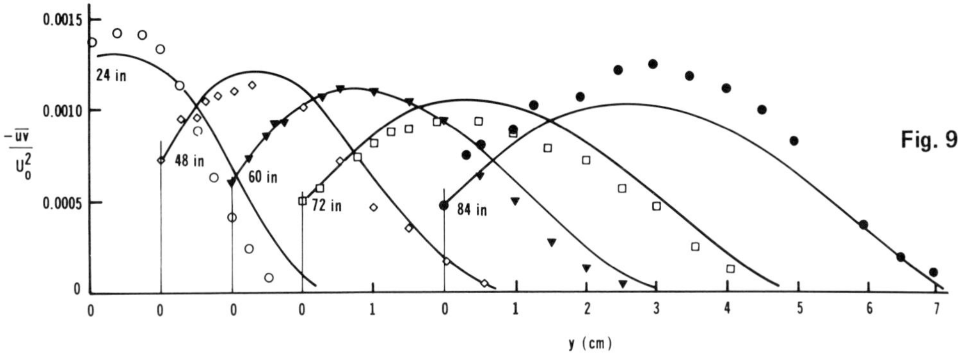

Fig. 9

and *Clauser*'s (1954) pressure gradient parameter β level out. The present model, however, faithfully reproduces the continuous measured rise of β and the decrease of c_f. The predicted and measured profiles of velocity and shear stress are shown in Figs. 8 and 9; for clarity the single-scale results have not been included. The calculated mean velocity profile exhibits virtually complete agreement with the experiment while the shear stress values are generally within about 10% of the reported data (a degree of proximity that must lie nearly within the uncertainty limits of the hot-wire measurements). Discrepancies in shear stress are somewhat larger at the final station though here it must be said that the experimental values seem to be out of line since, up to that station, the measured and calculated shear stress levels exhibit a gentle decrease in magnitude with downstream distance. The calculated profile at 84 in. continues that trend whereas the experiments show a 35% jump in magnitude from the previous station.

4. Concluding Remarks

The comparisons with experiments in the preceding section have perhaps served to show the potential advantages of including some account of the spectral character of turbulence in traditional second-moment closures. There is, however, still plenty of scope for model refinement. For example, the present proposals lead to a predicted rate of spread for the asymptotic plane wake which is some 30% too low – a discrepancy comparable with that obtained with a single-scale formulation. A more extensive study of the possible dependence of the coefficients in the model equations on the spectrum shape parameters may well allow a more generally applicable closure. In this connexion a recent exploration by *Launder* and *Schiestel* (1979) obtained much better predictions for the plane wake. That study, however, considered *only* the plane wake and jet and the axisymmetric jet while it made all the coefficients in the ϵ_P and ϵ_T equations dependent on the energy ratio k_P/k_T, a practice we no longer favor.

One possibly advantageous, though major, change in the current methodology would be to abandon the simplification that the energy production is confined to the large scale motions (which is equivalent to assuming the energy in k_T is isotropically distributed). The turbulence energy in a shear flow could then be distributed more evenly between the transfer and production ranges (at present, 80–90% of the energy is to be found in the production range in a flow where $P_k/\epsilon \simeq 1$). If that step were followed one would need separate closure hypotheses for the two parts of the Reynolds stresses, $\overline{u_iu_j}_P$ and $\overline{u_iu_j}_T$, say. Similar types of closure would probably be adopted for each part though there would be detailed differences. For example, the transfer range may be expected to exhibit a much faster rate of return to isotropy than the motions in the production range. Thus, if the contribution of fluctuating pressures (in the absence of mean strain) is represented as

$$-c_\alpha \frac{\epsilon_\alpha}{k_\alpha} (\overline{u_iu_j}_\alpha - \tfrac{2}{3}\delta_{ij}k_\alpha)$$

(the subscript α denoting either P or T), we should expect c_α to be larger for $\overline{u_iu_j}_T$ than for $\overline{u_iu_j}_P$. Moreover the P range would be strongly modified by rigid boundaries but such effects on the T part of the spectrum could probably be neglected. It is our view that an algebraic-stress closure for the transfer part of the spectrum would be an adequate treatment [parallel, for example to that given by *Gibson* and *Launder* (1976)]. Such formal subdivision of the energy spectrum is attractive from the point of view of strengthening the physical basis for the closure. It would moreover appear to provide an interfacing with the sub-grid-scale closures. The latter schemes have no use for the closure for the "production range" since they resolve that part of the spectrum through direct numerical solution. They could however employ something very similar to the model for the transfer range in place of the present highly rudimentary sub-grid-scale turbulence models that are usually employed at present.

Acknowledgments. The work reported herein has been supported by NASA Ames Experimental Fluid Mechanics Division under research grants NSG 2256. The continued interest of the NASA Scientific Staff, particularly Dr. J. G. Marvin and Dr. M. W. Rubesin is appreciated.

Our thanks are due also to L. S. Majesky who prepared the manuscript with her customary precision.

Authors' names appear alphabetically.

References

Bradshaw, P.: The response of a constant-pressure turbulent boundary layer to the sudden application of an adverse pressure gradient. Aero. Res. Counc. Rep. & Mem. 3575 (1969)

Clauser, F. H.: Turbulent boundary layers in adverse pressure gradients. J. Aeronaut. Sc. **21**, 91 (1954)

Daly, B. J., Harlow, F. H.: Transport equations in turbulence. Phys. Fluids **13**, 2634 (1970)

Gibson, M. M., Launder, B. E.: On the calculation of horizontal turbulent, free shear flows under gravitational influence. J. Heat Transfer **98** C, 80 (1976)

Jones, W. P., Launder, B. E.: The prediction of laminarization with a two-equation model of turbulence. Int. J. Heat Mass Transfer **15**, 301 (1972)

Klebanoff, P. S.: "Characteristics of Turbulence in a Boundary Layer with Zero Pressure Gradient". NACA Rpt. 1277 (1955)

Kolmogorov, A. N.: Equations of turbulent motion of an incompressible turbulent fluid. Izv. Akad. Nauk SSSR Ser. Fiz. **VI** (1–2), 56 (1942)

Launder, B. E., Morse, A.: "Numerical Prediction of Axisymmetric Free Shear Flows with a Second-Order Reynolds Stress Closure", in *Turbulent Shear Flows I*, ed. by F. Durst, B. E. Launder, F. W. Schmidt, J. H. Whitelaw (Springer, Berlin, Heidelberg, New York 1979) p. 279

Launder, B. E., Schiestel, R.: "Sur l'utilisation d'échelles temporelles multiples en modélisation des écoulements turbulents". C. R. Acad. Sci. Ser. A: **286**, 709 (1978)

Launder, B. E., Schiestel, R.: "Application d'un modèle de turbulence à échelles multiples au calcul d'écoulements libres turbulents". C. R. Acad. Sci., Ser. B: **288** 127 (1979)

Launder, B. E., Reece, G. J., Rodi, W.: Progress in the development of a Reynolds stress turbulence closure. J. Fluid Mech. **68**, 537 (1975)

Robins, A.: "The Structure and Development of a Plane Turbulent Free Jet". Ph. D. Thesis, University of London (1971)

Rodi, W.: "A Review of Experimental Data of Uniform Density Free Turbulent Boundary Layers", in Studies in Convection, Vol. 1 (Academic Press, London 1975) p. 79

Patankar, S. V., Spalding, D. B.: *Heat and Mass Transfer in Boundary Layers,* 2nd ed. (Intertext, London 1970)

Pope, S. B.: "An Explanation of the Turbulent Round-Jet/Plane Jet Anomaly". Rpt. FS/77/12, Department of Mechanical Engineering, Imperial College (June 1977)

Prandtl, L.: Bericht über Untersuchungen zur ausgebildeten Turbulenz. ZAMM **5**, 136 (1925)

Taylor, G. I.: Eddy motion in the atmosphere. Philos. Trans. R. Soc. **215**, 1–26 (1915)

Uberoi, M. S.: Equipartition of energy and local isotropy in turbulent flows. J. Appl. Phys. **28**, 1165 (1957)

Zeman, O., Lumley, J. L.: "Buoyancy Effects in Entraining Turbulent Boundary Layers: A Second-Order Closure Study", in *Turbulent Shear Flows I,* ed. by F. Durst, B. E. Launder, F. W. Schmidt, J. H. Whitelaw (Springer, Berlin, Heidelberg, New York 1979) p. 295

Two-Point Closure Applied to a Passive Scalar in Decaying Isotropic Turbulence

M. Larcheveque[2], J. P. Chollet[1], J. R. Herring, M. Lesieur[1], G. R. Newman[3], and D. Schertzer[4]
National Center for Atmospheric Research[5], Boulder, CO, 80307, USA

Abstract

We apply two-point closure to the decay of a passive scalar convected by homogeneous turbulence. The particular methods investigated are the test field model (TFM) and the related eddy damped Markovian (EDM) model. The accuracy of the methods is first assessed by comparing predictions to experiments for a range of Reynolds and Prandtl numbers. We also employ the theory to estimate eddy viscosity and conductivity. We then study the decay of an initially random scalar field injected into a preexisting field of turbulence, with the goal of understanding the dependence of velocity and scalar time scales on their respective energy spectra, paralleling – insofar as possible – the recent experimental study of *Warhaft* and *Lumley* [1]. Our results agree reasonably well with the trends predicted by the experiments. As in the experimental results, we find – for certain initial conditions – an anomalous scalar "inertial range". The theoretical results appear consistent with the interpretation that these cases have a higher scalar Reynolds number than those without an anomaly.

Nomenclature

$B(k, p, q)$	$\frac{1}{2}\pi \sin^2(p, q)\,[(p^2 - q^2)(k^2 - q^2) + p^2k^2]/k^3$
$B^s(k, p, q)$	$\pi kpq \sin^2(q, k)$
$C(k, p, q)$	$\pi/2\, kpq \sin^2(k, p) \sin^2(k, q)$
$D(k, p, q\,\vert\,t)$	Scalar triple moment relaxation factor
EDM $(\tilde{\lambda}, \tilde{\lambda}', \tilde{\lambda}'')$	Eddy Damped Markovian Model with time scales $\tilde{\lambda}, \tilde{\lambda}', \tilde{\lambda}''$. See Eqs. (A15–17) of Appendix
$E(k, t)$	Kinetic energy spectrum $\equiv 1/2\,(4\pi k^2)\,U(k, t, t)$
$E_\theta(k, t)$	Scalar energy spectrum $\equiv 1/2\,(4\pi k^2)\,\theta(k, t, t)$
g^2	Test field model scaling factor for velocity field
$g_\theta^2, \tilde{g}_\theta^2$	Test field model scaling factors for scalar field
k_s	Kolmogorov's wave number
(k, p, q)	Fourier wave numbers that form a triangle
L	Longitudinal integral scale

---

1 Institut de Mécanique de Grenoble, France.
2 Laboratoire de Météorologie Dynamique, Paris, France.
3 Dept. of Mechanical Engineering, Southeastern Massachusetts University, N. Dartmouth, MA 02747, USA.
4 Direction de la Météorologie, Paris, France.
5 The National Center for Atmospheric Research is sponsored by the National Science Foundation.

P_{λ_θ}	Microscale Peclet number $\equiv u\lambda_\theta/\gamma$
$\overline{q^2}$	Ensemble mean squared speed = $\overline{u_1^2} + \overline{u_2^2} + \overline{u_3^2}$
r	Ratio of velocity to scalar decay time scales $(\dot{\overline{q^2}}/\epsilon)/(\dot{\overline{\theta^2}}/\epsilon_\theta)$
R_L	Longitudinal integral scale Reynolds number
R_λ	Taylor microscale Reynolds number
$S(t)$	Velocity derivative skewness
$S'_\theta(t)$	Mixed derivative scalar skewness
t	Time
TFM$(g_v, g_\theta, \tilde{g}_\theta) \equiv$	Test field model with scaling parameters $g_v, g_\theta, \tilde{g}_\theta$. See Eqs. (A8–14) of Appendix.
$T(k)$	Kinetic energy transfer function
$T_\theta(k)$	Scalar energy transfer function
u_i	Cartesian ith component of velocity field, $\boldsymbol{v}$
$U_{ij}(\boldsymbol{k}, t, t')$	Fourier transform of $U_{ij}(\boldsymbol{x}, t; \boldsymbol{x}', t')$ with respect to $(\boldsymbol{x} - \boldsymbol{x}')$
$U_{ij}(\boldsymbol{x}, t; \boldsymbol{x}', t')$	Ensemble mean ($\langle\rangle$), two point covariance $\equiv \langle u_i(\boldsymbol{x}, t) u_j(\boldsymbol{x}', t')\rangle$
$\boldsymbol{v}$	Velocity field vector
α	(One-dimensional) Kolmogorov constant
β	(One-dimensional) Oboukhov-Corrsin constant
γ	Thermal diffusivity
ϵ	Kinetic energy dissipation
ϵ_θ	Scalar energy dissipation
$\mu(k)$	Eddy damped Markovian relaxation frequency
$\eta^c(k)$	Compressive test field relaxation frequency
$\eta^s(k)$	Solenoidal test field relaxation frequency
θ	Scalar concentration
$\overline{\theta^2}$	Ensemble mean scalar variance
$\Theta(k)$	Scalar variance Fourier spectrum
Θ	Ensemble mean, two point covariance of θ: $\equiv \langle\theta(\boldsymbol{x}, t)\, \theta(\boldsymbol{x}', t')\rangle$
λ	Taylor microscale length $\equiv (5\, \nu\, \overline{q^2}/\epsilon)^{1/2}$
$(\tilde{\lambda}, \tilde{\lambda}', \tilde{\lambda}'')$	Parameters of the eddy damped Markovian model
λ_θ	Microscale length for scalar field $\equiv (6\, \gamma\, \overline{\theta^2}/\epsilon_\theta)^{1/2}$
ν	Kinematic viscosity
ψ	Velocity normalized decay rate $\equiv -\dot{\epsilon}\, \overline{q^2}/\epsilon^2$
ψ_θ	Scalar normalized decay rate $\equiv -\dot{\epsilon}_\theta \overline{\theta^2}/\epsilon_\theta^2$.

1. Introduction

Two-point closures should provide a useful tool in the study of the decay of turbulence and, in particular, enable one to compute the "universal" rate coefficients which occur in the second order modeling. Alternately, they should provide sufficient details of the flow to see in what instances certain of these rate coefficients are not constant at all. Unfortunately, the more plausible of the closures [for example, the direct interaction approximation (DIA) (*Kraichnan* [2]) or one of its Lagrangian generalizations (LHDI) (*Kraichnan* [3]; *Kraichnan* and *Herring* [4]] prove to be somewhat intractable computational tools, and one is led to consider simpler but perhaps less secure theories for practical calculations. The test field model (TFM) (*Kraichnan* [5]; *Herring* and *Kraichnan* [6]) represents an attempt to retain reasonable physical description of the triple moment relaxation, as suggested by the LHDI, while at the same time reducing the amount of numerical computation to an acceptable level. The procedure replaces the rather complicated memory effects in the LHDI by plausibly motivated relaxation times computed by examining the effects of pressure scrambling. An even more drastic reduction in computational labor may be obtained through the eddy damped Markovian (EDM) procedure, which replaces the memory times by plausible and dimensionally consistent time scales related to large-scale straining. Such a method introduced by *Orszag* [7] has been employed by *André* and *Lesieur* [8] and by *Lesieur* and *Schertzer* [9] in their study of high Reynolds number turbulence.

The two methods (TFM and EDM) are actually closely related, and a more systematic comparison of their predictions will be made the topic of a subsequent paper. Our goal here is to establish the utility of these methods, having first demonstrated that they can give reasonable accuracy at high Reynolds number. With respect to this last point, we shall show that the combined inertial-dissipation range shapes of the velocity and scalar spectra are well-comprehended by the procedure, including the "bump" on the scalar spectrum recently discussed by *Hill* [10]. Reasonable predictions are also obtained for the limits of very high and low Prandtl numbers.

We should mention here briefly the possible role of two-point closures as a tool in understanding turbulent flows. Generally, such methods occupy a position intermediate in complexity between direct numerical simulation (with sub-grid-scale modeling) and single point closure (second order modeling), of the sort that has recently seen widespread application. The value of two-point closures lie in their relative simplicity – as compared to large eddy simulation on the one hand and their increased flexibility and universality as compared to the usual single point closure on the other hand. Thus they may be used to establish valid sub-grid scale procedures – as described by *Leslie* and *Quarini* [11]. They also have the potential of identifying proper coefficients to be used in the single point methods, as has been done for the Rotta return to isotropy term by *Herring* [12] and *Schumann* and *Herring* [13].

Another advantage such closures have over large eddy simulation methods lies in the fact that they can reach very large Reynolds numbers with relative ease. Thus, we have reported computations at $R_\lambda \sim 10^4$, far beyond the computational limits of full scale spectral simulation ($R_\lambda \sim 200$).

Additionally, they may be profitably employed to examine flows whose rapid development would seem to defy an accurate single point description, such as the development of an initial flow all of whose energy lies at large scale, a problem examined in some detail by *André* and *Lesieur* [8].

The usefulness of two-point closures is much vitiated when it comes to inhomogeneous flows. Indeed, the general two-point covariance equations (e.g., *Kraichnan* [3] or *Herring* [14] appear much more formidable than the Navier-Stokes equations, and considerable analytic reduction needs to be done to render them tractable. However, it may be that their complexity may be reduced in the direction of deriving single point closures, and such a program has been proposed in some detail by *Leslie* [15].

2. Equations of TFM, EDM and DIA: Large R Results

The TFM, DIA and EDM equations of motion for the modal velocity covariance, $U(k, t)$ and the modal scalar covariance $\Theta(k, t)$ are recorded in the appendix. For a complete physical interpretation of these equations, we defer to *Kraichnan* [2, 5], *Orszag* [7], *Sulem* et al. [16], *André* and *Lesieur* [8], and *Newman* and *Herring* [17]. Briefly, the TFM may be considered as a two-step abridgment and modification of the DIA: first, the memory effects represented by integrals over the past of two-time covariances, $\int U(k, t, s)ds$, are replaced by $\int \exp [-\eta(k)(t-s)] U(k, t, t)ds$. Then, instead of deriving equations for $\eta(k)$ from the DIA-$G(k, t, s)$ equations, new equations of similar structure are obtained by taking $\eta(k)$'s as proportional to the rate of change of the solenoidal component of a vector field – convected by the turbulence – interacting with its compressive component. Similarly, $\eta^c(k)$ is then introduced as the rate of change of the compressive component of the test-vector field, which is coupled to the η^s-component. The new equations for $\eta(k)$ eliminate the spurious dependence of the triple moment transfer on large scales, as found in the DIA. The EDM more simply takes the $\eta(k)$ [called $\mu(k)$ to distinguish it from TFM] as proportional to the rms strain of scales of wave number k by scales larger than $(2\pi k)^{-1}$.

Both TFM and EDM introduce phenomenological constants which specify the efficiency of large-scale strain in decorrelating velocity and scalar fields. The particulars of how these constants enter the energy transfer functions are given in the appendix. In the TFM we need one constant (called g_v) to specify the kinetic energy transfer and two (g_θ and $\tilde{g}_\theta$) to specify the scalar energy transfer. These correspond to ($\tilde{\lambda}$, $\tilde{\lambda}'$ and $\tilde{\lambda}''$) in the EDM. From the discussion given in the appendix, we note that $\tilde{g}_\theta = 1$ is indicated if we insist on a parallelism between the DIA and TFM with regard to the decorrelation of the scalar field. We shall here frequently (for simplicity of discussion) set $\tilde{g}_\theta = 1$, noting from numerical studies that its variation would produce some (mostly nonessential) variation in the scalar spectrum.

A straightforward method to determine the TFM value of (g, g_θ, $\tilde{g}_\theta$) or ($\tilde{\lambda}$, $\tilde{\lambda}'$, $\tilde{\lambda}''$) for EDM is by a best-fit of theory to a given experiment. This is the method followed here. Shown in Fig. 1 is a comparison of the one-dimensional spectra ϕ and ϕ_θ, with the data of *Champagne* et al. [18]. The values of (g_v^2, g_θ^2, $\tilde{g}_\theta^2$) are (1.17, 0.5, 1). We recall that the one-dimensional spectra ϕ and ϕ_θ are defined by

$$\phi(k) = \frac{1}{2} \int_k^\infty dp/p\, E(p)\, [1 - (k^2/p^2)] \tag{2.1}$$

and

$$\phi_\theta(k) = \int_k^\infty dp\, E_\theta(p)/p. \tag{2.2}$$

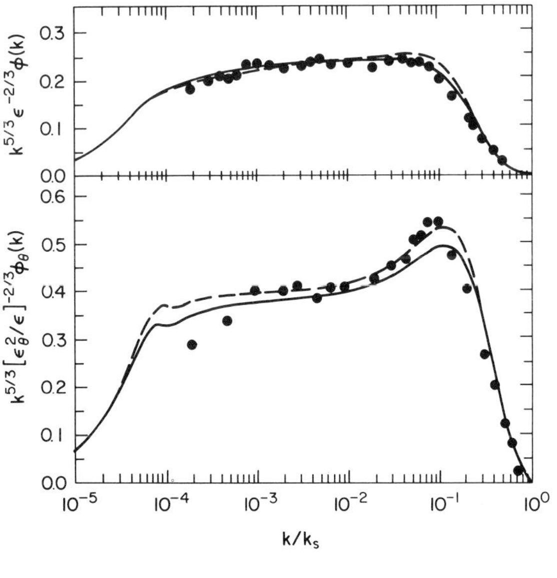

Fig. 1. One-dimensional kinetic energy spectrum $[\phi(k) = \frac{1}{2} \int_k^\infty dp(1 - k^2/p^2)E(p)dp]$ and scalar spectrum $[\phi_\theta(k) = \int_k^\infty dpE_\theta(p)/p]$, compared to data of *Champagne* et al. [18] *(points)*. *Solid curves,* TFM ($g_v^2 = g_\theta^2 = 0.5$, $\tilde{g}_\theta^2 = 1$), *dashed curves,* EDM ($\tilde{\lambda} = 0.36$, $\tilde{\lambda}' = 0$, $\tilde{\lambda}'' = 3.61\,\tilde{\lambda}$)

Here $E(k)$ and $E_\theta(k)$ are the three-dimensional kinetic energy and scalar variance spectrum whose initial shapes for the present runs were

$$E_v(k, 0) = E_\theta(k, 0) = 2\pi k^4(0.005 + k)^{-17/3}. \tag{2.3}$$

Other parameters of the run are: viscosity $\nu = 0.002$ and computational wavenumber span (0.001, 200). The ($dpdq$)-convolution domain entering the evaluation of the energy transfer function, $T(k)$ (see Eqs. A.1 and A.5), is a 45° tilted rectangular strip with (p, q)-axes intercepts equal to k, and longitudinal length $k_{max} = 200$. This procedure accurately conserves energy only if the length of the strip $k_{max} \gg 2k_s$, where $k_s = (\epsilon/\nu^3)^{1/4}$. An overall estimate of numerical accuracy is found by computing the energy flux function

$$\Pi(k) = \int_k^{k_{max}} T(p)dp \tag{2.4}$$

and comparing $\Pi(0)$ with the maximum of $\Pi(k)$, $k \leqslant k_{max}$. For the present case, such errors are $\lesssim 2\%$. The spectra $E(k, t)$ and $E_\theta(k, t)$ are allowed to decay until full self-similarity is obtained for Kolmogorov-scaled variables. The computed value of R_λ is 2,500. This is approximately equal to the observed value ($\simeq$ 2,400) provided the latter is based on the vertical velocity fluctuations. Figure 1 shows the compensated energy and variance spectra (α and β plots), defined as $k^{5/3}\epsilon^{-2/3}\phi(k)$ and $k^{5/3}(\epsilon_\theta^3/\epsilon)^{-1/3}\phi_\theta$, respectively. We see here a fairly accurate matching of TFM and experiment from just above the energy containing scales ($k/k_s \simeq 2 \times 10^{-4}$) through the dissipation range ($k/k_s = 1$). The dashed curves here show EDM results for the ($\tilde{\lambda}, \tilde{\lambda}', \tilde{\lambda}''$)-parameters given by (0.36, 0.0, 1.30). The TFM and EDM are seen to be in close accord throughout the universal range of wave numbers. We note that both TFM and EDM predict a "bump" on the scalar spectrum as k tends toward dissipation wave numbers (a much enhanced outer peak at $k/k_s \simeq 0.1$). The slightly higher "bumps" on the EDM compensated spectra may be understood in terms of the fact that the EDM relaxation frequency $\mu(k)$ goes to a constant in the dissipation range, where the corresponding $\eta(k)$ for the TFM go to zero. *Hill* [10] has interpreted this peak as a "high Pr" effect; the velocity spectrum ends near $k = 0.04\,k_s$, and consequently the scalar field at larger k feels dominantly a large-scale strain and hence tends momentarily toward k^{-1} before lapsing into the scalar dissipation range (beyond $k/k_s = 0.1$).

3. Arbitrary Prandtl Number

As yet, we have only shown that the values of (g, g_θ) chosen to match the inertial range constants appear to give reasonable shapes for the compensated spectra over the universal portion of k/k_s. We may next inquire as to how well the theory does at very large and small Prandtl number Pr. *Newman* and *Herring* [17], have noted earlier that the TFM gives qualitatively sensible results at arbitrary Pr; at low Pr it lapses over into the theory of *Batchelor* et al. [19], and for Pr $\to \infty$, it becomes identical to Batchelor's theory: $E_\theta(k) = C_B(\epsilon/\nu)^{-1/2}\epsilon_\theta k^{-1}$ for the viscous-convective range. For Pr $\to \infty$ the TFM yields

$$C_B = (5/2)\,(\epsilon/\nu)^{+1/2} \left\{ \int_0^\infty p^2 dp E(p)/[\eta^s(p) + 2g_\theta^2\eta^c(\infty) + \nu p^2] \right\}^{-1}. \tag{3.1}$$

Our TFM calculations (with $g_v^2 = 1.17, g_\theta^2 = 0.5, \tilde{g}_\theta^2 = 1$) of Sect. 2 yield $C_B = 0.78$ if we assume $\eta^c(\infty) = 0$, as indicated by the TFM. This value is less than *Gibson*'s [20] lower bound of $3^{1/2}$, and is similar to Kraichnan's ALH estimate. TFM (1.06, 0, 1.9) or its near equivalent EDM (0.36, 0, 1.3) yield C_B = 2.2 and 2.6. The latter choices also give Kolmogorov and Oboukhov constants of (1.4, 0.667), respectively. We note here that if we set $\nu = 0$ in (3.1) EDM ($\tilde{\lambda}$, 0, $\tilde{\lambda}''$) yields

$$C_B = 5\sqrt{2}\,\tilde{\lambda}''/4 = 2.298 \tag{3.2}$$

Newman and *Herring* [17] reported for $(g_v, g_\theta, \tilde{g}_\theta) = (2.25, 0.25, 1.0)$ $C_B = 1.63$, but this value is obtained only if $\eta^c(\infty) \sim (\epsilon/2\,\nu)^{1/2}$, an invalid extrapolation of an inertial range asymptotic result into the dissipation range.

At small Pr, both TFM and EDM lapse over into a (Markovianized) quasi-normal approximation, and hence in this limit we obtain the result of *Batchelor* et al. [19].

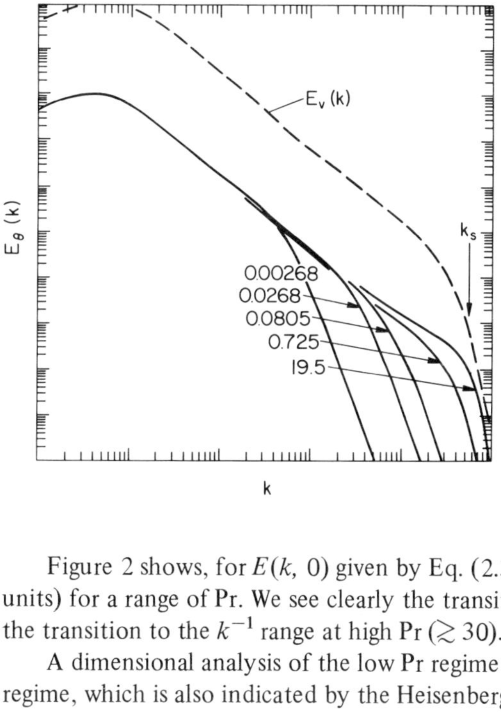

Fig. 2. Scalar energy spectrum $E_\theta(k)$, for initial data given by Eq. (1). Curves are labeled by Prandtl number, Pr. *Dashed curve* gives kinetic energy spectrum, $E(k)$. Only TFM (g_v^2, $g_\theta^2 = 0.5$, $\tilde{g}_\theta^2 = 1$) shown

Figure 2 shows, for $E(k, 0)$ given by Eq. (2.3), the developed $E(k, t = 20.8)$ (in arbitrary units) for a range of Pr. We see clearly the transition to the $k^{-17/3}$ range at low Pr, as well as the transition to the k^{-1} range at high Pr ($\gtrsim 30$).

A dimensional analysis of the low Pr regime suggests the possible existence of a $k^{-13/3}$ regime, which is also indicated by the Heisenberg theory (*Granatstein* et al. [21]; *Ogura* [22]). For this to occur, the conductive dissipation must balance the positive random stirring production, as for the $k^{-17/3}$ range, but the duration of the random stirring must be limited by the velocity and scalar decorrelation times rather than diffusion. In fact, a more complete analysis of the TFM (or EDM) for the low Pr regime suggests that

$$E_\theta(k) = (1/3)E(k)\epsilon_\theta\gamma^{-2}k^{-2}[g_\theta^2\eta^c(k) + \eta^s(k) + \gamma k^2]^{-1}. \tag{3.3}$$

However, for the $\eta(k)$ terms to dominate γk^2 requires $k \lesssim (\mathrm{Pr})^{3/4}k_s$, which is the same order of magnitude as that k at which the $k^{-5/3}$ to $k^{-13/3}$ transition occurs, assuming here that $\eta(k) \sim \epsilon^{1/3}k^{2/3}$. Hence, a possible $k^{-13/3}$ regime is probably masked by the inertial-conductive transition. This surmise is reinforced by the numerical calculation shown in Fig. 2.

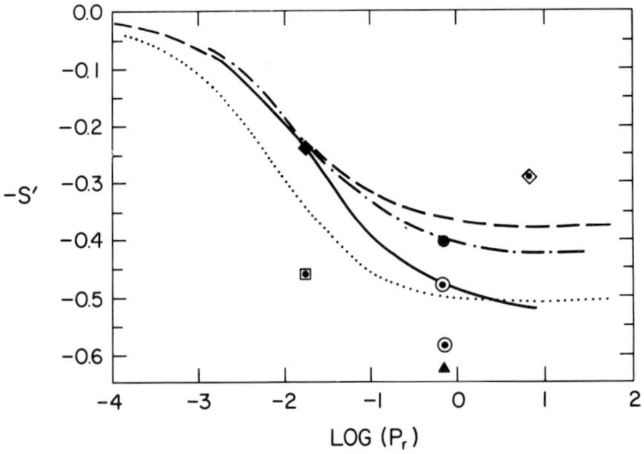

Fig. 3. Mixed scalar skewness [see Eq. (3.4)] as a function of Pr, from *Hill* [23]. Dashed line gives Hill's model #2; *dotted line* shows model #4. TFM ($g_v^2 = 1.17$, $g_\theta^2 = 0.5$, $\tilde{g}_\theta^2 = 1$) is *solid line* and *dot-dashed line* is EDM (0.36, 0, 1.3). Data are: ♦ Hill's computation from Howell's model (for mercury); □ *Clay*'s [35] mercury data; ○ Williams and *Pound*'s [36] -run average; • their low noise run; and ▲ *Champagne* et al. [18] data

A convenient small-scale measure of scalar energy transfer is the mixed scalar skewness, defined by

$$S'_\theta = \langle(\partial u/\partial x)(\partial\theta/\partial x)^2\rangle/\langle(\partial u/\partial x)^2\rangle^{1/2}\langle(\partial\theta/\partial x)^2\rangle. \tag{3.4}$$

Figure 3 presents TFM and EDM results for S' as a function of Pr, for $(g_\theta^2, g_v^2, \tilde{g}_\theta^2) = (0.50, 1.17, 1)$ and $\tilde{\lambda}, \tilde{\lambda}', \tilde{\lambda}'' = (0.36, 0, 1.3)$. The dotted and dashed lines are model #4 and #2 of *Hill* [10], as computed by *Hill* [23]. The points are experimental data, as indicated in the figure caption. Both TFM and EDM results appear reasonable. As $\mathrm{Pr}\to\infty$, $S'\to\sqrt{20/3}\,C_B^{-1}$ for either theory (see also *Hill* [23]). *Hill* notes that the data here are quite uncertain because of the unknown effects of intermittency.

4. Eddy Viscosity and Conductivity

Two-point closures may be used to obtain prescriptions of a type of eddy viscosity and conductivity by the following argument. Consider the evolution of a spectrum which contains a perturbation at very small wave number:

$$E(k,t) = E_0(k,t) + \delta E(k,t),$$

where $E_0(k,t)$ is some smooth spectrum which is zero below the lowest wave number k_c, and $\delta E(k,t)$ is very sharply peaked at some wave number k_0 much smaller than k_c. Then for $\delta E(k,t)$ both EDM and TFM gives (*Kraichnan* [24]; *Lesieur* and *Schertzer* [9])

$$d\delta E(k,t)/dt = -2(\nu+\nu_{\mathrm{eddy}})k^2\delta E(k,t), \tag{4.1}$$

where (for TFM, for example)

$$\nu_{\mathrm{eddy}} = (2/15)\int_{k_c}^{\infty} dp/\eta_0^S(p)\cdot\{E_0(p) + 1/4\cdot d/dp[pE_0(p)]\}. \tag{4.2}$$

Similarly, for the scalar field

$$d\delta E_\theta(k,t)/dt = -2(\gamma+\gamma_{\mathrm{eddy}})k^2\delta E_\theta(k,t), \tag{4.3}$$

where

$$\gamma_{\text{eddy}} = (2/3) \int_{k_c}^{\infty} E_0(p)/[\eta_0^s(p) + g_\theta^2 \eta_0^c(p)]dp. \tag{4.4}$$

In Eqs. (4.2) and (4.4), the relaxation frequencies η_0^c and η_0^s correspond to the spectrum $E_0(k)$. We have taken k_0, the wave number centroid of $\delta E(k)$, sufficiently small so that $\eta(k)$ may be neglected compared to $\eta(p)$. We should stress here that Eqs. (4.2) and (4.4) apply, strictly speaking, only for the diffusion of momentum and heat in homogeneous and isotropic flows. However, we may conjecture that they have validity for inhomogeneous and anisotropic flows since possible classes of ensembles have individual realizations in which the large scales $\delta E(k)$ are strongly anisotropic, with the orientation of the flow randomly distributed among the ensembles. From Eqs. (4.2) and (4.4) it is possible to evaluate an eddy Prandtl number. Assuming $E_0(k) \sim k^{-5/3}$, $k \geqslant k_c$; $E_0(k) = 0$, $k < k_c$, then we obtain for the EDM

$$\text{Pr}_{\text{eddy}} = (\tilde{\lambda}' + \tilde{\lambda}'')/6\tilde{\lambda} \tag{4.5}$$

The values (0.36, 0.0, 1.30) for the parameters $(\tilde{\lambda}, \tilde{\lambda}', \tilde{\lambda}'')$ then give

$$\text{Pr}_{\text{eddy}} = 0.60$$

This may be compared to (0.7–0.8) quoted by *Blom* [25] as an appropriate value for the turbulent part of the wall region for pipes or ducts (in this region Pr_{eddy} is observed to be approximately constant). The experiments are for air; the formula (4.5) is independent of the molecular Pr, at apparent variance with earlier results of *Ogura* [22] who employed Heisenberg's theory.

In fact the eddy Prandtl number depends also on the precise shape of the energy spectrum $E_0(k)$ in the neighbourhood of k_c. Let us integrate Eq. (4.2) by parts to obtain

$$\nu_{\text{eddy}} = (2/15) \int_{k_c}^{\infty} E_0(p)dp/\eta_0^s(p)\,[1 + d\ln \eta_0^s(p)/4\, d\ln p] \tag{4.2a}$$

for the TFM, which yields

$$\text{Pr}_{\text{eddy}} = 0.2\,[\overline{g_\theta^2} + \overline{g_\theta^2}(\overline{\eta_0^c/\eta_0^s})]\,(1 + \overline{d\ln \eta_0^s/4\, d\ln p}). \tag{4.5a}$$

Here the overbars indicate appropriate mean values as implied by Eqs. (4.2a) and (4.4). An examination of the equations for $\eta^s(k)$ and $\eta^c(k)$ suggests $\eta^c/\eta^s \cong 2$. For $k \to 0$, $\eta(k) \tilde{\to} k^2$, while at $k \to \infty$ (but $k \ll k_s$) $\eta(k) \to \epsilon^{1/3} k^{2/3}$. Since Eq. (4.2a) appears to weight small k more heavily, it would seem that an appropriate value of the logarithmic derivative $d\ln \eta_0^s/d\ln p$ in (4.5a) should be ~ 2. In this case, $\text{Pr}_{\text{eddy}} \sim 0.6$. More accurate numerical evaluations suggest this value is approximately correct. Finally we note here that the EDM yields

$$\text{Pr}_{\text{eddy}} = 0.2\,[(\tilde{\lambda}' + \tilde{\lambda}'')/\tilde{\lambda}]\,(1 + \overline{d\ln \mu/4\, d\ln p}) \tag{4.5b}$$

instead of Eq. (4.5a).

An examination of the scaling of (4.2) suggests that it be rewritten as

$$\nu_{\text{eddy}} = a u_0 L, \tag{4.6}$$

s	$a(s)$
2	1.58
3	1.24
4	1.11

Table 1. TFM eddy velocity parameter [see Eq. (4.6)] a as a function of s, where $E(k) \to k^s$

where $u_0 = (\overline{q^2}/3)^{1/2}$, L is the (longitudinal) integral scale, and a is a parameter depending only on the shape of the kinetic energy spectrum at small k [i.e., $a(s)$, where $E(k) \to k^s$, $k \to 0$]. Table 1 summarizes $a(s)$ for $s = 2, 3, 4$, for TFM ($g_v^2 = 1.17$). Values of $a(s)$ for EDM are about twice as large as those obtained from TFM; the reason lies in the different expressions for $\eta(k)$ and $\mu(k)$ as $k \to 0$.

Table 1 values may be compared to the results of *Weinstock* [26], who obtained 0.3761 $= 2/(3\,\pi^{1/2})$ for $\gamma_{\text{eddy}}/u_0 L$, independent of s.

Generally the real situation is somewhat more complex, since it is not really possible to define a clear cut separation between the characteristic spatial scales of the shear and of the turbulent fluctuations. Formulas (4.2) and (4.4) still hold but then the eddy viscosity and diffusivity depend on the energy spectrum below k_c. Nevertheless, for a $k^{-5/3}$ energy spectrum extending from zero to infinity, it is easy to show that Eq. (4.5) is still valid. It turns out that the eddy Prandtl number which models the damping action of small scales $p, q \gg k_c$ on large scales $k \ll k_c$ is relatively independent of the ratio k/k_c and of the possible existence of a gap in the energy spectrum between k and k_c. *Kraichnan* [24] has shown that, due to local interactions, the eddy viscosity has a cusp behaviour when $k \to k_c$. Nevertheless preliminary calculations of *Lesieur* and *Chollet* [27] have shown that an eddy viscosity based on (4.2) with an appropriate boundary condition at the cutoff wavenumber gave satisfactory results for the sub-grid-scale modeling of energy transfer.

5. Moderate R Results: Comparison with Wind Tunnel Experiments

A problem of considerable recent interest is to understand the decay of a passive scalar at moderate R_λ ($\simeq 45$) and, in particular, to explain why scalar decay seems to lack those universal aspects which the kinetic energy possesses. The problem is simplest put in terms of the decay exponents (n, m) defined by $n = -t\dot{\overline{q^2}}/\overline{q^2}$ and $m = -t\dot{\overline{\theta^2}}/\overline{\theta^2}$. Since the experiments of *Comte-Bellot* and *Corrsin* [28], most wind tunnel measurements of $\overline{q^2}(t)$ yield $n \simeq 1.4$.

However, measurements of m have ranged over (3.09, 0.87). This point has recently been discussed by *Warhaft* and *Lumley* [1], who stress the need to examine the E_θ-spectra in order to understand the reported differences in m. These authors observed in their experiments that the value of m increases with k_θ/k_v, where k_θ and k_v are the peak wave numbers of the spectra $E_\theta(k)$ and $E(k)$, respectively. To avoid unwanted heat fluxes, the above authors, in certain of their experiments, injected temperature fluctuations in an already decaying field of turbulence by a device they called a *mandoline*. The scale of the temperature field could then be set up independent of the turbulence scale. Somewhat the same initial conditions may also be realized in the closures by allowing the turbulence to decay toward self-similarity, and then at some time t_1 introduce the scalar field with any arbitrary spectrum, but having zero triple moments [i.e., $D^\theta(k, p, q, t_0) = 0$ in Eq. A.8]. Initial data which evolve approximately as the Warhaft-Lumley experiment are as given in Table 2, Run #1.

Table 2. $(k_{min}, k_{max}) = (0, 100)$, $\nu = 10^{-3}$. For both runs Constant c is such that total energy is normalized to unity. Here t_0 = time of injection of random scalar field

Run #	t_0	$E_v(k)$	$T_v(k)$	$E_\theta(k)$	$T_\theta(k)$
1	–	$ck^4e^{-0.8k}$	0	$ck^4e^{-0.8k}$	0
2	11.25	$E_v(k, t_0, \#1)$	$T_v(k, t_0, \#1)$	$ck^4e^{-0.5k}$	0

Run 2 represents a random scalar field injected into the evolved turbulence field at $t_1 = 11.25$. Initial spectral shapes and parameters of the run are specified in the figure captions. We recall that the Warhaft and Lumley experiment has $R_\lambda \simeq 45$, and values of P_λ range from ~ 50 for the heated grid experiments to $\sim 43-27$ for the mandoline experiments.

The course of $n(t) \equiv -t\dot{\overline{q^2}}(t)/\overline{q^2}(t)$ and $m(t) \equiv -t\dot{\overline{\theta^2}}(t)/\overline{\theta^2}(t)$ for the runs are given in Fig. 4. Note that beyond $t = 18$, for practical purposes the gross energetics are fully asymptotic $n \simeq 1.397$. [Note, however, $n(t)$ shows a long-term secular drift toward larger values

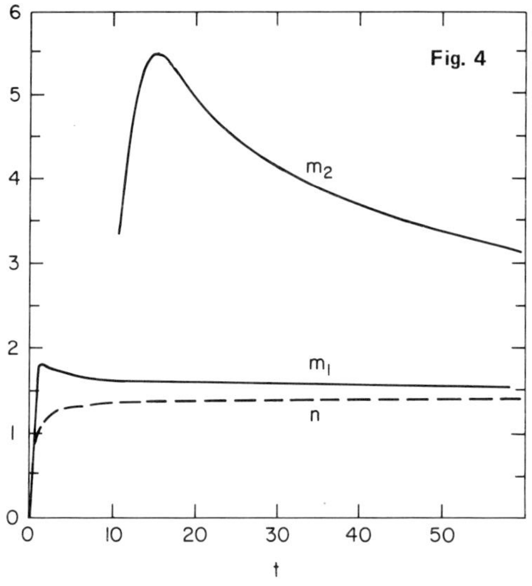

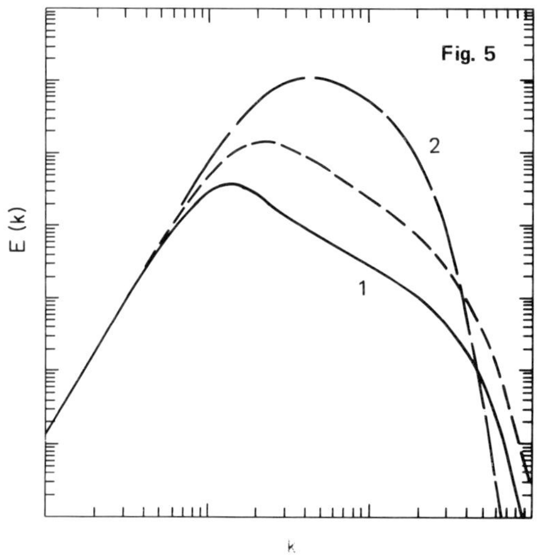

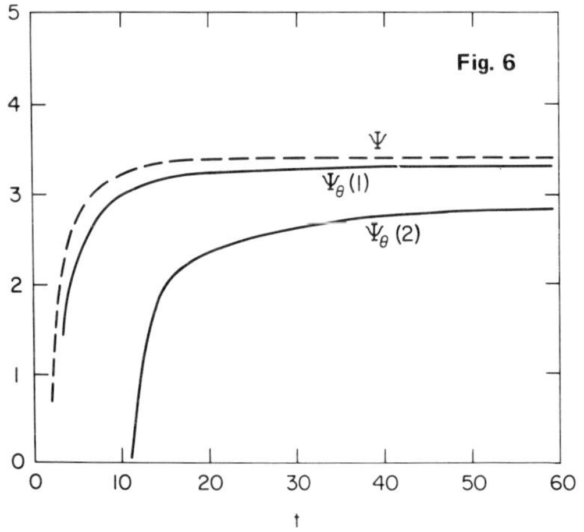

Fig. 4. Decay exponents for kinetic energy, n, and scalar energy m. Run numbers are indicated by subscripts. Model depicted is TFM ($g_v^2 = 1.17$, $g_\theta^2 = 0.5$, $\tilde{g}_\theta^2 = 1$)

Fig. 5. Three-dimensional kinetic energy (*dashed line*) and scalar energy (*solid lines*). Run numbers are indicated on Figure. For Run 1, $R_\lambda = 45$, $P_\lambda \equiv u\lambda\theta/\gamma = 37$, while for Run 2, $R_\lambda = 37.4$, $P_\lambda = 20$. (Runs for TFM ($g_v^2 = 1.17$, $g_\theta^2 = 0.5$, $\tilde{g}_\theta^2 = 1$)

Fig. 6. Normalized energy and scalar decay rates, ψ, and ψ_θ, respectively. Runs for TFM ($g_v^2 = 1.17$, $g_\theta^2 = 0.5$, $\tilde{g}_\theta^2 = 1$)

as R_λ decreases.] This value found for n is in good agreement with numerical results of *Lesieur* and *Schertzer* [9], *Tatsumi* et al. [29], and *Frisch* et al. [30]. The large-scale eddy circulation time, $\tau_0 \equiv L/(\overline{q^2}/3)^{1/2} \sim 0.5\, t^{1.058}$, $t \gtrsim 5$. The small-scale spectra appear to become self-similar much quicker; for example, $S(t)$ reaches its asymptotic value of 0.53 at $t = 3.0$ for both runs.

Figure 5 gives a log-log plot of the energy spectrum $E(k,\ 18)$ and $E_\theta(k,\ 18)$ for Runs 1 and 2. Note that for Run 1, $E_\theta(k)$ appears to have an anomalous inertial range, even at this modest value of $P_\lambda (\simeq 37)$. Run 2 does not have this range, but has $P_\lambda \simeq 18$. Such an anomaly has also been observed by *Warhaft* and *Lumley* [1] and by *Yeh* and *Van Atta* [31]. The former authors attribute the anomaly to the effects of heat fluxes generated near the heated grid. We suggest the difference may be partly attributable to the decrease in P_λ for the *mandoline* data. In terms of the TFM ingredients, it appears that this "inertial" range is more extensive for smaller g_θ. An examination of the theory suggests that it may actually be a somewhat damped vestigal k^{-1} range, similar to the large Prandtl number case.

Figure 6 gives $\psi(t)$ and ψ_θ, defined by

$$\psi \equiv -\dot{\epsilon}\overline{q^2}/\epsilon^2, \qquad \psi_\theta \equiv -\dot{\epsilon}_\theta\overline{\theta^2}/\epsilon_\theta^2. \tag{3.1}$$

These are important parameters for single-point models (e.g., *Newman* et al. [32]). We shall note later (Sect. 6) the asymptotic $(t \to \infty)$ limits for ψ and ψ_θ.

Finally, Fig. 7 shows $r(t)$ for all runs as a function of k_θ/k_v, where k_v and k_θ are wave numbers at which the energy spectra for velocity and scalar fields peak. The points here are the *mandoline* data of *Warhaft* and *Lumley.* The correlation of wave number ratio with $r(t)$ is similar to the experiments, but the TFM ($g_\theta^2 = 0.5$) curve is ~20% larger than the data, while the EDM (0.36, 0, 1.3) is ~10% too large. TFM ($g_\theta^2 = 1$) results are more nearly in accord with the data, as is indicated by the dashed line. This suggests a Peclet-number dependence of g_θ, when we recall that g_θ was picked to match experiments at large Peclet number.

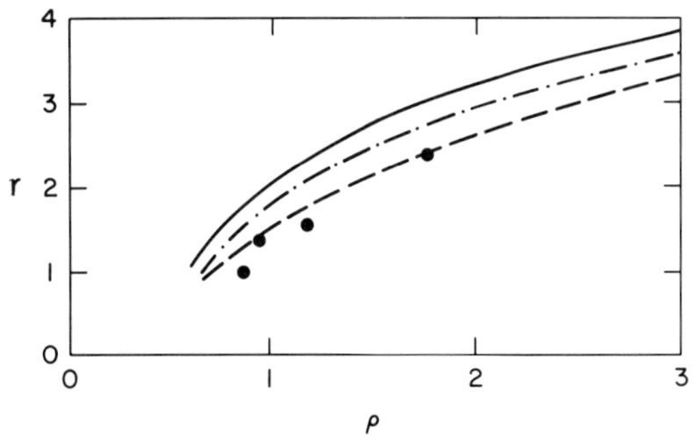

Fig. 7. Decay time scale ratio, r, as function of ratio of scalar energy peak wave number to kinetic energy peak wave number. TFM ($g_v^2 = 1.17$, $g_\theta^2 = 0.5$, $\tilde{g}_\theta^2 = 1$), *solid line;* TFM ($g_v^2 = 1.17$, $g_\theta^2 = 1$, $\tilde{g}_\theta^2 = 1$), *dashed line.* EDM ($\tilde{\lambda} = 0.36$, $\tilde{\lambda}' = 0$, $\tilde{\lambda}'' = 1.3$), *dot dashed line. Mandoline* data of *Warhaft* and *Lumley* [1] are points

6. Analytic Results for the Decay Exponents

Much of our discussion here of the behaviour of the indices $[n(t), m(t)]$ may be understood analytically in terms of the general context of two-point closures. The key points needed to determine (n, m) are based on the analytic behavior of $T(k)$ and $T_\theta(k)$ as $k \to 0$, and the inertial range shapes of $E(k)$ and $E_\theta(k)$ at large k. As $k \to 0$, we have from the analysis of *Lesieur* and *Schertzer* [9],

$$T(k) \to Ak^4 - 2(\nu + \nu_{\text{eddy}})k^2E(k), \tag{6.1}$$

and

$$T_\theta(k) \to A_\theta k^4 - 2(\gamma + \gamma_{\text{eddy}})k^2 E_\theta(k). \tag{6.2}$$

Here A and A_θ are functionals of $E(k)$ and $E_\theta(k)$. Suppose that now

$$E(k) \to C_s(t)k^s,$$

and

$$E_\theta(k) \to C_{s'}(t)k^{s'},$$

as $k \to 0$. Then Eqs. (6.1) and (6.2) imply that C_s and $C_{s'}$ are constants of motion, provided $[1 \leqslant (n, m) < 4]$. At large R_λ, and for Pr $\sim$ 1, we then have (approximately),

$$E(k) = C_s k^s, \; k < k_0; \quad E(k) \sim \epsilon^{2/3} k^{-5/3} \tag{6.3}$$

$$E_\theta(k) = C_{s'} k^{s'}, \; k < k_0; \quad E_\theta(k) \sim \epsilon_\theta \epsilon^{-1/3} k^{-5/3}, \; k > k_0. \tag{6.4}$$

We have here assumed that the spectral profile of either E or E_θ is accurately described by only two parameters (C_s and k_0). Then, using continuity and $\dot{E} = -\epsilon, \dot{E}_\theta = -\epsilon_\theta$ implies (see also *Comte-Bellot* and *Corrsin* [28]),

$$n = 2(s+1)/(s+3) \tag{6.5}$$

$$m = 2(s'+1)/(s+3), \tag{6.6}$$

which in turn imply

$$r = m/n = (s'+1)/(s+1), \tag{6.7}$$

$$\psi = 2(n+1)/n = (3s+5)/(s+1) \tag{6.8}$$

$$\psi_\theta = 2(m+1)/m = (2s' + s + 5)/(s'+1) \tag{6.9}$$

We note that $s = 4$ yields $n = 10/7$, a result previously derived by *Kolmogorov* [33], assuming the invariance of the Loitsyanskii integral. However, Eq. (6.1) is inconsistent with the invariance of such an integral, since it implies $dC_4/dt \neq 0$. We may include the case of $s = 4$ within the purview of our discussion by introducing the parameters Γ and Γ'

$$d \ln C_4/d \ln t = \Gamma, \quad d \ln C_4'/d \ln t = \Gamma'. \tag{6.10}$$

Then,

$$n(s=4) = 2(5-\Gamma)/7, \tag{6.5$'$}$$

and

$$m(s'=4) = 5(2+\Gamma)/7 - \Gamma' \tag{6.6$'$}$$

with corresponding changes in Eqs. (6.7–9). Numerical estimates of r at very large R_λ show it to be rather small, $r = 0.16$.

At moderate R_λ, the hypotheses under which these formulas were derived are no longer valid; however, our numerical results of Sect. 5 indicate their approximate validity even for $R_\lambda \simeq 20$.

7. Summary and Discussion

The two-point closures employed here to investigate the case of a passive scalar convected by isotropic turbulence account for the major features of this problem fairly well. At low Prandtl number, we recover the spectra of *Batchelor*, *Howels* and *Townsend* and predict the course of the mixed scalar skewness with Pr in a reasonable way (see Figs. 2 and 3). At Pr ~ 1, the *shapes* of the inertial-dissipation ranges for both velocity and scalar fields are satisfactory. At very large Pr, we find a Batchelor k^{-1} subrange; here, however, a satisfactory value of $C_B(\sqrt{3} < C_B \lesssim 6)$ [see Eq. (3.1)] depends sensitively on the choice of eddy relaxation parameters; for either TFM (1.06, 0.0, 3.61) or EDM (0.36, 0.0, 1.30) $C_B \cong 2.6$. From the numerical comparisons presented here, these appear for practical purposes as equivalent procedures. The choice also appears the most reasonable at lower Pr. Finally it gives an eddy Prandtl number of ~ 0.6, in good agreement with experimental values.

At moderate R_λ, we have attempted to shed some light on the question of why the scalar index m is frequently not universal in experiments. Here we are able to recover results similar to the Warhaft-Lumley experiment in that the value of m substantially larger than the $(t \to \infty)$ asymptotic value (~ 1.55) is obtained if the scalar initially resides at smaller scales than the velocity field. However, the theoretically predicted scalar decay is not – in reality – a power law; rather $m(t)$ decreases secularly until the peak scalar wave number catches that of the velocity field. We note that this secular decrease of $m(t)$ is sufficiently slow so that an experimental observation of $\overline{\theta^2}(t)$ (with typical scatter) decreasing by only a factor of ~ 10 would be consistent with a large but constant value of m. *Kerr* and *Nelkin* [34] have estimated (through a simple cascade model) that the time for the scalar to catch the velocity field is so long that the flow would exit the wind-tunnel first. Our calculations here are consistent with theirs on this point.

The correlation of $m(t)$ with the peak wave number of the scalar is predicted by TFM and EDM as given in Fig. 7 and is similar to the experimental values, but the values of the predicted peak k_θ is 10–20% too large, unless the eddy relaxation rates are allowed to depend on R_λ. The trend shown in this figure $(d\rho/dr > 0)$ holds only if the scalar is injected at smaller scales than the velocity field; it would be interesting to check this point experimentally.

Finally, we note that the large R_λ time asymptotic analysis of *Lesieur* and *Schertzer* [9] appears to hold – at least approximately – at moderate R_λ also, despite the fact that the assumptions of the analysis are not well-justified at low R_λ.

Acknowledgements. The authors are indebted to J. C. André and U. Frisch for helpful comments.

Appendix

We record here the direct interaction approximation (DIA) (*Kraichnan* [2]); the test field model (TFM) (*Kraichnan* [5]) (*Newman* and *Herring* [17]); and the eddy damped Markovian (EDM) approximation (*Orszag* [7]), (*André* and *Lesieur* [8]) for the passive covariance $\theta(\boldsymbol{k}, t, t')$ and velocity field covariance, $U_{ij}(\boldsymbol{k}, t, t')$. We are interested in isotropic statistics for which $U_{ij} = \frac{1}{2}(\delta_{ij} - k_i k_j/k^2)\, U(k, t, t')$. The DIA equations are:

$$\left(\frac{\partial}{\partial t} + \gamma k^2\right) \Theta(k, t, t') + \int_{-\infty}^{t} \hat{\eta}(k, t, s)\, \Theta(k, s, t')\, ds$$

$$= + \int_{-\infty}^{t'} ds \int_{\Delta} B^{s}_{kpq}\, G_{\Theta}(k, t', s)\, U(q, t, s)\, \Theta(p, t, s)\, dp\, dq \qquad t \geqslant t' \tag{A.1}$$

$$\left(\frac{\partial}{\partial t} + 2\gamma k^2\right) \Theta(k, t, t) + 2 \int_{-\infty}^{t} \hat{\eta}(k, t, s)\, \Theta(k, s, t)\, ds$$

$$= 2 \int_{-\infty}^{t} ds \int_{\Delta} B^{s}_{kpq}\, G_{\Theta}(k, t, s)\, U(q, t, s)\, \Theta(p, t, s)\, dp\, dq \tag{A.2}$$

$$\hat{\eta}(k, t, s) \equiv \int_{\Delta} B^{s}_{kpq}\, G_{\Theta}(p, t, s)\, U(q, t, s)\, dp\, dq \tag{A.3}$$

Here the scalar Green's function $G_{\Theta}(\boldsymbol{k}, t, t')$ satisfies

$$\left(\frac{\partial}{\partial t} + \gamma k^2\right) G_{\Theta}(k, t, t') + \int_{t'}^{t} \hat{\eta}(k, t, s)\, G_{\Theta}(k, s, t') \quad ds = 0 \; t > t'$$

with

$$G_{\Theta}(k, t, t') = 1, \quad G(k, t, t') = 0 \quad t < t'.$$

The DIA equations for $U(k, t, t')$ are

$$(\partial/\partial t + 2\nu k^2)\, U(k, t, t) + 2 \int_{0}^{t} \eta(k, t, s)\, U(k, s, t)\, ds \tag{A.5}$$

$$= 2\pi \int_{0}^{t} ds \iint_{\Delta} B_{kpq}\, G(k, t, s)\, U(p, t, s)\, U(q, t, s)\, pq\, dp\, dq$$

and for $G(k, t, t')$

$$(\partial/\partial t + \nu k^2)\, G(k, t, t') + \int_{t'}^{t} \eta(k, t, s)\, G(k, s, t')\, ds = 0 (t \geqslant t') \tag{A.6}$$

$$G(k, t', t') = 1$$

with

$$\eta(k, t, s) = \int_\Delta\int B_{kpq}\, G(p, t, s)\, U(q, t, s)\, dp\, dq \tag{A.7}$$

The TFM equations are

$$\left(\frac{d}{dt} + 2\,\gamma k^2\right) \Theta(k, t, t) = \int_\Delta B^s_{kpq}\, D^\theta_{kpq}\, U(q, t, t) \times [\Theta(p, t, t) - \Theta(k, t, t)]\, dp\, dq \tag{A.8}$$

$$dD^\theta_{kpq}/dt = 1 - \{g_\theta^2[\eta^c(k) + \eta^c(p)] + \tilde{g}_\theta^2\eta^s(q) + \gamma(k^2 + p^2) + \nu q^2\}\, D^\theta(k, p, q)$$

with

$$D^\theta(k, p, q \mid t = 0) = 0 \tag{A.9}$$

and

$$\left(\frac{d}{dt} + 2\,\nu k^2\right) U(k, t, t) = \int_\Delta B_{kpq} D_{kpq}\, [U(p, t, t) - U(k, t, t)]\, U(q, t, t)\, dp\, dq \tag{A.10}$$

$$\frac{dD_{kpq}}{dt} = 1 - [\eta^s(k) + \eta^c(p) + \eta^c(q) + \nu(k^2 + p^2 + q^2)]\, D_{kpq} \tag{A.11}$$

$$\frac{d}{dt} D^c_{kpq} = 1 - [\eta^c(k) + \eta^c(p) + \eta^s(q) + \nu(k^2 + p^2 + q^2)]\, D^c_{kpq} \tag{A.12}$$

with

$$\eta^c(k) \equiv \int_\Delta C_{kpq} D^c_{kpq}\, U(q, t, t)\, dp\, dq \tag{A.13}$$

and

$$\eta^s(k) \equiv \int_\Delta C_{kpq} D^s_{kpq}\, U(q, t, t)\, dp\, dq. \tag{A.14}$$

The EDM equations are the same as Eqs. (A.8), (A.10), with (A.9) and (A.11) replaced by

$$dD^\theta_{kpq}/dt = 1 - \{\tilde{\lambda}'[\mu(k) + \mu(p)] + \tilde{\lambda}''\mu(q) + \gamma(k^2 + p^2) + \nu q^2\}\, D^\theta_{kpq} \tag{A.15}$$

$$dD_{kpq}/dt = 1 - \{\tilde{\lambda}\, [\mu(k) + \mu(p) + \mu(q)] + \nu(k^2 + p^2 + q^2)\}\, D_{kpq} \tag{A.16}$$

where

$$\mu^2(k) = \int_0^k E(p) p^2 dp. \tag{A.17}$$

In the TFM and EDM the coefficients $(g, g_\theta, \tilde{g}_\theta)$ and $(\tilde{\lambda}, \tilde{\lambda}', \tilde{\lambda}'')$ must be fixed through an appeal to experiments. The TFM values used here, as determined through a comparison with experiments, as explained in Sect. 2 are

$$g^2 = 1.12 \tag{A.18}$$

$$g_\theta^2 = 0.5 \tag{A.19}$$

$$\tilde{g}_\theta = 1.0. \tag{A.20}$$

In general, the TFM and EDM give slightly different evolution equations for $U(k, t)$ and $\Theta(k, t)$; however, in the inertial range, the difference is entirely a matter of adjusting the $\tilde{\lambda}$'s and g's. In this case the choice of λ's corresponding to Eqs. (A.18–20) is:

$$\tilde{\lambda} = 0.360 \tag{A.21}$$

$$\tilde{\lambda}' = 0.320 \tag{A.22}$$

$$\tilde{\lambda}'' = \tilde{\lambda}. \tag{A.23}$$

The above choice is not unique; one may increase λ'' and decrease λ' and obtain the same Oboukhov-Corrsin constant, while not upsetting the agreement of theory and experiment at large Prandtl number. The above choice blends the theories with the DIA predictions at low Reynolds number; one may with equal logic blend in the scalar theory with the Lagrangian history approximation (*Kraichnan* [3] in which case $\tilde{\lambda}' = 0$. In this case $\tilde{\lambda}''/\tilde{\lambda} = 3.61$, in order to obtain a Oboukhov-Corrsin constant of 0.667.

References

1. Z. Warhaft, J. L. Lumley: "An experimental study of the decay of temperature fluctuations in grid-generated turbulence". J. Fluid Mech. **88**, 659 (1978)
2. R. H. Kraichnan: "The structure of isotropic turbulence at very high Reynolds numbers". J. Fluid Mech. **5**, 497–543 (1959)
3. R. H. Kraichnan: "Lagrangian history closure approximation for turbulence". Phys. Fluids **8**, 575 (1965)
4. R. H. Kraichnan, J. R. Herring: "A strained based Lagrangian-history turbulence theory". J. Fluid Mech. **88**, 355–367 (1978)
5. R. H. Kraichnan: "An almost-Markovian Galilean-invariant turbulent model". J. Fluid Mech. **47**, 513–524 (1971)
6. J. R. Herring, R. H. Kraichnan: "Comparison of some approximations for isotropic turbulence", in *Statistical Models and Turbulence,* Lecture Notes in Physic, Vol. 12, ed. by M. Rosenblatt and C. Van Atta (Springer, Berlin, Heidelberg, New York 1972) pp. 148–194
7. S. A. Orszag: Analytical theories of turbulence. J. Fluid Mech. **41**, 363–386 (1970)
8. J. C. André, M. Lesieur: Influence of helicity on the evolution of isotropic turbulence at high Reynolds number. J. Fluid Mech. **81**, 187 (1977)
9. M. Lesieur, D. Schertzer: "Amortissement auto similaire d'une turbulence à grand mombre de Reynolds". J. Mec. **17**, 609–646 (1978)
10. J. R. Hill: Models of the scalar spectrum for turbulent advection. J. Fluid Mech. **88**, 541 (1978)
11. D. C. Leslie, G. L. Quarini: The application of turbulence theory to the formulation of subgrid modeling procedure. J. Fluid Mech. **91**, 65–91 (1978)
12. J. R. Herring: Approach of axisymmetric turbulence to isotropy. Phys. Fluids **17**, 859–972 (1974)
13. U. Schumann, J. R. Herring: Axisymmetric homogeneous turbulence: A comparison of direct spectral simulations with the direct- interaction approximation. J. Fluid Mech. **76**, 755–782 (1976)
14. J. R. Herring: "Statistical Turbulence Theory and Turbulence Phenomenology". *Proc. Langley Working Conf. Free Turbulent Shear Flows,* NASA SP321 (1973), Langley Research Center, Langley, VA, (available from NTIS as N73-2815415GA)

15. D. C. Leslie: *Developments in the Theory of Turbulence* (Clarendon Press, Oxford 1973)
16. P. L. Sulem, M. Lesieur, U. Frisch: "Le test-field model interprété comme méthode de fermeture des équations de la turbulence". Ann. Geophys. **31**, 487–495 (1975)
17. G. R. Newman, J. R. Herring: "A test field model study of a passive scalar in isotropic turbulence". J. Fluid Mech. **94**, 163–194 (1979)
18. F. H. Champagne, C. A. Friehe, J. C. LaRue, J. C. Wyngaard: "Flux measurements, flux estimation techniques and fine-scale turbulence measurements in the unstable surface layer over land". J. Atmos. Sci. **34**, 515 (1977)
19. G. H. Batchelor, I. D. Howels, A. Townsend: "Small scale variations of convected quantities like temperature in turbulent fluid". J. Fluid Mech. **5**, 134 (1959)
20. C. H. Gibson: "Fine structure of scalar fields mixed by turbulence. II. Spectral theory". Phys. Fluids **11**, 2316 (1968)
21. V. L. Granatstein, S. J. Buchsbaum, D. S. Bugmolo: "Fluctuation spectrum of a plasma additive in a turbulent gas". Phys. Rev. Lett. **12**, 504 (1966)
22. Y. Ogura: "The dependence of the eddy diffusivity on the fluid Prandtl number". Adv. Geophys. **6**, 175 (1959)
23. R. J. Hill: "Solution of Howell's model of the scalar spectrum and comparison with experiment". Preprint, 2 June 1978)
24. R. H. Kraichnan: "Eddy viscosity in two and three dimensions". J. Atmos. Sci. **33**, 1521 (1976)
25. J. Blom: "An Experimental Determination of the Turbulent Prandtl Number in a Developing Temperature Boundary Layer". Ph. D. Thesis, Eindhoven University of Technology (1970)
26. J. Weinstock: "Analytic approximations on the theory of turbulent diffusion". Phys. Fluids **21**, 2129 (1977)
27. M. Lesieur, J. P. Chollet: "Introduction aux théories statistiques de la turbulence pleinement développée" in *Bifurcations et Problèmes Non Linéaires,* ed. by C. Bardos, C. Schatzmann, J. M. Lasry (Springer, Berlin, Heidelberg, New York, to be published)
28. G. Comte-Bellot, S. Corrsin: "The use of a contraction to improve the isotropy of grid-generated turbulence". J. Fluid Mech. **24**, 657 (1966)
29. T. Tatsumi, S. Kida, J. Mizushima: "The multiple scale cumulant expansion for isotropic turbulence". J. Fluid Mech. **85**, 97–142 (1978)
30. U. Frisch, M. Lesieur, D. Schertzer: "Comment on the quasinormal Markovian approximation for fully developed turbulence". J. Fluid Mech. **97**, 181–192
31. T. T. Yeh, C. W. Van Atta: "Spectral transfer of scalar and velocity fields in heated-grid turbulence". J. Fluid Mech. **58**, 233 (1973)
32. G. R. Newman, B. E. Launder, J. L. Lumley: "Modeling the behavior of homogeneous scalar turbulence". (Preprint to be submitted to J. Fluid Mech.)
33. A. N. Kolmogorov: "Decay of isotropic turbulence in incompressible viscous fluids". Dokl. Akad. Nauk SSSR **31**, 538–541 (1941)
34. R. M. Kerr, M. Nelkin: "The decay of temperature fluctuations convected by isotropic turbulence". (Preprint)
35. J. P. Clay: "Turbulent Mixing of Temperature in Water, Air and Mercury". Ph. D. Thesis in Engineering Science (Engr. Phys.), University of California at San Diego (1973)
36. R. M. Williams, Jr.: "High Frequency Temperature and Velocity Fluctuations in the Atmospheric Boundary Layer". Ph. D. Thesis, Oregon State University, Corvallis (1974) p. 81

Vortex Simulation of Three-Dimensional Spotlike Disturbances in a Laminar Boundary Layer

A. Leonard

Ames Research Center, NASA, Moffett Field, CA 94035 USA

Abstract

The growth of a turbulent spot in a laminar boundary layer, as the spot evolves from a localized disturbance in the layer, is simulated numerically using a three-dimensional vortex filament description of the vorticity field. The filaments are marked with a sequence of node points which are tracked in a Lagrangian reference frame. Velocity computation is done by Biot-Savart integration. Although some discrepancies with experiment appear to exist in the near wall region, the gross properties of the spot, including the velocities of the leading and trailing edges and the velocity perturbations away from the wall, are in good agreement with experiment.

Introduction

Studies of the turbulent spot in a laminar boundary layer data back to the theoretical and experimental work of *Emmons* [1] and *Mitchner* [2], who were concerned with the role of spots in the transition process. Such spots are now recognized as interesting flow phenomena in their own right and, in addition, spots or spotlike objects have been suggested as the coherent flow structure responsible for much of the turbulent transport in a fully developed *turbulent* boundary layer [3, 4]. Recent experiments by *Wygnanski* et al. [5] and *Cantwell* et al. [6] confirmed earlier observations of *Schubauer* and *Klebanoff* [7] and provided much new information on the structure of turbulent spots. In particular, entrainment rates in the plane of symmetry and the mean flow field within the spot were obtained by ensemble averaging over a large number of spots. In another recent experiment, *Gaster* and *Grant* [8] studied the evolution of a wave packet in a laminar boundary layer produced by a localized perturbation at the wall boundary. The amplitude of the perturbation was very small so that the linear development of the resultant disturbance could be observed for some time after its initiation. Indeed, model calculations by *Gaster* [9] based on linear theory agree well with the experimental measurements of *Gaster* and *Grant* [8] during the early stages, but some qualitative differences were noted in the later stages of growth, presumably due to nonlinear mechanisms.

Despite these significant advances in our understanding of turbulent spots, a number of questions remain, such as the interaction of two or more spots, the formation of sublayer streaks, the three-dimensional vorticity field within the spot, the nonlinear mechanics of growth, and the birth of new spots.

In this paper, the growth of a turbulent spot in a laminar boundary layer, as the spot evolves from a localized disturbance in the layer, is simulated numerically using a vortex filament description of the vorticity field. The attempt is to provide additional insight into the structure and growth of a turbulent spot. In many respects, the method employed [10, 11] is an extension to three dimensions of the two-dimensional vortex method [12, 13] and is done in the spirit of the large-eddy simulation technique [14] in which small-scale turbulence is not simulated directly but is modeled. In an earlier numerical study *Hama* [15], motivated by his experimental observations of boundary-layer transition, investigated the motion of a single vortex loop.

Previous three-dimensional numerical simulations of wall-bounded shear flows have included channel flow simulations by *Deardorff* [16] and *Schumann* [17] using an artificial law-of-the-wall boundary condition and the recent improved simulation by *Moin* et al. [18] using the no-slip condition. In addition, transition phenomena have been simulated numerically by *Orszag* [19] and by *Wray* et al. [20].

Numerical Method

For numerical purposes, a turbulent spot is viewed as a localized region of boundary-layer vorticity that has undergone three-dimensional displacements, strains, and rotations from its original monodirectional laminar state (see Fig. 1). Each tube of vorticity in the boundary layer is decomposed into its straight, unperturbed contribution and into a loop representing its contribution to the perturbed vorticity field, as shown in Fig. 2. The velocity field is decomposed, accordingly, as

$$u(x) = U(y)\hat{e}_x + u'(x)$$

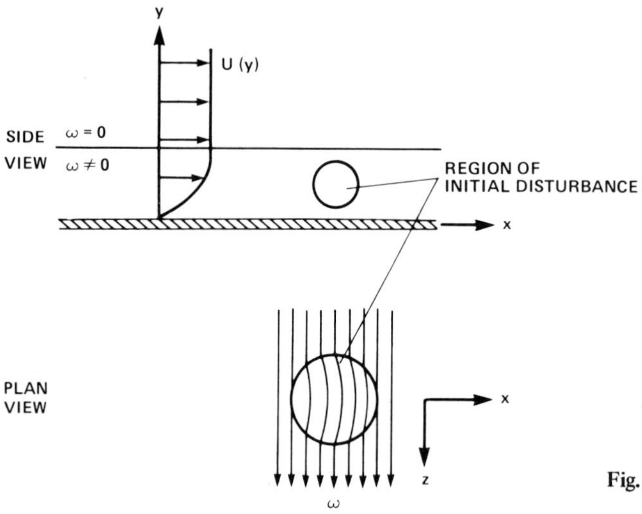

Fig. 1. Spot geometry

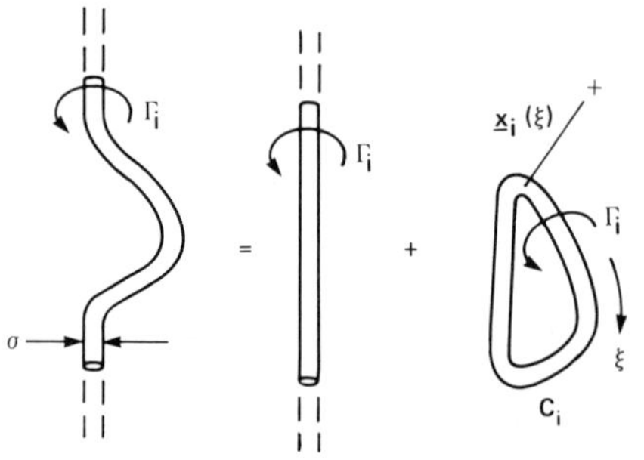

Fig. 2. Decomposition of vorticity field

where $U(y)$ is the laminar profile and $\boldsymbol{u}'$ is given by a sum of Biot-Savart integrations over all the filament curves, C_i, plus their images. The image contributions ensure tangency of the flow at the wall. In this initial study we ignore the generation of vorticity at the wall boundary due to the disturbance, i.e., we use the inviscid boundary condition at the wall. Biot-Savart integration yields an exact representation of the boundary conditions at infinity in all directions.

Each filament is represented by a space curve $\boldsymbol{x}_i(\xi, t)$, where ξ is a parameter along the curve; by a circulation Γ_i; and by an effective core radius σ_i, which parameterizes the assumed Gaussian vorticity distribution within the filament. Thus the perturbed vorticity field is given by

$$\omega(\boldsymbol{x}, t) = \sum_i \Gamma_i \oint_{C_i} \exp\left(-|\boldsymbol{x} - \boldsymbol{x}_i|^2/\sigma_i^2\right)/(\sqrt{\pi}\,\sigma_i)^3 \frac{\partial \boldsymbol{x}_i}{\partial \xi}\, d\xi$$

A filament or curve represents a collection of vortex lines so that the filament velocity must be computed as an appropriate average over the filament core. We use the following equation of motion for the space curves

$$\frac{\partial \boldsymbol{x}_i}{\partial t} = -\frac{1}{4\pi} \sum_j \Gamma_j \oint_{C_j} \frac{(\boldsymbol{x}_i - \boldsymbol{x}_j) \times \dfrac{\partial \boldsymbol{x}_j}{\partial \xi'}\, d\xi'}{\left[|\boldsymbol{x}_i - \boldsymbol{x}_j|^2 + \alpha(\sigma_i^2 + \sigma_j^2)\right]^{3/2}}$$

$$+ \hat{e}_x U(y_i) + \text{image contributions.}$$

The equation is seen to be the Biot-Savart induction law for an infinitesimal filament but corrected in the near field for finite core size. Using $\alpha = 0.2065$ we recover the correct speed of a ring vortex in the limit $\sigma_i \ll$ ring radius.

For numerical purposes, the filaments are marked with a sequence of node points which are tracked in a Lagrangian reference frame. The space curve $\boldsymbol{x}_i(\xi, t)$ is assumed piecewise linear, passing through each node point. The spacing between node points is $0(\sigma_i)$ to minimize discretization error. For example, satisfying this constraint yields a computed speed of a ring vortex that is essentially independent of the number of node points.

Discretization in the directions normal to the vorticity vector must also be treated carefully. If the nominal spacing between vortices is so large that adjacent cores do not overlap, the smooth vorticity field that exists in the undisturbed regions near the spot is poorly approximated. Given a number of horizontal layers of vortex filaments, equally spaced in the vertical direction, an optimum core size as well as a set of optimum circulations for each horizontal layer is determined that minimizes the mean square error in approximating the desired vorticity distribution. The result of this procedure is displayed in Fig. 3a for the case of eight layers, whereas the effect of choosing a smaller value of core size is shown in Fig. 3b.

For many incompressible flows in which the vorticity is concentrated in small regions, vortex methods have the conceptual advantage that computational points need only appear in the vorticity-containing regions. In this problem only the region of perturbed vorticity requires points. Thus, in this simulation, points are added automatically in the streamwise and, in some cases, the spanwise direction, as the spot grows. More specifically, the maximum perturbation of $\boldsymbol{x}_i$ in the farthest upstream and farthest downstream filaments in each layer is monitored. As soon as that perturbation reaches a critical value (usually 0.004 of the displacement thickness, δ^*) a new, adjacent filament is added. This feature, which helps to minimize the number of computational points, is particularly important because $0(N^2)$ operations per time step are required to compute the velocity at each of N vortex node points. For reason-

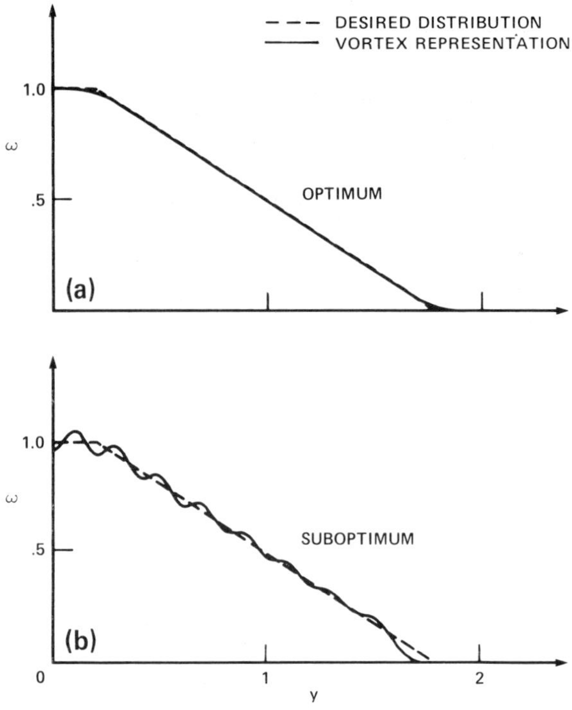

Fig. 3a, b. Vorticity distributions. (**a**) Optimum vortex representation. (**b**) Suboptimum choice of core size

able total run times (1–2 cpu hr), the maximum number of points representing one-half the spot is < 2,000 on the CDC 7600 and < 8,000 on the Illiac IV.

The range of computed scales of motion is quite large; e.g., in the streamwise direction the ratio of the largest to the smallest computed scale is ~ 125 to 1. Nevertheless, this range is not large enough to span those scales found experimentally (see, e.g., aluminum particle photographs in Ref. [6]). Thus, the effect of the smaller or subgrid scales on the computed scales must be modeled. The usual procedure used in finite-difference simulations is to use a subgrid viscosity dependent on the local strain rate [14]. In this study we simply maintain a constant filament core size. Indeed, this procedure is basically equivalent to using a subgrid viscosity dependent on the local rate of vortex stretching.

Results

Figure 4 displays the computed grid of vortex filaments as the spot develops. All the results given in this paper were obtained on the Illiac IV, although development of the algorithm was done on a CDC 7600. As shown at $t = 0$, the initial perturbation imposed on the boundary layer consists of a localized deformation of the vortex lines in the streamwise direction with a maximum displacement equal to 0.3. The net effect of this deformation is equivalent to adding a ringlike blob of vorticity to the laminar boundary layer, oriented so that the blob is initially moving upward away from the plate. The maximum streamwise velocity perturbation in this initial disturbance is $-0.07\, U_\infty$. Velocities and lengths are normalized so that $U_\infty = 1$ and $\partial U/\partial y\,|_{\text{wall}} = 1$. The unperturbed vorticity profile is that shown in Fig. 3a and the corresponding δ^* is 1.21. Eight levels of filaments were used in the y-direction normal to the plate and were spaced 0.2 apart. The same spacing was used in the streamwise (x) direc-

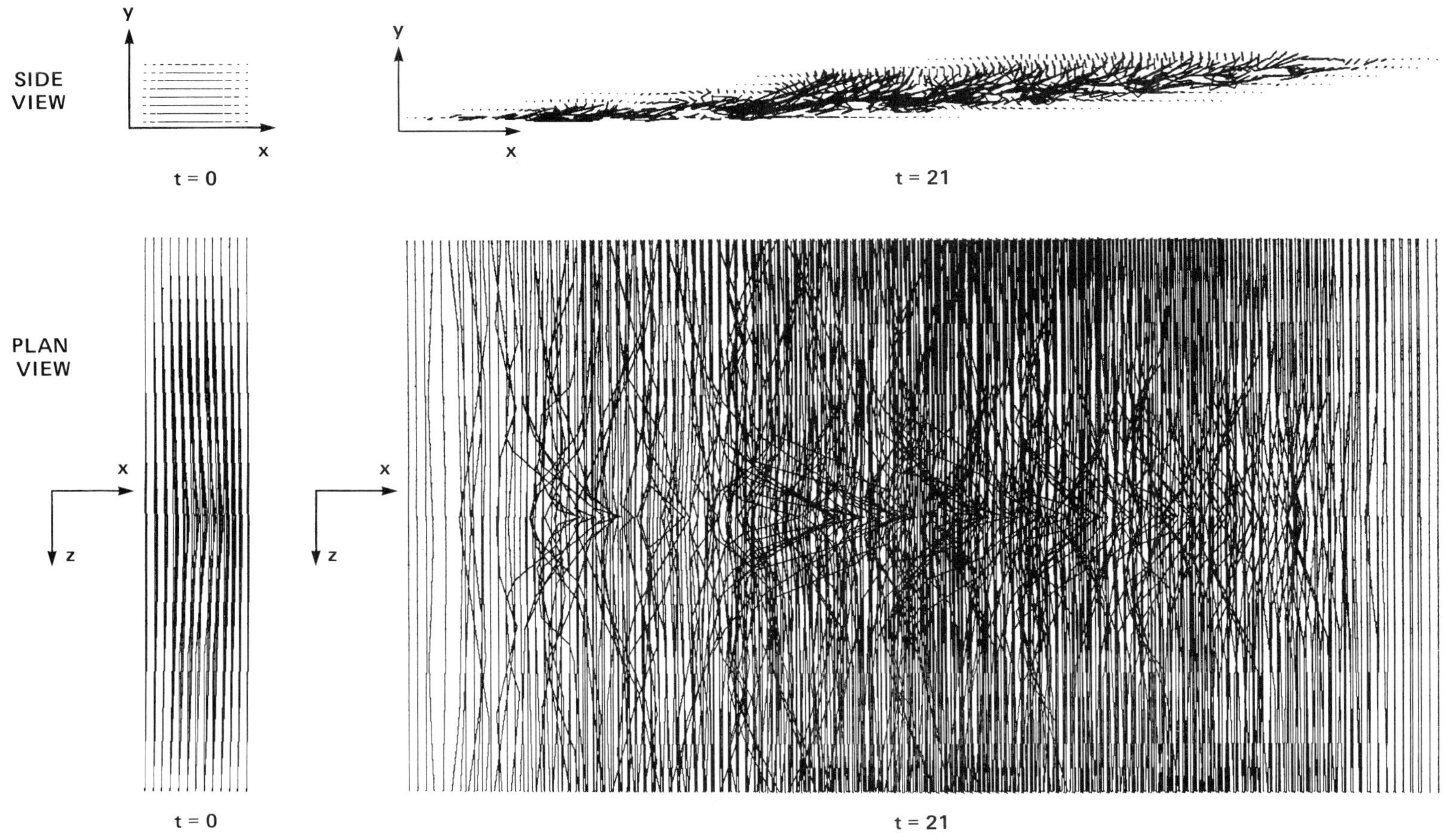

Fig. 4. Computation grid of vortex filaments

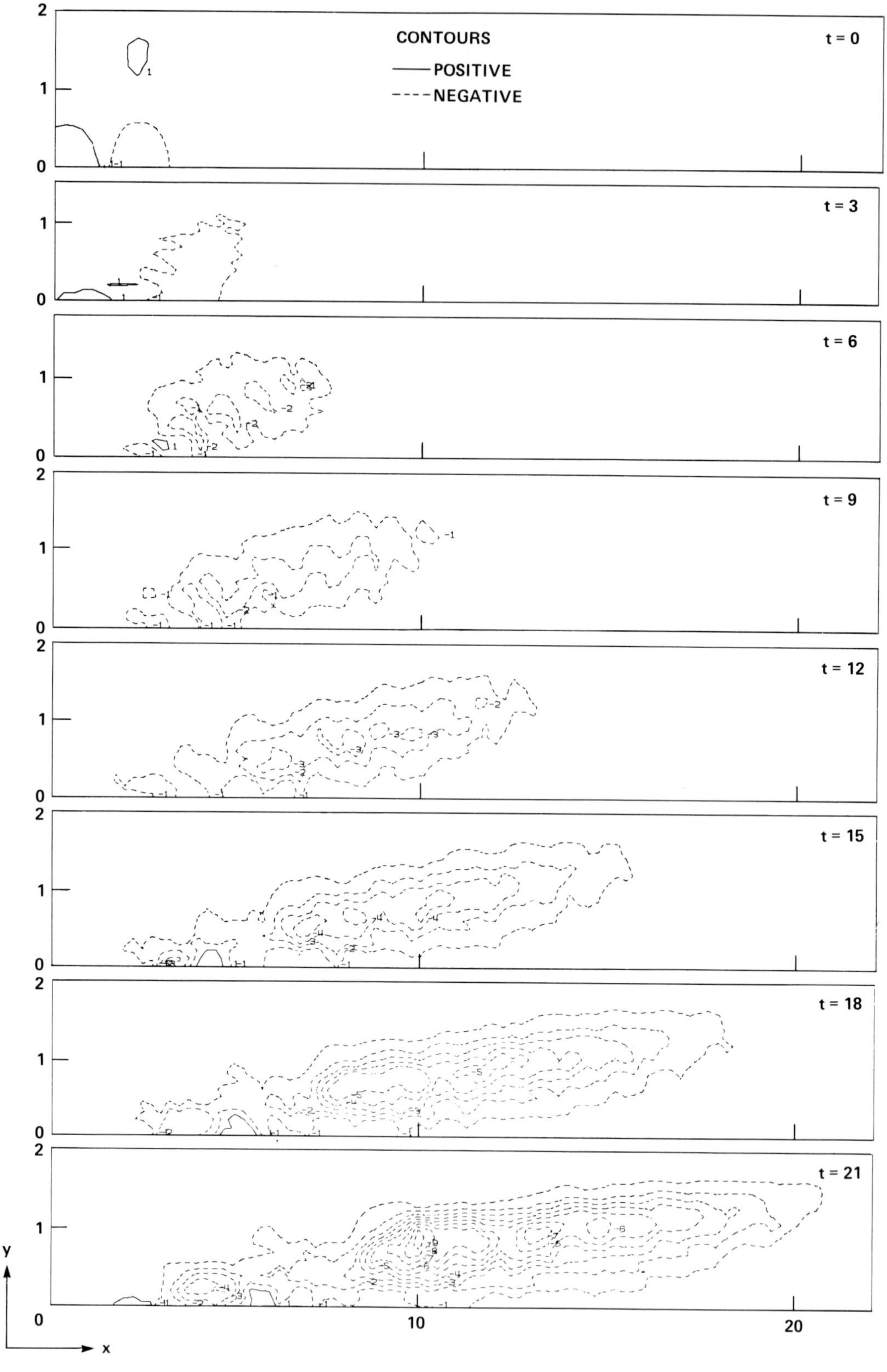
CONTOURS
POSITIVE
NEGATIVE
t = 0
t = 3
t = 6
t = 9
t = 12
t = 15
t = 18
t = 21
0
10
20
x
y

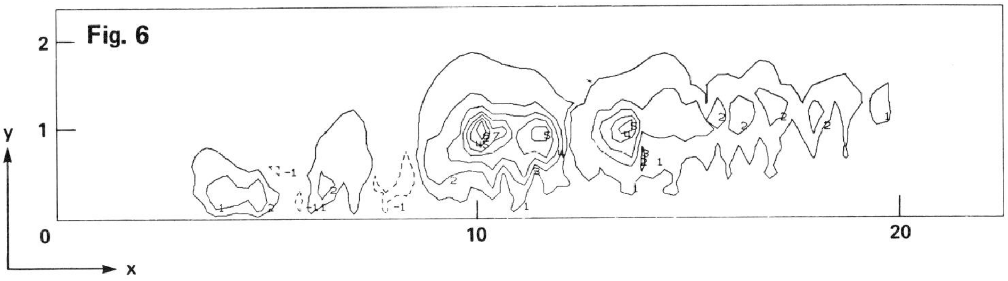

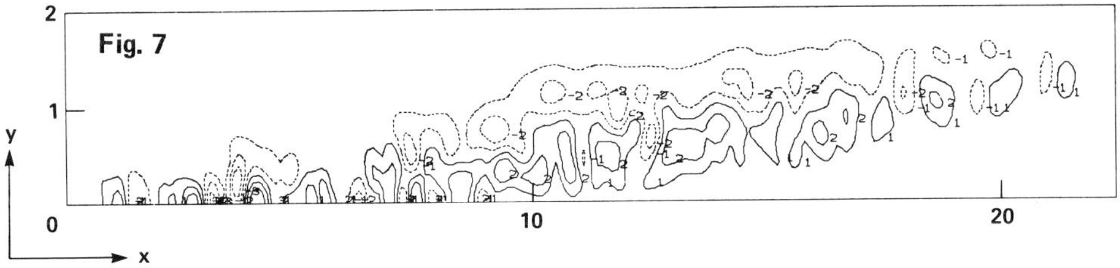

Fig. 6. v' contour plot; same contour levels as figure 5; $t = 21$; $z = 0$

Fig. 7. ω_z' contour plot; contours 0.1, 0.3, 0.5, . . .; $t = 21$; $z = 0$

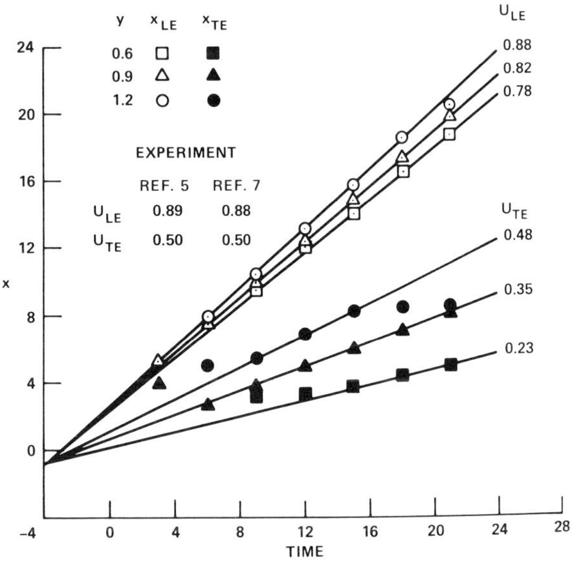

Fig. 8. Trajectories of leading and trailing edges

tion due to the isotropy of the velocity field induced by each filament. The initial spacing of points along each filament was 0.67. As the simulation proceeded, however, points were automatically redistributed to provide a higher density of points in regions of high curvature. The lifting of vortex lines is evident in the side view of Fig. 4. The stretching of vortex lines and the development of streamwise vorticity is quite evident in the plan view. In the simulation

◄

Fig. 5. u' contour plots, y scale expanded by 2; contours ± 0.02, ± 0.06, ± 0.10, . . .; $z = 0$

shown in Fig. 4, 2,184 points in 104 filaments were used at $t = 0$. At the end of the run of 71 steps ($\Delta t = 0.3$, first-order Euler explicit time differencing), there were 12,222 points in 582 filaments representing the spot. The spot was assumed to be symmetric with respect to the $z = 0$ plane. Therefore, the number of computational points requiring nodal velocities was only approximately half the total number representing the disturbance.

In Fig. 5, a time sequence of contour plots is shown for the perturbed streamwise velocity in the plane of symmetry. The plots are dominated by a growing patch of streamwise velocity deficit with a peak of $-0.37\ U_\infty$ at the final time, $t = 21$. Small regions of positive perturbation are evident near the wall in the later stages. Corresponding plots of vertical velocity and perturbed spanwise vorticity are shown in Figs. 6 and 7, respectively. As shown, the vertical velocities are predominately positive. The vorticity plot shows an elongated region of negative vorticity riding on a similar region of positive vorticity.

To compare with experiment, profiles of the spot boundary were determined from plots similar to those in Fig. 5 according to the criterion of Ref. [5]: 2% deviation of the streamwise velocity from its free-stream value. Shown in Fig. 8 are trajectories of the leading and trailing edges measured in the plane of symmetry at $y = 0.6$, 0,9, and 1.2. (These y values have undisturbed laminar velocities of $0.55\ U_\infty$, $0.75\ U_\infty$, and $0.89\ U_\infty$, respectively.) The trajectories for $y = 1.2$ yield leading – and trailing-edge velocities of $0.88\ U_\infty$ and $0.48\ U_\infty$, respectively. These values compare well with those of *Wygnanski* et al. [5], who found $U_{LE} = 0.89\ U_\infty$ and $U_{TE} = 0.50\ U_\infty$, and with those of *Schubauer* and *Klebanoff* [7], who reported $U_{LE} = 0.88\ U_\infty$ and $U_{TE} = 0.50\ U_\infty$. Trailing-edge velocities, however, show a significant variation with y. The plot also reveals the virtual origin (x_0, t_0) of the spot.

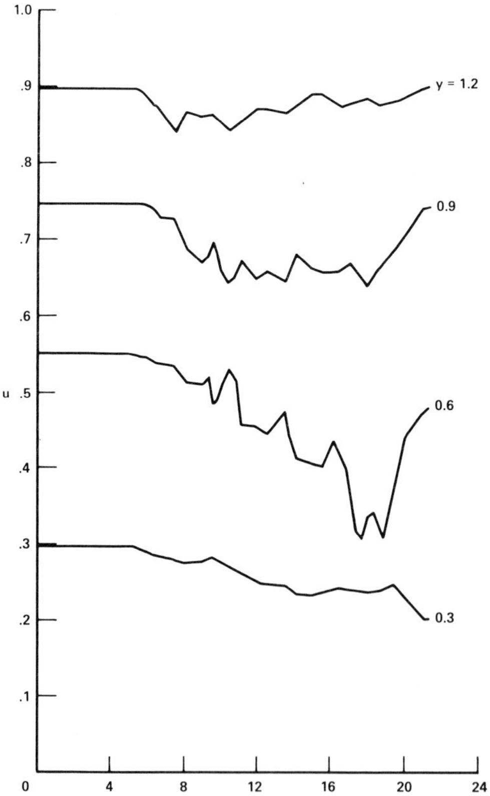

◀ **Fig. 9.** u vs time, $x = 8.0$, $z = 0.0$

▼ **Fig. 10.** v' vs time, $x = 8.0$, $y = 0.9$, $z = 0.0$

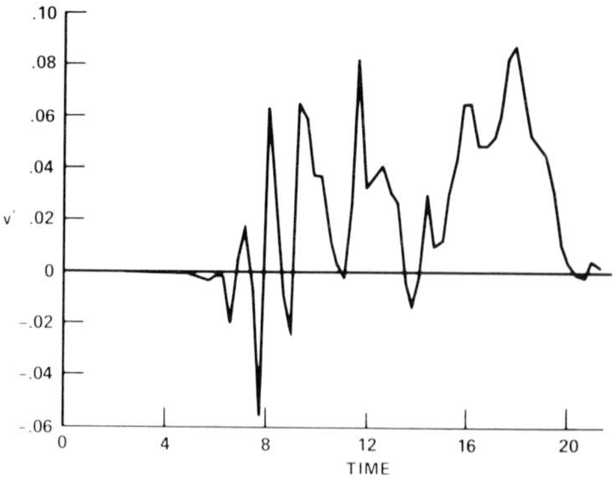

Velocity histories at fixed points are shown in Figs. 9 and 10. The u-velocity records away from the wall show the characteristic dip and recovery in agreement with the measurements of Refs. [5, 6] based on ensemble means over large numbers of spots. The simulated u-velocities in the plane of symmetry near the wall show no indication of the high-speed rise which is observed experimentally. The normal velocity record shown in Fig. 10 has an initial dip followed by a rise to positive values in agreement with the ensemble-averaged results of Ref. [5].

Streamwise velocity perturbations in three horizontal planes at $t = 18$ are shown in Fig. 11. Although the velocity histories taken in the plane of symmetry (Fig. 9) showed no evidence of high-speed fluid near the wall, the contour plot for $y = 0.3$ shows extensive regions of positive perturbation away from the plane of symmetry.

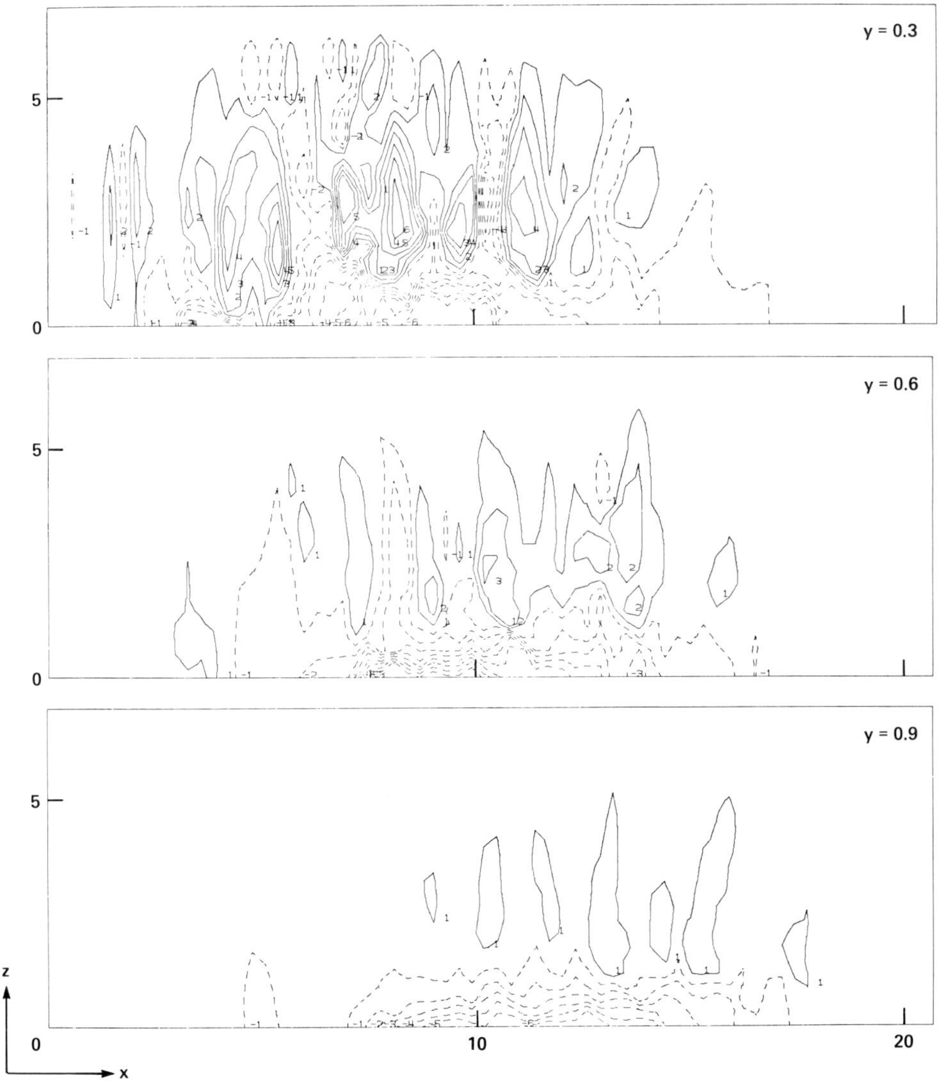

Fig. 11. u' contour plots in horizontal planes; contour levels ± 2%, ± 6%, ± 10%, . . .

Conclusions

The numerical simulation of a spotlike disturbance in a laminar boundary layer using vortex filaments has provided insight into the development of the three-dimensional vorticity and velocity fields within the spot. Gross properties of the spot away from the wall, including the velocities of the leading and trailing edges and the velocity perturbations, are in good agreement with experiment.

Discrepancies with experiment appear to exist in the near wall region, especially at the trailing edge. This suggests that future simulations should include the effect of vorticity production at the wall and an improved representation of viscous dissipation. Another source of disagreement may be the relatively short history of the computed spot. Experimental results (5–7) are based on spots that have traveled several hundred δ^* whereas the computed spot has traveled about 20 δ^*. Another possibility is that the experimental data were obtained from ensemble averaging over a large number of spots, a procedure that possibly filters out important coherent motions on the scale of several δ^* or more in addition to unwanted "noise".

Achnowledgment. The author thanks Dr. R. S. Rogallo for reprogramming the original CDC 7600 code for use on Illiac and for many helpful suggestions.

References

1. H. W. Emmons: The laminar-turbulent transition in a boundary layer, Part I. Aeronaut Sci. **18** (7), 490–498 (1951)
2. M. Mitchner: Propagation of turbulence from an instantaneous point disturbance. Aeronaut Sci. **21**, (5), 350–351 (1954)
3. D. Coles, S. J. Barker: "Some Remarks on a Synthetic Turbulent Boundary Layer", in *Turbulent Mixing in Nonreactive and Reactive Flows* (Plenum Press, New York 1975)
4. G. R. Offen, S. J. Kline: Combined dye-streak and hydrogen-bubble visual observations of a turbulent boundary layer. J. Fluid Mech. *62* (2), 223–239 (1974)
5. I. Wygnanski, M. Sokolov, D. Friedman: On a turbulent "Spot" in a laminar boundary layer. J. Fluid Mech. **78**, (4), 785–819 (1976)
6. B. Cantwell, D. Coles, P. Dimotakis: Structure and entrainment in the plane of symmetry of a turbulent spot. J. Fluid Mech. **87** (4), 641–672 (1978)
7. G. B. Schubauer, P. S. Klebanoff: "Contributions on the Mechanics of Boundary-Layer Transitions". NASA TN 3489 (Sept. 1955)
8. M. Gaster, I. Grant: An experimental investigation of the formation and development of a wave packet in a laminar boundary layer. Proc. R. Soc. London Ser. A: **347**, 253–269 (1975)
9. M. Gaster: A theoretical model of a wave packet in the boundary layer on a flat plate. Proc. R. Soc. London Ser. A: **347**, 271–289 (1975)
10. A. Leonard: "Numerical Simulation of Interacting, Three-Dimensional Vortex Filaments". *Proceedings of the Fourth International Conference on Numerical Methods in Fluid Dynamics* (Springer, Berlin, New York, Heidelberg 1975) pp. 245–250
11. A. Leonard: "Simulation of Three-Dimensional Separated Flows with Vortex Filaments". *Proceedings of the Fifth International Conference on Numerical Methods in Fluid Dynamics* (Springer, Berlin, New York, Heidelberg 1976) pp. 280–284
12. A. J. Chorin: Numerical study of slightly viscous flow. J. Fluid Mech. **57** (4), 785–796 (1973)
13. W. T. Ashurst: "Numerical Simulation of Turbulent Mixing Layers via Vortex Dynamics", in *Turbulent Shear Flows I,* ed. by F. Durst, B. E. Launder, F. W. Schmidt, J. H. Whitelaw (Springer, Berlin, Heidelberg, New York 1979) p. 402
14. J. H. Ferziger: Large eddy numerical simulations of turbulent flows. AIAA J. **15**, 1261–1267 (1977)
15. F. R. Hama: Progressive deformation of a curved vortex filament by its own induction. Phys. Fluids 5 (10) 1156–1162 (1962)

16. J. W. Deardorff: A numerical study of threedimensional turbulent channal flow at large Reynolds number. J. Fluid Mech. **41** (2), 453–480 (1970)
17. U. Schumann: Subgrid scale model for finite difference simulations of turbulent flows in plane channels and annuli. J. Comput. Phys. **18** (4), 376–404 (1975)
18. P. Moin, W. C. Reynolds, J. H. Ferziger: "Large Eddy Simulation of Incompressible Turbulent Channel Flow". TF-12, Dept. of Mechanical Engineering, Standford Univ., Standford California (May 1978)
19. S. A. Orszag, L. C. Kells: "Transition to Turbulence in Plane Poiseuille and Plane Couette Flow". Cambridge Hydrodynamics Rpt. **10** (May 1978)
20. A. A. Wray, M. Y. Hussaini, D. Degani: "Numerical Simulation of Transition to Turbulence". *Proceedings of Second Gamm Conference on Numerical Methods in Fluid Mechanics.* DFVLR, Köln, Fed. Rep. of Germany (1977) pp. 247–254

Direct Numerical Simulations of the Turbulent Wake of an Axisymmetric Body

James J. Riley and Ralph W. Metcalfe

Flow Research Company, Kent, Washington 98031, USA

Abstract

We present comparisons of results of direct numerical simulations of turbulence with both laboratory data and self-similarity theory for the case of the turbulent wakes of towed, axisymmetric bodies. In general, the agreement of the simulation results with both the laboratory data and the self-similarity theory is good, although the comparisons are hampered by inadequate procedures for initializing the numerical simulations.

Nomenclature

C_D	Drag coefficient
D	Maximum body diameter
$E(t)$	Total integrated mean flow kinetic energy
E_0	$\equiv E(t)$ at $t = 0$
$e(t)$	Total integrated turbulence kinetic energy
$I(r)$	Form function used in initialization scheme
k	Wave number vector
l_x	Characteristic axial length scale of mean quantities
l_r	Characteristic radial length scale of mean quantities
n	Annulus number
p	Pressure
$r_m(t)$	Radial distance from x-axis to where U achieves one-half of its peak value
$r_m^*(t)$	Best estimate of $r_m(t)$ from simulations
$r_T(t)$	Radial distance from x-axis to where u' achieves one-half of its peak value
$\mathscr{R}$	$\equiv r_m(t)$ at $t = 0$
R_D	$\equiv U_0 D/\nu$
t	Time
T	$\equiv t\mathscr{U}/\mathscr{R}$, a nondimensional time
$U(r, t)$	Axial mean velocity
$U_m(t)$	Maximum U at a fixed time
$U_m^*(t)$	Best estimate of $U_m(t)$ from simulations
U_0	Body speed
$\mathscr{U}$	$\equiv U_m(t)$ at $t = 0$
(u, v, w)	Axial, circumferential, and radial components of turbulence velocity vector
(u', v', w')	Axial, circumferential, and radial components of rms turbulent fluctuations
$u_m(t)$	Maximum u' at a fixed time
$u_m^*(t)$	Best estimate of $u_m(t)$ from simulations
(x, θ, r)	Axial, circumferential, and radial coordinates
ϵ	Turbulent energy dissipation
ν	Kinematic viscosity
ρ	Density

Introduction

This paper reports a study of the turbulent wakes of axisymmetric bodies using direct numerical simulations.[1] Direct numerical simulations (sometimes referred to as large eddy simulations) involve the numerical solution of the detailed evolution of the large-scale structure of the complex turbulent velocity field. Using very efficient numerical methods (e.g., pseudospectral methods), we can solve directly the fully nonlinear low-pass filtered equations of motion, so closure assumptions are only necessary (if at all) for the smaller scale motions, which have been filtered out. Statistical data are obtained by performing spatial, temporal, and/or ensemble averages over the computed flow fields.

This procedure is analogous to performing experiments in the laboratory. However, it has the advantages that (I) much more statistical information of interest can be obtained (since the entire flow field is known at every time step), (II) parameters can be varied easily, and (III) experimental conditions are more controllable. The technique also offers the advantage of circumventing the closure problems that have plagued turbulence theory. Because of these advantages, direct numerical simulations of turbulence can serve a variety of purposes. For example, they can be usefully employed to test theories from the more exotic (e.g., the direct interaction approximation) to Reynolds stress closure models. They can also be very useful in obtaining a better understanding of the underlying physical processes. And these simulations may be ultimately used as a predictive technique for applied problems. However, before this approach can be used with confidence, especially when applied to high Reynolds number flows, its validity and accuracy under various conditions need to be determined through careful comparison of the results of the simulations with experimental data.

Some work has been carried out with the purpose of establishing the validity and accuracy of numerical simulations of turbulence. For example, research at Stanford and NASA Ames (*Mansour* et al. [2]; *Ferziger* et al. [3]) has led to the successful prediction of the large-scale features of homogeneous turbulence decay. There has also been some success in the comparison of simulation results with channel flow data (*Deardorff* [4]; *Grötzbach* and *Schumann* [5]). However, in these studies the flow in the region near the wall, where many of the physical processes of current interest occur, was modeled. Recent work by *Moin* et al. [6] has examined removing this latter limitation by computing the flow in the entire boundary layer. While significant contributions have been made in establishing the validity and accuracy of numerical simulations, it is apparent that much more work needs to be done to establish the capabilities of the method.

The study of free shear flows may offer an excellent vehicle for determining the capabilities of direct numerical simulations to treat turbulent shear flows. While possessing many of the dynamic processes of all turbulent flows, free shear flows are generally easier to treat numerically than bounded flows. This is because, first of all, the boundary conditions are usually easier to implement numerically, and secondly, the resolution requirements are much less strigent. In particular, the powerful pseudospectral numerical methods can be easily applied to some free shear flows.

We have chosen to study the wakes of axisymmetric, slender, towed bodies. This choice was made for several reasons. First, there are extensive data available for this case (e.g., *Chevray* [7]; *Pao* and *Lin* [8]; *Bukreev* et al. [9]). Secondly, as long as (I) the mean velocity deficit is small with respect to the body speed, and (II) the lateral scale of nonhomogeneity is small with respect to the axial length scale, then it can be shown that this flow can be treated

[1] For a more detailed discussion of this work, see *Riley* and *Metcalfe* [1].

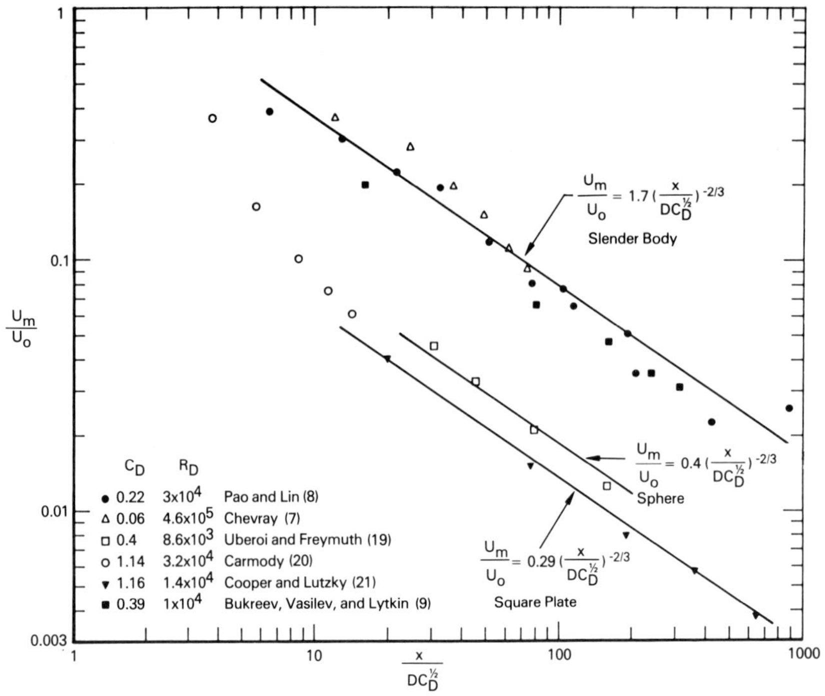

Fig. 1. Decay of maximum mean velocity U_m/U_0 (from *Pao* and *Lin* 1973)

as a time-dependent problem (with time interpreted as axial variation in the data) which is statistically homogeneous in the axial direction. This leads to significant simplifications in the numerical procedures, and also gives us the opportunity to study a statistically time-dependent flow. The two conditions described above are satisfied in the late wake in the experiments. Another advantage of late wakes of axisymmetric bodies is that they appear to be approximately self-preserving, so that the flow develops in a fairly simple manner. For example, Fig. 1 is a plot from *Pao* and *Lin* showing the collapse of several sets of data in the late wake, and also the agreement of the data with the self-preserving decay rate.

The main objective of the work presented is the determination of how well direct numerical simulations can model a simple turbulent shear flow. Our method of approach is to initialize our calculations with available wake data at a specific location downstream of the body, and then compute the subsequent development of the wake. The results of the numerical simulations are then compared with data downstream of the original plane to determine how well the wake has been simulated. The effects of a subgrid-scale model (or lack of one), of the initialization procedure, of the numerical algorithms and other features of the simulations are examined.

Method of Approach

Pseudospectral numerical methods are used to solve the governing equations of motion. (See *Orszag* and *Pao* [10], for a discussion of these methods.) In our case this involves expanding the dependent variables in a truncated Fourier series, and evaluating the derivatives in Fourier space. The equations are solved on a 32 x 32 x 33 point grid in physical space. The computations were performed on the NASA Ames CDC 7600. They require about 3 s/time step for the 32 x 32 x 33 computational mesh. The numerical algorithm uses fast Fourier transforms to evaluate transforms and inverse transforms, and uses leapfrog time differencing on the nonlinear terms, and Crank-Nicolson (implicit) time differencing on the viscous terms.

In the calculations reported in this paper, a sub-grid scale model was not used. The Reynolds number was lowered sufficiently to resolve all of the scales accurately. Our initial turbulence Reynolds number (R_λ) was about 50, compared to values in the data ranging from 70 to several hundred. One objective of this study was to determine whether a sub-grid scale model was necessary for simulating this free-shear flow, or whether the flow is approximately Reynolds number similar even at these low Reynolds numbers.

We are interested in the far wake of an axisymmetric body, where the flow is approximately self-similar. In this region the following conditions are approximately satisfied (see *Tennekes* and *Lumley* [11] for a detailed discussion of self-similarity):

I. The axial variation of any statistical quantity is much smaller than its radial variation, i.e.,

$$l_r \ll l_x,$$

where l_x and l_r are characteristic length scales of axial and radial variations in mean quantities;

II. The maximum value over which the mean velocity varies, say U_m, is much smaller than the body speed, U_0,

$$U_m \ll U_0;$$

III. The Reynolds number of the flow is large,

$$R = U_m l_r / \nu \gg 1.$$

In the laboratory data, the coordinate system is chosen to be in a frame of reference moving with the body so that the flow is statistically steady in that frame. The rate-of-change of any statistical quantity is expressed in terms of the axial coordinate (x). When conditions (I) and (II) are satisfied, the derivatives of all mean quantities with respect to x can be neglected, except when they are multiplied by U_0. If x/U_0 is interpreted as time, then it can be easily shown that the resulting equations for any mean quantity are identical to the corresponding equations for a time dependent, statistically axisymmetric flow, which is also statistically homogeneous in the axial direction.

It is much easier numerically treat a problem which is statistically homogeneous in the axial direction and time-dependent, than to compute a flow which is statistically steady, but nonhomogeneous in the axial direction. In the latter case we have to deal with complex upstream inflow and downstream outflow boundary conditions. Also in the far field, the mean radial velocity is negligible with respect to the mean axial velocity, so that we need not be concerned with radial inflow boundary conditions. Thus, in the present simulations, we choose to compute the statistically axially homogeneous, time-dependent problem. Time, in our calculations, is interpretable as x/U_0 in the laboratory data.

In presenting the results of our numerical simulations, we make comparisons with predictions from self-similarity theory. In particular, this theory predicts that

I. Velocity scales should decay as $x^{-2/3}$;
II. Length scales should increase as $x^{1/3}$;
III. Total energies should decay like $x^{-2/3}$.

When a mixing length assumption is made, it is found that the mean velocity profile satisfies

$$U(r,t) = U_{\mathrm{m}}(t) \exp\left[-0.693 \frac{r^2}{r_{\mathrm{m}}^2(t)}\right],$$

and the Reynolds stress profile is

$$\overline{uv}(r) = C_{uv} U_{\mathrm{m}}^2(t) \exp\left[-0.693 \frac{r^2}{r_{\mathrm{m}}^2(t)}\right],$$

where C_{uv} is a constant determined empirically (*Pao* and *Lin* [8]).

To initialize our calculation, we used *Pao* and *Lin*'s data at $x/D = 36$, which is in the self-preserving region. In cylindrical polar coordinates,

$$(x, \theta, r) \Leftrightarrow (u, v, w),$$

we chose the axial mean velocity $U(r)$ to be identical to *Pao* and *Lin*'s. In order to initialize the turbulence field, we used a method suggested by *Orszag* and *Pao* [10] to determine the initial wave number amplitudes. The method involves choosing a wave number spectrum, generating appropriate values of wave number amplitudes with a pseudo-random number generator, transforming to real space, multiplying by a form factor $I(r)$, transforming back to wave number space, and finally multiplying by ikx to ensure continuity. The form factor $I(r)$ is related to the axial turbulent intensity $u'(r)$ through a non-linear ordinary differential equation.

With this method, realizations with the proper turbulence intensity $u'(r)$ can be initialized. Also the axial velocity correlation (and integral scale) can be specified. The initial field, however, still has several drawbacks. First, the Reynolds stresses and various higher moments are initialized to zero, which is physically unrealistic. Secondly, there is no clear distinction between the turbulent and nonturbulent regions. This latter fact is reflected in the uniformity of the velocity flatness factor throughout the wake. One of the objectives of this study was to determine the ramifications of this initialization procedure, and whether alternative procedures were necessary.

The data presented is obtained by performing spatial averages over the computed flow field, taking advantage of the axial homogeneity and axisymmetry, and dividing the flow field into concentric annuli with axes along the x-axis. The gap width of each annulus is taken to be the (constant) mesh spacing. The average value of a quantity q in the nth annulus is obtained by summing up the values of q for each grid point in the annulus, and dividing by the number of points in the annulus (N_n). The statistical error is roughly proportional to $1/r$.

Numerical Results

In this section we give the results of our direct numerical simulations of the axisymmetric turbulent wake of a towed body. We present the results of the simulations for two independent realizations, generated using different random number seeds in the initialization process.

The combined results show both the capabilities of the numerical simulations to treat turbulent wakes, as well as provide an estimate of the statistical scatter in the results. We show first some constant contour plots of the instantaneous velocity and vorticity fields to help give a qualitative picture of the wake evolution. We then present plots of the decay of characteristic velocity and length scales, as well as total energies, with time. Finally, we present radial plots of various statistical quantities, and establish the degree to which the self-similar decay is modeled.

Contour Plots

Although statistical results are necessary to determine the capabilities of the simulations, it is useful to examine contour plots to see qualitatively the development of various features in the

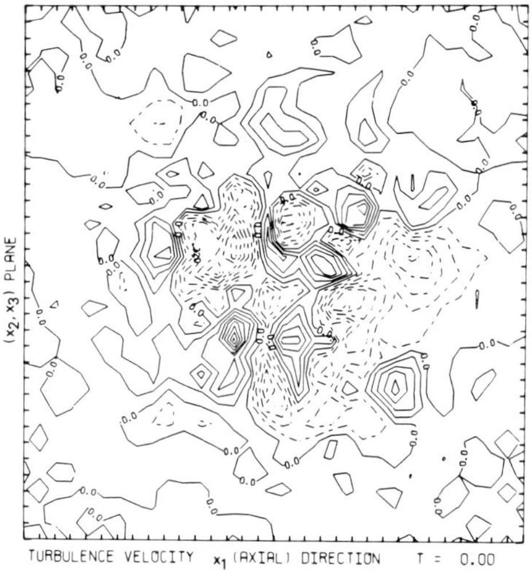

Fig. 2. Velocity field contour plot

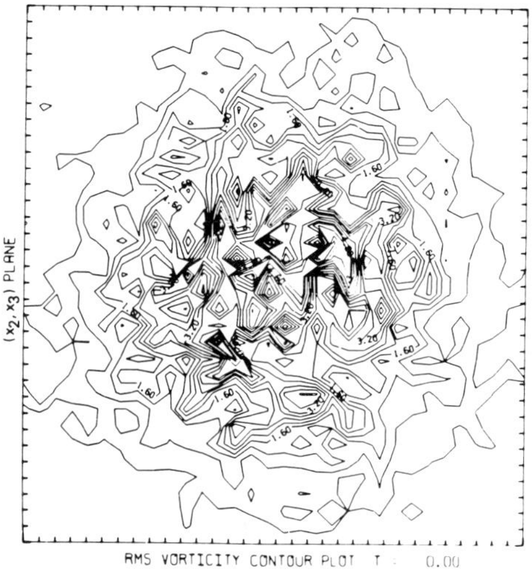

Fig. 3. Vorticity magnitude contour plot

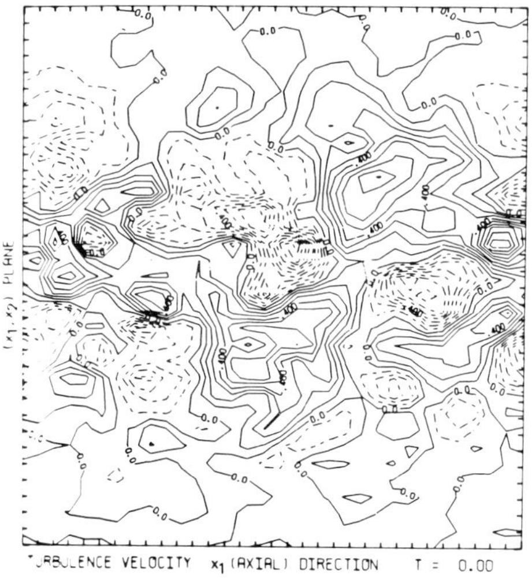

Fig. 4. Velocity field contour plot

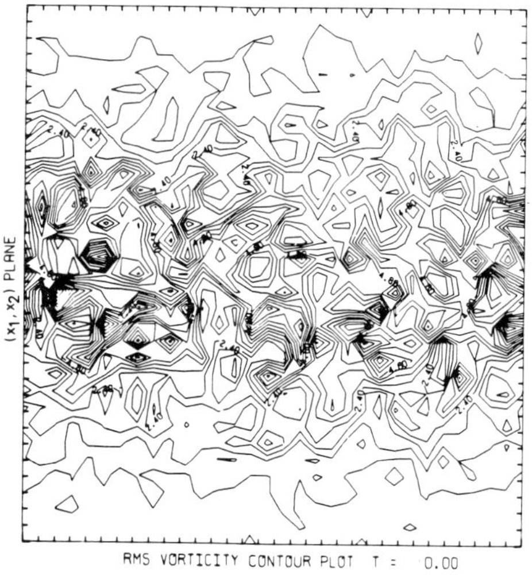

Fig. 5. Vorticity magnitude contour plot

flow field. Figures 2 and 3 are contour plots of the longitudinal velocity fluctuation (u) and the magnitude of the fluctuating vorticity ($\omega = \sqrt{\omega_i \omega_i}$) respectively, in a typical plane perpendicular to the mean motion (x or x_1), at time zero. The vorticity plot especially shows the localization of the turbulence in the cylindrical region in the center. Note the axisymmetry of the vorticity field, as well as the manner in which it goes rather smoothly to zero with radial distance. These are both a result of the use of the smooth form function in the initialization. Figures 4 and 5 are contour plots of u and ω in a plane parallel to and passing through the x_1 axis, at time zero. Again note the lack of bulges and other nonuniformities at the wake edge, especially in the vorticity field. Also, there appears to be no directional preference in the contours. This reflects the fact that the Reynolds stresses are initially approximately zero.

Figures 6 and 7 are similar contour plots of u and ω in a plane perpendicular to the x_1 axis at nondimensional time T of 15.46, two-thirds of the way through our calculation. Note the growth of the wake, the increased convolution of its boundary, and the loss of small scale

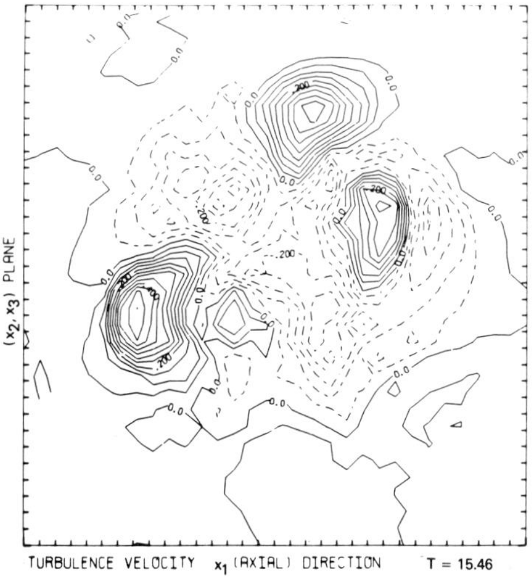

Fig. 6. Velocity field contour plot

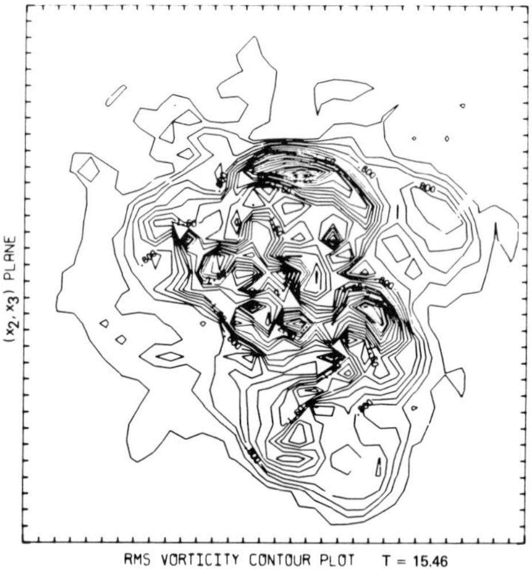

Fig. 7. Vorticity magnitude contour plot

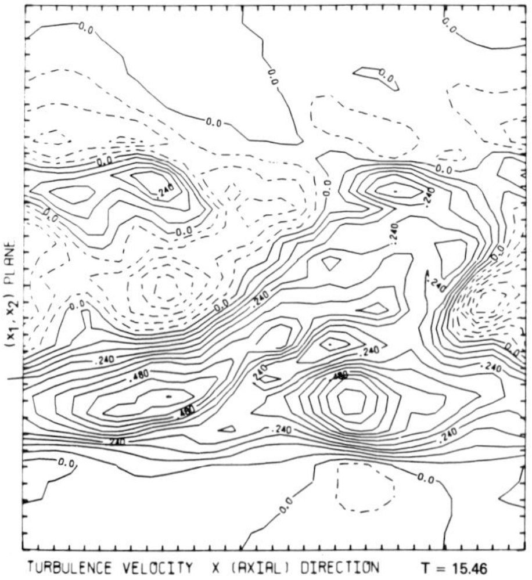

Fig. 8. Velocity field contour plot

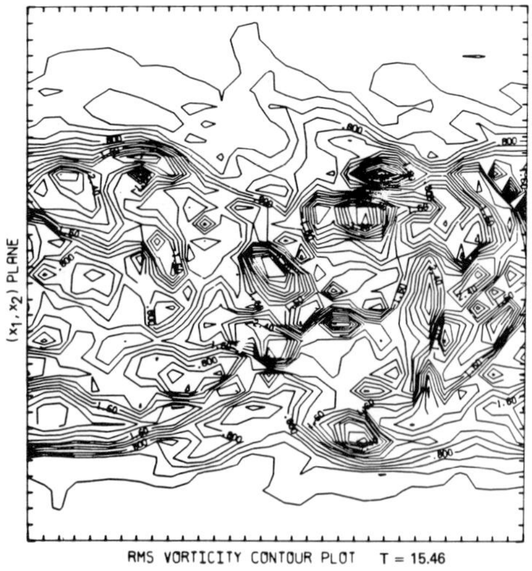

Fig. 9. Vorticity magnitude contour plot

features to molecular viscosity. Figures 8 and 9 are contour plots of u and ω in the x_1x_2 plane through the x_1 axis at the same time, $T = 15.46$. Again note the convoluted nature of the wake boundary, and the growth of the wake. The extent of convolution of the boundary regions of the wake will show up as increased values of the velocity (and velocity derivative) skewness and kurtosis in those regions. It is also apparent that the velocity field has developed a directional orientation. In the upper-half-plane, the constant contours tend to be oriented with their major axes in the second and fourth quadrants, whereas in the lower half-plane the contours are oriented with their major axes in the first and third quadrants. This can be shown to correspond to a positive Reynolds stress in the upper-half-plane and a negative stress in the lower-half-plane. Note that the mean velocity is positive in the direction of increasing x_1. The mean velocity gradient has kinematically distorted the initially isotropic-like field into one which has directional preferences.

Temporal Behavior. Figure 10 is a plot of the temporal behavior of $U_m(x)$, the maximum axial mean velocity deficit, and $u_m(x)$, the maximum rms turbulent axial velocity, for the two different realizations. In presenting the simulation results, we show the horizontal axis in terms of both x/D (as in *Pao* and *Lin*), and also the nondimensional time $T = t\mathcal{U}/\mathcal{R}$. $T = 0$ corresponds to $x/D = 36$, and $T - T_0 = -3.76$ corresponds to $x/D = 0$. Consider U_m first. Initially, the decay rate of U_m is less than predicted by theory, or observed in the data. The dynamic equation for U is:

$$\frac{\partial}{\partial t} U + \frac{1}{r}\frac{\partial}{\partial r}(r\,\overline{uw}) = \nu \frac{1}{r}\frac{\partial}{\partial r}\left(r\frac{\partial U}{\partial r}\right). \tag{1}$$

The second term on the LHS represents the radial diffusion of mean momentum due to the Reynolds stress $\overline{uw}$. The term on the RHS is the molecular diffusion term, and is small in this case. Initially, the mean velocity peak does not decay since $\overline{uw} = 0$. The Reynolds stress builds up rapidly, however, and by $x/D = 90$, U_m is approximately following the $t^{-2/3}$ behavior for both realizations. Note that there is some difference in the results for the two realizations.

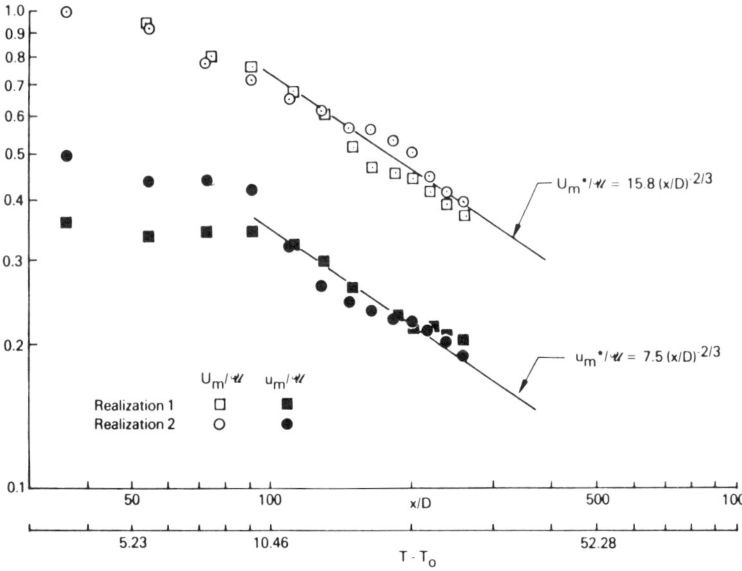

Fig. 10. Decay of maximum mean velocity (U_m) and maximum axial turbulent intensity (u_m)

The behavior of u_m is similar. The equation satisfied by $\overline{u^2}/2$ is:

$$\frac{\partial}{\partial t}\frac{1}{2}\overline{u^2} + \overline{uw}\frac{\partial U}{\partial r} + \frac{1}{r}\frac{\partial}{\partial r}\left(\frac{r}{2}\overline{u^2 w}\right) = -\frac{1}{\rho}\overline{u\frac{\partial p}{\partial x}} + \nu\frac{1}{r}\frac{\partial}{\partial r}\left[r\frac{\partial}{\partial r}\left(\frac{1}{2}\overline{u^2}\right)\right]$$

$$-\nu\left[\overline{\left(\frac{\partial u}{\partial r}\right)^2} + \frac{1}{r^2}\overline{\left(\frac{\partial u}{\partial \theta}\right)^2} + \overline{\left(\frac{\partial u}{\partial x}\right)^2}\right]. \qquad (2)$$

The dynamics of u' are governed by a complex balance between energy feed from the mean flow ($\overline{uw}\,\partial U/\partial r$), the diffusion of u' by turbulence $1/r\,\partial/\partial r\,(r\,\overline{u^2 w}/2)$, the redistribution of u' between w' and v' ($-1/\rho\,\overline{u\,\partial p/\partial x}$), and the viscous dissipation of u'. Initially the generation terms and diffusion terms are both zero, due to the initialization scheme. However, by x/D of about 110, these effects appear to have built up, and u_m decays at roughly the proper rate for the two realizations. Some idea of the relative error can again be seen by comparing the results for the two realizations.

Figure 11 is a plot of the temporal behavior of r_m, the mean velocity profile half-width, and of r_T, the turbulent velocity profile half-width. Considering first r_m, we see that it displays the same effect as does U_m. Initially it does not grow, but apparently after the Reynolds stress has built up, it increases at approximately the proper rate. The same holds for r_T. Initially there is no growth, but by x/D equal to about 110 it is growing at approximately the right rate in correspondence with the decay of u_m.

Figure 12 contains plots of the integrated mean flow energy, E, and the integrated turbulent energy, e, as functions of time. Multiplying Eq. (1) by ρU and integrating over the crossplane gives the equation for E, i.e.,

$$\frac{d}{dt}E = 2\pi\rho\int_0^\infty r\,\overline{uw}\,\frac{\partial U}{\partial r}\,dr - 2\pi\rho\nu\int_0^\infty r\left(\frac{\partial U}{\partial r}\right)^2 dr \qquad (3)$$

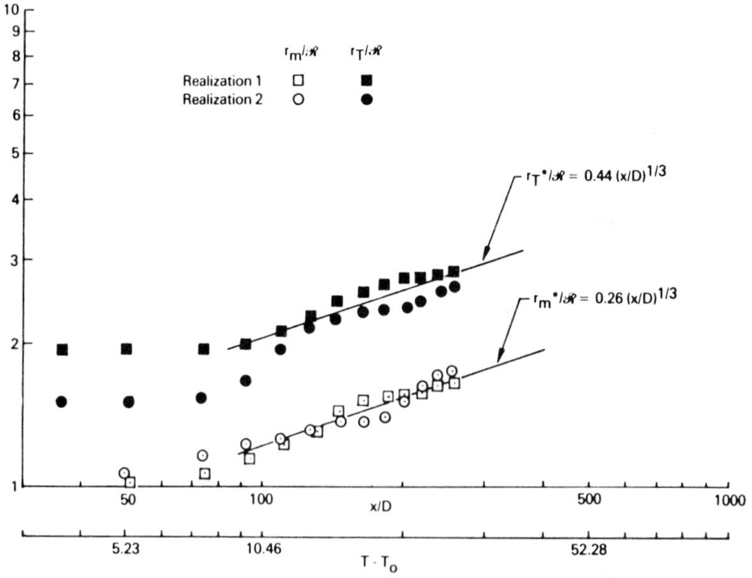

Fig. 11. Growth of mean wake radius (r_m) and turbulent wake radius (r_T)

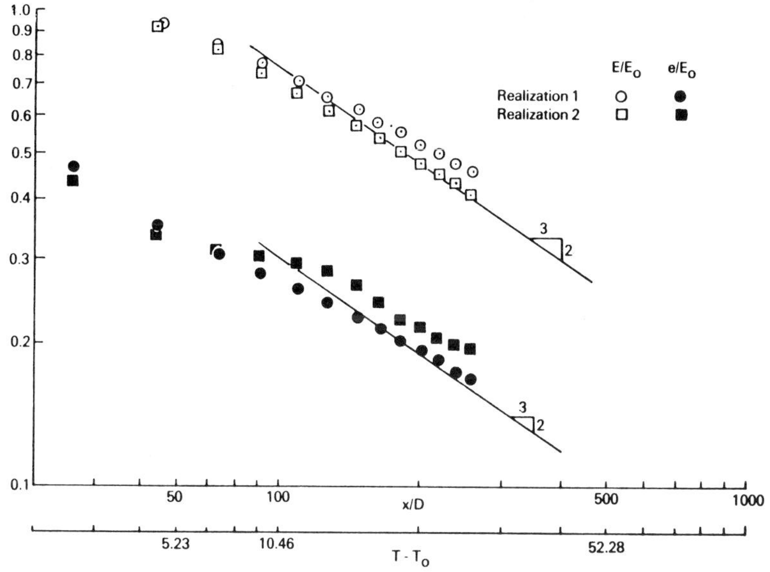

Fig. 12. Decay of integrated mean and turbulent energies

The first term on the RHS represents the loss of mean kinetic energy to turbulence, while the second term is the molecular dissipation term, which is small. Mean flow energy is lost to the turbulence through the action of the Reynolds stress $\overline{uw}$ on the mean velocity gradient. The decay of E has established close to its proper behavior by about $x/D = 100$. The equation for e is obtained by integrating the sum of equation (2) and its counterparts for $\overline{w^2}$ and $\overline{v^2}$, and is

$$\frac{d}{dt} e = -2\pi\rho \int_0^\infty r\,\overline{uw}\,\frac{\partial U}{\partial r}\,dr - \epsilon, \tag{4}$$

where ϵ is the viscous dissipation. Thus the time rate of change of e is a balance between energy feed from mean flow and viscous dissipation. Again, the proper decay rate is approximately obtained by x/D about 110. The slightly slower decay rate can be improved by a change in the virtual origin.

Although the approximate proper decay rate has been attained by these quantities, their ratios have changed from the initial conditions. From *Pao* and *Lin* at $x/D = 36$,

$$\frac{u_m}{U_m} = 0.32, \quad \frac{r_T}{r_m} = 2.0, \tag{5}$$

and these ratios are maintained in their laboratory data throughout the self-similar region. In our simulations, however, although we started the values of the above ratios properly (to within statistical scatter), the values in the self-similar range became

$$\frac{u_m}{U_m} = 0.47, \quad \frac{r_T}{r_m} = 1.66. \tag{6}$$

These values are somewhat different from the data, and are another reflection of our inadequate method of initialization. These ratios do vary, however, for different shaped obstacles. Our wake may be more representative of body different than what *Pao* and *Lin* used.

To summarize, initially the time development of the flow does not correspond to the self-similarity theory and the data due to inadequate initialization, especially of the Reynolds stress. After the flow has adjusted, however, and the Reynolds stresses have built up, these quantities develop in time at approximately the proper self-similar rate, over a distance of about 100 to 270 diameters (or a time of $23.2\,\mathcal{U}/\mathcal{R}$). In the meantime the ratios u_m/U_m and r_T/r_m change from their initial values, so that our calculations may be more representative of the wake of a different body than that used by *Pao* and *Lin*.

Spatial Behavior. In this section we present the radial distribution of various statistical quantities for different decay times. Self-similarity theory predicts that if the velocities are normalized by U_m (or u_m), and the lengths are scaled by r_m (or r_T), then the radial distribution of a particular statistical quantity, say $U(r/r_m)/U_m$, should collapse for different times. The ability of this scaling to collapse the data, and the degree to which the resulting profiles agree with corresponding profiles from laboratory data are good checks of the capability of the numerical simulations to correctly model the physical processes occurring in the wake. To normalize the data, we use best fit estimates which are given by the solid lines in Figs. 10 and 11.

Figure 13 gives data from the first realization (R1) for U/U_m^* vs r/r_m^* for the locations $x/D = 150$, 205, and 260; U_m decreases by about 30 percent and r_m increases by 20 percent over this range. We see, first of all, that the scaling collapses the data very well, except near the origin, where the statistical scatter is expected to be the greatest. The laboratory results tend to follow the self-similar result rather well, although their scatter is fairly large for $r/r_m > 1.5$, so that it is difficult to conclude that the agreement is good in this region. Previous measurements of the wakes of two-dimensional bodies (see, e.g., *Townsend* [12]) have consistently shown that the theory predicts somewhat larger values than are observed for $r/r_m > 1.5$. This is usually attributed to decreased turbulent diffusion near the wake edge, due to the intermittent nature of the flow there, which is not accounted for in the theory. We see that the simulation results tend to be slightly low for small r/r_m^*. This is due to the fact that both realizations were taken into account in the normalization. Data from the second realization lie slightly above the theoretical curve in this same region. The simulation results also lie below the theoretical curve for $r/r_m^* > 1.5$. The fact that U_m initially started as an exact fit to the similarity result may indicate that the numerical simulations have added a realistic feature.

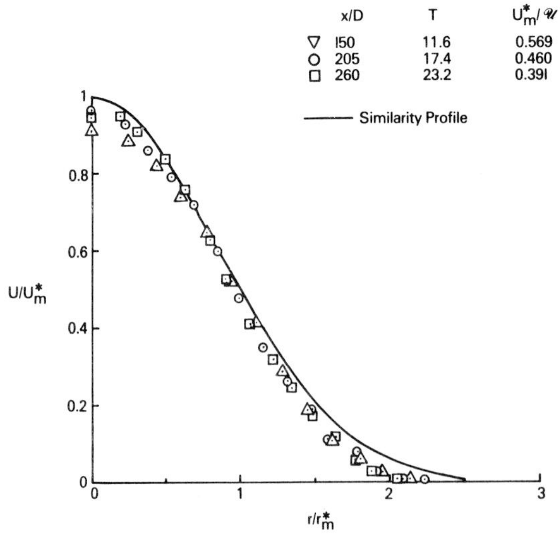

Fig. 13. Normalized profile of axial mean velocity for realization 1

Fig. 14

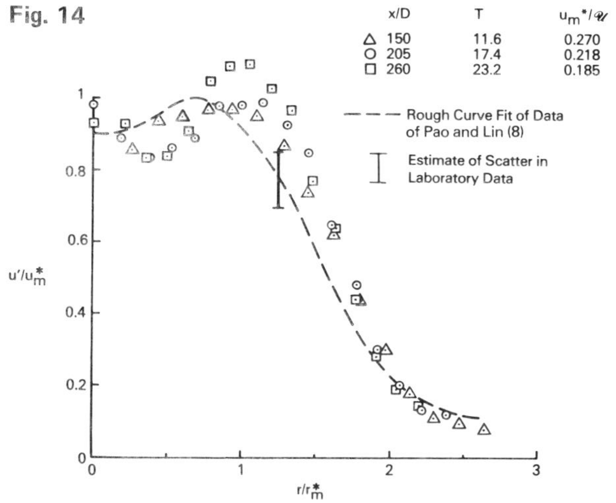

Fig. 15

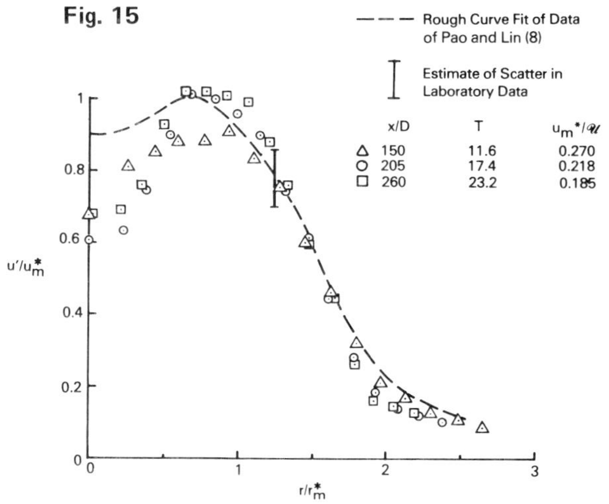

Fig. 14. Normalized profile of axial turbulent intensity for realization 1

Fig. 15. Normalized profile of axial turbulent intensity for realization 2

Fig. 16. Normalized profile of Reynolds stress for realization 1

Fig. 16

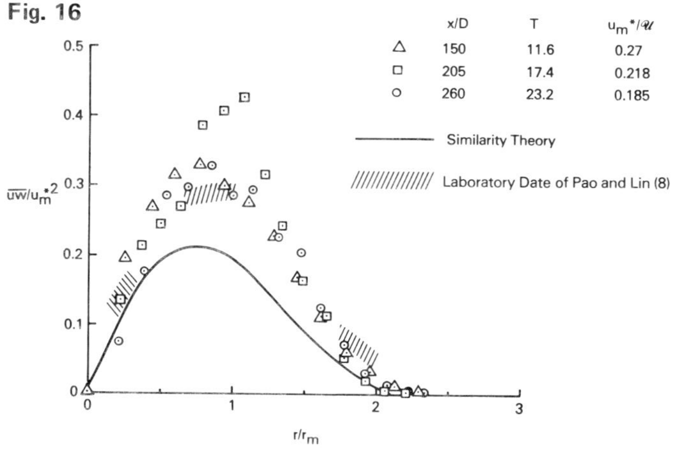

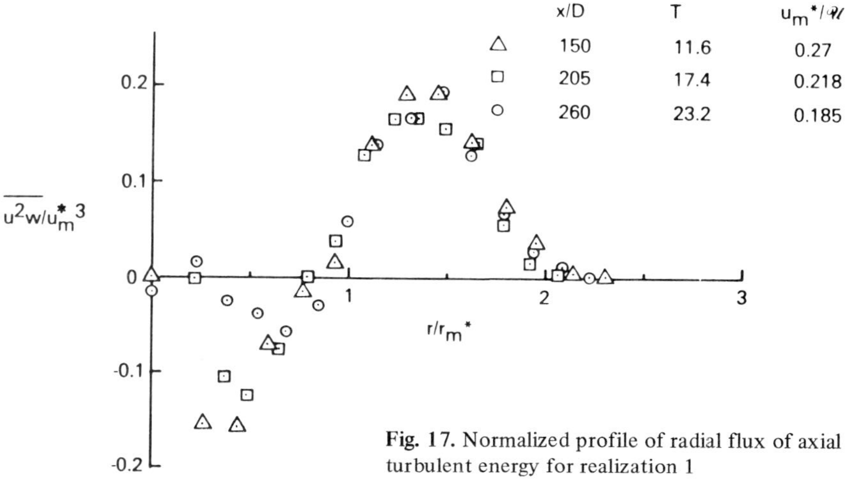

Fig. 17. Normalized profile of radial flux of axial turbulent energy for realization 1

Figure 14 is a plot of u'/u_m^* vs r/r_m for the downstream locations $x/D = 150$, 205, and 206 for R1. The dashed line is a rough fit of the laboratory data of *Pao* and *Lin*. The laboratory data curve has been compressed so that r_T/r_m for the data takes on the same value given by Eq. (6) instead of Eq. (5) i.e., 1.66. By plotting all the data vs r/r_m^*, we can compare the simulation results with laboratory data, and simultaneously compare u' with U and other variables. The collapse of the simulation data is again fairly good. The scatter in the data for u' is expected to be significantly larger than that for U because the sampling error increases monotonically with the order of the statistical quantity. For the choice of u_m^* and r_m^* given by the best fit estimates, the agreement with the laboratory data is only fair. If this data were matched independent of R2, a better fit could be obtained.

Figure 15 is a similar plot for R2. Again the collapse of the data is fairly good except near the origin. This realization agrees much better with the laboratory data than the first realization. This is partly due to the choice of u_m^* and r_m^*. When the data for u'/u_m^* are superimposed for the two realizations, distinct differences in the two sets can be noted. This gives some idea of the magnitude in the scatter in the data to be expected in the simulation.

Figure 16 is a plot of the normalized Reynolds stress $\overline{uw}/u_m^{*2}$ vs r/r_m^* for $x/D = 150$, 205, and 260 in R1. The solid curve gives the prediction of the similarity theory, and the hatched lines give the data of *Pao* and *Lin.* The collapse is fairly good considering the error expected in computing the Reynolds stress. Furthermore, the agreement with the laboratory data is reasonable. Recall that the Reynolds stress is zero in our simulations at $T = 0$.

Figure 17 is a plot of $\overline{u^2w}/u_m^{*3}$ vs r/r_m^* for $x/D = 150$, 205, and 260 for R1. The radial flux of $\overline{u^2}$ is represented by $\overline{u^2w}$ and is an essential part of the dynamics of $\overline{u^2}$. The collapse is reasonable, except near the origin where the scatter in the data is expected to be greatest. We see that the data have the proper qualitative behavior; $\overline{u^2}$ is being diffused by the turbulence away from its peak towards the origin for $r/r_m^* \lesssim 0.9$, and away from the origin for $r/r_m^* \gtrsim 0.9$.

The intermittency of the wake turbulence is reflected in the flatness factor of the velocity field. *Townsend* [13] has shown for wakes, and *Wygnanski* and *Fiedler* [14] for mixing layers, that the velocity kurtosis is approximately equal to the inverse of the intermittancy factor. Figure 18 is a plot of the radial distribution of the velocity kurtosis for R1 at the same three downstream locations. At the start of the calculation, the profile is flat, due to the method of initialization, indicating there is no intermittency in the flow. As the wake devel-

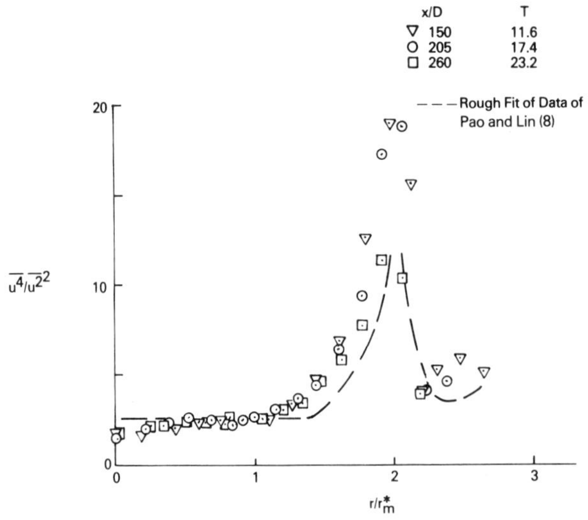

Fig. 18. Profile of velocity kurtosi for realization 1

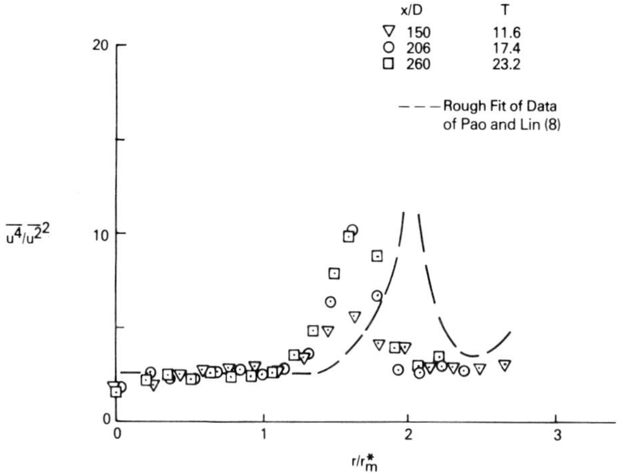

Fig. 19. Profile of velocity kurtosis for realization 2

ops, however, the outer regions of the wake become highly contorted, and the velocity becomes highly intermittent, which is reflected in the sharp increase in the kurtosis at the wake edge. The dashed line is a rough fit of the data of *Pao* and *Lin*, and the agreement between the simulations and laboratory data is reasonable.

Figure 19 is the same plot for R2. While the collapse of the data is good, the agreement with the laboratory data and with R1 is poor, especially in the location of the peak value. Apparently in R2 the wake edge is neither as contorted nor as far out as in R1. This again is an indication of the deviations to expect from one realization to another. To compute the kurtosis accurately, we may need to compute several realizations and use ensemble averaging.

To summarize the radial plots, in general the collapse of the data using the self-similarity scaling is very good. Also the agreement with laboratory data is generally good, except for the turbulence intensity (u') profile, where the ratio r_T/r_m has changed from its initial value, Eq. (5). However, when u' is plotted against r/r_T^*, or, equivalently, when the ratio of r_T/r_m given by Eq. (6) is used to plot the laboratory data, the agreement is then fairly good.

Sub-Grid Scale Modelling

Almost all studies up to the present have modeled the sub-grid scale in physical space (but see *Love* and *Leslie* [15]). This is because most studies have employed finite-difference numerical methods, so that the sub-grid scale models have been most naturally applied in physical space. However, much that is known about turbulence, especially the energy transfer from the large to the smaller scales, can better be stated in Fourier space. The spectral methods used in the present work are better suited to treat sub-grid effects in wave number space. We are presently developing models in Fourier space for the energy transfer from wave numbers on our computational grid to those which have been truncated. These models have been made consistent with various spectral theories, e.g., *Kovasznay*'s [16], *Ellison*'s [17], and *Heisenberg*'s [18].

Conclusions

From the numerical simulation data presented in the preceding section, we can draw a number of conclusions about the capabilities of the direct numerical simulation of turbulence. In general, the agreement of the numerical simulation results with the laboratory data and the self-similarity theory was good. In terms of the temporal behavior of the simulated flow over a span of time of about 15 $\mathcal{U}/\mathcal{R}$,

I. the maximum mean and turbulent velocities decayed at approximately the proper rate $((x/D)^{-2/3})$;
II. both the mean and turbulent wake radii increased at approximately the proper rate $((x/D)^{1/3})$;
III. the total integrated mean and turbulent energies decayed at approximately the proper rate $((x/D)^{-2/3})$.

Furthermore, over this same interval of time, using self-similarity scaling, we were able to obtain reasonably good collapse, and also reasonable agreement with laboratory data, for

I. the mean velocity profile,
II. the turbulent intensity profile,
III. the Reynolds stress profile,
IV. the profile for $\overline{u^2 w}$, and
V. the profiles for skewness and kurtosis.

In addition, several realistic features, which were not in our initial conditions, subsequently appeared later in time in the numerical simulations. They include

I. the development and maintenance of the Reynolds stress,
II. the development and maintenance of higher order correlations, e.g., $\overline{u^2 w}$;
III. the development of intermittency near the wake edge, which qualitatively shows up as bulges in the wake, and is reflected by the development of the skewness and kurtosis to reasonable values.

It also appeared that a subgrid-scale model is not a primary necessity for computing this free shear flow, although such a model may prove to be necessary in the future when more careful comparisons with data, especially for the decay of turbulent energy, are made.

The comparisons were hampered by two inadequacies in the present calculations. The first was the initialization process (which we are in the process of improving). The turbulent field, especially in terms of the Reynolds stresses, was not adequately initialized, which limited the conclusions that we could draw concerning numerical simulations. Secondly, there was a significant amount of scatter in our data, both in an individual realization, and from one realization to another.

Given these limitations, the present work greatly increases our confidence in the capabilities of direct numerical simulations to treat the physics of turbulent shear flows accurately. The turbulence in the axisymmetric wake is governed by a complex balance among the following mechanisms: generation of energy from the mean flow, diffusion of turbulence radially by the turbulence, intercomponent transfer of energy, and dissipation. All of these processes have to be represented faithfully for the resulting statistics to possess the proper behavior. (It should be noted that there are no adjustable constants in our simulations.) Thus, we feel that the outlook is good for using direct numerical simulations to predict statistical results which depend on the large-scale features of free turbulent shear flows.

Acknowledgements. This work was supported by the NASA Ames Research Center, Contract No. NAS2-9855.

References

1. J. J. Riley, R. W. Metcalfe: The direct numerical simulations of turbulent wakes of axisymmetric bodies – An interim report. Flow Research Rpt. **135** (March, 1979) (also NASA-CR-152282)
2. N. N. Mansour, P. Moin, W. C. Reynolds, J. H. Ferziger: "Improved Methods for Large Eddy Simulations of Turbulence", in *Turbulent Shear Flows I*, ed. by F. Durst, B. E. Launder, F. W. Schmidt, J. H. Whitelaw (Springer, Berlin, Heidelberg, New York 1979) p. 386
3. J. H. Ferziger, V. B. Mehta, W. C. Reynolds: (Submitted to the 1st International Symposium on Turbulent Shear Flows, Pennsylvania, USA 1977)
4. J. W. Deardorff: J. Fluid Mech. **41** (2), 453–480 (1970)
5. G. Grötzbach, V. Schumann: "Direct numerical Simulation of Turbulent Velocity, Pressure, and Temperature Fields in Channel Flows", in *Turbulent Shear Flows I*, ed. by F. Durst, B. E. Launder, F. W. Schmidt, J. H. Whitelaw (Springer, Berlin, Heidelberg, New York 1979) p. 370
6. P. Moin, W. C. Reynolds, J. H. Ferziger: Large eddy simulation of incompressible turbulent channel flow. Rpt. TF-**12** Standford Thermosciences Div. (May, 1978)
7. R. Chevray: J. Basic Engin. **90**, 275–284 (1968)
8. Y.-H. Pao, J.-T. Lin: Velocity and density measurements in the turbulent wakes of a towed slender body in stratified and nonstratified fluids. Flow Research Rpt. **12** (Dec., 1973)
9. V. I. Bukreev, O. F. Vasilev, Yu. M. Lytkin: Proc. Acad. Sci. USSR **207**(4), 805–807 (1972)
10. S. A. Orszag, Y.-H. Pao: Adv. Geophys. **18A**, 225–236 (1974)
11. H. Tennekes, J. L. Lumley: *A First Course in Turbulence* (MIT 1972)
12. A. A. Towsend: *The Structure of Turbulent Shear Flow,* 2nd ed. (Cambridge University Press. 1976)
13. A. A. Townsend: Proc. Roy. Soc. Ser. A: **197**, 124 (1949)
14. I. Wygnanski, H. E. Fiedler: J. Fluid Mech. **41**(2), 327–361 (1970)
15. M. D. Love, D. C. Leslie: "Studies of Subgrid Modelling with Classical Closures and Burgers Equation", in *Turbulent Shear Flows I,* ed. by F. Durst, B. E. Launder, F. W. Schmidt, J. H. Whitelaw (Springer, Berlin, Heidelberg, New York 1979) p. 353
16. L. S. G. Kovasznay: Spectrum of locally isotropic turbulence. J. Aeronaut. Sci. **15**(12), 745–753 (1948)
17. J. H. Ellison: "The Universal Small-Scale Spectrum of Turbulence at High Reynolds Number", in *Mechanique De La Turbulence,* ed. by CNRS, Paris (1962) pp. 113–121
18. W. Heisenberg: Z. Physik **124**, 628–657 (1948)
19. M. S. Uberoi, P. Freymuth: Turbulent energy balance and spectra of the axisymmetric wake. Phys. Fluids **13**, 2205–2210 (1970)
20. T. Carmody: Establishment of the wake behind the disk. J. Basic Eng. **86**, 869–882 (1964)
21. R. D. Cooper, M. Lutzky: Exploratory investigation of the turbulent wakes behind bluff bodies. The David Taylor model Basin, Rpt. **963** (Oct. 1955)

Part II

Wall Flows

Introductory Remarks

J. C. Rotta

Deutsche Forschungs- und Versuchsanstalt für Luft- und Raumfahrt e. V., Aerodynamische Versuchsanstalt Göttingen, Bunsenstraße 10, D-3400 Göttingen, Fed. Rep. of Germany

Wall flows, i. e. turbulent flows along solid walls, arise in flows around bodies (external flows) or in flows through pipes and ducts (internal flows). They present a large group of turbulent shear flows that pose many engineering problems. They, therefore, form an important topic of research. Simple cases such as the fully developed flow through a cylindrical pipe or the boundary layer developing over a flat plate were very early subject to intensive investigations. Lateron, interest centred on two-dimensional boundary layers in variable pressure distributions and on flows through ducts with variable cross section. In the course of time, very many of these flows with practical importance have been studied. Present day investigations, therefore, concentrate generally on more complex flow situations such as three-dimensional layers in which effects of streamline curvature, buoyancy forces as well as heat and mass transfer are of importance. Various contributions to these topics were presented at the 2nd Turbulent Shear Flow Symposium in London and a selection of these is presented in this chapter.

The papers chosen for this chapter are generally of an experimental nature. This does not mean that there are no theoretical acivities in the field of wall flows; however, the theoretical research on wall flows appears in this book in the chapters on Turbulence Modelling and Complex Flows. The experimental investigations given in this chapter include, in most of the papers, the determination of details of the turbulence structure without providing information on coherent structures. The latter are again the topic of another chapter of the present book.

Two of the five papers in this chapter are concerned with the effects of streamline curvature on turbulent shear flows, a subject which has been fully reviewed up to 1972 by *Bradshaw* [1]. The investigation of *Nakamura, Yamashita,* and *Furuya,* supplement the available data on turbulent boundary layers on rotating bodies in axial flow, as published e.g. by *Parr* [2], *Furuya* and *Nakamura* [3], and *Bisonnette* and *Mellor* [4]. These studies of shear flows with skewed velocity profiles have their practical application in the flow on hubs of axial turbomachines. The important aspect of the present investigations of the flow on a rotating body with the radius decreasing in downstream direction is the occurrence of additional body forces and transfer of turbulent energy.

Gillis and *Johnston* report on experiments made in a nominally two-dimensional turbulent boundary layer over a convex wall followed by a flat plate. The turbulent motion is strongly affected by the convex curvature of the wall and recovers slowly to the flat plate boundary layer. Such history effects on turbulence is one of the problems which usually cannot be predicted accurately with available turbulence models; even multi-equation schemes often fail to give satisfactory results. The authors propose an empirical lag equation as an extension of *Bradshaw's* curvature corrections. These investigations are interesting in view of the recent experiments and analysis of *Smits* et al. [5], which show how large changes in the turbulent properties take place in short bends with large turning angle, and how the structure recovers in the straight part of the duct. Investigations of this type will help to complete modelling of streamline curvature effects on turbulent flows.

Two of the papers in this chapter are concerned with the passive diffusion of heat in two-dimensional turbulent boundary layers. *Belorgey, Paranthoen* and *Trinite* present detailed measurements of the diffusion of heat from a line source in a wall boundary layer. The paper considers the intermittent nature of the contact of the hot-fluid with the wall and postulates the validity of the sublayer renewal theory as described in the literature. In order to apply this theory, the contact of the hot fluid with the wall is needed and this information was obtained from measurements. Further downstream of the flow, the authors claim that boundary layer type of calculations give good agreement with their experimental results.

The paper by *Senda, Suzuki* and *Sato* presents a detailed study on the turbulent structure of a wall boundary layer over a porous plate with uniform injection of heated air. Detailed measurements of various turbulence properties are reported that seem to suggest that a turbulent Prandtl number approach would be satisfactorily to predict the turbulent transport of heat from computed $\overline{u'v'}$-correlations. The data provided by the authors are detailed and may serve as a useful test for turbulence models for heat transfer such as those reviewed by Launder [6].

The effect of stable and unstable density stratification on the nominally two-dimensional turbulent flow through a channel is the subject of the paper by *Petukov, Polyakov* and *Tsypulev,* who present detailed measurements on the structure of turbulence. The topic of this paper is closely related to the two papers mentioned above, as there exists an analogy between the effects of streamline curvature and buoyancy on turbulent flow, which *Bradshaw* [7] has exploited to derive certain characteristic parameters.

References

1. P. Bradshaw: "Effects of streamline curvature on turbulent flow". AGARDograph **169** (1973)
2. O. Parr: Untersuchungen der dreidimensionalen Grenzschicht an rotierenden Drehkörpern bei axialer Anströmung. Ing. Arch. **32,** 393–413 (1963)
3. Y. Furuya, I. Nakamura: Velocity profiles in the skewed boundary layers on some rotating bodies in axial flows. J. Appl. Mech. **37,** 17–24 (1970)
4. L. R. Bissonnette, G. L. Mellor: Experiments on the behaviour of an axisymmetric turbulent boundary layer with a sudden circumferential strain. J. Fluid Mech. **63,** 369–413 (1974)
5. A. J. Smits, S. T. B. Young, P. Bradshaw: The effect of short regions of high surface curvature on turbulent boundary layers. J. Fluid Mech. **94,** 209–242 (1979)
6. B. E. Launder: "Heat and Mass Transport", in *Turbulence,* ed. by P. Bradshaw, Topics in Applied Physics, Vol. 12 (Springer, Berlin, Heidelberg, New York 1978) pp. 231–287
7. P. Bradshaw: The analogy between streamline curvature and buoyancy in turbulent shear flow. J. Fluid Mech. **36,** 177–191 (1969)

Experiments on Turbulence Properties in the Thick Turbulent Boundary Layer on a Rotating Conical Body of Radius Decreasing Toward Downstream

I. Nakamura, S. Yamashita, and Y. Furuya

Department of Mechanical Engineering, Nagoya University
Furocho, Chikusaku, Nagoya, Japan

Abstract

A study has been carried out of a skewed turbulent boundary layer over a spinning cone with decreasing radius placed in a uniform flow. Consideration is given to the mechanism of turbulent energy transfer. Measurements of mean velocity profiles and of Reynolds stress components are presented. The rotating cone generates two body forces in the boundary layer which act as an adverse pressure gradient. Mean flow and turbulent fields show clearly the effect of this gradient. Distribution of novel production terms and transfer terms display significant effects in the turbulent energy budget. An account is given of the change in the mixing length using the flux Richardson number. Evaluation of terms of z-momentum equation indicates that the turbulence terms should be included.

Nomenclature

A	Radius of the principal curvature of the cone, Fig. 1
a	Radius of the body, Fig. 1
l	Mixing length
P	Static pressure
$\overline{q^2}/2 = (\overline{u^2} + \overline{v^2} + \overline{w^2})/2$	Kinetic energy of turbulence
$R = A + z$	Radius of the principal curvature of the $x - y$ surface, Fig. 1
$R_e = U_m a_m/\nu$	Reference Reynolds number
R_f	Flux Richardson number, Eq. (19)
U, V, W	Mean velocity components in the x-, y- and z-directions, respectively
u, v, w	Fluctuating velocity components in the x-, y- and z-directions, respectively
u', v', w'	RMS values of u, v and w, respectively
x	Meridian coordinate; meridian distance from the foward stagnation point, Figs. 1 and 3
y	Azimuthal angle, Fig. 1
z	Normal coordinate; normal distance from the body surface, Fig. 1
δ	Boundary layer thickness
$\Theta_x = \int_0^\delta \frac{U}{U_e}\left(1 - \frac{U}{U_e}\right)\frac{R}{A}\,dz$	Momentum thickness of the thick boundary layer in the x-direction
$\Theta_{xy} = \int_0^\delta \frac{U}{U_e}\frac{V}{V_0}\left(\frac{R}{A}\right)^2 dz$	Momentum thickness of the thick boundary layer in the y-direction

$\lambda = \partial_x R$ Gradient of the cone
$\Omega_m = a_m \omega / U_m$ Reference speed ratio
ω Angular velocity of the body

Subscripts

e Outer edge of the boundary layer
m Reference value measured at $\bar{x}$ = 138 mm, Fig. 3
o Surface of the body
x, y Components in the x- and y-directions

Introduction

Although the velocity vector in a turbulent boundary layer developing on a body of revolution and spinning in an axial stream exhibits a completely three-dimensionally skewed profile, the rotational symmetry imposed on the flow makes the problem of this boundary layer more tractable to experimental studies than that of general three-dimensional flows. However, the structure of the turbulent boundary layer of this kind is still too complex. It is difficult to deduce the character of this flow from knowledge of the two-dimensional turbulent boundary layer. For example, the thick turbulent boundary layer on a rotating cylinder in a constant-pressure axial flow, which is the simplest boundary layer of this kind, is influenced by both the effects of the transverse curvature and of the centrifugal force. Both these effects are not observed in the case of two-dimensional plane flow. Evidently these effects become significant in relation to the boundary layer thickness, and this turbulent boundary layer may qualify as a full-fledged member of Bradshaw's complex turbulent flows [1].

The authors have already reported the characteristics of skewed turbulent boundary layers on rotating bodies in papers dealing with the features of the mean velocity profiles [2, 3]. Turbulent fields on rotating cylinders in an axial flow were studied by *Bissonnette* and *Mellor* [4] and *Lohmann* [5]. However, the research aims of these authors, namely the clarification of the effect of a sudden rate of strain imposed on the boundary layer, were rather different from ours.

In this report, the results of mean flow and turbulence measurements in the thick turbulent boundary layer on a spinning cone, which has a radius decreasing in the main flow direction, are presented. The main flow pressure has been held approximately constant using a tapered sleeve in a wind tunnel. The shape of the body has been selected to clarify the role of the forces produced by the rotation: the centrifugal force and "Coriolis-like force" (see next section). The structure of the mean flow and turbulence in rotating flow fields are discussed as influenced by these forces. These discussions are bored on the boundary layer equations and the energy equations, and results are presented in relation to the flux Richardson number.

Basic Equations and Consideration on Turbulent Energy

Let us agree to take a conical coordinate system as shown in Fig. 1. Application of the boundary layer approximation to the Reynolds equations expressed in this coordinate system yields the following equations:

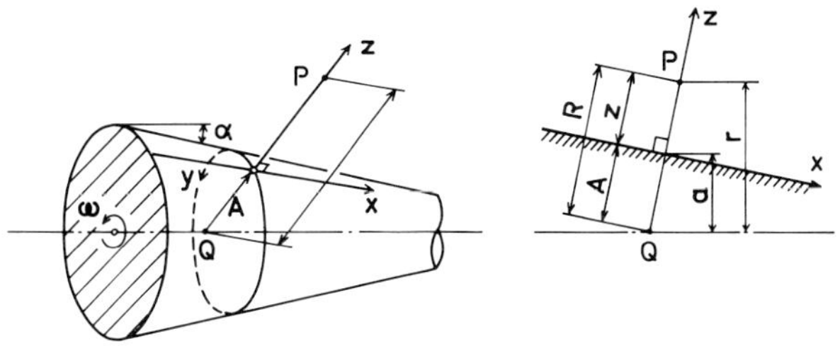

Fig. 1. Coordinate system

$$\rho \mathscr{D} U = -\partial_x P - \rho\lambda V^2/R + \partial_z(R\tau_x)/R, \tag{1}$$

$$\rho(\mathscr{D} V + VW/R) = \rho\lambda UV/R + \partial_z(R^2\tau_y)/R^2, \tag{2}$$

$$\rho V^2/R = \partial_z P + \rho[\partial_z(\overline{w^2}) + (\overline{w^2} - \overline{v^2})/R], \tag{3}$$

$$\partial_x(RU) + \partial_z(RW) = 0, \tag{4}$$

where $\mathscr{D} \equiv U\partial_x + W\partial_z$, and

$$\tau_x = \mu\partial_z U - \rho\overline{uw}, \quad \tau_y = \mu(\partial_z V - V/R) - \rho\overline{vw}. \tag{5}$$

Boundary conditions are

$$\left.\begin{aligned} z &= 0; \quad U = W = u = v = w = 0, \quad V = V_o = a\omega, \\ z &\to \infty; \quad U \to U_e, \quad V, u, v, w \to 0. \end{aligned}\right\} \tag{6}$$

Reduction of the cone radius toward downstream affects the fluid motion through two terms: the second term on the right-hand side of Eq. (1) and the first term on the right-hand side of Eq. (2). The term $-\rho\lambda V^2/R$ is the meridian component of the centrifugal force per unit volume of the fluid and it decelerates the fluid in the boundary layer in the meridian direction. On the other hand, the term $\rho\lambda UV/R$ accelerates the fluid in the azimuthal direction. In what follows, the latter will be called a "Coriolis-like force" because of its nature. These forces were already observed in the previous experiment [2] but their directions were the opposite of the present ones. Integration of the above equations gives two momentum integral equations which have parameters Θ_x and Θ_{xy} as defined in the nomenclature. They will be used in the presentation of experimental results.

The approximate turbulent energy equation of the present problem is

$$\mathscr{D}(\overline{q^2}/2) = -\overline{uw}\,\partial_z U - \overline{vw}\,R\partial_z(V/R) + D - \epsilon, \tag{7}$$

where D and ϵ stand for the turbulent diffusion and the dissipation respectively. Two production terms appear in the right-hand side of Eq. (7), the first and the second; the latter is lacking in the case of a stationary body.

Now the transfer of the kinetic energy from the mean flow to the fluctuation and its conversion between the fluctuating components will be discussed. Equations describing the kinetic energy components of mean flow are,

$$\left.\begin{aligned}\mathscr{D}(U^2/2) &= -\lambda UV^2/R - U\partial_x P/\rho + \overline{uw}\,\partial_z U \\ &\quad -\partial_z(\overline{uw}RU)/R - \nu(\partial_z U)^2 + \nu\partial_z(R\partial_z U^2)/2R, \\ \mathscr{D}(V^2/2) &= \lambda UV^2/R - V^2W/R + \overline{vw}R\partial_z(V/R) \\ &\quad -\partial_z(\overline{vw}RV)/R - \nu[R\partial_z(V/R)]^2 \\ &\quad +\nu\partial_z[R^3\partial_z(V/R)^2]/2R.\end{aligned}\right\} \qquad (8)$$

The terms corresponding to the first and the second production terms in Eq. (7), $-\overline{uw}\,\partial_z U$ and $-\overline{vw}R\partial_z(V/R)$, appear in the first and the second equations of Eq. (8), respectively, in opposite sign. Hence the second production term plays an important role for turbulent intensity in the rotating boundary layer.

The first term on the right-hand side of the equation for $U^2/2$ represents the rate of work done by the centrifugal force per unit mass in the x-direction; this makes a negative contribution to the U-energy. The corresponding term appears in the equation for $V^2/2$ in opposite sign. Therefore the kinetic energy extracted from the U-component by the centrifugal force is transferred directly to the V-component; the latter receives the energy. This is consistent with the earlier statement that the fluid in this boundary layer is decelerated in the x-direction and accelerated in the y-direction.

If viscous transport is neglected and if the boundary layer approximation is applied so that the terms such as $\partial_x(*)$ can be neglected, the equations for $\overline{u^2}/2$, $\overline{v^2}/2$ and $\overline{w^2}/2$ are, respectively,

$$\left.\begin{aligned}\mathscr{D}(\overline{u^2}/2) &= -2\lambda\overline{uv}\,V/R - \overline{uw}\partial_z U + D_{11} + P_{11} - \epsilon_u, \\ \mathscr{D}(\overline{v^2}/2) &= 2\lambda\overline{uv}\,V/R - 2\,\overline{vw}\,V/R - \overline{vw}R\partial_z(V/R) + D_{22} + P_{22} - \epsilon_v, \\ \mathscr{D}(\overline{w^2}/2) &= 2\,\overline{vw}\,V/R + D_{33} + P_{33} - \epsilon_w,\end{aligned}\right\} \qquad (9)$$

where D_{ij} and P_{ij} are the turbulent diffusion and the pressure-strain rate correlation terms, respectively, and ϵ_u, ϵ_v and ϵ_w are the turbulent dissipation for u-, v- and w-fluctuations, respectively. The expressions of these terms are obtained as follows;

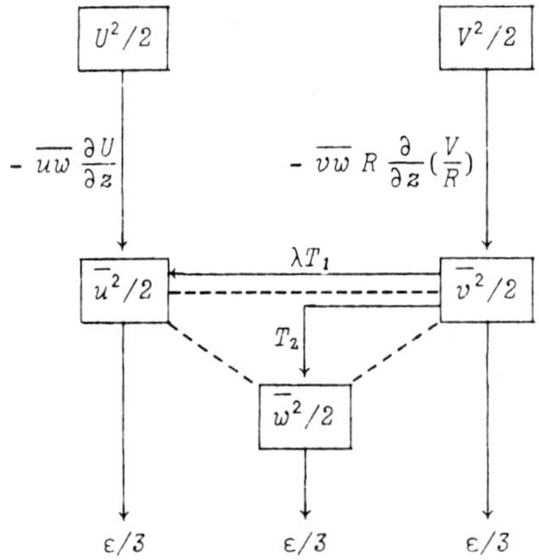

Fig. 2. Schematic diagram of energy transfer

$$\left.\begin{aligned} D_{11} &= -\partial_z(R\overline{u^2w}/2)/R, \\ D_{22} &= -\partial_z(R\overline{v^2w}/2)/R - \overline{v^2w}/R + (\lambda\overline{up} - \overline{wp})/\rho R, \\ D_{33} &= -\partial_z(R\overline{w^3}/2)/R + \overline{v^2w}/R - \partial_z(\overline{wp}/\rho), \end{aligned}\right\} \tag{10}$$

$$\left.\begin{aligned} P_{11} &= \overline{p\partial_x u}/\rho \\ P_{22} &= \sqrt{\lambda^2+1}\,\overline{p\partial_y v}/\rho R - (\lambda\overline{up} - \overline{wp})/\rho R \\ P_{33} &= \overline{p\partial_z w}/\rho \end{aligned}\right\} \tag{11}$$

$$\left.\begin{aligned} \epsilon_u &= \nu[2\overline{(\partial_x u)^2} + \overline{(\sqrt{\lambda^2+1}\ \partial_y u/R + \lambda v/R)^2} + \overline{(\partial_z u)^2} \\ &\quad + \overline{\partial_x v(\sqrt{\lambda^2+1}\ \partial_y u/R + \lambda v/R)} + \overline{\partial_z u \partial_x w}], \\ \epsilon_v &= \nu\,[\overline{(\partial_x v)^2} + \overline{(\partial_z v)^2} + 2\,\overline{(\partial_x u + \partial_z w)^2} \\ &\quad + \overline{\partial_x v(\sqrt{\lambda^2+1}\ \partial_y u/R + \lambda v/R)} \\ &\quad + \overline{\partial_z v(\sqrt{\lambda^2+1}\ \partial_y w/R - v/R)}], \\ \epsilon_w &= \nu[\overline{(\partial_x w)^2} + \overline{(\sqrt{\lambda^2+1}\ \partial_y w/R - v/R)^2} + 2\,\overline{(\partial_z w)^2} \\ &\quad + \overline{\partial_z v(\sqrt{\lambda^2+1}\ \partial_y w/R - v/R)} + \overline{\partial_x w \partial_z u}]. \end{aligned}\right\} \tag{12}$$

Of course, $\epsilon_u + \epsilon_v + \epsilon_w = \epsilon$, and it may be plausible to assume $\epsilon_u = \epsilon_v = \epsilon_w = \epsilon/3$ in the case of high Reynolds number.

When the body is stationary, that is $V = 0$, the kinetic energy is educed from the mean flow by $-\overline{uw}\partial_z U$ and enters directly into the streamwise fluctuation and is then redistributed to the other components. In this case the effect of curvature affects only D_{ij} and P_{ij}.

The rotation of the body complicates the flow of kinetic energy appreciably. *Townsend* [7] also discussed this problem in the case of two-dimensional flow along a curved surface. In the present case, the speed ratio and the inclination of the cone are additional factors affecting the turbulence mechanism. Comparison of Eqs. (8) and (9) shows that the energy extracted from $U^2/2$ and $V^2/2$ converts directly into $\overline{u^2}/2$ and $\overline{v^2}/2$ respectively. The right-hand sides of Eq. (9) have two novel terms:

$$-2\lambda\overline{uv}V/R = \lambda T_1, \quad 2\overline{vw}V/R = T_2. \tag{13}$$

Obviously, the summation of equations in Eq. (9) cancels out these terms and this fact reveals the new transfer mechanism of turbulent energy between fluctuating velocity components in this boundary layer. The term λT_1 appears in the equations of $\overline{u^2}/2$ and $\overline{v^2}/2$ with the opposite sign, and T_2 does in the equations of $\overline{w^2}/2$ and $\overline{v^2}/2$ with the opposite sign, hence λT_1 and T_2 perform a mutual transfer of energy. These terms are usually positive in the present case, so they are a sink for $\overline{v^2}/2$ and a source for $\overline{u^2}/2$ and $\overline{w^2}/2$. The process is depicted schematically in Fig. 2.

Experimental Apparatus and Procedures

The wind tunnel employed for the present experiments is a suction type with a test section inner diameter of 350 mm. The rotating conical body with a gradient of about 0.0783 and a maximum radius of 100 mm has been equipped in the tunnel, as shown in Fig. 3. As mentioned above, a tapered sleeve has been inserted in the tunnel; taking the displacement thickness into account, the effective cross-sectional area for the air-stream has been made nearly constant over the working section when the body was at rest.

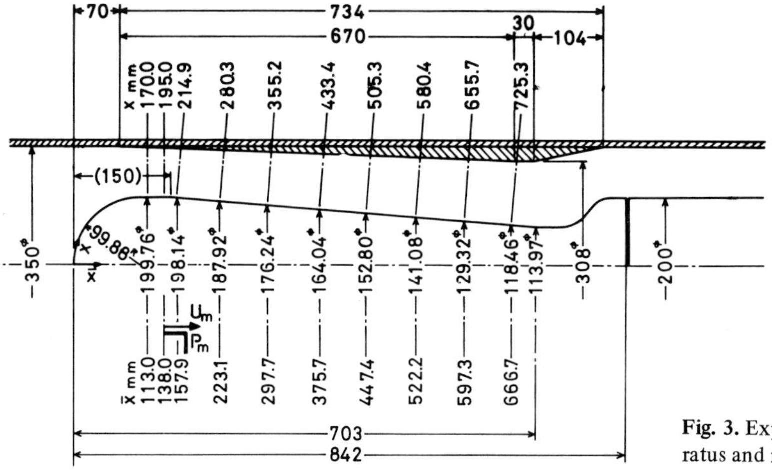

Fig. 3. Experimental apparatus and measuring stations

Experiments have been carried out with the Reynolds number R_e fixed at 8×10^4 and speed ratio Ω_m varied: 0, 1, 1.5 and 2. The skewed mean velocity profiles have been measured by means of a three-hole yaw probe and a static probe. Measurements of Reynolds stresses have been performed at four different stations: $x = 433.4$ mm to 655.7 mm, using the method of a rotated hot-wire [4]. The two types of sensors used were designed especially to prevent prong interference, one with the wire mounted normal to the axis of the stem and the other with it mounted at angle of about 45° to the axis as shown in Fig. 4. In the two sets of measurements, the sensor has been rotated about the mean flow angle that had been obtained beforehand by the three-hole yaw probe.

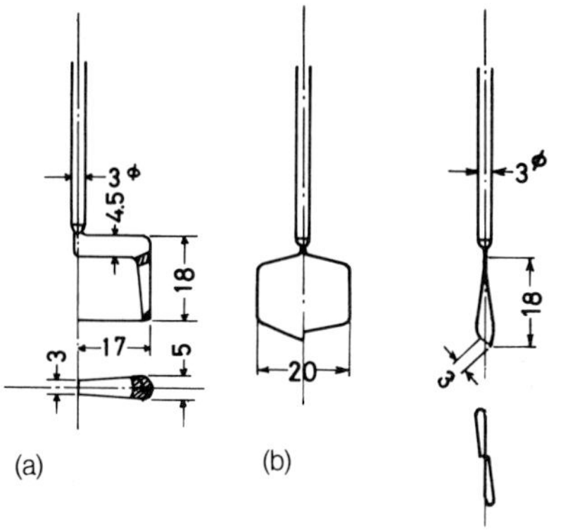

Fig. 4a, b. Hot-wire probes: (a) normal type, (b) inclined type

All six components of $\overline{u_i u_j}$ tensors have been computed from data taken from least-square fits to the ten different sets of experimental data obtained from the two sensors. For simplicity of data processing, it was assumed that the directional sensitivity of the hot-wire can be expressed by the cosine law. This assumption does not seem to lead to any serious error if the angle β between the flow direction and the plane normal to the wire axis is in the range $-45^\circ \lesssim \beta \lesssim 45^\circ$. This condition was always satisfied in the case of the inclined-type hot-wire, whereas the normal-type sensor, for which the rotation angle of the stem equals the angle β, has been rotated within this limit about the flow direction in the present experiment. Moreover, the assumption of the cosine law has a distinct advantage in that accurate preliminary knowledge of the angle of the wire-axis of the inclined type sensor is unnecessary; it can be computed from the output corresponding to the uniform flow according to the rotated wire method. The effect of the skewness of the mean velocity profile in the boundary layer on the turbulence measurements, i.e., the effect of the change in flow direction over the wire length of the inclined-type sensor, has been estimated at about 1% at most and is therefore negligible.

Results and Discussion

Mean Flow Measurements

The static pressure, measured when the body is stationary, is plotted in Fig. 5. This pressure is shown to be nearly constant in the region of x = 355.2 mm to 655.7 mm. The velocity distribution at all test sections without rotation of the body is presented in Fig. 6. The ordinate is non-dimensionalized by the momentum thickness Θ_x. The profiles in the turbulent boundary layer at the downstream stations are nearly similar.

Figure 7 shows the meridian and azimuthal velocity profiles in the boundary layer on the conical part of the body rotating at $\Omega_m = 2$. The rotation of the body promotes the development of the boundary layer and hence the boundary-layer transition, and the boundary layer can be deemed fully turbulent even at x = 280.3 mm.

Figure 8 shows the typical profiles in the turbulent boundary layer compared with those on a rotating circular cylinder [3]. Although a quantitative comparison may not be made because the experimental condition in Ref. [3] is slightly different from the present one, the differences between the mean flow fields on the rotating conical body and on the rotating cylinder are clearly seen; the cone geometry has a significant effect on the mean flow structure. Although the fluid particles are accelerated in the y-direction, the change in the y-component velocity profiles or their gradient across the layer is similar to that caused by adverse pressure gradient in the main flow direction. This is also clear when we observe the profiles in the frame fixed to the rotating body. Hence, it can be thought that the cone geometry has an effect similar to the adverse pressure gradient on the whole flow field.

The ratio of the centrifugal force to the Coriolis-like force is,

$$(\rho\lambda V^2/R)/(\rho\lambda UV/R) = V/U \sim \Omega_m. \tag{14}$$

The effect of the speed ratio Ω_m on the velocity profile can be explained by use of the above expression. That is, as the speed ratio becomes higher the role of the centrifugal force becomes relatively more important than that of the Coriolis-like force and then the fluid particle in the boundary layer is more decelerated in the x-direction than it is accelerated in the y-direction. This variation of the velocity distribution will be made clear using a polar diagram of mean velocity vectors rather than such usual plots of velocity profiles as Figs. 6–8. The veloc-

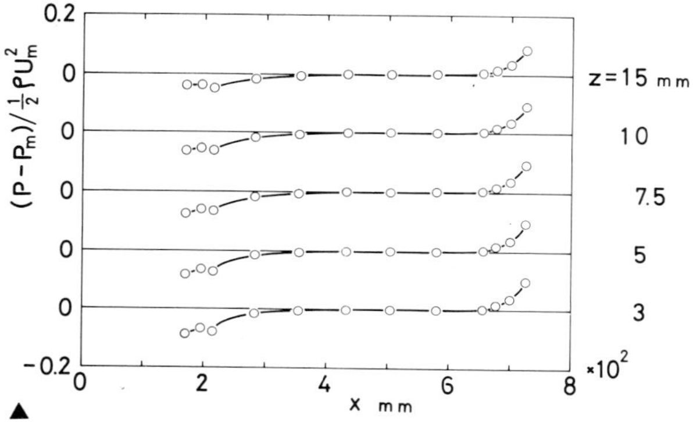

Fig. 5. Static pressure distribution with no rotation

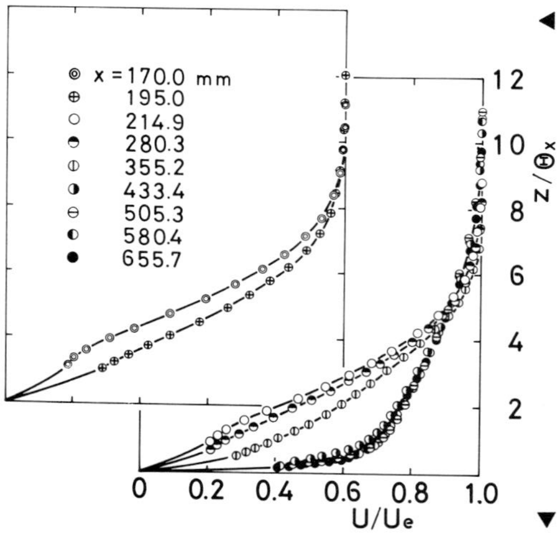

Fig. 6. Velocity profiles with no rotation. Θ_x is the momentum thickness of the thick boundary layer

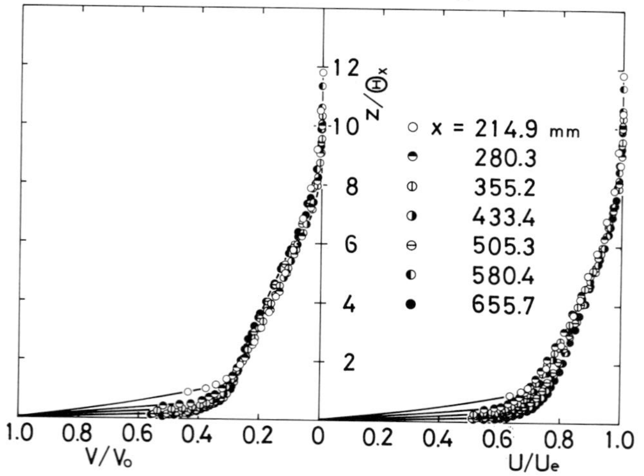

Fig. 7. Velocity profiles with rotation: $\Omega_m = 2$

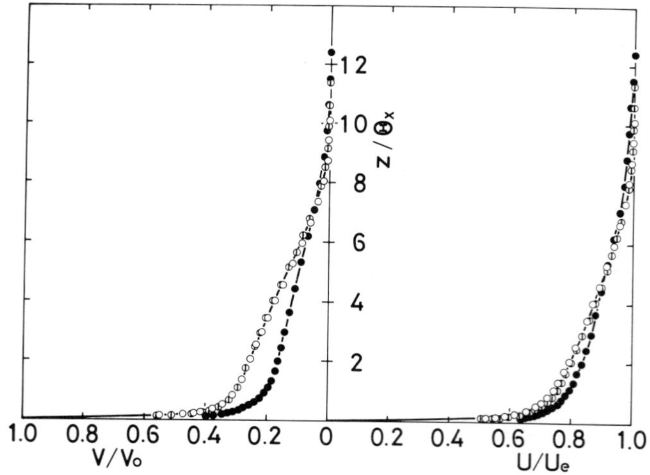

Fig. 8. Comparison of velocity profiles: $\Omega_m = 2$. ○, ⦶, present experiment: $R_e = 5 \times 10^4$, ○, $x = 505.3$ mm; ⦶, 580.4 mm. ●, experiment on a rotating cylinder [3]: $R_e = 5 \times 10^4$, $x = 549$ mm

ity distribution of the present flow may be said as quasi-collateral when it satisfies the following condition;

$$1 - U/U_e = (A/R)\,V/V_o. \tag{15}$$

If the profile is quasi-collateral and if it is observed from the rotating frame fixed to the body, the wall-parallel component of the relative velocity vector in the boundary layer coincides in its direction with the relative velocity vector of the uniform inviscid flow whose magnitude is U_e in the stationary frame. Equation (15) reduces to $1 - U/U_e = V/V_o$ used in [2] when the boundary layer is thin compared with the radius of the body. A diagram taking the abscissa as U/U_e and the ordinate as $(A/R)V/V_o$ on the basis of the Eq. (15) may be termed a polar or hodograph diagram as was done in the earlier study [2], although *Bissonnette* and *Mellor* [4] used the different expression in their ordinate, i.e., $(r/a)V$ (their test model is a cylinder of constant radius a; r, radial coordinate); but a basis provided for their expression is not clear.

Figure 9 shows the polar plots of the velocity profiles at $\Omega_m = 1$ and 2. While the velocity profiles are quasi-collateral at the stations upstream within the regions that the boundary layers are turbulent, say $x = 433.4$ mm at $\Omega_m = 1$ and 280.3 mm at $\Omega_m = 2$, the profiles tend to deviate upward in this figure from the quasi-collateral condition in the downstream stations. The degree of the deviation is smaller at $\Omega_m = 2$ than at $\Omega_m = 1$. This feature can be explained by Eq. (14); U/U_e is more decelerated than V/V_o is accelerated as the speed ratio is higher. If the speed ratio becomes still greater, the polar plots of the velocity profiles would bend downward. Then the boundary-layer separation might occur.

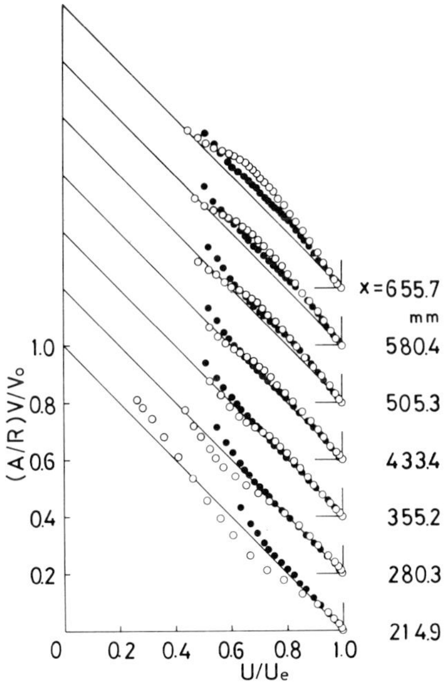

Fig. 9. Polar plots of velocity profiles. ○, $\Omega_m = 1$; ●, 2

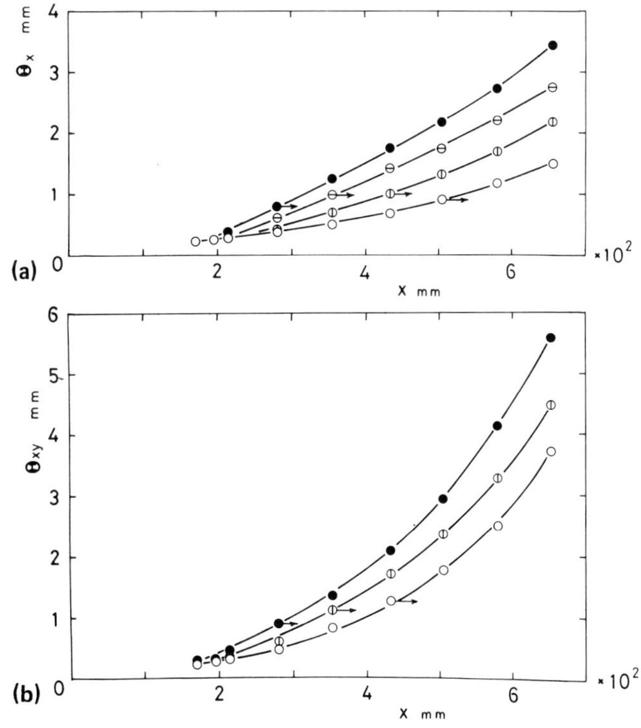

Fig. 10a, b. Developments of momentum thicknesses. (a) Θ_x: ○, $\Omega_m = 0$; ⦶, 1; ⊖, 1.5; ●, 2. (b) Θ_{xy}: ○, $\Omega_m = 1$; ⦶, 1.5; ●, 2. Arrows designate the turbulent zone

Figure 10a and b show the development in the momentum thicknesses of the boundary layer, Θ_x and Θ_{xy}, respectively, which remain the simplest and most important parameters of the boundary layer for the description of its mean growth.

Turbulent Field

Reynolds Stress Components. The turbulent field in the boundary layer on the rotating cone will manifest the effect of inclination of the rotating surface, i.e., the effect similar to an adverse pressure gradient as mentioned in the previous section.

In Fig. 11 the distribution of non-dimensional turbulent intensity in the various directions exhibit a definite increase in turbulence with the speed ratio. The intensification of azimuthal component should be noted in particular. A comparison with the case of a stationary cone indicates that in the rotating case the rather extensive region of the boundary layer has intense turbulent energy.

Reynolds shear stress components are shown in Figs. 12 and 13a–c in non-dimensional form. The distribution of the shear stress $-\overline{uw}/U_e^2$ on the stationary body shown in Fig. 12

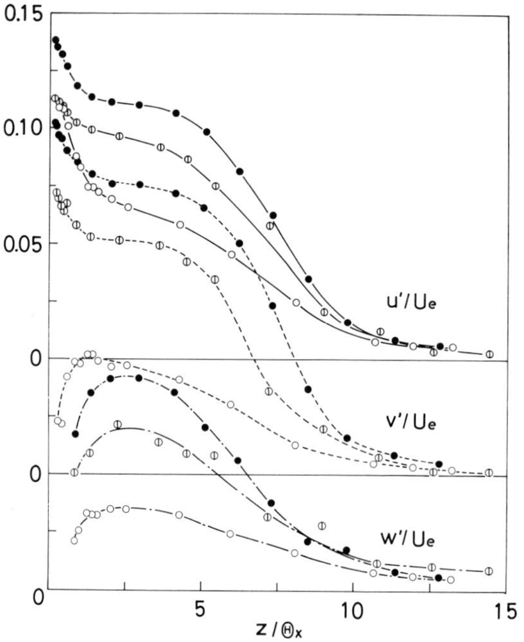

Fig. 11. Distribution of turbulent intensity components: x = 580.4 mm. ○, Ω_m = 0; ◐, 1.5; •, 2

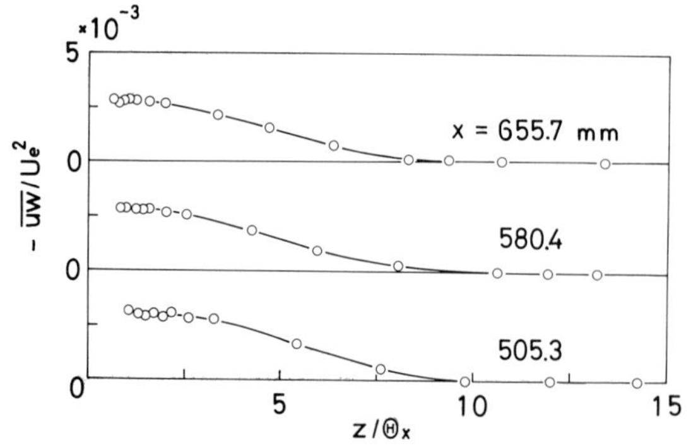

Fig. 12. Distribution of Reynolds shear stress with no rotation

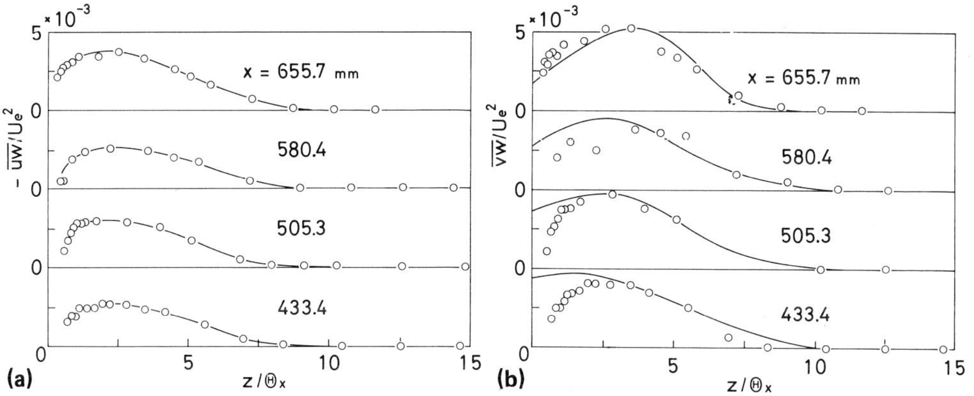

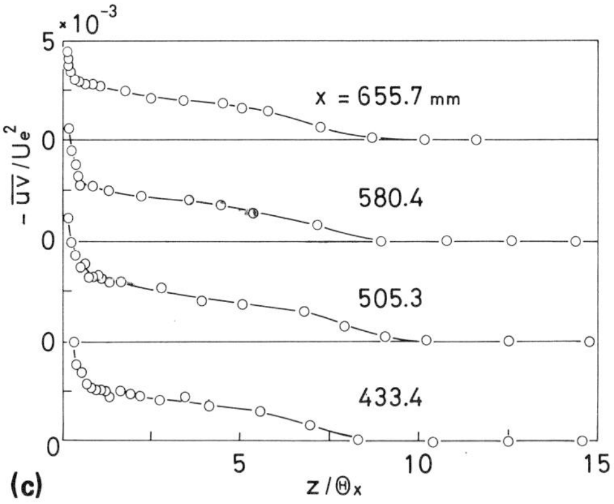

Fig. 13a–c. Distributions of Reynolds shear stresses: $\Omega_m = 1.5$. **(a)** $-\overline{uw}/U_e^2$, **(b)** $\overline{vw}/U_e^2$; —, $-\tau_y/\rho U_e^2$ computed from the basic equations using the measured mean velocity, **(c)** $-\overline{uv}/U_e^2$

almost coincides with *Klebanoff*'s result for a flat-plate flow [8]. The rotation of the cone produces all of the non-zero shear stress components. Figure 13a exhibits the streamwise development of the component $-\overline{uw}/U_e^2$ and it deviates from the profile in the case $\Omega_m = 0$, that is, each profile has a peak rather far from the wall. The distribution of $\overline{vw}/U_e^2$ is shown in Fig. 13b, and a large peak is also evident. Studies of the effect of adverse pressure gradient on the turbulent structure of the two-dimensional boundary layer suggest that the above-mentioned profiles appear due to the action of the centrifugal and Coriolis-like forces. Solid lines in Fig. 13b are drawn by use of the basic equations and the measured mean velocity distribution. Except for near the cone surface, there is good agreement between direct measurements and results calculated from the mean velocity. Profiles of $-\overline{uv}/U_e^2$ shown in Fig. 13c show rather large values very near the wall and decrease monotonically toward the edge of the layer.

Turbulent Energy Production. One of the most important terms of the turbulent energy budget is the production terms, which are the first and second terms on the right-hand side of Eq. (7). The distribution of the production with no rotation shown in Fig. 14 is similar in the streamwise direction and coincides almost with the result for a flat-plate flow [8].

The representative profiles of production terms in the boundary layer on the rotating cone are presented in Fig. 15. Non-dimensional value of the first production term $-\overline{uw}\partial_z U$ is slightly larger than that of the stationary cone. The value of the second production term, $\overline{vw}R\partial_z(V/R)$, coming from the rotation of the cone amounts to about twice as much of the value of $-\overline{uw}\partial_z U$ in non-dimensional form. The profiles of the second production term exhibit a remarkable peak in the central part of the layer, and the profiles of the first pro-

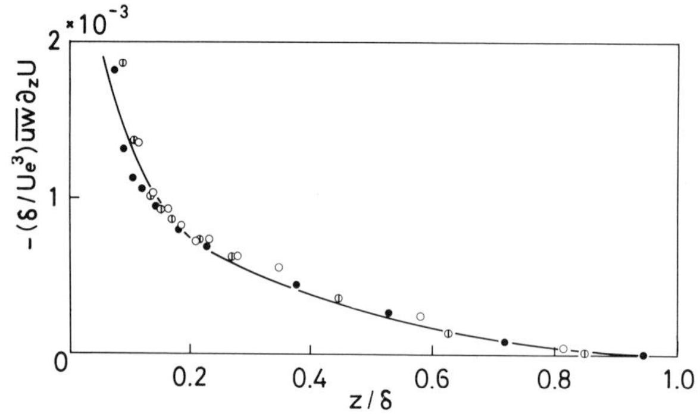

Fig. 14. Turbulent energy production with no rotation. ○, x = 505.3 mm; ◐, 580.4 mm; •, 655,7 mm; —, data of *Klebanoff* [8]

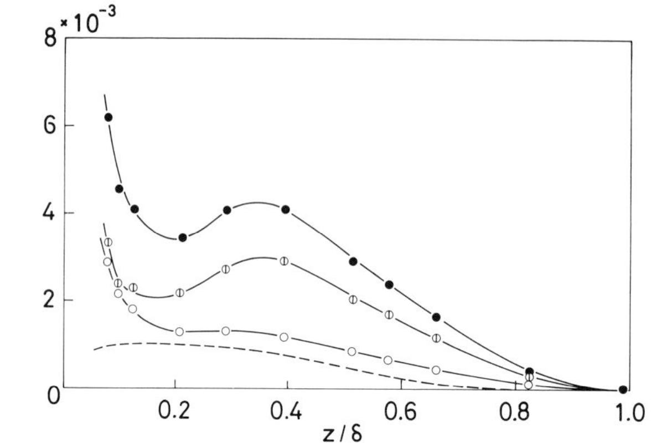

Fig. 15. Distribution of energy production terms and a transfer term: Ω_m = 1.5, x = 655.7 mm. ○, $-(\delta/U_e^3) \times \overline{uw}\,\partial_z U$; ◐, $-(\delta/U_e^3)\,\overline{vw}R\partial_z(V/R)$; •, $-(\delta/U_e^3)\,[\overline{uw}\,\partial_z U + \overline{vw}R\partial_z(V/R)]$; ---, $(\delta/U_e^3)T_2$

duction term maintain a large value over the central part of the layer in comparison with those in the non-rotating case.

The turbulent energy generated by the second production term appears directly as $\overline{v^2}/2$, and then it is redistributed to $\overline{u^2}/2$ and to $\overline{w^2}/2$. The result of the process shows the increase in v'/U_e and other components of turbulent intensity as shown in Fig. 11. The mechanism of the turbulent energy transfer of the flow has two characteristic terms λT_1 and T_2 as discussed earlier, and the latter may never be neglected. As illustrated, about 40% of the turbulent energy produced by $-\overline{vw}R\partial_z(V/R)$ in v-fluctuation is transported to w-fluctuation. The transport of the turbulent energy from $\overline{v^2}/2$ through λT_1 and T_2 to $\overline{u^2}/2$ and $\overline{w^2}/2$ prevents the intensification of v'/U_e as conjectured from the value of $-\overline{vw}R\partial_z(V/R)$. These features of the production and the transfer of the turbulent energy, chiefly caused by the effect of the rotating cone geometry, must affect the overall budget of the turbulent energy.

Eddy Viscosity and Mixing Length. From a phenomenological viewpoint, the structural differences of this turbulent boundary layer from two-dimensional flow should show up quantitatively in the mixing length or eddy viscosity distributions. In order to show this we apply the formulation of *Monin* and *Yaglom* [9], that is,

$$-\rho\overline{u_i u_j} \;=\; -\rho\overline{q^2}\delta_{ij}/3 + \rho(\nu_{Ti\alpha}S_{\alpha j} + \nu_{Tj\alpha}S_{\alpha i}) \tag{16}$$

where $S_{ij} = \partial_j U_i + \partial_i U_j$ is the deformation tensor, and ν_{Tij} the eddy viscosity tensor. Perhaps this form is one of the most simple representation of the non-isotropic eddy viscosity. Invoking boundary layer approximations in the present coordinate system, we can simplify the

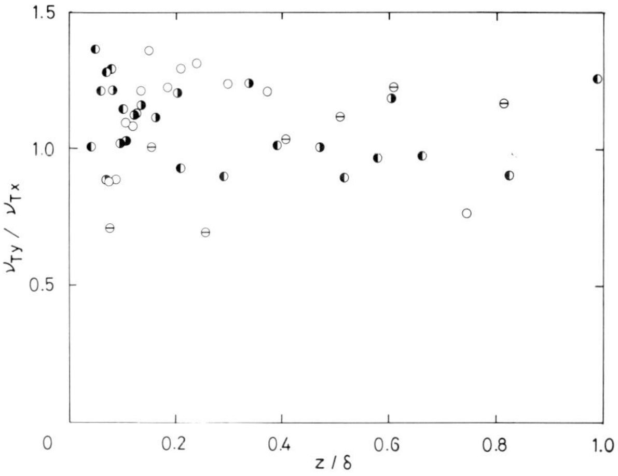

Fig. 16. Ratio of eddy viscosity coefficients: $\Omega_m = 1.5$. ○, $x = 433.4$ mm; ◑, 505.3 mm; ⊖, 580.4 mm; ◐, 655.7 mm

deformation tensor. Further if we assume that the eddy viscosity tensor deviates only a small amount from the isotropic form, that is $\nu_{Tij} = \nu_T \delta_{ij} + \epsilon_{ij}$, then Eq. (16) yields,

$$\left.\begin{aligned}
\overline{u^2} &= \overline{w^2} = \overline{q^2}/3, \quad \overline{v^2} = \overline{q^2}/3 + 2(\nu_T + \epsilon_{22})\lambda U/R, \\
\overline{uv} &= -(2\nu_T + \epsilon_{11} + \epsilon_{22})\lambda V/2R, \\
\overline{uw} &= -(2\nu_T + \epsilon_{11} + \epsilon_{33})\partial_z U/2 = -\nu_{Tx}\partial_z U, \\
\overline{vw} &= -(2\nu_T + \epsilon_{22} + \epsilon_{33})R\partial_z(V/R)/2 = -\nu_{Ty}R\partial_z(V/R).
\end{aligned}\right\} \quad (17)$$

In Fig. 16, the ratio ν_{Ty}/ν_{Tx}, computed from the measured values using above relations, is plotted for all x stations in the turbulent region when $\Omega_m = 1.5$. The scatter is somewhat large, owing to the errors in measured stresses and the numerical derivatives. The mean value appears to be slightly larger than 1.

For swirling flows, neglecting terms of the order of λ^2, the relationships between the components of the eddy viscosity and mixing length are written by

$$\nu_{Ti} = l_i^2 \{(\partial_z U)^2 + [R\partial_z(V/R)]^2\}^{1/2}, \quad i = x, y. \quad (18)$$

In Fig. 17, the value of l_y/δ, calculated from Eq. (18) together with Eq. (17), is plotted for two x stations. It can be seen that the mixing length is considerably greater than in the tur-

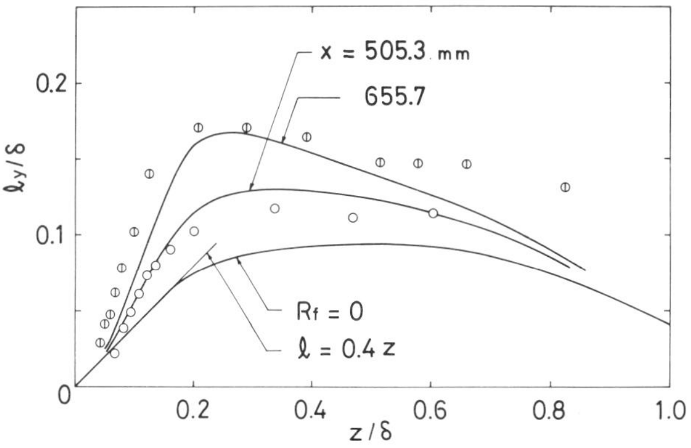

Fig. 17. Variation of mixing length: $\Omega_m = 1.5$. ○, $x = 505.3$ mm; ⦶, 655.7 mm; —, computation from the Monin-Oboukhov formula (21), $\beta = 3$. The curve for $R_f = 0$ is the dissipation length parameter of *Bradshaw* [6]

bulent boundary layer on a flat plate and, in particular, even in the near wall region it is not similar to the mixing length of the flat-plate flow. Hence, the logarithmic velocity distribution law related to the mixing length near the wall would be no longer valid in the form generally accepted for flat-plate flow; this law requires a modification. At the same time, the wall variable, that is, the argument in the logarithmic term must be altered to include the curvature effect, as was done in the case of a rotating cylinder [3].

It can be thought that the mixing length is increased through the agency of the transport term T_2 because the w-fluctuation is intensified through this term. Thus, if we adopt a Richardson number as a measure of the effect of the curvature or rotation, as suggested by *Bradshaw* [6], (an appropriate one is the "flux" Richardson number), we can express this fact; the "flux" Richardson number is related to the production and transport terms in the turbulent energy equations. In swirling boundary layer the flux Richardson number R_f may be expressed as:

$$R_f = (2\,\overline{vw}V/R)/[\overline{uw}\partial_z U + \overline{vw}\partial_z(RV)/R]. \tag{19}$$

This relationship is derived from the turbulent energy equations, Eq. (9), and is analogous to but different from that in two-dimensional curved flow given by Bradshaw. Figure 18 shows an example of the distribution of the flux Richardson number computed from the experimental data.

Using the flux Richardson number, the turbulent energy equation, Eq. (7), becomes

$$\mathscr{D}(\overline{q^2}/2) = [-\overline{uw}\partial_z U - \overline{vw}\partial_z(RV)/R](1 - R_f) + D - \epsilon. \tag{20}$$

If transport of turbulent energy by advection and diffusion is not significant, as it is usually the case in the inner layer, the change in the dissipation length parameter, and hence the mixing length, may be estimated qualitatively by the flux Richardson number.

The variation of mixing length is expressed by the following semi-empirical formula, i.e., the Monin-Oboukhov formula [6],

$$l/l_o = 1 - \beta R_f, \tag{21}$$

where l_o is the mixing length for the flat-plate flow, in which $R_f = 0$. Using this formula and the dissipation length parameter L [6] in place of l_o, the variation of the mixing length has been evaluated. In Fig. 17, the predicted profiles of the mixing length, together with the dissipation length parameter at $R_f = 0$, are presented in full lines. The best fitting value of β has been equal to 3. It can be seen that the estimates from Eq. (21) with $\beta = 3$ show fairly good

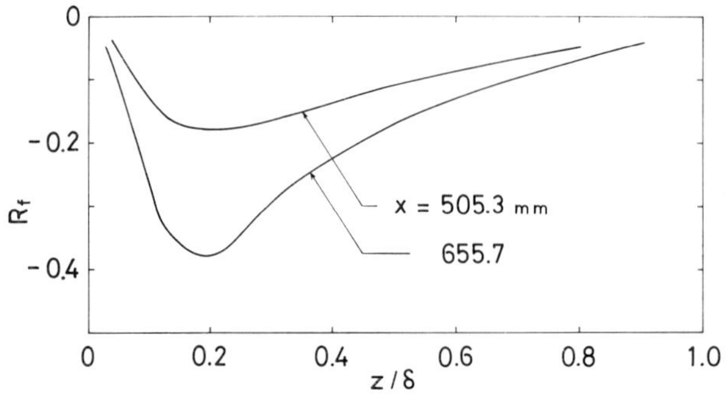

Fig. 18. Variation of the flux Richardson number: $\Omega_m = 1.5$

agreement with experimental results. In particular, it should be noted that the near wall behavior of the mixing length is predicted very well by this formula.

It was inferred from the properties of the present boundary layer flow that the main effect of rotation was simply to increase the length of the helical path travelled by the fluid from one end of the cylinder to the other [6]. This inference might be true as long as the boundary layer is so thin that the effects of curvature or rotation are negligible. However, it has been shown that, even in the laminar boundary layer on a stationary cylinder (e.g., [10, 11]) as well as the laminar and turbulent boundary layers on a rotating cylinder [3], the curvature or rotation effects are important. The drastic deviations of the mixing length shown in Fig. 17 and the mean velocity from those the expected distributions for two-dimensional flow on a flat plate, illustrate the significant effect of rotation on the turbulent boundary layer on a rotating cylindrical body in an axial stream.

Momentum Equation in the z-Direction

The order estimation method for the terms in Reynolds equations leading to the boundary layer equations is not always explicit. The method given by *Hinze* [12] provides the Eqs. (1) to (5). Several recent prediction methods for swirling turbulent boundary layers (e.g., [13, 14]) have neglected all the fluctuation terms in the momentum equation in the normal direction to the wall, Eq. (3); it was assumed that the centrifugal force was in equilibrium only with the force generated by the mean static pressure gradient across the layer. Consideration is made of its equilibrium using measured values.

It is easier to use an integrated form of Eq. (3) for the purpose. Integration of Eq. (3) from some point z to the outside of the boundary layer yields the following equation in non-dimensional form;

$$\frac{1}{A}\int_z^\delta \frac{A}{R}\left(\frac{V}{V_o}\right)^2 dz = \frac{P_e - P}{\rho V_o^2} + \frac{\overline{w_e^2} - \overline{w^2}}{V_o^2} + \frac{1}{A}\int_z^\delta \frac{A}{R}\,\frac{\overline{w^2} - \overline{v^2}}{V_o^2}\, dz. \tag{22}$$

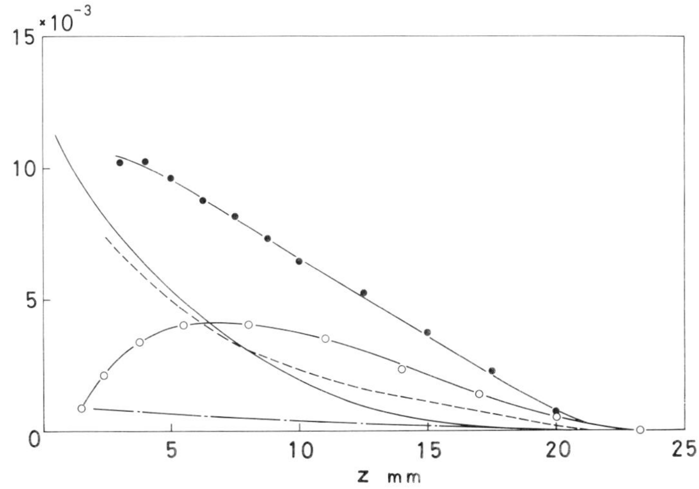

Fig. 19. Balance of z-momentum equation: $\Omega_m = 2$, $x = 580.4$ mm. ——, left-hand side of Eq. (22); – – –, right-hand side of Eq. (22); –•–, $(P_e - P)/\rho V_o^2$; –○–, $-(\overline{w_e^2} - \overline{w^2})/V_o^2$; –·–, $-\frac{1}{A}\int_z^\delta \frac{A}{R}\frac{\overline{w^2} - \overline{v^2}}{V_o^2}\, dz$

Figure 19 shows the distribution of the value of each term of Eq. (22) evaluated from the measurements. The upper boundary of the integration was determined from the variation of fluctuating velocity. It can be seen that the normal component of the centrifugal force due to the mean velocity is not in equilibrium only with the mean pressure gradient term. On the other hand, inclusion of the fluctuating terms in Eq. (22) serves to restore the momentum balance in the z-direction as shown in Fig. 19. The role of the fluctuation terms in equilibrium of z-momentum is never negligibly small but is the same order as the mean pressure term. From these results it may be conjectured that the pressure gradient term in Eq. (1) also contains the effect of the fluctuating velocities.

Concluding Remarks

The measurements and the consideration of basic equations for a thick turbulent boundary layer on a spinning cone with the radius decreasing in the downstream direction in a uniform axial flow, lead to the following conclusions:

First, the centrifugal force and Coriolis-like force have an upstream component and an azimuthal component, respectively, along the cone surface. They affect the boundary layer similar to an adverse pressure gradient, although these forces are body forces in contrast to the pressure force. The measured mean velocity profiles and Reynolds stress distributions clearly show their effects. Secondly, new mechanisms of turbulent energy production and turbulent energy transport which are lacking in the normal two-dimensional flow appear in the boundary layer. The measurements of each term reflect their importance in terms of the turbulent energy budget. Thirdly, the distribution of the mixing length evaluated from the Reynolds stress and mean flow measurements is significantly higher when compared to the flat-plate flow. This feature can be well predicted by use of the flux Richardson number and the Monin-Oboukhov formula. Lastly, the boundary layer momentum equation in the normal direction to the surface is not satisfied only by the centrifugal force due to the mean velocity and the mean pressure gradient in that direction, but it requires the complement of the turbulent terms.

References

1. P. Bradshaw: REVIEW – Complex turbulent flows. J. Fluids Eng. **97**, 146–154 (1975)
2. Y. Furuya, I. Nakamura: Velocity profiles in the skewed boundary layers on some rotating bodies in axial flow. J. Appl. Mech. **37**, 17–24 (1970)
3. Y. Furuya, I. Nakamura, S. Yamashita: The laminar and turbulent boundary layers on some rotating bodies in axial flows. Mem. Fac. Eng. Nagoya Univ. **30** (1), 1–58 (1978)
4. L. R. Bissonnette, G. L. Mellor: Experiments on the behaviour of an axisymmetric turbulent boundary layer with a sudden circumferential strain. J. Fluid Mech. **63**(2), 369–413 (1974)
5. R. P. Lohmann: The response of a developed turbulent boundary layer to local transverse surface motion. J. Fluids Eng. **98**, 354–363 (1976)
6. P. Bradshaw: The analogy between streamline curvature and buoyancy in turbulent shear flow. J. Fluid Mech. **36**(1), 177–191 (1969)
7. A. A. Townsend: *The Structure of Turbulent Shear Flow,* 2nd ed. (Cambridge University Press, London 1976) pp. 393–398
8. P. S. Klebanoff: Characteristics of turbulence in a boundary layer with zero pressure gradient. NACA Rpt. **1247**, 1135–1153 (1955)
9. A. S. Monin, A. M. Yaglom: *Statistical Fluid Mechanics,* Vol. 1 (MIT Press, Cambridge, Mass. 1971) pp. 388–393

10. R. A. Seban, R. Bond: Skin-friction and heat-transfer characteristics of a laminar boundary layer on a cylinder in axial incompressible flow. J. Aeronaut. Sci. **18**(10), 671–675 (1951)
11. M. B. Glauert, M. J. Lighthill: The axisymmetric boundary layer on a long thin cylinder. Proc. R. Soc. London Ser. A: **230**, 188–203 (1955)
12. J. O. Hinze: *Turbulence,* 2nd ed. (McGraw-Hill, New York 1975) pp. 589–593
13. M. L. Koosinlin, F. C. Lockwood: The prediction of axisymmetric turbulent swirling boundary layers. AIAA J. **12**(4), 547–554 (1974)
14. B. E. Launder, C. H. Priddin, B. I. Sharma: The calculation of turbulent boundary layers on spinning and curved surfaces. J. Fluids Eng. **99**, 231–239 (1977)

Experiments on the Turbulent Boundary Layer Over Convex Walls and Its Recovery to Flat-Wall Conditions

J. C. Gillis, and J. P. Johnston

Department of Mechanical Engineering, Stanford University,
Stanford, CA 94305, USA

Abstract

The experiments were performed at $\delta/R \simeq 0.09$ on a convex wall, where the curvature effects are strong. The data show that a flat-wall boundary layer is strongly affected by curvature only a few boundary layer thicknesses downstream of the start of convex curvature. On the other hand, in flow over a flat surface downstream of convex curvature, the effects of curvature persist downstream for twenty or more boundary layer thicknesses.

Over the convex surface, the boundary layer seems to have two regions. In the inner layer the shear stress is reduced from comparable flat-wall values but is still appreciable. In the outer layer the shear stress is negligible. Measurements show that in the inner layer the ratio of shear stress to turbulent kinetic energy, which is approximately constant over a flat wall, becomes a strong function of n/δ. Mixing lengths computed from the data indicate that in the inner layer, l is equal to κn very near the wall but that it takes on a constant value at about 0.07δ.

The data from the recovery region show that the shear stress, turbulence intensity and mean velocity profiles are slow to recover. However, the structural coefficient assumes its flat-wall value very quickly.

Nomenclature

β	Empirical constant in mixing length correction
δ	Thickness at which velocity is 99% of potential flow velocity
δ^*	Displacement thickness, $\int_0^\infty (1 - U/U_p)dn$
ϵ	Dissipation rate
κ	Karman constant, 0.41
ν	Kinematic viscosity
θ	Momentum thickness, $\int_0^\infty (U/U_p)(1 - U/U_p)dn$
ρ	Density
τ_0	Surface shear stress
$C_f/2$	Skin friction coefficient, $\tau_0/\rho U_{pw}^2$
l	Mixing length
l_0	Flat-wall mixing length
$\dot{m}$	Boundary layer mass flux per unit spanwise width
n	Distance normal to surface
n^+	Non-dimensional distance normal to surface, $n\sqrt{\tau_0/\rho}/\nu$
p	Static pressure
$\frac{1}{2}q^2$	Turbulence kinetic energy
R	Radius of surface curvature
R_{eff}	Effective radius of curvature [see Eq. (4)]
R_I	Richardson number
s	Streamwise coordinate, measured along *surface*
U	Mean velocity
u	Fluctuating streamwise velocity
u'	Root mean square value of u

U_p	Potential flow velocity	v'	Root mean square value of v
U_{pw}	Potential flow velocity at the wall	w	Fluctuating velocity parallel to surface in spanwise direction
U_∞	Flat-plate freestream velocity	w'	Root mean square of w
v	Fluctuating velocity normal to surface	y	Distance normal to flat wall

Introduction

The sensitivity of turbulent boundary layers to wall curvature has been known for some time and at one time was the subject of experiments by Ludwig Prandtl and his students. However, it has really been only in the last decade that widespread attention has been paid to this phenomenon. Lately, several curved boundary layer experiments have been performed at various values of δ/R. Both *Meroney* and *Bradshaw* [5] and *Ramaprian* and *Shivaprasad* [7,8] have done experiments which yielded mean and turbulent data for very slight surface curvature. Their results showed noticeable changes in structure, even though the values of δ/R were about 0.01. *So* and *Mellor* [9] have provided much data at $\delta/R \simeq 0.10$, and *Smits* et al. [11] have explored exceptionally strong curvature effects at $\delta/R \simeq 0.2$.

With the exception of the last, all of these experiments examined the transformation from flat-wall to curved-wall boundary layer flow, but ignored the reverse process, the recovery to flat-wall conditions following longitudinal curvature. If one thinks of areas where the effects of curvature will have practical consequences (airfoils, turbomachine blades), one almost always finds that there are flat surfaces or regions of low surface curvature behind curved surfaces, and so the relaxation of curvature effects is an area worth study. The experiment to be described here was designed to provide additional data on flow with strong curvature, $\delta/R \simeq 0.09$, and to explore the relaxation process on a flat wall downstream of the curved region.

The Experimental Apparatus

In our experiment we sought to isolate the effects of curvature on turbulent boundary layers by eliminating from the flow field any other flow conditions which are known to affect turbulence structure. Most such conditions, such as surface roughness, were easily eliminated, but the elimination of longitudinal pressure gradients in the wall layers required an unusual tunnel design. A streamwise cross section of this tunnel and its pertinent dimensions are shown in Fig. 1.

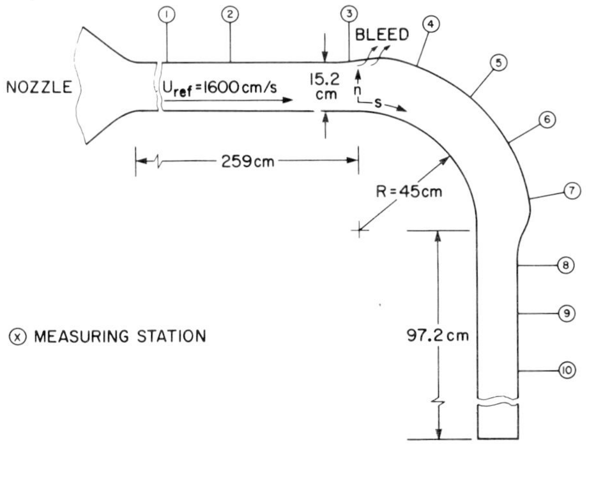

Fig. 1. Wind tunnel schematic

Upstream of the curved test section is a long, straight duct (see [6]) of rectangular cross section. The flow in this duct was set to zero pressure gradient simply by adjusting the top wall. To maintain zero pressure gradient on the lower wall (test surface) in the neighborhood of the point where the surface changes from flat to curved required a widening of the tunnel cross section. This geometry produced a strong adverse pressure gradient on the upper wall, and consequently strong suction had to be applied to keep the upper wall (non-test) boundary layer attached. At the end of the curved region, the tunnel width had to be reduced to avoid an adverse pressure gradient on the test surface. The final configuration for the tunnel was established by changing the shape of the upper wall and measurement of static pressure distribution along the test (convex) wall. The static pressure coefficient,

$$C_p = \frac{P_{\text{wall}} - P_{\text{wall},s=0}}{\frac{1}{2}\rho U_{\text{pw}}^2}, \tag{1}$$

was less than 0.029 at all points on the test surface after the final iteration.

In the region of curvature, the boundary layers on the parallel side walls are subjected to spanwise (i.e., radial) pressure gradients which cause secondary flows down the side walls and onto the convex test surface. Since the tunnel has low aspect ratio (the test surface width was approximately 12 boundary layer thicknesses), steps had to be taken to minimize the effects of these secondary flows. At the entrance to the curved region, the side-wall boundary layers were skimmed off. Further, two sets of boundary layer fences were mounted near the side walls, perpendicular to the test surface and parallel to the centerline. Measurements of the off-centerline mean velocity profiles showed that the wall shear and the integral quantities ($C_f/2$, δ^*, θ) varied no more than ±2.5% over the center 13 cm of the tunnel. Surveys taken of the mean velocity flow direction with a two-hole Conrad probe showed the flow to be less than ±2 deg out of parallel to the centerline over a 13 cm spanwise width of the test surface. The skew angles reached a maximum of ±4 deg near the side walls at the end of the curved section.

For the test conditions chosen, the free stream speed upstream of curvature was 1,600 cm/s. Surveys of the boundary layer in the flat pre-tunnel showed it to have normal, zero-pressure-gradient, turbulent boundary layer profiles in the mean velocity and all Reynolds stresses. The boundary layer's momentum thickness Reynolds number was 4,620 at the start of curvature, the point where distance s, measured along the surface, was designated to be zero. The thickness of the boundary layer varied from 2.95 cm at the first measuring station, 71.1 cm upstream of the curvature, to 4.88 cm at the end of the section, 44.25 cm downstream of the curvature. In the curved section, the ratio of the 99% boundary layer thickness to radius of curvature varied from 0.085 near the start to 0.097 at the end.

All mean velocity measurements, with the exception of those at station 3, were taken with pitot tubes. At station 3, near the start of curvature, hot-wire measurements were necessary because the static pressure distribution was unknown. Measurements of the longitudinal component of turbulence intensity were made using a DISA 55M01 Constant Temperature Anemometer with a DISA 55P01 hot-wire probe. Other turbulence measurements were taken using a DISA 55P51 x-wire probe.

Results

Three profiles representative of those taken in the upstream (flat-wall), curved, and recovery sections of the flow are plotted in Fig. 2. These three typical mean velocity profiles show that

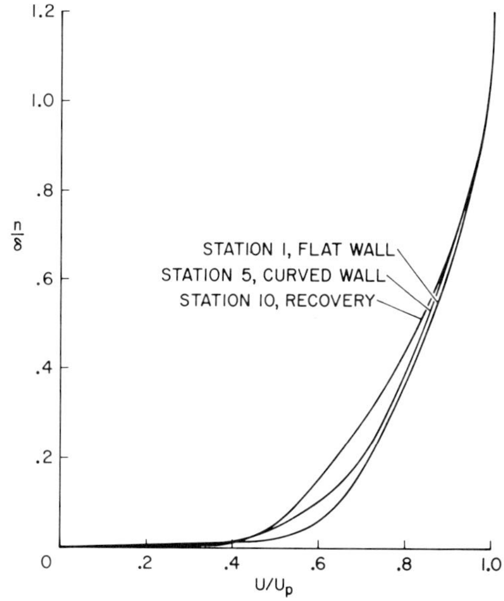

Fig. 2. Mean velocity profiles at representative stations

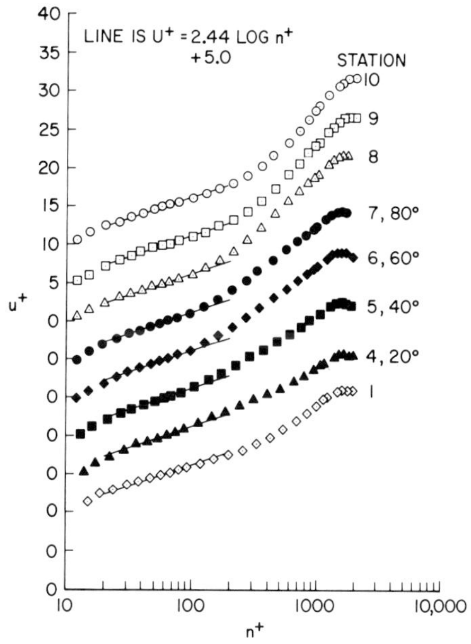

Fig. 3. Mean velocity profiles plotted in wall coordinates

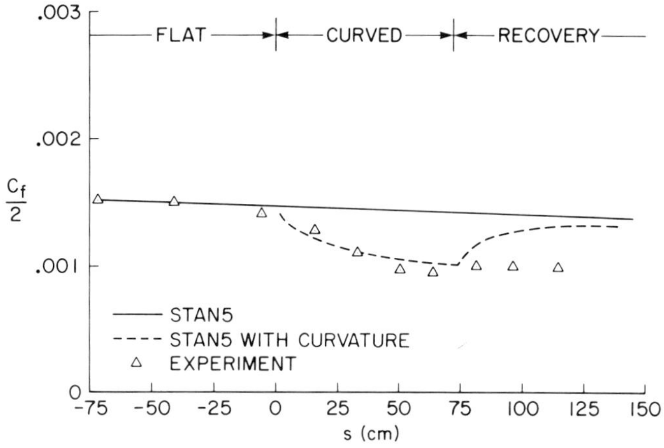

Fig. 4. Skin friction (calculated by Clauser plot) vs distance in flow direction

profiles taken downstream of the start of curvature follow a lower exponent power law than the flat-wall profiles measured upstream of the curvature. This has been observed before [5]. Fig. 2 also shows that the profiles in the recovery zone have not begun to recover their sharpness at the last measuring station. This second observation is surprising – at the start of the curved section, the response of the profiles to the sudden appearance of curvature is quite fast. One might expect the response to the disappearance of curvature to be almost as quick. Fig. 3 presents the mean velocity profiles replotted in wall coordinates. They show an easily discernible logarithmic layer, although the profiles become non-logarithmic at a lower value of n^+ in the curved region. In the profiles from 60° (station 6) on, it is also possible that there is a *second* logarithmic zone, from $n^+ \simeq 200$ to $n^+ \simeq 1{,}500$. The profiles at stations 6, 7, and 8 show this phenomenon most clearly; the second logarithmic zone seems to be breaking down at stations 9 and 10.

Since the data show a region where the law of the wall fits the data, it is appropriate to calculate an approximate skin friction using a Clauser plot. Fig. 4 is a graph of $C_f/2$, determined in this manner, as a function of distance s, along the centerline.

For purposes of comparison, we have included a solid line showing the flat-wall skin friction distribution predicted by the boundary layer code STAN5 [2]. Also shown is a dashed line which is the prediction of a version of STAN5 modified by *Johnston* and *Eide* [3] to account for the effects of curvature. The Johnston-Eide model uses a curvature Richardson number correction to the mixing length, as suggested by *Bradshaw*:

$$l = l_0(1 - \beta R_\mathrm{I}), \tag{2}$$

where R_I is the curvature Richardson number, l_0 is the standard mixing length, and β is a constant obtained from the data. The Richardson number is defined as

$$R_\mathrm{I} = \frac{2\,U/R_\mathrm{eff}}{\partial U/\partial n}. \tag{3}$$

The effective radius of curvature, R_eff, is computed from a first-order lag equation, which simulates the effect of a changing radius of curvature.

$$\frac{d(1/R_\mathrm{eff})}{ds} = \frac{1}{10\,\delta}\left(\frac{1}{R} - \frac{1}{R_\mathrm{eff}}\right) \tag{4}$$

The measured skin friction is decreased by the curvature about 33% – a large effect, but not too different from what might be expected from the data of other investigators of the curvature effect [1]. However, it is surprising to see that the skin friction has not begun to approach its flat-wall level at a distance approximately ten boundary layer thicknesses downstream of the end of curvature. The data and the predictions are in good agreement in the curved region, but the model predicts a quicker recovery than is observed.

Figure 5 is an isometric plot of the streamwise turbulence intensity plotted vs n/δ and s. The curvature clearly affects the profile shape – the turbulence intensity declines quickly after the start of curvature. The last three profiles in the curved section (at 40, 60, and 80 deg) are very similar and seem to indicate that some degree of self-similarity or equilibrium has been reached. After 90 deg, the recovery from curvature appears to take place first at the wall and then to diffuse outward slowly. At the last recovery station, the "bulge" of increased turbulence has reached only about halfway across the layer. This is consistent with the earlier observations that the mean velocity profiles are slow in recovering.

The shear stress profiles plotted in Fig. 6 show how drastically the curvature affects the turbulence structure. The stress levels at stations 4, 5, 6, and 7 are dramatically reduced relative to the profiles taken upstream. Once again, the slow recovery of the profiles downstream of the curvature is apparent. Fig. 6 also shows how the skin friction calculated from the mean profiles compares with the extrapolations of the turbulent shear stress to the wall. There is good agreement at all stations except station 4, where the boundary layer has just suffered a large change in turbulence structure and is probably far from equilibrium. The fact that there is good agreement in the curved region between the extrapolated shear stress and the wall shear computed from the law of the wall is perhaps a surprise. Many analytical derivations of

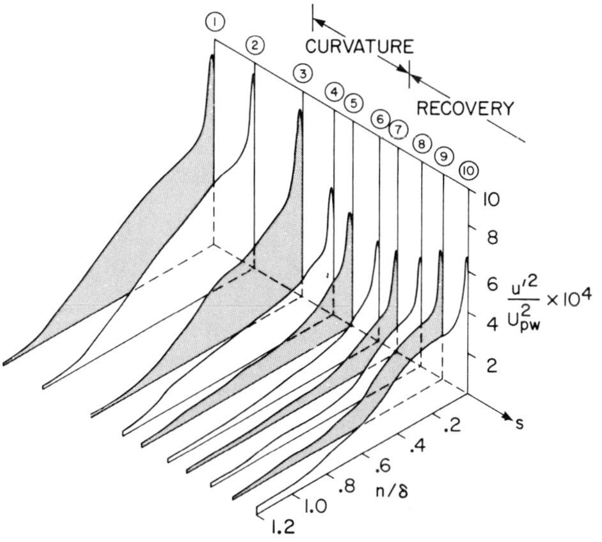

Fig. 5. Isometric plot of streamwise turbulence intensity profiles vs distance in flow direction

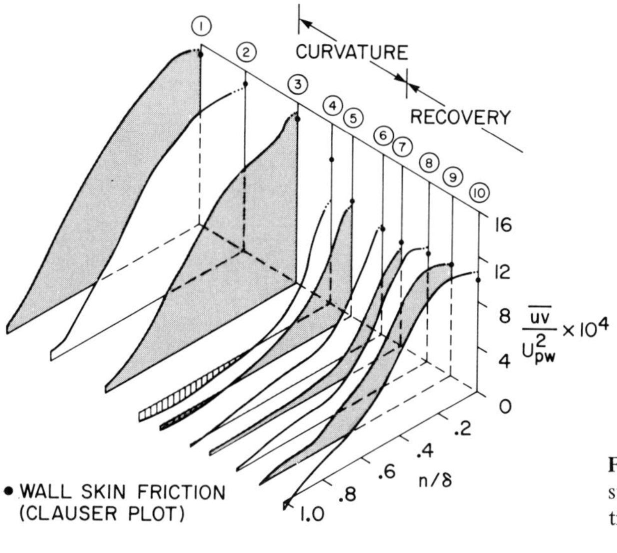

Fig. 6. Isometric plot of turbulent shear stress profiles vs distance in flow direction

the law of the wall assume that there is a region in the fully turbulent zones near the wall where the shear stress is nearly independent of y and equal to the wall shear stress. The curved profiles do not show such a region.

Figure 7 shows three representative shear stress profiles for the flat, curved, and recovery regions, respectively. The profile at 40 deg (which is very similar to the 60 and 80 deg profiles) shows that shear stress drops to negligible values in the outer portion of the layer; the data of *So* and *Mellor* [9] also show this result. In the recovery region, the shear stress profiles recover in the same way the turbulence intensity recovered – a slow diffusion and growth of a "bulge" out from the near-wall region towards the freestream.

In Fig. 8 we have plotted the shear stress profiles (non-dimensionalized on U_{pw}^2) for the stations at 40, 60, and 80 deg of the curve. There is a region between $n/\delta = 0.10$ and $n/\delta = 0.35$ where all the dimensionless profiles collapse. While this is not a large fraction of δ, it

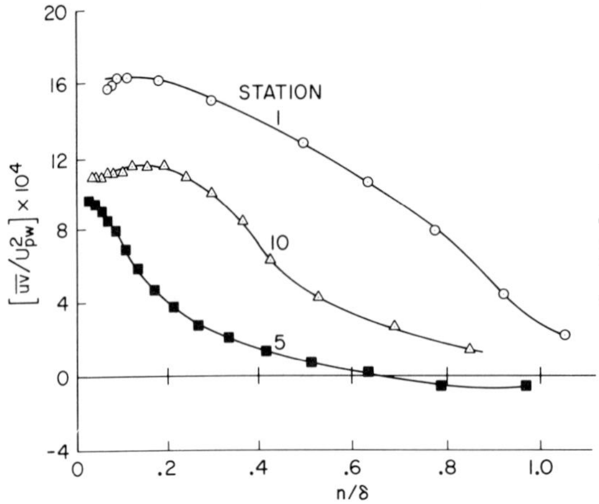

Fig. 7. Turbulent shear stress vs distance from the wall for representative stations

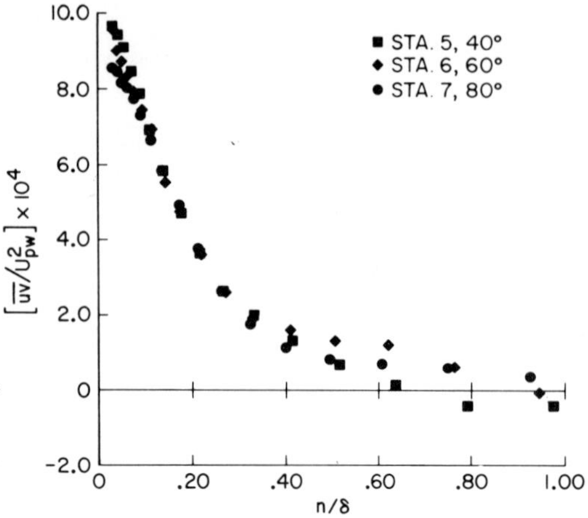

Fig. 8. Partial collapse of shear stress profiles for three curved wall stations

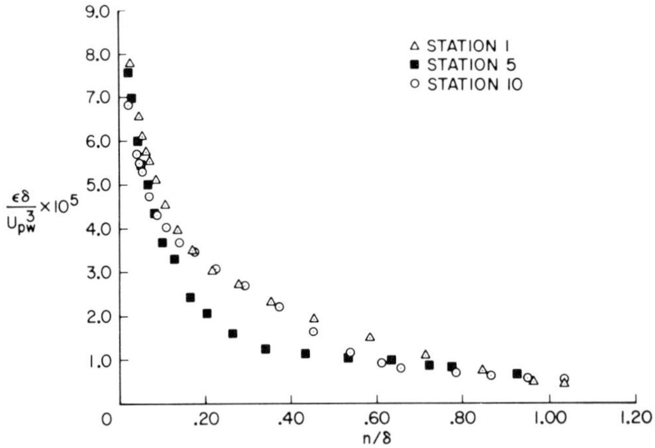

Fig. 9. Dissipation profiles at three stations

is a significant fraction of the region where $-\overline{uv}$ is large. We are currently investigating the meaning of this collapse.

We have also made approximate measurements of the dissipation rate. In doing so we have used the usual assumption that the small-scale motions which are responsible for the dissipation are isotropic, and so

$$\epsilon \simeq 15\,\nu \overline{\left(\frac{\partial u}{\partial s}\right)^2}. \tag{5}$$

Using Taylor's hypothesis, this becomes

$$\epsilon = \frac{15\,\nu}{U^2} \overline{\left(\frac{\partial u}{\partial t}\right)^2}, \tag{6}$$

and the dissipation rate was then obtained from the differentiated horizontal wire signal. As pointed out by Klebanoff, this method of measuring dissipation is not completely accurate because of assumption of isotropy. However, the measurements are useful for comparison

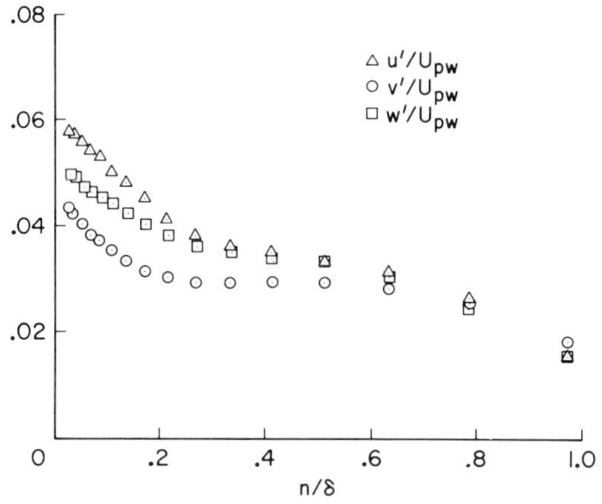

Fig. 10. Turbulent normal stress profiles after 40 degrees of curvature

between curved and flat boundary layers. Profiles taken at stations 1, 5, and 10 are shown in Fig. 9, non-dimensionalized on U_{pw} and δ. The profiles show that the dissipation rate over the curved surface is not too different from the dissipation rate over the flat surface. It is reduced, but not to the extent that the shear stress is reduced. The profile at station 10 shows a two-layered structure, not unlike the shear stress and turbulence intensity profiles at the same station.

In Fig. 10 we have plotted the three normal stress intensities vs n/δ at 40 deg. For the region $n/\delta > 0.40$ where the shear stress is nearly zero, the turbulence becomes quasi-isotropic. This is reasonable because the main production term in the turbulence kinetic energy equation, $-\overline{uv}(\partial U/\partial n)$, is zero there. The turbulence in the outer part of the layer is then "debris" from the preceding flat-wall boundary layer, which simply decays as it is convected downstream. To test this conjecture, we made the following calculations. First, we analyzed the mean flow data and located the n position of mean flow streamlines as a function of s. If the turbulence in the outer parts of the boundary layer is indeed quasi-isotropic, then its decay, as it is convected downstream, should be predictable. One model of the decay of isotropic turbulence was given by *Reynolds* [10] as

$$q^2 = q_0^2 \left(1 + \frac{5\, t\epsilon_0}{3\, q_0^2}\right)^{-6/5} \tag{7}$$

where q_0 is the initial value of q^2 and ϵ_0 is the initial value of turbulent dissipation, which is taken from the dissipation measurements. Using Taylor's hypothesis to replace t by x/U and with the conditions at station 5 as initial conditions, we have calculated the decay of the turbulence along a streamline.

Table 1. Comparison of calculation and experiment

Station	$S - S_0$ [cm]	q^2 Expt. [cm^2/s^2]	q^2 Calc. [cm^2/s^2]
5	0.0	6,230 = q_0^2	
6	17.8	4,900	4,400
7	30.5	4,200	3,600
8	42.2	4,900	3,100

The model and experiment show fair agreement in the curved region. The agreement is bad in the recovery region, but this must be so because of the appearance of significant $\overline{uv}$ in the outer region during recovery. The agreement is about the same for other streamlines which exist between the place where the shear stress becomes negligible ($y/\delta \simeq 0.4$) and where intermittency effects become important ($y/\delta \simeq 0.7$). This calculation is intended only to show that the turbulence decays at roughly the predicted rate in the curved region.

In the recovery region, the turbulence energy ($\frac{1}{2}q^2$) profiles show the same trends as the shear stress and u' profiles, a quick recovery near the wall followed by slow diffusion outwards. The profiles for u', $-\overline{uv}$, or q^2 seem to have barely started to recover toward equilibrium flat-plate structure at the last measuring station, approximately ten boundary layer thicknesses downstream of the end of curvature.

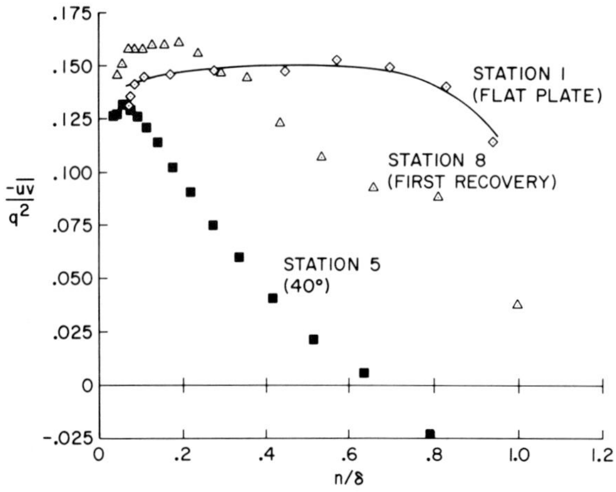

Fig. 11. Effect of curvature on structural coefficient

Since all the non-zero Reynolds stresses have been measured, the structural coefficient $a\ (-\overline{uv}/q^2)$ can be computed. a is plotted vs n/δ for three stations in Fig. 11. Upstream of the curvature, a is constant over almost all of the boundary layer and is about equal to 0.145. For the curved boundary layer, however, a becomes a strong function of position from the wall. In the recovery region, a very surprising trend is noticed. The recovery of the structural coefficient is very quick (compared to the very sluggish recovery of the $\overline{uv}$ and u'^2 profiles).

We have also computed, from the mean velocity profiles, the boundary layer mass flux per unit width at each of the measurement stations. In the flat regions, boundary layer mass flux non-dimensionalized on station 1 conditions is given by

$$\frac{\dot{m}}{(\rho U_\infty \delta)_1} = \frac{\delta - \delta^*}{\delta_1} \tag{8}$$

where

$$\delta^* = \int_0^\infty \left(1 - \frac{U}{U_\infty}\right) dn \tag{9}$$

and δ is the n value where U is 99% of freestream velocity. In the curved region, the corresponding formula is (to good approximation):

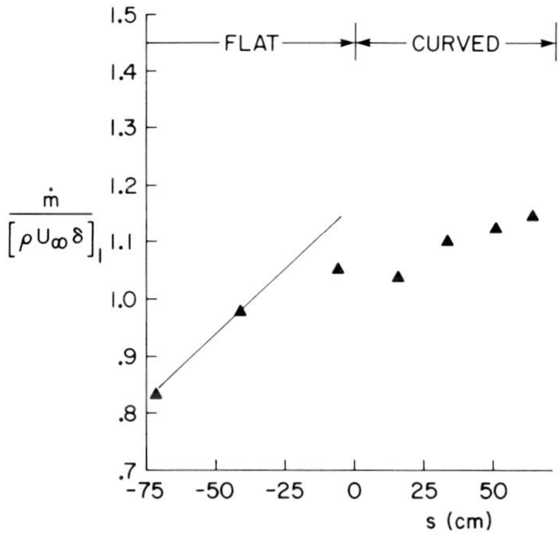

Fig. 12. Variation of boundary layer mass flux with streamwise distance

$$\frac{\dot{m}}{(\rho U_{pw}\delta)_1} = \frac{1}{(\delta)_1}\left(\delta - \delta^* - \frac{\delta^2}{2R}\right) \tag{10}$$

where

$$\delta^* = \int_0^\infty \left(1 - \frac{U}{U_p}\right) dn \tag{11}$$

and δ is the value of n where U is 99% of the potential flow velocity U_p.

In Fig. 12 the mass flux is plotted as a function of s. The line drawn on Fig. 12 has slope calculated from the known correlations for flat-plate boundary layers:

$$\theta = 0.037\, s \left(\frac{\nu}{U_\infty s}\right)^{0.2} \tag{12}$$

and

$$\delta = \frac{\theta}{0.097}, \quad \delta^* = \frac{\theta}{0.776}. \tag{13}$$

The rate of growth of the mass flux is a measure of the entrainment, and Fig. 11 shows that the entrainment rate is reduced in the curved region. Data on entrainment rate during recovery is not clear at this time.

From the shear stress and mean velocity profiles, it is possible to calculate the mixing length, and in Fig. 13 we have plotted mixing length profiles for the three stations used previously to represent flat, curved, and recovery profiles. The mean velocity gradients needed to calculate the mixing length were obtained by fitting a cubic spline close to the data points and then differentiating it. Fortunately, very little smoothing was needed to fit the spline. In the profiles of Fig. 13, the great reduction in the length scale of the turbulence which has been observed by all workers in this field is evident. Apart from this simple observation, however, there are many ways in which Fig. 13 may be interpreted.

One possible interpretation is that, even in the curved region, there is a layer close to the wall where the classical mixing length distribution ($l = \kappa n$) holds. In the curved region, how-

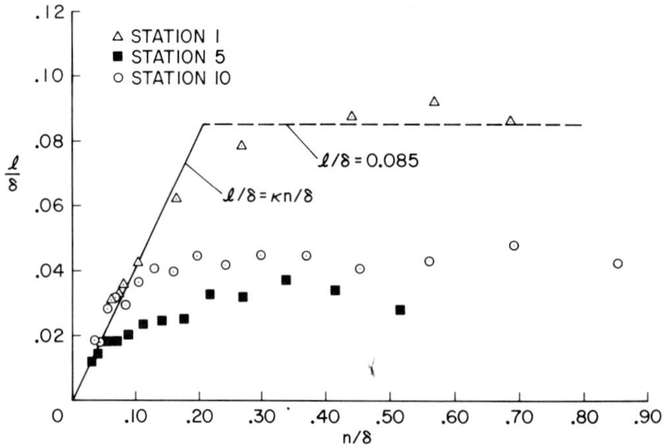

Fig. 13. Mixing length profiles at representative stations

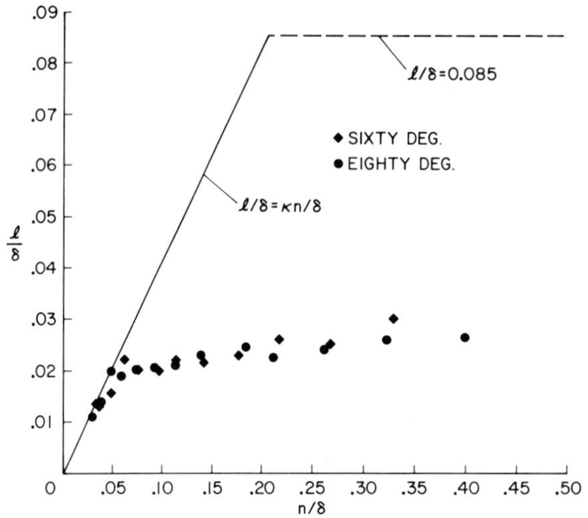

Fig. 14. Mixing length profiles at 60 and 80 degrees

ever, the mixing length becomes constant closer to the wall. This view is supported by Fig. 14, in which we have plotted l/δ vs n/δ at 60 and 80 deg. (Mixing lengths calculated from data taken far out in the boundary layer are not shown because, far from the wall, both the shear stress and velocity gradient are small and the computed mixing lengths scatter.) The plots show profiles which seem to have two regions: $l = \kappa n$ for n less than $0.07\,\delta$ and $l = 0.022\,\delta$ for n greater than $0.07\,\delta$. The recovery profiles suggest that the point at which the mixing length becomes constant moves slowly back toward the usual value of approximately $0.085\,\delta$. This picture is consistent with the idea that the main effect of the curvature is to confine the turbulent motions close to the wall and to destroy the large-scale motions. The slow regrowth of the length scale in recovery is probably associated with the slow reappearance of large-scale structures.

All the data reported above were taken on the rig shown schematically in Fig. 1. The main new result is the slow recovery back to flat-wall conditions. As part of a parallel experimental program on the effects of curvature on heat transfer, a second experimental facility has been constructed. It is very similar to the rig in Fig. 1, except that the recovery section is about twice as long. At the time this paper is being written, only mean velocity data are available from this new rig. Figure 15 is a plot similar to Fig. 4, showing a comparison of new rig

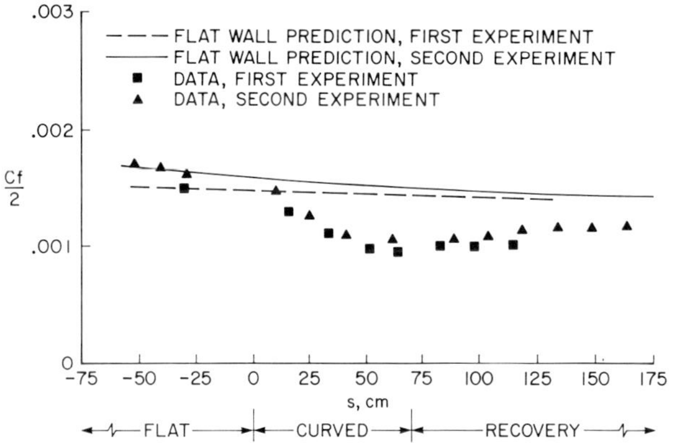

Fig. 15. Comparison of skin friction data for two similar experiments

and old rig data. The free stream speed for the second experiment is 5% less than for the first experiment; the inlet boundary layer is slightly thinner, but in all other respects the experiments are identical. Fig. 15 also shows, for purposes of comparison, the predicted flat-wall skin friction for the two experiments. The data are in good agreement, and the curvature effects are *still* appreciable at the end of the longer recovery section.

Summary and Discussion

This paper presents results of experiments designed to isolate the effects of convex surface curvature on turbulent boundary layers. The data show how a flat-wall boundary layer changes to a curved-wall boundary, and indicate how the reverse process takes place.

As noted in experiments of others, the mean velocity profiles follow the law of the wall, although not out as far from the wall as in a flat-wall boundary layer. The skin friction shows a sharp dropoff, which is quantitatively consistent with previous theories. From the measured turbulence profiles, it seems clear that, after the onset of curvature, the outer portion of the boundary layer quickly splits into two regions. There is an inner layer characterized by turbulence with non-zero mean turbulent shear stress and non-zero mean turbulence kinetic energy, and an outer layer with insignificant mean shear stress but significant turbulence kinetic energy. In the outer layer, the three normal stresses become nearly equal in value, and the turbulence kinetic energy there decays at the rate quantitatively consistent with the decay of isotropic turbulence. This suggests that the outer layer has been cut off from the influence of the wall. Measurements of the dissipation rate show that it, too, is reduced, but not as much as in the shear stress.

In the curved region the structural coefficient a becomes a strong function of distance from the wall in the region where $(-\overline{uv}) \neq 0$. This may be an indication that the region characterized by large $-\overline{uv}$ has an intermittent structure of its own.

Although the effects of curvature develop quickly, they persist many boundary layer thicknesses downstream of the end of the curvature. Even 20 δ downstream of the end of the curved wall, the skin friction continues to be much lower than would be expected for a flat-wall. In addition, the turbulence profiles (although there is clearly some recovery process under way as the fluid moves downstream) do not begin to resemble flat-wall profiles for at least ten boundary layer thicknesses.

Acknowledgments. This study is supported by NASA-Lewis Laboratories under Grant NSG-3124. It is part of an ongoing research project to examine the effects of surface curvature on turbulent convective heat transfer. The authors would also like to acknowledge the contributions made to this project by Russell Westphal and Paul Youssefmir, who assisted in the acquisition and analysis of the data reported here.

References

1. P. Bradshaw: Effects of streamline curvature on turbulent flow. AGARD-ograph **169** (1973)
2. M. E. Crawford, W. M. Kays: "STAN5 – A Program for Numerical Computation of Two-Dimensional Internal and External Boundary Layer Flows". NASA CR-2742 (1976)
3. J. P. Johnston, S. A. Eide: Turbulent boundary layers in centrifugal compressor blades: Prediction of the effects of surface curvature and rotation. J. Fluids Eng. **98**(I), 374–381 (1976)
4. P. S. Klebanoff: "Characteristics of Turbulence in a Boundary Layer with Zero Pressure Gradient". NACA Rpt. **1247** (1955)
5. R. N. Meroney, P. Bradshaw: Turbulent boundary layer growth over a longitudinally curved surface. AIAA J. **13**(11), 1448–1453 (1975)
6. R. J. Moffat, W. M. Kays: The turbulent boundary layer over a porous plate: Experimental heat transfer with uniform blowing and suction. Int. J. Heat Mass Transfer **11** (11), 1547–1565 (1968)
7. B. R. Ramaprian, B. G. Shivaprasad: Mean flow measurements in turbulent boundary layers along mildly curved surfaces. AIAA J. **15**(189) 189–196 (1977)
8. B. R. Ramaprian, B. G. Shivaprasad: The structure of turbulent boundary layers along mildly curved surfaces. J. Fluid Mech. **85**, Part 2, 273–303 (1978)
9. R. M. C. So, G. L. Mellor: Experiment on convex curvature effects in turbulent boundary layers. J. Fluid Mech. **60**, Part 1, 43–62 (1973)
10. W. C. Reynolds: Computation of turbulent flows. Annu. Rev. Fluid Mech. 8 (1976)
11. A. J. Smits, S. T. B. Young, P. Bradshaw: The effects of short regions of high surface curvature on turbulent boundary layers. (To be published in J. Fluid Mech.)

Diffusion from a Line Source in a Turbulent Boundary Layer with Transfer to the Wall

M. Belorgey, A. D. Nguyen, M. Trinite

Laboratoire de Thermodynamique (L.A. au C.N.R.S. n° 230)
Faculté des Sciences et des Techniques de Rouen
B.P. 67 76130 Mont-Saint-Aignan France

Abstract

A thermal field generated by a line source inside a turbulent boundary layer is studied when "flux at the wall" problems exist. Characteristics of the field are presented, and two methods of calculation are developed in order to determine the wall flux $\overline{\phi_p(x)}$ (x from the source). Far downstream of the source, the thermal field is sufficiently homogeneous to allow the prediction of $\overline{\phi_p(x)}$ by using a gradient type transport equation. Using in this case a calculation method based on *Spalding*'s theory calculated results are in good agreement with experiments. On the other hand, inside the region where the thermal field reaches the plate, high intermittency occurs in its edge and the calculation required the renewal sublayer theory formulated by *Danckwerts* and completed by *Hanratty*. In this case, predictions could be obtained which agreed with experimental results by developing experimental methods to obtain the contact time of hot structures.

Nomenclature

- $a = \lambda/\rho C_p$ Thermal diffusivity [m²/s]
- $b = \sqrt{\lambda \rho C_p}$ effusivity [J/m² K s^{1/2}]
- C_p Specific heat at constant pressure of fluid [J/kg K]
- Q Heat quantity for unit area [J/m²]
- T Fluid temperature [K]
- t Time [s]
- U Fluid local velocity [m/s]
- x Distance downstream from the source [m]
- y Distance from and normal to the wall [m]
- y_s Source height from the wall [m]
- θ Temperature difference $T - T_0$ [K]
- λ Thermal conductivity [W/mK]
- ν Kinematic viscosity [m²/s]
- ρ Density [kg/m³]
- τ Contact time [s]
- ϕ_p Wall heat flux [W/m²]
- $U_* = U_o \sqrt{C_f/2}$ Friction velocity [m/s]
- $y^+ = y \dfrac{U_o \sqrt{C_f/2}}{\nu}$ Normalized distance
- $\theta^+ = \dfrac{\rho C_p \bar{\theta} U_o \sqrt{C_f/2}}{\bar{\phi}_p}$ Normalized temperature

Greek Symbols

- δ^0 Boundary layer thickness at source level [m]
- δ_1^0 Displacement thickness at source level [m]

Subscripts

- o Corresponding to free flow
- p Corresponding to the wall
- s Corresponding to the source
- m Maximal

Superscripts

$\bar{}$	Mean value
$'$	Fluctuating component
$+$	Dimensionless

Introduction

Dispersion of a passive contaminant by turbulence is very important as much for practical aspects (atmospheric pollution, chemical reaction, dispersion, mixing . . .) as for fundamental intrinsic aspects (tracer).

Few experiments have been performed in a boundary layer under the controlled conditions of the laboratory. Often only mean values of contaminant concentrations (heat or mass) were measured (*Skramstad* and *Schubauer* [1], *Wieghart* [2], *Poreh* and *Cermak* [3]. Investigations with fluctuation measurements have been made by *Shlien* [4] and *Paranthoen* [5].

More recently this diffusion problem has been approached by considering a perfect conductive plate (*Bélorgey* [6]) with a heated line source normal to the flow and parallel to the wall. The main aim of this work was to relate heat exchange mechanisms between the flow and the plate with the characteristics of the scalar and dynamic field. Determination of the local heat transfer rate to the wall is not possible by classical methods in the region where the temperature field reaches the wall intermittently. Precise investigations of turbulent heated structures are then necessary to explain the large value of flux in such regions (frequency, probability distribution of pulse and gap widths of the temperature signal, temperature level). Applications of this transfer aspect are numerous at the present time (turbulent combustion with heat transfer from a flame front to a wall, chemical reactions . . . .)

Analysis of the mean field in the case of an adiabatic and in the case of a conductive plate will be reported on the first part of this paper. The second part will consider fluctuation measurements and the last part to the wall flux. Surface renewal and penetration theory will be adapted to the latter.

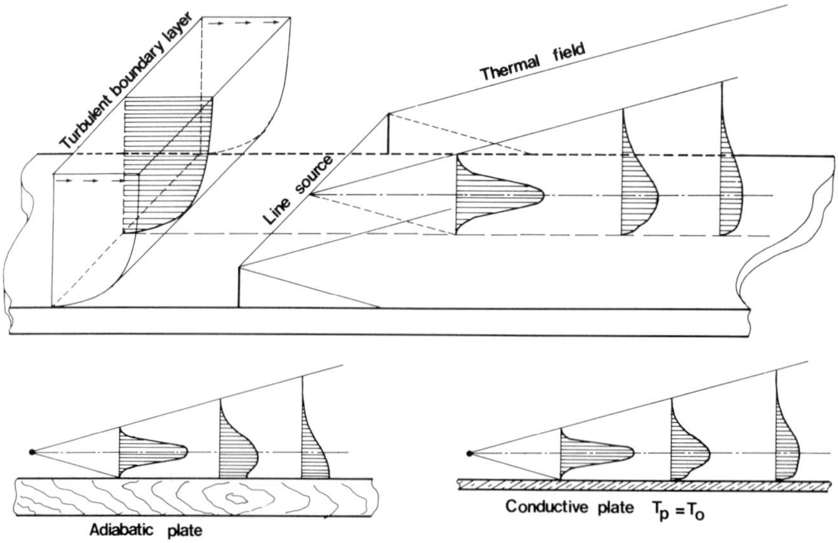

Fig. 1. Experimental configuration and boundary conditions

1. Longitudinal Evolution of the Thermal Field

Non-dimensional parameters characterising the different cases investigated, are summed up in the Table 1.

Distances are in relation with δ_1^0, which is the displacement thickness of the boundary layer at the source point.

The two cases which have been studied in different wind tunnels are:

adiabatic case:

$U_0 = 28.5$ m/s $\quad \delta_0 = 8.5$ mm $\quad \delta_1^0 = 1.32$ mm

conductive case:

$U_0 = 7.35$ m/s $\quad \delta_0 = 110$ mm $\quad \delta_1^0 = 12.5$ mm

It has been possible to study the thermal field nearer the source in the second case.

When the plate is adiabatic, the mean temperature profile shows, for the considered distance from the source, a maximum that quickly reaches the vicinity of the plate and stays

Table 1. Parameters of investigated cases

	Source height		Distance downstream the source
	y_0^+	y_0/δ_1^0	$\Delta x/\delta_1^0$
Adiabatic plate	67	0.62	15–125
Conductive plate	125	0.50	1.6–40

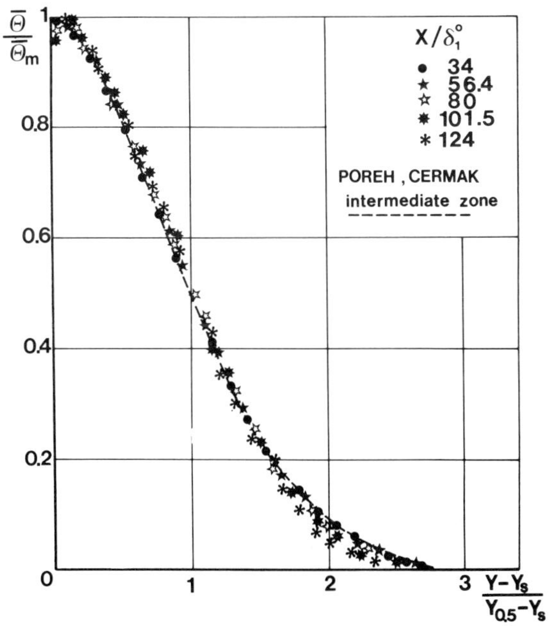

Fig. 2. Dimensionless profiles of mean temperature (adiabatic case)

there with increasing distance from this source. Similar profiles to those of a linear source located at the wall are quickly reached.

These results are presented in Fig. 2: The difference in temperature is normalized with the peak mean temperature difference and the height is normalized with the height $y_{0.5}$ for which

$$\frac{\overline{T} - T_0}{\overline{T}_m - T_0} = 0.5$$

In the case of conductive plate, the temperature peak remains nearly centered on the source height for short distances downstream. It cannot reach the wall however, because of the particular temperature gradient that settles in the lower part of the thermal field (Fig. 3). In this case three phases can be distinguished in the mean thermal field evolution (Fig. 4):

a) before the thermal field reaches the plate, the profile remains centered on source position;
b) when the thermal field reaches the plate the profile remains largely centered on the source position, except in the close vicinity of the wall. Wall boundary conditions involve a "connection" profile in the lowest part of the thermal field. This "connection" profile increases with the distance from the source;
c) in this phase, the "connection" profile extends to the maximum of the temperature profile and significantly influences it.

Phase b) corresponds to the stage where absorption by the plate is most influential.

With increasing distance from the source, the temperature gradient imposed by the wall slowly dominates the preexistent gradient, and the "connection" profile evolves to a boundary layer type profile, as is shown by the results in Fig. 5. This figure shows evolution of θ^+ vs y^+ for different locations behind the source.

Notice that far from the source, the points corresponding to the lower part of the thermal field, fit on some "Law of the wall" type curve.

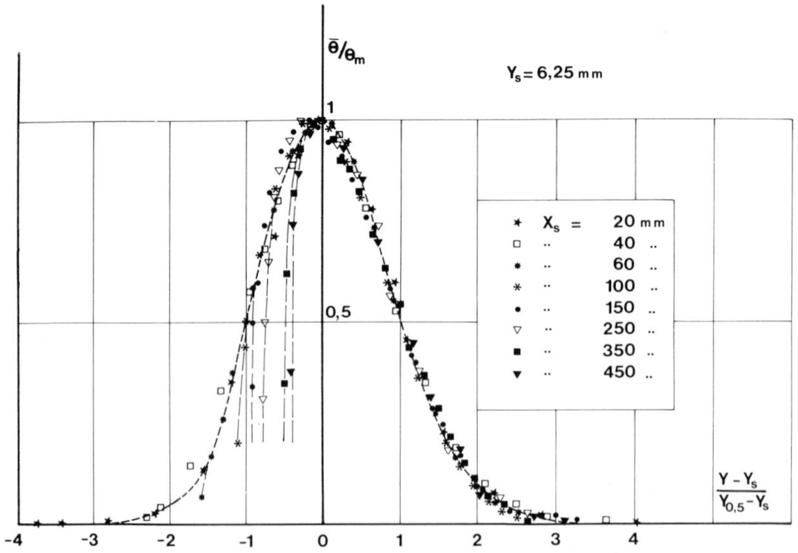

Fig. 3. Dimensionless profiles of mean temperature (conductive case)

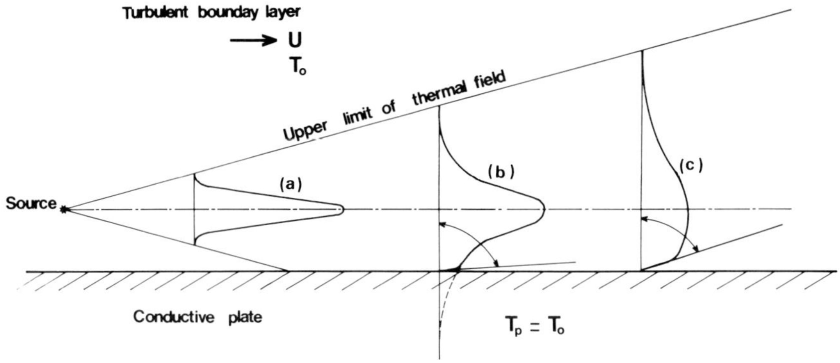

Fig. 4. The phase of the evolution of the thermal field

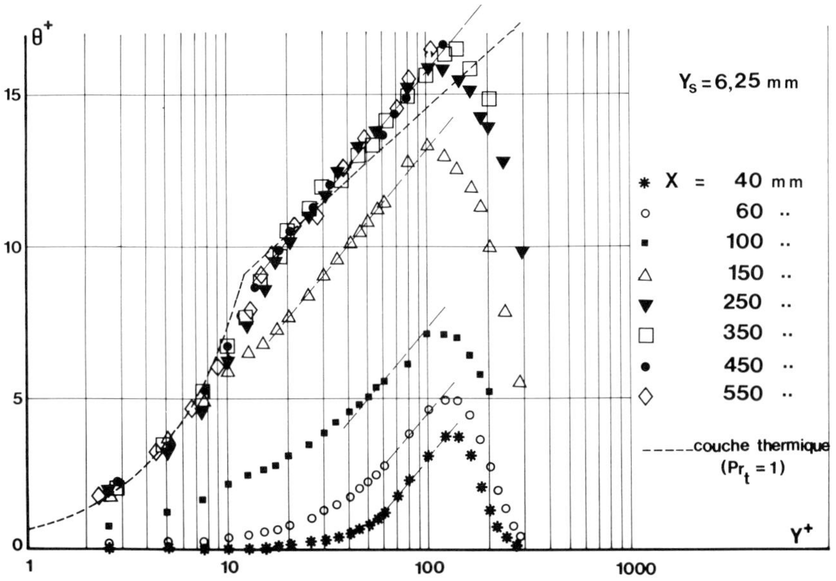

Fig. 5. Evolution of θ^+ vs y^+ for different distances downstream of the source

Within the sublayer $\theta^+ = y^+ \mathrm{Pr}$

Within the turbulent layer $\theta^+ = A \log y^+ + B$

RMS temperature fluctuations have been measured downstream of the source. In both cases longitudinal evolution of $\overline{\theta'^2}$ and the existence of maxima can be explained by considering the importance of the production term in temperature variance equation [7].

On Fig. 6 the results corresponding to the conductive plate are presented. RMS values are normalized with the local mean temperature of the considered profile, and are plotted against non-dimensional height.

It must be noticed that an important thermal turbulence intensity exists on the scalar field sides near the source. In particular, this significant intensity near the plate in connection with the intermittent nature of the temperature signal allows the heat flux at the wall within this region to be explained.

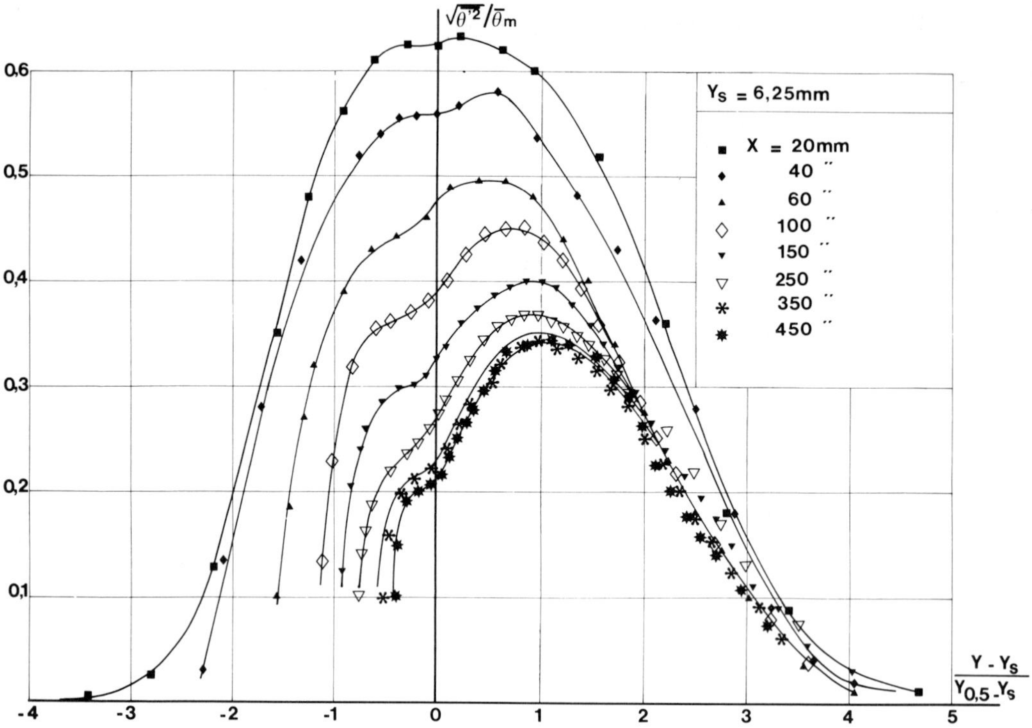

Fig. 6. Evolution of temperature RMS values for different distances downstream of the source

2. Determination of the Parietal Flux

Experimentally, the mean heat flux at the wall $[\overline{\phi_p(x)}]$ has been determined from mean temperature values by two different methods; by the integral method (more precise within the region where the thermal field initially reaches the plate and the temperature longitudinal variation is very significant), and by measurement of the temperature gradient at the wall (determination of this gradient is more precise for sections far downstream of the source).

$$\overline{\phi_p(x)} = \frac{d}{dx}\left(\int_0^\infty \rho C_p \overline{U}\overline{\theta} dy\right) = \lambda \left(\frac{\partial T}{\partial y}\right)_{y=0}$$

In the first method, the longitudinal derivative of the turbulent heat flux

$$\left(\frac{d}{dx}\int_0^\infty \rho C_p \overline{u'\theta'} dy\right)$$

has been neglected.

Study of $\overline{\phi_p(x)}$ far Downstream of the Source

The results concerning longitudinal evolution of the mean temperature values lead to the hypothesis that downstream of the source in the lower part of the thermal field, "a pseudo local thermal boundary layer" increases in thickness.

Characteristics of this boundary layer depend on the considered section, and are:

$U_\infty = U_0$: free flow velocity

$\theta_\infty = \theta_m(x)$: mean temperature maximum of the considered profile.

Thus, the prediction of the heat flux at the wall, far downstream of the source can be made using a thermal-boundary-layer-type method, in which the fictive origin X_o is adjusted [8].

The determination of X_o is not easy. Considerations that lead to its definition are the following:

- the transfer can be considered as a boundary layer type transfer only if the turbulent mixing at the wall is developed enough, that is to say that a heat loaded particulate will have forgotten the characteristics of its origine.
- the distance X_o must depend on the source height.

Thus, it seems possible to connect X_o to the "memory" distance associated with the "memory time" defined by *Batchelor* [9].

$$\left(t_1 = \frac{y_s}{U_*}\right).$$

A calculation method was developed based on *Spalding*'s formulation [10], in which the mean heat flux at the wall far downstream of the source is given by the following relation:

$$\overline{\phi_p(x)} = \rho C_p \overline{U}_o \overline{\theta}_m S_T \left[(x - X_o)^+, \Pr\right] \frac{\sqrt{C_f/2}}{\Pr}$$

where (S_T) is Spalding's function, and $U_o, \overline{\theta}_m$ are characteristics of the local pseudo-thermal boundary layer.

Pr is the Prandtl number, C_f is the friction coefficient.

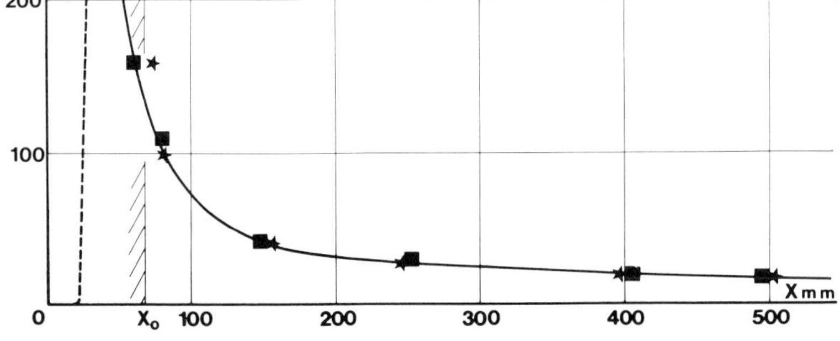

Fig. 7. Flux at the wall: longitudinal evolution

Results given by the calculation defined above are compared to experimental results, in Fig. 7.

The prediction method for the flux just defined clearly cannot be used within the region where the thermal field initially reaches the plate, (i.e. within the region of large intermittency defined in Sect. 2). Due to the intermittency of the thermal field, it is therefore not recommended to define $\overline{\phi_p(x)}$ using mean values within this region. It is the region where the mean flux at the wall is the most important.

3. Flux Study Within the Region Where the Thermal Field Initially Reaches the Wall

It is possible in this case to approach the calculation by means of the renewal surface theory, based on *Fage* and *Townsend*'s [11] and *Popovich* and *Hummel*'s [12] observations.

This model proposed by *Higbie* [13] and completed by *Dankwerts* in the context of turbulent mass transfer associated with a shear surface between two fluids, was extended to a contact surface between a solid and a fluid by *Einstein* and *Li* [14], and by *Hanratty* [15].

These authors proposed a three phases mechanism for the sublayer.

1) Sudden movement of a fluid mass from the turbulent region core towards the wall. ("sweep phase").
2) Very short period of stability for some structures within the wall region.
3) Return of the fluid mass towards the turbulent region.

Transfer of heat between the fluid and the wall is performed during phase 2) by molecular diffusion during the hot fluid structure contact time.

The equation of conduction inside this structure (considered as homogeneous with constant physical properties, and having an initial uniform temperature) can be written as:

$$\frac{\partial T}{\partial t} = a \frac{\partial^2 T}{\partial y^2} \qquad a = \frac{\lambda}{\rho C_p} \text{ thermal diffusivity}$$

Taking into account experimental conditions the problems can be considered as a problem of a semi-infinite wall whose entrance face is suddenly raised to a constant temperature T_o (the isothermal plate of copper has an effusivity ($b_c = \sqrt{\lambda \rho C_p}$) 6,000 times larger than air, therefore it keeps a constant temperature, independant of the structure temperature).

Thus, the solution of the diffusion equation leads to the following expression for the instantaneous heat flux through the contact surface:

$$\phi_p = \frac{b}{\sqrt{\pi}} \cdot \frac{\theta}{\sqrt{t}} \qquad \theta = T - T_o$$

It is then possible to calculate the mean flux through the contact surface for a temperature structure in contact with the plate during a time τ:

$$\phi(\theta, \tau) = \frac{2b}{\sqrt{\pi}} \cdot \frac{\theta}{\sqrt{\tau}}$$

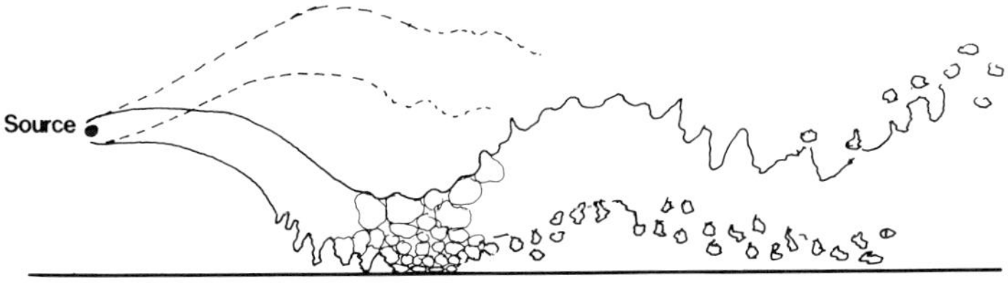

Fig. 8. Schematic of hot fluid stream

Therefore an expression can be written for the mean flux at a distance x from the source:

$$\bar{\phi}_p = \int_0^\infty \int_0^\infty \phi_p(x, \theta, \tau)\, p(\theta, \tau)\, d\theta d\tau$$

where $p(\theta, \tau)$ is the probability density that at a distance x from the source a structure with the temperature compared to the plate temperature, remains in contact with it during a time τ.

Until now, this theory was only applied to uniform temperature distributions, as in the case of an established thermal boundary layer. Thus the studies were mainly centered on the probability density of structures contact [$p(\tau)$]. In the particular problem of a source thermal field, however a distinction must be made between fluid particles, hot structures and dynamic structures.

That necessitates the use of a Lagrangian description for the movement of heat loaded particles downstream of the source.

At the source, the fluid particles marked with heat are all close to be other; the difference of the initial scale being of the same magnitude as the wire source diameter ($\overline{\Delta y_0} \simeq d$). Since the heat source is inside the wall law region of the boundary layer, the mean eddy size is large. In this case, the motion of these particles is a motion as a whole, characterized by a single particle of the *Batchelor*'s theory [16].

Thus the totality of these particles, behaves like a "hot fluid line stream" that waves downstream of the source, under influence of big eddies. In this case, the mean temperature profile corresponds to the passing probability density in this hot fluid stream at a given point.

This hot fluid line stream expands progressively (molecular diffusion) and when it enters a region of the boundary layer where the eddy size has the same magnitude as its own thickness it is dissociated by the eddies associated with the velocity gradient (two-particle aspect of Batchelor's theory). These structures, dissociated into smaller structures, are not seperated from one another by a sufficient distance to be thermally distinguishable with a hot wire.

Thus, within the region where the thermal field initially reaches the plate, the hot structures going to the wall, have portions of the hot fluid stream core, captured by small dynamic structures. Far removing from the source, the hot fluid stream core expands and dissociates itself more and more, thus the thermal field identifies itself more and more with the dynamic field, and in the vicinity of the wall, every dynamic structure is then loaded with heat and the thermal boundary layer aspect of the phenomenon, is observed.

Calculation of Heat Transfer at the Wall

- When the "hot fluid stream core" reaches the wall, two assumptions are used to calculate heat flux.
- This fluid stream consist of small structures which have all the same initial temperature when the contact occurs (the temperature of the hot fluid stream core.)

– Inside the hot fluid stream core a renewal phenomenon occurs for the small structures in parietal contact. Every small structure being in contact during a time τ', while the phenomenon occurs during the hot fluid stream contact time τ.

Consequently:

– The mean flux during the contact time τ of a small structure of temperature θ_k, can be expressed by:

$$\phi_p(\theta_k, \tau') = \frac{2\,b\theta_k}{\sqrt{\pi}\sqrt{\tau'}}$$

– Heat exchanged during the transit time τ of the corresponding hot fluid stream core is:

$$Q(\theta_k, \tau', \tau) = \frac{2\,b\theta_k}{\sqrt{\pi}\sqrt{\tau'}}\,\tau$$

Thus the mean flux at a distance x from the source is equal to the amount of heat yielded during unit time by all hot fluid stream cores reaching the plate at the source distance. That is to say

$$\overline{\phi_p(x)} = \sum_{\theta_k} Q(\theta_k, \tau, \tau')\,N(\theta_k)$$

where $N(\theta_k)$ is the number of hot fluid streams of temperature θ_k in contact with the plate during unit time.

Such a calculation method requires knowledge of τ, τ', $N(\theta_k)$. These parameters were not available in the present measurements however in order to verify the hypothesis, the mean contact time for different temperature levels have been determined.

The measurements of τ' cannot be obtained for two reasons:

– within the dynamic boundary layer the time τ' is about 100 µs (see below, and is therefore indistinguishable from noise.
– close to the wall where velocities are low the signal is attenuated by the time constant of hot wires used for the measurements [17].

The measurements were performed with a DISA anemometer equiped with a temperature bridge.

Two different wires were used (1.2 µm and 0.6 µm).

Inside the intermittency region, the temperature signal appears as illustrated in Fig. 9 and on Photos 1 and 2. It is a succession of peaks corresponding to hot fluid stream core crossings, followed by long pauses corresponding to dynamic structures of unheated fluid.

From the output of the DISA anemometer the signal is connected to a Hewlett Packard timer (5345 A) provided with two triggers A and B of variable level, for the start and the stop of the time measurement. (To avoid errors due to noise the trigger levels are different.)

By uncoupling the signal levels with regard to the trigger levels, it is possible to measure the time during which the signal level is greater or equal to a fixed value (θ_i). (Fig. 9).

Recording of the mean values allows the contact time probability density of the hot fluid stream cores which have a temperature level equal or greater to a given value θ_i to be obtained.

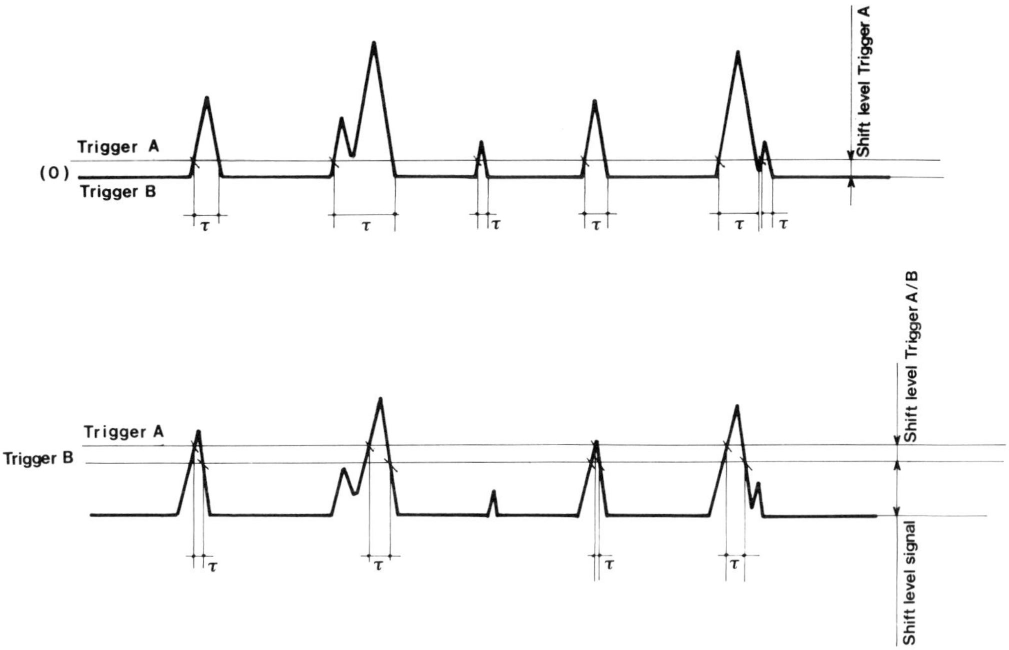

Fig. 9. Measurement method of the structures' contact time according to their temperature level

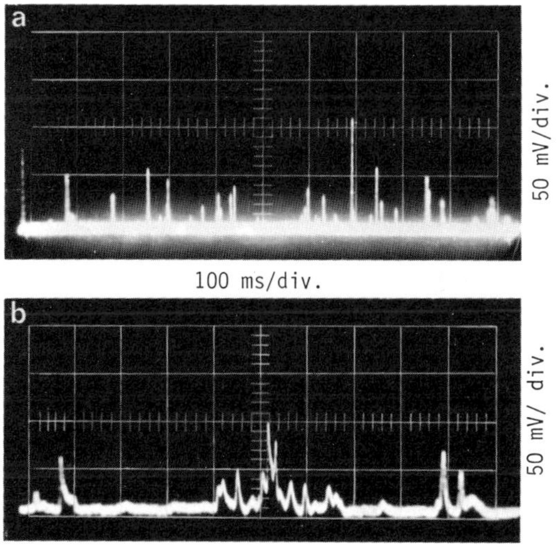

Photo 1. Visualization of the temperature signal: x = 40 mm, y = 0.45 mm (two sweeps)

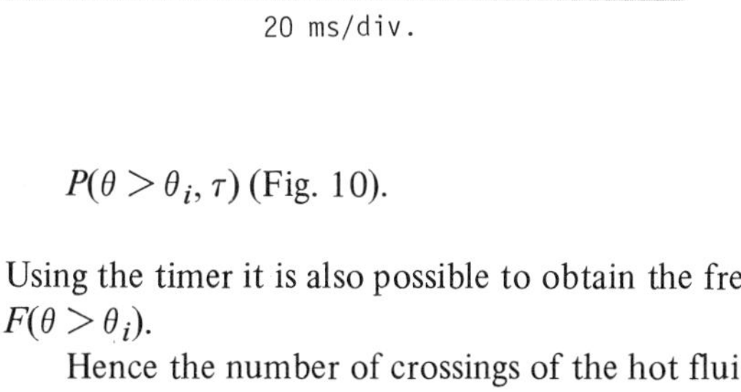

Photo 2. Visualization of the temperature signal: x = 70 mm, y = 0.125 mm (one sweep)

$P(\theta > \theta_i, \tau)$ (Fig. 10).

Using the timer it is also possible to obtain the frequency of occurance of the detected signals: $F(\theta > \theta_i)$.

Hence the number of crossings of the hot fluid stream cores having a temperature θ_k or greater is:

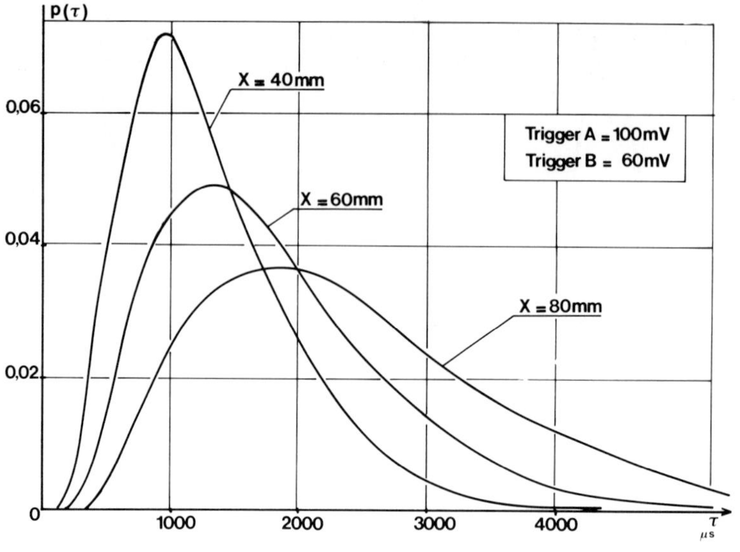

Fig. 10. Probability density of the structures' contact time

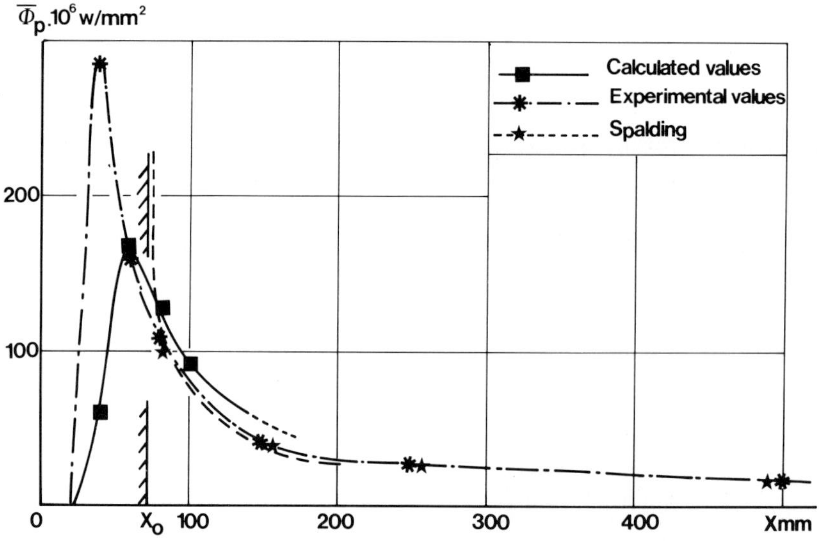

Fig. 11. ϕ_p longitudinal evolution inside the region where the thermal field reaches the plate

$$N(\theta_k) = F(\theta > \theta_i) - F(\theta > \theta_{i+1}) \quad \text{with} \quad (\theta_k) = \frac{\theta_i + \theta_{i+1}}{2}$$

As we are not able to determine exact value. An approximative value of was determined corresponding to the following conditions:

- The scalar field is carried by the dynamic field, i.e. hot structures in the neighbourhood of the plate are dynamic structures loaded with heat. Consequently hot structures' contact time is dynamic structures' contact time.

– If in the vicinity of the plate, the turbulence is presumed tobe nearly isotrope, it is possible to use Kolmogoroff's theory to define the order of magnitude of the structures' size in this region and consequently a crossing time (or contact time).

$\eta = \epsilon^{-1/4} \nu^{3/4}$
η is Kolmogoroff's microscale
ϵ is the turbulent dissipation rate
Hence $\tau' \simeq \eta \bar{U}$

Applications corresponding to the present dynamic boundary layer leads to times of τ' about 100 μs.

This magnitude is consistent with the flux far downstream of the source taking into account the renewal sublayer in stable rate.

On Fig. 11 the data given by the calculation is compared to the experimental results.

In the region where the thermal field reaches the wall, ϕ_p max. is distinct, and the magnitude of the prediction agrees very well.

Concluding Remarks

For this problem of transfer to the wall in a boundary layer, two regions must be distinguished.

Far downstream of the source when the turbulent mixing is well established, prediction of parietal flux can be obtained by boundary-layer type methods.

Near the source in the intermittent region of the scalar field, these methods cannot be used. A complete statistical analysis of the signal with conditional sampling associated with the application of the renewal theory, provides a good approximation to parietal flux problems.

The problem of intermittent transfer can appear in numerous applications such as jet edges and flame edges.

References

1. Skramstad, Schubaüer: The application of thermal diffusion to the study of turbulent air flow. Phys. Rev. **53**(11), 927 (1938)
2. Wieghardt: Über Ausbreitungsvorgänge in turbulenten Reibungsschichten. Z. Angew. Math. Mech. **27**, No. 11/12 (1948)
3. Poreh, Cermak: Study of the diffusion from a line source in a turbulent boundary layer, Int. J. Heat Mass Transfer 7 (1964)
4. Shlien, Corrsin: Dispersion measurements in a turbulent boundary layer. Int. J. Heat Mass Transfer **19** (1976)
5. Paranthoën: "Etude expérimentale de la diffusion de la chaleur en aval d'une source linéaire placée dans une couche limite turbulente". Thèse de Doctorat ès Sciences, Rouen (1978)
6. Belorgey: "Etude du flux à la paroi à l'aval d'une source linéaire de chaleur placée dans une couche limite turbulente". Thèse de Doctorat ès Sciences, Rouen (1978)
7. Paranthoën, Trinite: Profils adimensionnels de température en eval d'une ligne source de chaleur. C. R. Acad Sci. Ser. A: **276**, 1473–1476 (1973)
8. Belorgey, Trinite: Etude du flux à la paroi à l'aval d'une source linéaire de chaleur placée dans une couche limite turbulente. Lett. Heat Mass Transfer **6**, No. 1 (1979)
9. Batchelor: "Diffusion from sources in a turbulent boundary layer". Arch. Mech. Stosow. **3**, 16 (1964)
10. Spalding: "Contribution to the theory of heat transfer across a turbulent boundary layer". Int. J. Heat Mass Transfer 7, 443–461 (1964)

11. Fage, Townsend: An examination of turbulent flow with an ultramicroscope. Proc. Roy. Soc. Ser. A: **135**, No. 828 (1932)
12. Popovich, Hummel: Experimental study of the viscous sublayer in turbulent pipe flow. AIChE J. 854 (Sept. 1967)
13. Higbie: Trans. Amer. Inst. Chem. Eng. **31**, No. 65 (1935)
14. Einstein, Li: The viscous sublayer along a smooth boundary. J. Eng. Mech. Div. **82**, No. 293 (1956)
15. Hanratty: Turbulent exchange of mass and momentum with a boundary. AIChE J. **2**, No. 3 (1956)
16. Batchelor: "The Theory of Homogeneous Turbulence". Cambridge Monographs on Mech. and Appl. Path. (Cambridge University Press 1967)
17. Millon, Paranthoën, Trinite: Influence des échanges thermiques entre le capteur et ses supports sur la mesure des fluctuations de température dans un écoulement turbulent. Int. J. Heat Mass Transfer **21**, No. 1 (1978)

Turbulence Structure Related to the Heat Transfer in a Turbulent Boundary Layer with Injection

M. Senda, K. Suzuki, and T. Sato
Department of Mechanical Engineering,
Kyoto University, Kyoto, Japan

Abstract

The mechanism of turbulent heat transfer can be made clear through the measurement of temperature fluctuations and related correlations. For the purpose, an analog circuit solving a set of simultaneous thermal equations for the cold and hot wires was applied to a turbulent boundary layer with uniform injection of heated air and the effects of injection on turbulence structure was studied. The main effects of injection on turbulence structure can be summarized as follows.

(a) Turbulent intensities of both velocity and temperature fluctuations increase in the boundary layer with increasing the injection rate except very near the wall. (b) One-point double correlation coefficients are constant and depend slightly on the injection rate. (c) Bradshaw's structure functions are almost constant in the boundary layer, independent of the injection rate. (d) Integral length scales of streamwise velocity and temperature fluctuations decrease with increasing the injection rate.

Nomenclature

a	Bradshaw's structure function, $-\overline{uv}/q^2$
a_θ	Bradshaw's structure function, $\overline{v\theta}/\theta' \sqrt{\tau/\rho}$
B	Blowing parameter, $2F/c_f$
c_f	Skin friction coefficient
E_u	Spectrum of velocity fluctuation
E_θ	Spectrum of temperature fluctuation
E_{uv}	Spectrum of Reynolds shear stress
$E_{v\theta}$	Spectrum of turbulent heat flux
F	Injection rate, $(\rho V)_w/(\rho U)_e$
k_1	Wave number, $2\pi n/U$
n	Frequency
$\mathrm{Pr_t}$	Turbulent Prandtl number
q^2	Turbulent kinetic energy
R	Time scale ratio, $(\epsilon/q^2)/(\epsilon_\theta/\overline{\theta^2})$
Re_δ	Boundary layer thickness Reynolds number
Re_θ	Momentum thickness Reynolds number
T	Mean temperature
t	Time
Δt	Delay time
U, V	Mean velocity components in x and y directions
u, v	Fluctuating velocity components in x and y directions
x, y	Rectangular coordinates along and normal to flat plate
δ	Velocity boundary layer thickness
δ_T	Thermal boundary layer thickness
ϵ	Dissipation rate of velocity fluctuation
ϵ_θ	Dissipation rate of temperature fluctuation
θ	Fluctuating temperature

Λ_u	Turbulent integral scale of velocity
Λ_θ	Turbulent integral scale of temperature
ν	Kinematic viscosity
ρ	Density
τ	Shear stress
Φ	Parameter of velocity defect

Subscripts and Superscripts

e	Value at free stream
w	Value at wall
0	Value without injection
$'$	Root mean square value
$\overline{}$	Time average

1. Introduction

The turbulent boundary layer with injection and/or combustion is one of the important flows in engineering, as seen in transpiration cooling of turbine blades or in ablation cooling of high speed reentry vehicles. As it is necessary to know exactly the reduction of heat transfer or skin friction due to injection, there have been many investigations [1–7] of this flow. The mechanism of turbulent heat transfer can be made clear through the measurement of temperature fluctuations and related correlations, but there have been only a few publications on the turbulence structure in the transpired turbulent boundary layer; *Peterson* and *Isaacson* [8], and *Andersen* et al. [9] in an isothermal boundary layer with injection, *Pimenta* et al. [10] in a non-isothermal boundary layer with and without injection on a rough wall, *Wooldridge* and *Muzzy* [11] in a boundary layer with injection and combustion. An attempt has been made in this study to study the turbulence structure related to the heat transfer in the boundary layer with uniform injection of heated air.

Even in the case when the added heat can be treated as a passive contaminant, the signal from the hot wire anemometer is usually a mixture of the velocity and the temperature fluctuation signals. Separation of pure velocity and temperature signals from the mixed signal, therefore, is essential to obtain the temperature-normal velocity correlation, i.e. the turbulent heat flux $\overline{v\theta}$. For the purpose, an analog circuit, which enables one to measure the velocity and the temperature fluctuations simultaneously, was constructed and was applied to the turbulent boundary layer presently being studied. In this paper the results of the measurement of turbulent quantities, including the Reynolds shear stress and the turbulent heat flux, are described and the effects of injection on turbulence structure are discussed.

The following quantities are measured in this study. (a) The turbulent intensities of streamwise and transverse velocity fluctuations $\sqrt{\overline{u^2}}$ and $\sqrt{\overline{v^2}}$, and the intensity of temperature fluctuations $\sqrt{\overline{\theta^2}}$. (b) One-point double correlations $-\overline{uv}$, $-\overline{u\theta}$ and $\overline{v\theta}$, and their coefficients $-\overline{uv}/u'v'$, $-\overline{u\theta}/u'\theta'$ and $\overline{v\theta}/v'\theta'$. Here, $-\overline{uv}$ is the Reynolds shear stress and $\overline{v\theta}$ is the normal turbulent heat flux. (c) The auto-correlations $\overline{u(t)u(t+\Delta t)}$ and $\overline{\theta(t)\theta(t+\Delta t)}$, and their one-dimensional spectra $E_u(k_1)$ and $E_\theta(k_1)$. (d) The time-averaged velocity and temperature profiles and the turbulent Prandtl number, defined as $\overline{uv}(\partial T/\partial y)/\overline{v\theta}(\partial U/\partial y)$.

2. Experimental Procedure

2.1. Experimental Apparatus

The experimental apparatus presently used is the same one as that used previously in the experiments on a turbulent boundary layer with injection and combustion [12], except that air is used as an injectant in this study. Therefore, it is briefly described below.

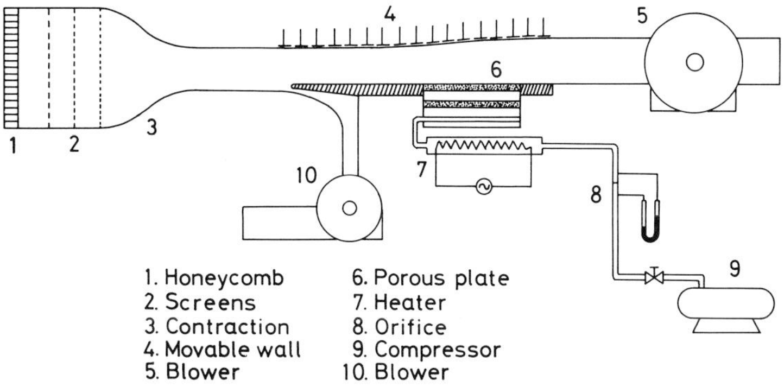

Fig. 1. Schematic diagram of experimental apparatus

Figure 1 shows the schematic diagram of the low speed wind tunnel used. The test section is 300 mm by 300 mm in cross section and 2,400 mm in length. A turbulent boundary layer developing on the bottom flat plate is studied. The porous part of the plate starts 1,200 mm downstream of the flat plate leading edge, and extends 900 mm downstream. Except in a preliminary, isothermal experiment to be mentioned just below, heated air is uniformly injected through the porous plate. The mass flux ratio of the injectant to the main flow, namely the injection rate F, is varied in three steps; 2.12×10^{-3}, 3.50×10^{-3} and 4.43×10^{-3}. The free stream velocity is kept the same at every run and is equal to 8.0 m/s. Prior to the experiments, the top wall of the wind tunnel is carefully adjusted, and the streamwise variation of the static pressure is minimized, i.e. within ±0.5% of the dynamic pressure of the main flow.

In the early stage of the work, the isothermal boundary layer was studied. In Fig. 2 are shown the velocity profiles obtained at the station 680 mm downstream from the starting position of injection. The ordinate Φ is defined by the following equation.

$$\Phi = \frac{\sqrt{2c_f}}{F}\left[(1+B)^{0.5} - \left(1 + B\frac{U}{U_e}\right)^{0.5}\right] \tag{1}$$

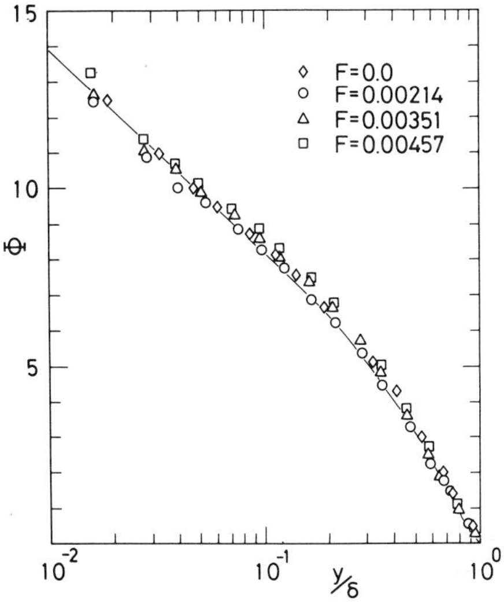

Fig. 2. Mean velocity profiles in the isothermal boundary layer

The present results agree well with the extended form of the Coles' wake law proposed by *Stevenson* [13]. It is shown by the solid line in the figure. The later non-isothermal experiments were performed at the same station. Unfortunately, however, the thermal boundary layer to be examined does not seem to be fully developed at the station. The thickness of the thermal boundary layer is thinner than that of the velocity boundary layer by about 20%, and this must be kept in mind when discussing the present results.

The injectant is as mentioned before the heated air and its temperature in the reservoir is controlled to be higher than the main flow temperature by about 30 K. Depending on the injection rate, the corresponding temperature difference between the wall and the main flow varies from 20 K to 25 K. From the value of the Richardson number evaluated, this level of temperature difference can be considered to be within the range where the added heat is a passive contaminant. Thus, the characteristics of the averaged and fluctuating velocity fields are expected to differ little from those in the isothermal boundary layer examined above.

2.2. Measurements of Turbulence

Even in the case when the added heat can be treated as a passive contaminant, both the velocity and the temperature fluctuations are mixed up in the output signal of a hot wire anemometer. The separation of the mixed signal into pure velocity and temperature signals is accomplished by applying an analog circuit similar to that proposed by *Hishida* and *Nagano* [14]. An extention was made in the present study such that two velocity components and temperature fluctuations can be measured. The basic idea of the present method is described below.

A probe with three wires is mainly used in this study. Two of them are hot wires and the remaining one is a cold wire. For a limited purpose, a combination of a hot wire and a cold wire is also used. In a constant temperature anemometer, the effective thermal inertia time lag is reduced considerably. Then, the instantaneous heat balance equation for each hot wire may be written as follows.

$$E \equiv \frac{V_h^2}{(T_0 - T_a)} = A + B\sqrt{U_{eff}} \tag{2}$$

where V_h is the output voltage from the constant temperature anemometer, T_0 the wire temperature, T_a the fluid temperature, U_{eff} the effective cooling velocity of the hot wire, and A and B the coefficients related to the heat transfer of a fine wire to be determined through the wire calibration. B does not vary so pronouncedly with the fluid temperature, but A does. Thus, the fluctuation of the fluid temperature affects the output signal V_h directly and indirectly. To get the pure velocity signal, the instantaneous fluid temperature must be measured separately.

A cold wire is operated in the constant current mode, and is used as a resistance thermometer. Different from the above case, the thermal inertia term should be retained in the instantaneous heat balance equation for the cold wire. The equation may be written as follows.

$$V_a = V_c + M\frac{dV_c}{dt} \tag{3}$$

where V_a is the true output voltage of the cold wire in a steady state at the temperature T_a, V_c the actual output voltage at the same temperature, M the thermal inertia time lag, and t the time. The time constant M depends on the fluid velocity U_{eff}. Thus, Eq. (2) for the hot

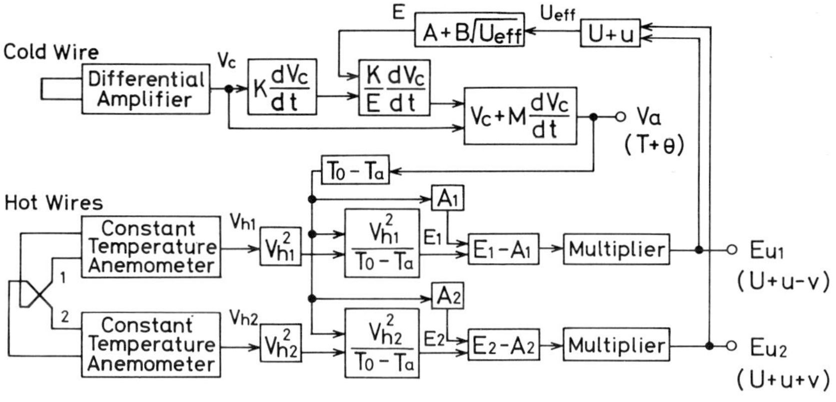

Fig. 3. Block diagram of analog circuit

wire and Eq. (3) for the cold wire should be solved simultaneously. For the purpose, an analog circuit is used and its block diagram is shown in Fig. 3. The coefficient A is assumed to vary linearly with the fluid temperature and its dependence is evaluated experimentally as well as the dependence of M on U_{eff}. The gain of the operational amplifier K is adjusted at a standard condition through a response test of the cold wire to a step heating such that K/E coincides with M at that condition.

The cold wire is operated at the electric current 1.0 mA, and each hot wire is operated at the same overheat ratio 0.40. The cold wire is placed always parallel to the wall and perpendicular to the main flow direction. In the case when a single hot wire is used, the cold wire is located 1.0 mm upstream of the hot wire, while in the case using X-type hot wires, it is located 1.0 mm upstream of the X-wire intersection. These sensing wires consist of 5 μm tungsten wire with copper plated ends, and the effective lengthes of the hot wire and the cold wire are respectively 1.0 mm and 2.0 mm long.

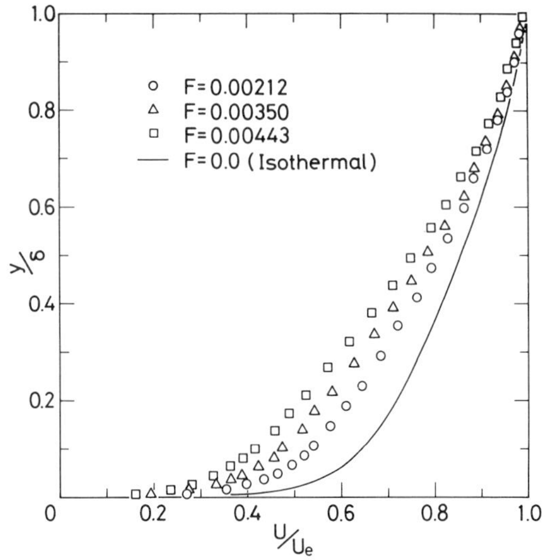

Fig. 4. Mean velocity profiles

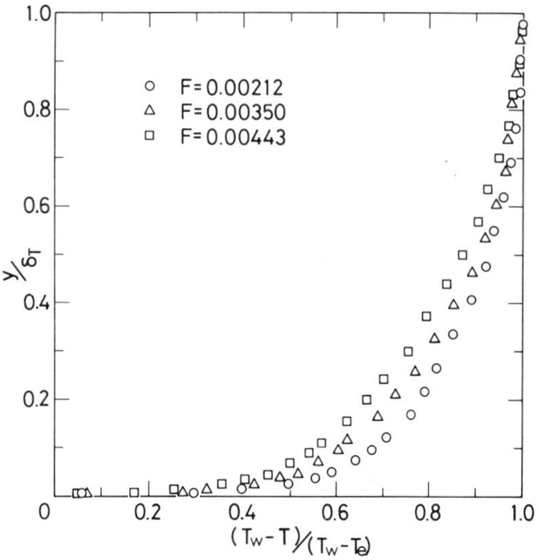

Fig. 5. Mean temperature profiles

3. Results and Discussion

3.1. Mean Profiles

The mean velocity and temperature distributions are shown in Figs. 4 and 5. As seen from the figures, the effect of injection on these mean profiles is such that the velocity and the temperature gradients near the wall decrease with increasing injection rate.

3.2. Turbulent Intensities

The turbulent intensities of streamwise and transverse velocity fluctuations $\sqrt{\overline{u^2}}$ and $\sqrt{\overline{v^2}}$, and the intensity of temperature fluctuations $\sqrt{\overline{\theta^2}}$ are shown in Figs. 6 and 7. For simplicity, u', v' and θ' are used respectively for $\sqrt{\overline{u^2}}$, $\sqrt{\overline{v^2}}$ and $\sqrt{\overline{\theta^2}}$ in the following. Both u' and θ' increase with increasing injection rate in the boundary layer except very near the wall. In the very near wall region, they seem to decrease with increasing injection rate. v' also increases with injection in the boundary layer. These results agree with those of *Peterson* et al. [8] and *Pimenta* et al. [10].

3.3. Reynolds Shear Stress and Turbulent Heat Flux

In Figs. 8 and 9 are shown the transverse distributions of the Reynolds shear stress $-\overline{uv}$ and the turbulent heat flux $\overline{v\theta}$ in the boundary layer. As pointed out by many investigators, these fluxes have maxima in the transpired boundary layer and the maximum value increases with increasing injection rate. This is in contrast to the reduction of the momentum and the heat fluxes at the wall. The transverse positions of both maxima tend to go outwards with the increase of the injection rate. In Fig. 8 are shown the Reynolds shear stress distributions calculated from the mean velocity profiles by using the method proposed by *Andersen* [9]. The method is based on the following equation for the Reynolds shear stress obtained by integrating the time-averaged boundary layer equations for incompressible flow.

$$\frac{-\overline{uv}}{U_e^2} = \frac{c_f}{2} + F\frac{U}{U_e} - \frac{\mathrm{Re}_\delta}{\mathrm{Re}_\theta}\left(F + \frac{c_f}{2}\right)\left[\frac{U}{U_e}\int_0^y\left(\frac{U}{U_e}\right)d\left(\frac{y}{\delta}\right) - \int_0^y\left(\frac{U}{U_e}\right)^2 d\left(\frac{y}{\delta}\right)\right] - \frac{\nu}{U_e^2}\frac{\partial U}{\partial y} \tag{4}$$

Here, the skin friction coefficient c_f is obtained by applying the equation in the small region close to the wall. The points on the ordinate of the figure are the skin friction coefficients so evaluated. Once the skin friction coefficient has been determined, the Reynolds shear stress distribution can be calculated from the mean velocity profiles observed. The agreement between the experimental results and the calculated profiles is fairly good. Also plotted are the skin friction coefficients evaluated from Kays' empirical formula Eq. (5) by obtaining the momentum thickness from the measured velocity profile,

$$\left.\frac{c_f}{c_{f0}}\right|_{\mathrm{Re}_\theta} = \left[\frac{\ln(1+B)}{B}\right]^{1.25}(1+B)^{0.25} \tag{5}$$

where $c_{f_0} = 0.026\,\mathrm{Re}_\theta^{-0.25}$. In view of the difficulty of estimating the wall shear stress, the agreement among the skin friction coefficients evaluated by these two methods may be considered tolerably good.

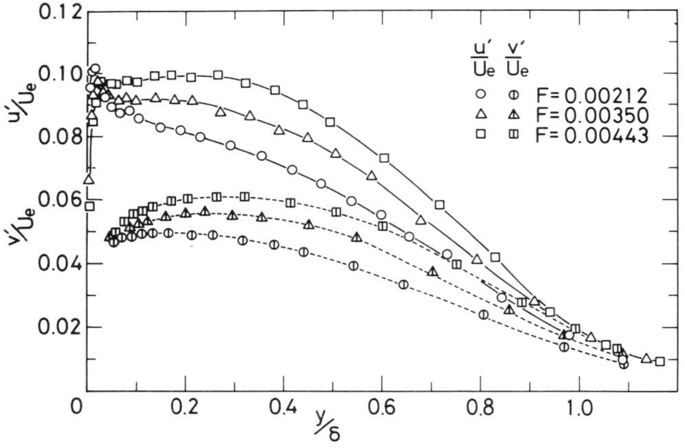

◀ **Fig. 6.** Intensities of velocity fluctuations

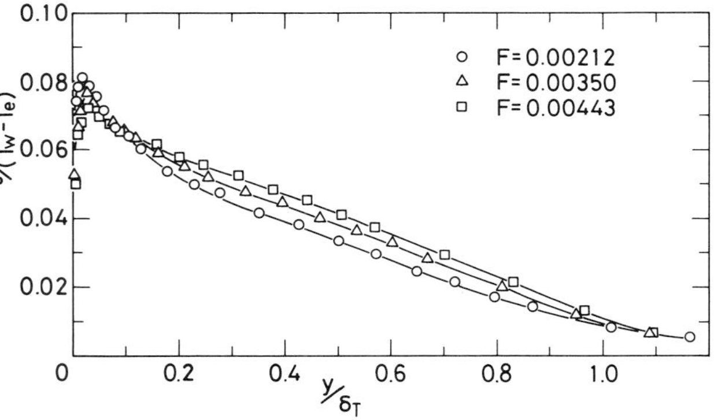

Fig. 7. Intensity of temperature fluctuation

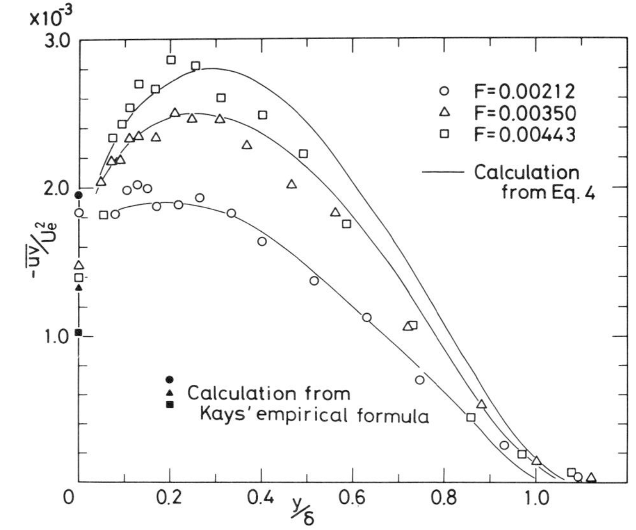

◀ **Fig. 8.** Distribution of Reynolds shear stress

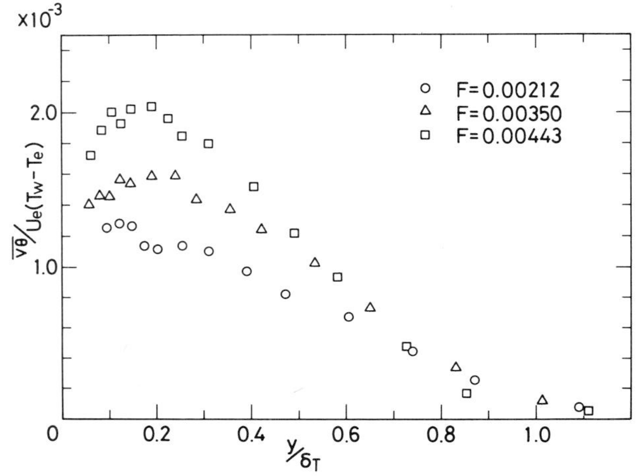

Fig. 9. Distribution of turbulent heat flux

3.4. One-Point Double Correlation Coefficients

Except near the wall region and near the boundary layer edge, the one-point double correlation coefficients $-\overline{uv}/u'v'$ and $\overline{v\theta}/v'\theta'$ are distributed almost uniformly over the entire region of the boundary layer, i.e. $-\overline{uv}/u'v' \simeq 0.50$ and $\overline{v\theta}/v'\theta' \simeq 0.48$. The dependence of their values on the injection rate is very slight. This can be seen in Fig. 10. In the figure are also shown the results of *Chen* and *Blackwelder* [15], $-\overline{uv}/u'v' \simeq 0.54$ and $\overline{v\theta}/v'\theta' \simeq 0.51$, for the flat plate boundary layer without injection and those of *Pimenta*, $-\overline{uv}/u'v' \simeq 0.45$ and $\overline{v\theta}/v'\theta' \simeq 0.60$, for the rough wall boundary layer with and without injection. The result of *Fulachier* [16] for the flat plate boundary layer is that $\overline{v\theta}/v'\theta' \simeq 0.60$. The present results for both $-\overline{uv}/u'v'$ and $\overline{v\theta}/v'\theta'$ are in agreement with the results of other investigators. Slightly lower present values of $\overline{v\theta}/v'\theta'$, however, might be partly characteristic of the present flow situation where the thermal boundary layer has not still attained the thickness matching the fully developed condition.

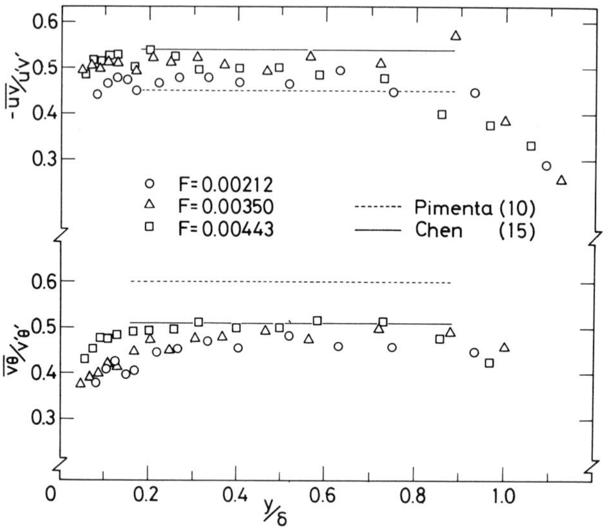

Fig. 10. One-point double correlation coefficients

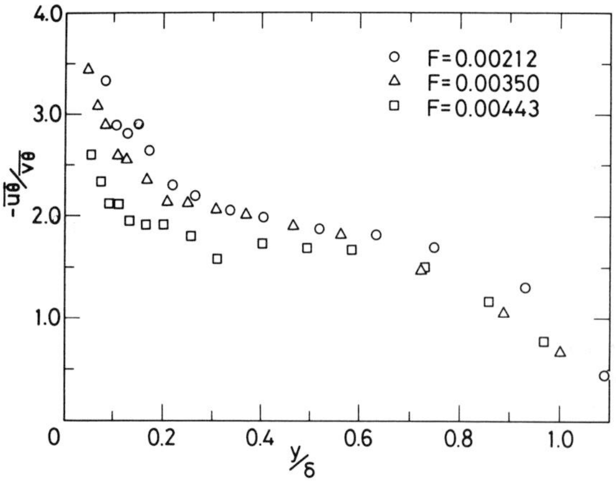

Fig. 11. Ratio of streamwise to transverse turbulent heat flux

3.5. Turbulent Heat Flux Ratio

The ratio of streamwise to transverse turbulent heat flux $-\overline{u\theta}/\overline{v\theta}$ is shown in Fig. 11. The streamwise turbulent heat flux $-\overline{u\theta}$ is of the same order of magnitude as $\overline{v\theta}$, but is greater than $\overline{v\theta}$. Although this result indicates that the streamwise velocity fluctuation couples strongly with temperature, especially near the wall, this does not mean that the streamwise gradient of $-\overline{u\theta}$ is significant. The streamwise to the transverse turbulent heat flux ratios $-\overline{u\theta}/\overline{v\theta}$ decrease with increasing injection rate near the wall. In the middle of the boundary layer, the ratio is almost constant and independent of the injection rate.

3.6. Bradshaw's Structure Functions

In Figs. 12 and 13 are shown the characteristic parameters a and a_θ introduced by *Bradshaw* and *Ferriss* [17, 18]. They are defined as follows.

$$a = \frac{-\overline{uv}}{q^2} \tag{6}$$

$$a_\theta = \frac{\overline{v\theta}}{\theta' \sqrt{\tau/\rho}} \tag{7}$$

where q^2 is the turbulent kinetic energy and was evaluated by approximating it to $(\overline{u^2} + 2\,\overline{v^2})$ in this study. *Bradshaw* and *Ferriss* assumed that a and a_θ are constant and equal to 0.15 and 0.45, respectively, across the boundary layer. Except at the boundary layer edge, a is nearly constant over most of the boundary layer and is independent of the injection rate. It is

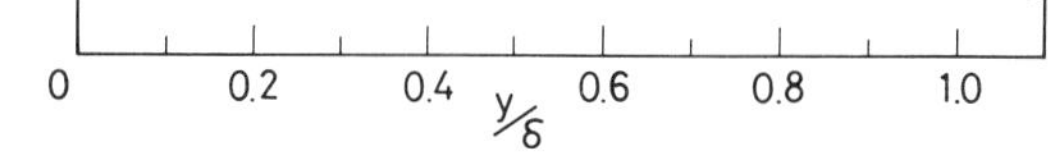

Fig. 12. Structure function a

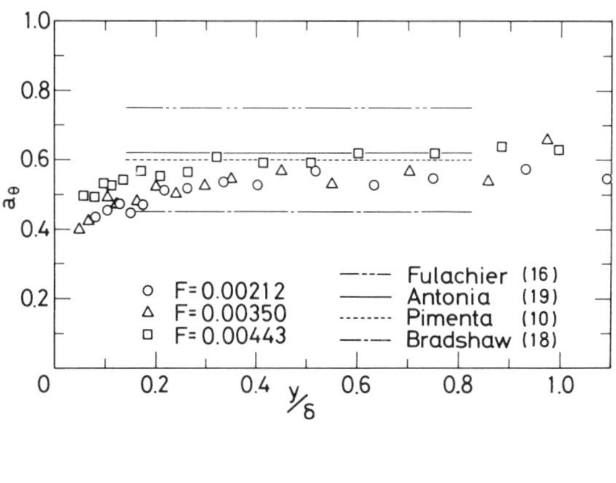

Fig. 13. Structure function a_θ

approximately 0.17. In Fig. 13, the results of other investigators are shown for comparison. Present results show that a_θ is almost constant across the boundary layer and that the dependency on the injection rate is slight. It is approximately 0.55, which is slightly smaller than that of *Antonia* et al. [19], 0.62 and that of *Pimenta* et al. [10], 0.60, but larger than that of *Bradshaw*.

3.7. Time Scale Ratio

Following is discussed the time scale ratio R for the decay of temperature and velocity fluctuations defined below,

$$R = \left(\frac{\epsilon}{q^2}\right) \Big/ \left(\frac{\epsilon_\theta}{\theta'^2}\right) \tag{8}$$

where ϵ and ϵ_θ are the dissipation rates of velocity and temperature fluctuations. With an assumption that both the velocity and the temperature fields are in local equilibrium, the above equation can be rewritten as follows with the Bradshaw's parameters a and a_θ, and the turbulent Prandtl number $\mathrm{Pr_t}$.

$$R = \frac{a}{a_\theta^2 \cdot \mathrm{Pr_t}} \tag{9}$$

By using the values obtained in this experiments (Prandtl numbers are given in Fig. 20), the time scale ratio becomes approximately 0.40, which is agreeable with the results of other investigators [20, 21].

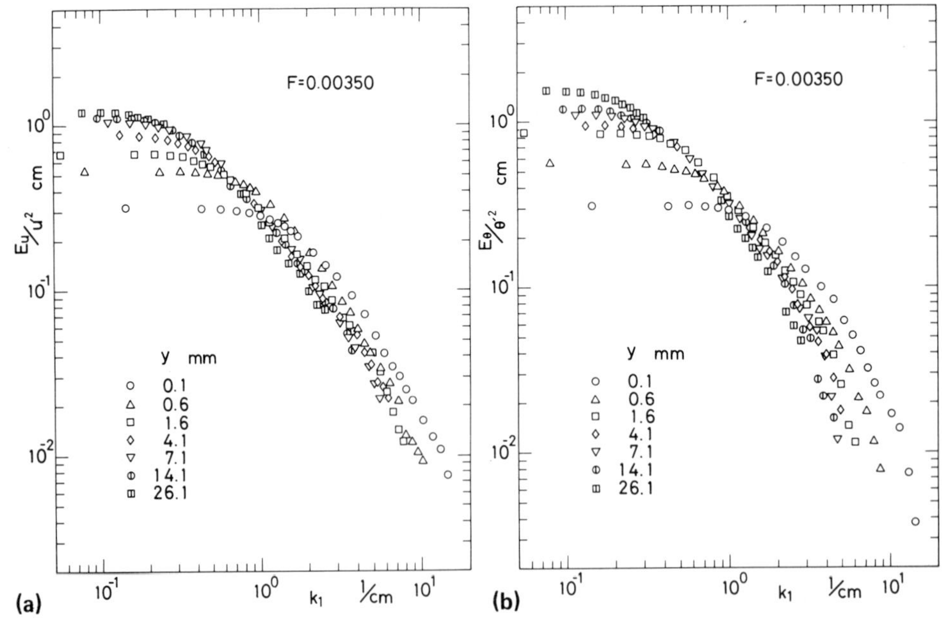

Fig. 14. **(a)** Spectrum of streamwise velocity fluctuation (F = 0.00350). **(b)** Spectrum of temperature fluctuation (F = 0.00350)

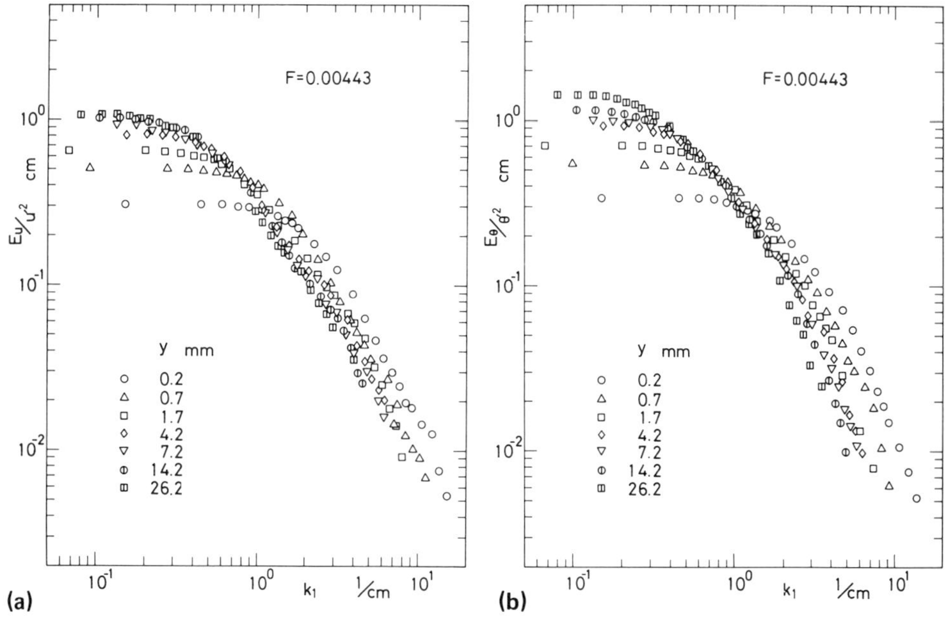

Fig. 15. (a) Spectrum of streamwise velocity fluctuation ($F = 0.00443$). (b) Spectrum of temperature fluctuation ($F = 0.00443$)

3.8. Spectra of $\overline{u^2}$ and $\overline{\theta^2}$

The one dimensional spectra $E_u(k_1)$ and $E_\theta(k_1)$ of the streamwise velocity and the temperature fluctuations are shown in Figs. 14 and 15, where k_1 is wave number evaluated by using Taylor's hypothesis. First, it can be recognized that the temperature spectrum falls off more rapidly in the high wave number region than the velocity one does, as clearly seen if one compares Fig. 14a, b, and Fig. 15a, b, with each other. This indicates that the dissipation of turbulent kinetic energy occurs in the higher wave number region than that of the intensity of temperature fluctuations, or that the microscale of temperature fluctuations is larger than the velocity one, as pointed out by Corrsin [22]. In Fig. 16a, b, the results of the integral scales of turbulence are plotted in the form non-dimensionalized with the boundary layer thickness.

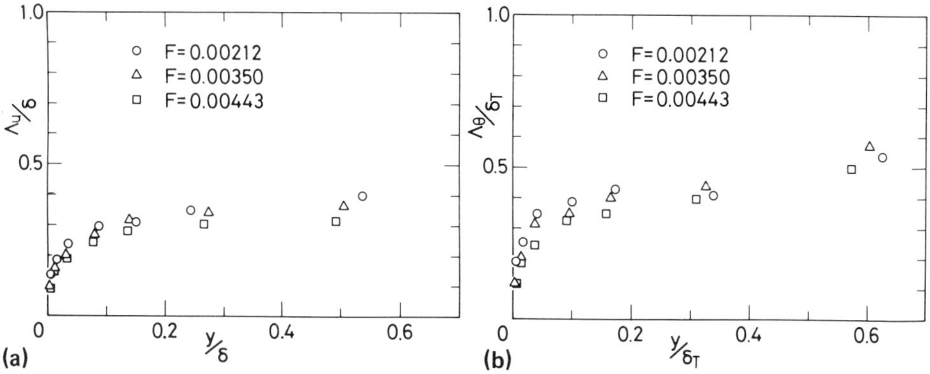

Fig. 16. (a) Integral length scale of velocity. (b) Integral length scale of temperature

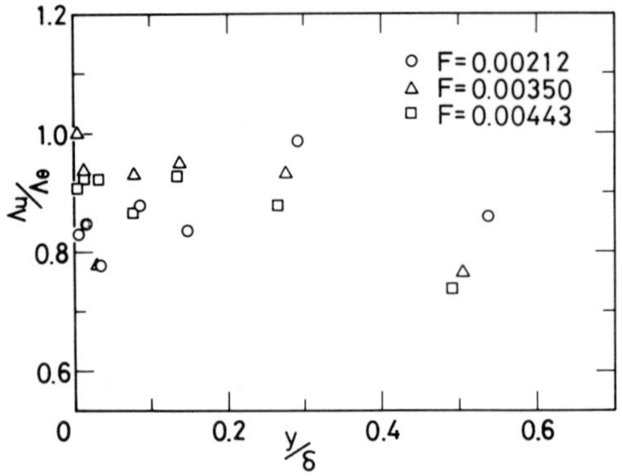

◀ **Fig. 17.** Integral scale ratio Λ_u/Λ_θ

Fig. 18. (a) Spectrum of Reynolds shear stress (F = 0.00350). **(b)** Spectrum of turbulent heat flux (F = 0.00350)

▼

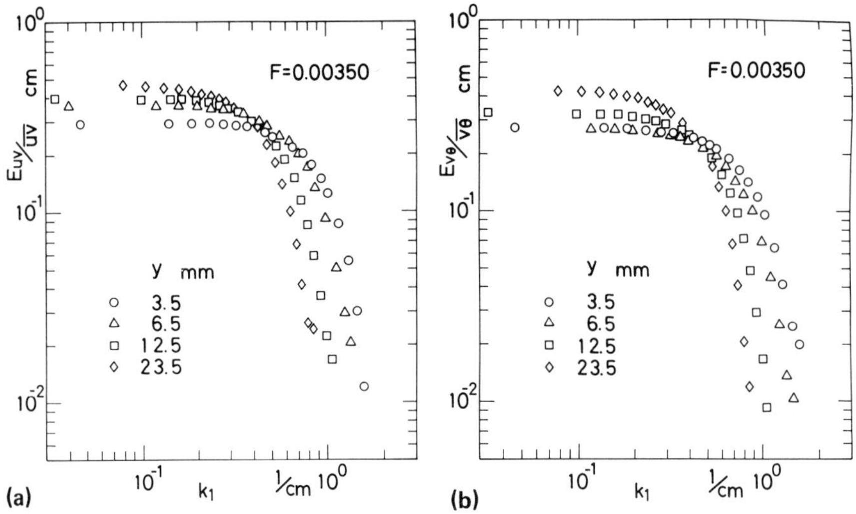

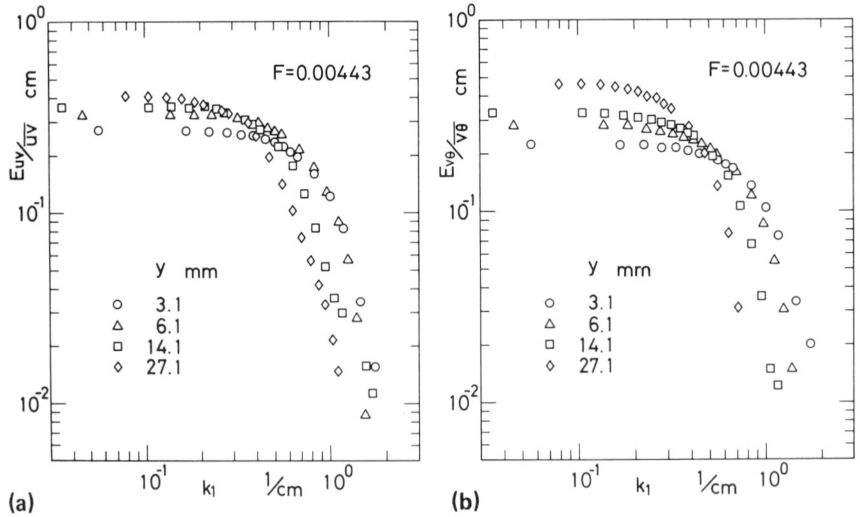

Fig. 19. (a) Spectrum of Reynolds shear stress (F = 0.00443). **(b)** Spectrum of turbulent heat flux (F = 0.00443)

The integral scales of turbulence of the velocity and the temperature fluctuations, Λ_u and Λ_θ, are deduced here from the following equations.

$$\Lambda_u = \frac{\pi}{2} \lim_{k_1 \to 0} \frac{E_u(k_1)}{u'^2} \qquad \Lambda_\theta = \frac{\pi}{2} \lim_{k_1 \to 0} \frac{E_\theta(k_1)}{\theta'^2} \tag{10}$$

As seen from the figures, both the velocity and the temperature integral scales decrease with the injection rate in the inner layer and this becomes more pronounced as the wall is approached. This may be related with the decreasing mean strain of the turbulence eddies in the near wall region caused by the reduction of the mean shear rate with injection. (see the mean velocity profiles). On the other hand, these scales remain unchanged in the outer layer.

In Fig. 17, the ratio of the integral length scale of velocity Λ_u to that of temperature Λ_θ is shown. It reveals clearly that the integral scale of temperature is larger than that of velocity, but the effect of injection on this scale ratio does not appear in the figure.

3.9. Spectra of $-\overline{uv}$ and $\overline{v\theta}$

The spectra of $-\overline{uv}$ and $\overline{v\theta}$ are shown in Figs. 18 and 19. Comparing these results with the spectra of velocity and temperature fluctuations, Figs. 14 and 15, the spectra of the Reynolds shear stress and the turbulent heat flux are observed to fall much faster as the wave number increases. This confirms that the turbulence in high wave number range tends to local isotropy and that the turbulent exchange of momentum and heat is borne mainly by the large scale eddies.

3.10. Turbulent Prandtl Number

The values of the turbulent Prandtl number, defined as the ratio $\overline{uv}(\partial T/\partial y)/\overline{v\theta}(\partial U/\partial y)$ of the momentum diffusivity to the thermal diffusivity, are shown in Fig. 20. That the turbulent Prandtl number is not constant across the boundary layer increasing slightly as the wall is approached is similar to the results found in reference [1], but the present results are quantitatively a little higher than those results. This could be related to the special feature of the present thermal boundary layer of thinner thickness. The quantitative discussion of the results will be postponed until the fully developed data is obtained with the present method.

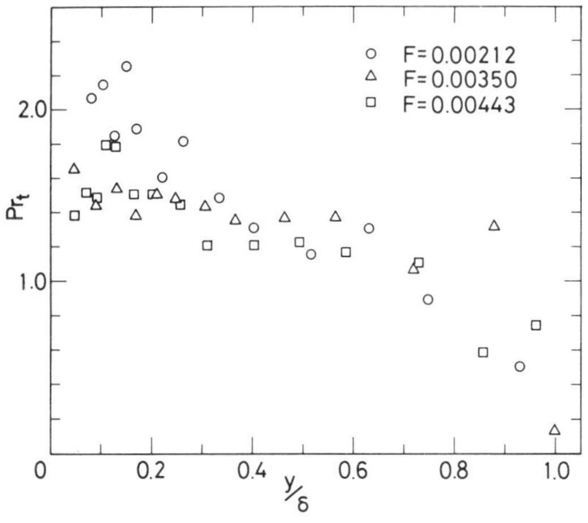

Fig. 20. Turbulent Prandtl number

4. Concluding Remarks

An analog circuit was constructed to extract the pure velocity and temperature fluctuation signals from the mixed signal of the hot wire anemometer. This was applied to a turbulent boundary layer with uniform injection of heated air. The various turbulent quantities, among others the turbulent heat flux, were measured. The quantities evaluated with these signals were discussed and the effects of injection on the turbulence structure were investigated. Although the measurements have been made in the boundary layer possibly still developing thermally, most of the results obtained are in agreement with other results in the turbulent boundary layer with and without injection.

The main effects of injection on the turbulence structure obtained in this study can be summarized as follows.

a) Turbulent intensities of both velocity and temperature fluctuations increase in the boundary layer with increasing injection rate except very near the wall.
b) One-point double correlation coefficients, $-\overline{uv}/u'v'$ and $\overline{v\theta}/v'\theta'$, depend slightly on the injection rate and are constant across the boundary layer.
c) Bradshaw's structure functions, a and a_θ, are almost constant in the boundary layer and independent of the injection rate.
d) Integral length scales of streamwise velocity and temperature fluctuations decrease with increasing injection rate and this may be related to the reduction of the mean shear rate.

The value of turbulent Prandtl number obtained, however, shows a consistently higher value. For further discussion of it, experiments in fully developed thermal boundary layer are desired.

References

1. W. M. Kays, R. J. Moffat: "The Behaviour of Transpired Turbulent Boundary Layers", in *Studies in Convection,* Vol. 1 (Academic Press, London 1975)
2. A. I. Leont'ev: "Heat and Mass Transfer in Turbulent Boundary Layers", in *Advances in Heat Transfer,* Vol. 3 (Academic Press, London 1966)
3. K. Suzuki: Non-isothermal turbulent boundary layer with foreign gas injection (1st Rpt. A phenomenological theory predicting surface skin-friction). Trans. JSME **42,** No. 353, 226–235 (1976) (in Japanese)
4. K. Torii, N. Nishiwaki, M. Hirata: "Heat Transfer and Skin Friction in Turbulent Boundary Layer with Mass Injection". 3rd Int. Heat Transfer Conference, Vol. 3 (1966) pp. 34–48
5. R. J. Baker, B. E. Launder: The turbulent boundary layer with foreign gas injection – I. Measurements in zero pressure gradient. Int. J. Heat Mass Transfer **17**(2), 275–291 (1974)
6. R. H. Pletcher: Prediction of transpired turbulent boundary layers. J. Heat Transfer **96**(1), 89–94 (1974)
7. R. L. Simpson: Characteristic of turbulent boundary layer at low Reynolds number with and without transpiration. J. Fluid Mech. **42,** Part 4, 769–802 (1970)
8. R. L. Peterson, L. K. Isaacson: "Statistical Properties of a Turbulent Boundary Layer with Surface Mass Injection". AFOSR-TR-71-1880 (1971)
9. P. S. Andersen, W. M. Kays, R. J. Moffat: Experimental results for the transpired turbulent boundary layer in an adverse pressure gradient. J. Fluid Mech. **69,** Part 2, 353–375 (1975)
10. M. M. Pimenta, R. J. Moffat, W. M. Kays: "The Turbulent Boundary Layer: An Experimental Study of the Transport of Momentum and Heat with the Effect of Roughness", Rpt. HMT-21, Mech. Engng. Dept., Stanford Univ. (1975)
11. C. E. Wooldridge, R. J. Muzzy: Boundary layer turbulence measurements with mass addition and combustion. AIAA J. **4**(11), 2009–2016 (1966)

12. M. Senda, K. Suzuki, T. Sato: Study on turbulent boundary layer with injection and combustion. Mem. Fac. Eng., Kyoto Univ. **38**, Part 1, 21–36 (1976)
13. T. N. Stevenson: Turbulent boundary layers with transpiration. AIAA J. **2**(8), 1500–1502 (1964)
14. M. Hishida, Y. Nagano: Simultaneous measurements of velocity and temperature in nonisothermal flows. J. Heat Transfer **100**(2), 340–345 (1978)
15. C.-H. P. Chen, R. F. Blackwelder: Large-scale motion in a turbulent boundary layer: A study using temperature contamination. J. Fluid Mech. **89**, Part 1, 1–31 (1978)
16. L. Fulachier: Thèse Docteur ès Sciences, Université de Provence (1972)
17. P. Bradshaw, D. H. Ferriss, N. P. Atwell: Calculation of boundary-layer development using the turbulent energy equation. J. Fluid Mech. **28**, Part 3, 593–616 (1967)
18. P. Bradshaw, D. H. Ferriss: "Calculation of Boundary Layer Development Using the Turbulent Energy Equation: IV. Heat Transfer with Small Temperature Differences". Nat. Phys. Lab., Aero Rpt. 1271 (1968)
19. R. A. Antonia, H. Q. Danh, A. Prabhu: Response of a turbulent boundary layer to a step change in surface heat flux. J. Fluid Mech. **80**, Part 1, 153–177 (1977)
20. B. E. Launder: "Heat and Mass Transport", in *Turbulence,* ed. by P. Bradshaw, Topics in Applied Physics, Vol. 12 (Springer, Berlin, Heidelberg, New York 1976)
21. C. Béguier, I. Dekeyser, B. E. Launder: Ratio of scalar and velocity dissipation time scales in shear flow turbulence. Phys. Fluids **21**, 307–310 (1978)
22. S. Corrsin: On the spectrum of isotropic temperature fluctuations in an isotropic turbulence. J. Appl. Phys. **22**, 469–473 (1951)

Peculiarities of Non-Isothermal Turbulent Flows in Horizontal Flat Channels at Low Reynolds Number and Under Significant Buoyancy Forces

B. S. Petukhov, A. F. Polyakov, Yu. V. Tsypulev

Institute of High Temperatures, Moscow, U.S.S.R.

Abstract

An experimental study has been carried out for an air flow in a flat horizontal channel. The effects of stable and unstable density stratifications on the distributions of mean and fluctuating velocity and temperature values have been studied. The study took into account spectra and amplitude information in order to reveal some features of the structure of the near wall turbulent flow under strong influences of buoyancy forces. Detailed information on the experimental data is provided.

Nomenclature

b	Channel width [m]
C_p	Specific heat at constant pressure [J/kgK]
h	Channel height [m]
f	Frequency [1/s]
g	Gravitational acceleration [m/s^2]
$G(f)$	Spectrum of temperature fluctuation [s]
$G(K_1) = \frac{u}{2\pi} G(f)$	[m]
$K_1 = 2\pi f/U$	one-dimensional wave number [1/m]
q	Heat flux [J/m^2 s]
t	Time-averaged temperature, mean temperature [K]
t'	Temperature fluctuation [K]
u	Mean velocity [m/s]
u'	Velocity fluctuation [m/s]
$\bar{u}$	Cross-section-averaged velocity, averaged velocity [m/s]
$\overline{\rho u}$	Averaged mass weighted velocity [kg/m^2 s]
$v_* = \sqrt{\tau_w/\rho}$	Shear velocity [m/s]
x	Axial coordinate [m]
y	Distance from the heated wall [m]
β	Coefficient of thermal expansion [1/K]
λ	Thermal conductivity [J/msK]
μ	Dynamic viscosity [kg/ms]
ν	Kinematic viscosity [m^2/s]
ρ	Density [kg/m^3]
$\sigma_t = \sqrt{\overline{t'^2}}$	Intensity of temperature fluctuation [K]
$\sigma_u = \sqrt{\overline{u'^2}}$	Intensity of velocity fluctuation [m/s]
$E_t = (\overline{t'^4}/\sigma_t^4) - 3$	Excess coefficient of temperature fluctuations
$Gr = g\beta q_w(2h)^4/\lambda\nu^2$	Grashof number
Pr	Prandtl number
$Re = \overline{\rho u}\, 2h/\mu$, $Re_* = v_*\, 2h/\nu$	Reynolds number
$S_t = (\overline{t'^3})/\sigma_t^3$	Asymmetry coefficient of temperature fluctuations
$T^+ = \frac{t_w - t}{q_w} \rho C_p v_*$	Dimensionless temperature
$U^+ = u/v_*$	Dimensionless velocity

$Y = y/h$ Dimensionless coordinate

$\eta = \frac{v_*}{\nu} y$ Dimensionless coordinate

$\mathrm{Ri}_w = \mathrm{Gr}_w/\mathrm{Re}_*^4$ Richardson number

Subscripts

cr Critical value

cw Refers to cold wall

eqv Equivalent value

l Limit value

t Refers to temperature characteristic

u Refers to velocity characteristic

w Refers to hot wall

The strong effects of the gravitational field on forced turbulent flow and heat transfer in channels become more pronounced with increasing heat transfer rates and increasing geometrical dimensions of the heating surfaces. A detailed review has been provided in [1] and presents quite a complete report of the state of the art in this field. It is established both theoretically and experimentally that at relatively large Grashof numbers (Gr), and not so high Reynolds numbers (Re), thermo-gravitational forces result in variations of the velocity and temperature fields and influence turbulent transport parameters, heat transfer rates, etc. The influence of strong buoyancy forces on the turbulent transport of momentum and heat is not small and is comparable with the influence on the mean flow. In heated horizontal round pipes the effect of buoyancy on the mean flow and the turbulence properties results in the formation of secondary flows in a plane normal to the pipe axis [2], which substantially complicates the analysis of the buoyancy effects. These secondary flows are less significant in two-dimensional channel flows and in this connection the study of the two-dimensional turbulent flow in a gravitational field seems to be important because in this case the secondary flow is not developed. The present work is devoted to the experimental study of the various effects of the gravitational field on the turbulent flow on a horizontal, flat, asymmetrically heated channel. As it is shown below, the studies revealed influence of buoyancy on the structure of the near wall turbulence.

The present experimental investigation is carried out for air flowing in a channel with a $40 \times 180\ \mathrm{mm}^2$ cross-section. The work aimed at detailed studies of the peculiarities of the stable and unstable density stratification on the turbulent structure. Asymmetric heating was applied by heating one of the channel surfaces by an electric current, while the other consisted of an adiabatic wall, i.e. the thermal boundary conditions were q_w = const. at the wall and $q_{cw} = 0$ at the other. The conditions of stable and unstable density stratification were provided by means of a 180° rotation of the channel around the longitudinal axis. The measurements in the flow were performed at a distance $x = 46.5$ h from the start of the heat supplying surface in the vertical and horizontal planes. At two-dimensional flow region was analytically and experimentally proved to exist in the central region of the channel, both hydrodynamically and thermally. Its size was 30–40% of the channel width. All the main measurements were carried out within this zone; the flows within the corner zones, in which the secondary flows existed, were not considered in the present study.

To perform flow measurements, a 55M-DISA hot-wire anemometer was employed and the resulting signals were processed with correlation and probability analysis electronics. The measuring technique and experimental equipment were described in previous publications, [3, 4]. No further details are required here.

The experiments required non-isothermal flow measurements with special attention given to the near wall region where high velocity and temperature gradients exist. Strong temperature variations in the flow and the requirements of near wall measurements made the design and manufacturing of special probes necessary. Commercial hot-wire sensors do not permit local, simultaneous measurements of both velocity and temperature at distances less than 1 mm from the wall. Special probe designs are necessary for near wall records of velocity and temperature fluctuations.

The prongs of the specially designed hot-wire sensor were L-shaped, as the prongs of boundary layer probes usually are, in order eliminate influences of the velocity and temperature gradients in the near wall region. The prongs provided the support for two identical tungsten wires of 3 μm in diameter or two platinum wires of 5 μm in diameter. The wire length was 1.5 to 2 mm in length and varied within these limits from one probe to the other. Both wires were based along a straight line with a common central point of zero potential. One of the wires was working as the hot-wire sensor and the other was employed for temperature compensation. The two wires were placed parallel to the wall taking into account the two-dimensionality of the flow in the central region of the channel.

To represent the experimental results for the region near the heated wall in universal coordinates, it is necessary to know the wall shear stress τ_{hw} and wall heat flux q_{hw}. The values for τ_{hw} and q_{hw} were obtained on the basis of the velocity and temperature distributions in the viscous sublayer region where these distributions are linear.

The effects of buoyancy forces are given by the Grashof number ratio, Gr/Gr_l, where Gr_l is given by the following relationship:

$$Gr_l = 1.3 \cdot 10^{-4} \frac{Pr\, Re^{2.75}[Re^{1/8} + 2.4\,(Pr^{2/3} - 1)]}{\left[lg(Re) + 1.15\, lg\left(\frac{5\,Pr + 1}{30}\right) + 0.5\,Pr\right]^{-1}} \tag{1}$$

It is shown in [6] that this Grashof number determines the starting of buoyancy effects on the heat transfer in horizontal turbulent wall boundary layers.

In the present experiments the parameters Re and Gr/Gr_l could not be changed independently so that large values of Gr/Gr_l correspond to small values of the Reynolds number. Nevertheless, as the measurements under isothermal conditions showed, the flow was turbulent even for the low Reynolds number studies ($Re = 4 - 8 \cdot 10^3$). This is in accordance with findings in Ref. [7] where it was found that channels with aspect ratios varying from 1:1 to 1:10 showed laminar to turbulent transition in the Reynolds number range from 2.2 to $2.5 \cdot 10^3$. The hydrodynamic entrance region for channels with aspect ratios of 1:5 and Reynolds numbers larger than $4 \cdot 10^3$ was found to be smaller than 20 equivalent diameters.

The results of velocity and temperature measurements are shown in normalized coordinates in Figs. 1 and 2 for various ratios of Gr/Gr_l. The results show that for $Gr/Gr_l = 5$ a slight effect of buoyancy is observed on the mean quantities. The $U^+(\eta)$ and $T^+(\eta)$ profiles are slightly distorted near the channel axis. With increasing values of Gr/Gr_l the distortion of the profiles $U^+(\eta)$ and $T^+(\eta)$ becomes much more pronounced and the influence of the buoyancy effects shifts towards the wall. The velocity and temperature profiles become more elongated under stable density stratification, thus approaching the shape characteristic to laminar flows. On the contrary, under unstable stratification the profiles become more flat. As the experimental results show, for $Gr/Gr_l > 100$ practically all the flow region is strongly affected by buoyancy forces except the viscous sublayer ($\eta < 5$).

The present experimental data have been used in connection with the Monin-Obukhov-theory [9–12] which has been developed for turbulent flows in stratified media. It should be realized, however, that this theory was developed assuming near-earth atmospheric layer as some ideal model for a stratified medium. Because of this, the theory of Monin-Obukhov is limited in its applicability to internal flows as studied by the authors. Nevertheless, some

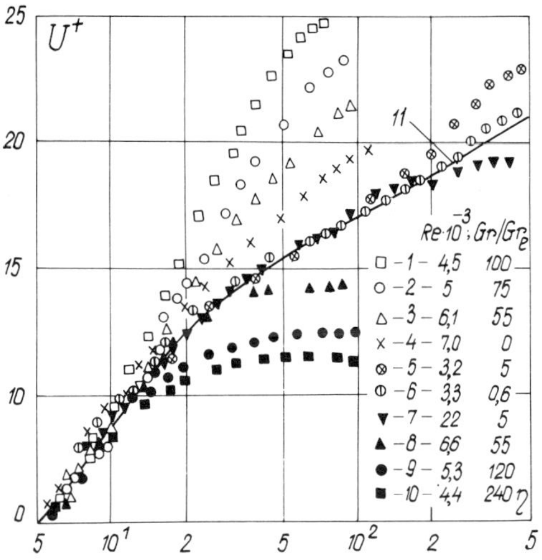

Fig. 1. Dimensionless velocity distribution as a function of the dimensionless wall distance:
1–3, 5: stable stratification;
4, 6: neutral stratification;
7–10: unstable stratification;
11: calculation from Reichardt equation [5]

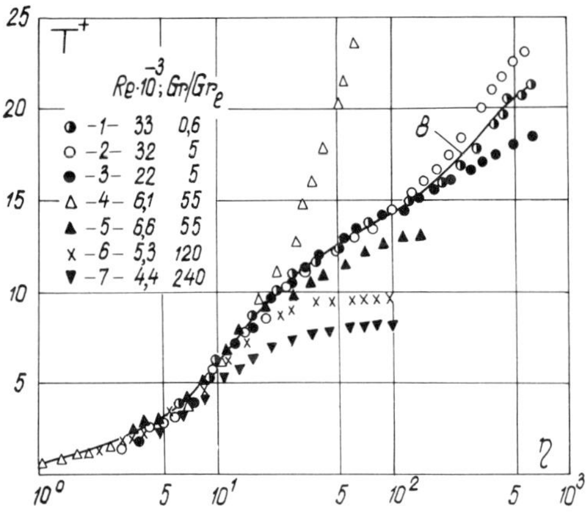

Fig. 2. Dimensionless temperature distribution as a function of the dimensionless distance from the heated wall:
1: neutral stratification;
2, 4: stable stratification;
8: calculation from the equation [8] for neutral stratification;
3, 5–7: unstable stratification

common influences to internal and external flows exist and can be understood from the Monin-Obukhov theory.

In [9–12] the logarithmic region of wall boundary layer flows (τ = const, q = const.) is considered. With the help of dimensional analysis and by introducing some simplifying assumptions, Monin-Obukhov received a single independent variable which affects the turbulence characteristics in stratified flows. It is the so-called stability parameter:

$$\xi = \frac{y}{L}, \quad \text{where} \quad L = -\frac{v_*^3}{\kappa}\left(\frac{g}{T_b} \cdot \frac{q_w}{\rho c_p}\right)^{-1} \tag{2}$$

The negative sign in the above equation is necessary in order to obtain $L > 0$ for the stable thermal stratification. For distances $y \ll |L|$ the turbulence transport properties are only slightly affected by buoyancy forces and correspond to those measured in isothermal flows. Therefore the L-scale defines the thickness of the near wall layer within which the buoyancy influences are small. In this region the flow is mainly controlled by the convective terms in the fluid mechanics equations and the energy equation [11, 12].

Monin and *Obukhov* called their layer dimension, L, the dynamic turbulence sublayer height. The actual thickness of this height is, however, only proportional to L and the proportionality constant is not known. Atmospheric measurements showed [5], that the proportionality constant is of the order of 10^{-2} but the exact magnitude could not be deduced. The present results permit the proportionality constant to be computed. For this purpose the value of:

$$L = -\frac{v_*^3}{0.4}\left(\frac{g}{T_b} \cdot \frac{q_w}{\rho c_p}\right)^{-1} \tag{3}$$

was computed using T_b (bulk temperature) as reference temperature. The dynamic turbulence

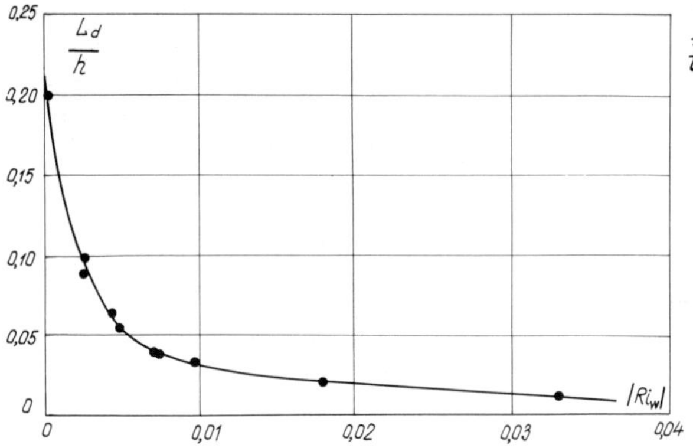

Fig. 3. Dependence of the dynamic turbulence sublayer thickness on the magnitude of the Richardson number

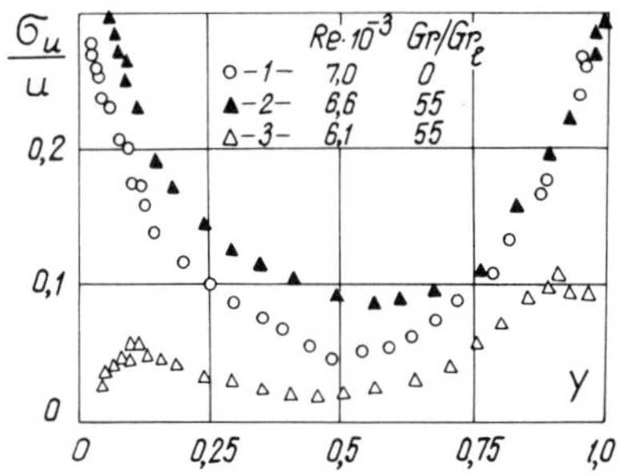

Fig. 4. Distribution of intensity of the velocity fluctuation across the channel:
1: neutral stratification;
2: unstable stratification;
3: stable stratification

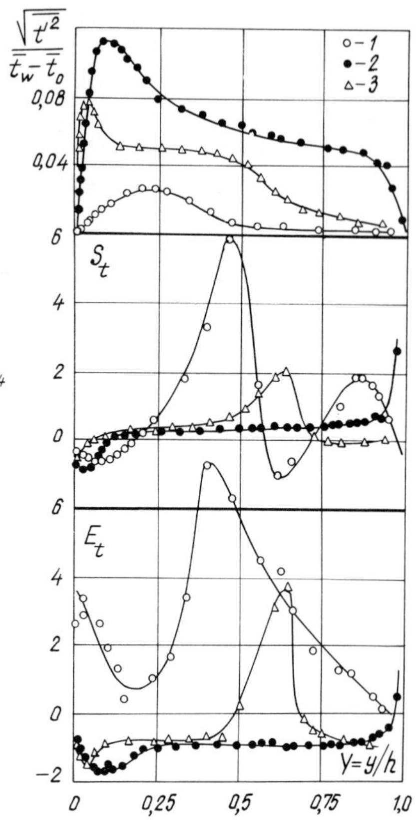

Fig. 5. Distributions of intensities, asymmetry and excess coefficients (skewness and flatness factors) of the temperature fluctuations across the channel:
1: stable stratification, Re = $6.1 \cdot 10^3$, $Gr/Gr_l = 55$;
2: unstable stratification, Re = $6.6 \cdot 10^3$, $Gr/Gr_l = 55$;
3: neutral stratification, Re = $3.3 \cdot 10^4$, $Gr/Gr_l = 0.6$

sublayer thickness was defined as the distance from the wall where the velocity and temperature profiles begin to deviate from the values measured for the isothermal flow case. It was found that the proportionality constant is equal to 0.03, a value which is close to the estimation in [5]. The dependence on the dynamic turbulent sublayer thickness, $L_d = kL$ is shown in Fig. 3 as a function of the Richardson number, Ri_w. The numerical values for L_d are given in Fig. 3 for various Richardson numbers, Ri_w. The values of L_d are normalized with the channel height.

The results of measurements of velocity and temperature fluctuations, σ_u and σ_t are shown in Figs. 4 and 5. It is evident from these data that the stable stratification results in significant decreases of σ_u and σ_t in the entire channel cross section. The characteristic near wall maximum of these quantities is being broadened and shifted away from the heated wall to the channel axis. The decrease of the fluctuations for stable stratification can be explained by the energy consumption required for the displacement of turbulent eddies in the stably stratified field.

As the results show, the intensities of velocity and temperature fluctuations, σ_u and σ_t, increase in the entire flow region under conditions of unstable stratification. On the one hand, buoyancy forces lead for unstable stratification to an increase of velocity and temperature gradients in the area close to the heated wall, thus intensifying in this region the turbulent energy production and the energy of the temperature fluctuations due to the mean flow. On the other hand, the unstable stratification increases the turbulent diffusion which is responsible for the increase of the temperature fluctuations over the entire flow region. It is this increased diffusion that causes the increase in temperature fluctuations near the unheated wall. A very strong increase of the temperature boundary layer is also observed near the heated wall.

The significant variations of the mean velocity and temperature fields and the turbulent transport properties, naturally leads to the observed changes of such integral parameters as friction factor and heat transfer coefficient near the heated wall. It has been shown by the present experiments that under unstable stratification the heat transfer and wall friction increase, while in the case of stable stratification both the wall friction and heat transfer diminish to the values specific to laminar flow. Thus, one could speak of laminarization of the turbulent flow under stable stratification.

In addition to the measurements of mean velocity and mean temperature, and the intensities of velocity and temperature fluctuations, the probability density distributions of the temperature and velocity fluctuations were measured for conditions on which considerable buoyancy effects were present. The second, third and fourth moments of these distributions were computed and compared with previous data for neutral stratification conditions, e.g. see [13]. This distributions revealed the effect of stratifications given above.

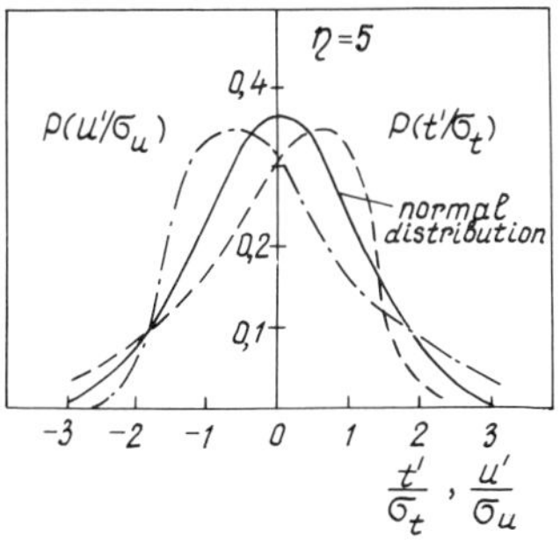

Fig. 6. Probability density distributions of fluctuation amplitudes of temperature and velocity in the viscous sublayer, Re = $3.2 \cdot 10^4$, $Gr/Gr_l = 0.6$

Some examples of the measured probability density distribution of the velocity and temperature fluctuations $P(u'/\sigma_u)$ and $P(t'/\sigma_t)$ are presented in Fig. 6 for a measured point inside the viscous sublayer. The figure also shows a normal (Gaussian distribution) and indicates that the probability density functions are asymmetric. The maximum value of $P(u'/\sigma_u)$-distribution is shifted to negative values of the velocity fluctuations, u', whereas the maxima of $P(t'/\sigma_t)$-distribution is shifted to positive values of the temperature fluctuations, t'. This difference in the most probable velocities and the mean velocity as well as the most probable temperature and the mean temperature is reflected by the third moments of the measured probability density distributions to be not equal to zero. This asymmetry is caused by the proximity of the solid surface and the differences in the signs are due to the fact that the wall is impermeable to the velocity field but provides a constant heat flux, (compare with [14]). The long right hand branch of the $P(u'/\sigma_u)$-function and the elongated branch of the $P(t'/\sigma_t)$-function correspond to the rare, although relatively strong, positive fluctuations of the velocity and negative fluctuations of the temperature which are determined by the inrushes of the faster colder fluid from the centre of the channel. This fluid occasionally reaches the viscous sublayer and can be recorded there.

The experiments also show that the intensity of the velocity fluctuations, reaches its maximum value in the region where the turbulent energy production reaches its maximum near the heated wall ($\eta \approx 15$). In this region the asymmetry coefficient S_u is equal to zero and the excess coefficient E_u takes on its minimum. Similar relationships exist for corresponding parameters of the temperature distribution for values of $\eta = 20-22$. This location is dependent on the Prandtl number.

As pointed out, the asymmetry coefficient of the velocity and temperature fluctuations are of different signs ($S_u > 0, S_t < 0$). This difference is caused by the fact that near the heated wall the temperature of the fluid decreases while its velocity increases with increasing distance from the wall. Thus, turbulent vortices creating positive velocity fluctuations near the wall produce a negative temperature fluctuations.

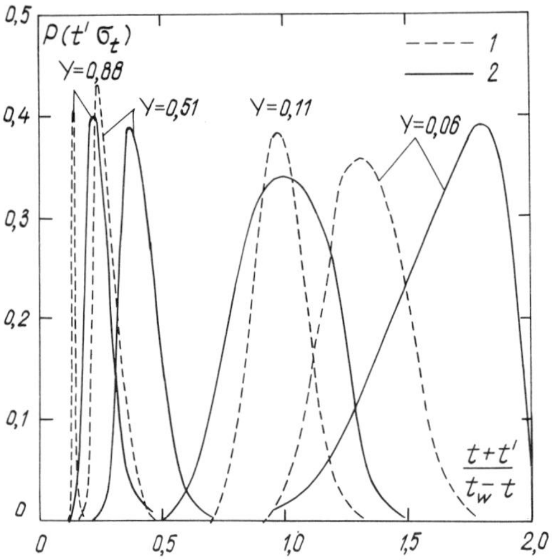

Fig. 7. Probability density distributions of temperature fluctuations:
1: stable stratification, Re = $6.1 \cdot 10^3$, $Gr/Gr_l = 55$;
2: unstable stratification, Re = $6.6 \cdot 10^3$, $Gr/Gr_l = 55$

It is apparent from Fig. 7, that the $P(t'/\sigma_t)$-distribution undergoes significant changes as the buoyancy forces become more dominant. Figure 7 clearly shows that the $P(t'/\sigma_t)$-distributions are much broader for the unstable stratification than for the stable one. In accordance with this observation, only positive excess coefficients are observed under the stable stratification, e.g. see Fig. 5. The data indirectly show that the probability of the occurrence of the large temperature fluctuations decrease under conditions of stable stratification due to a suppression of the large scale pulsations. Figure 5 shows that there are interconnections between the changes in the second, the third and fourth moment. Specific values are reached for these quantities at a distance from the heated wall $y < 0.25$. These specific values occur at a larger distance for conditions of stable stratification.

The data provided in Fig. 5 document that the viscous layers are thickening under conditions of stable stratification.

The asymmetric thermal boundary conditions lead to significant distinctions between thermal and hydrodynamic parameters. The abrupt decrease in the intensity of the temperature fluctuations is observed close to the outer edge of the temperature boundary layer, i.e. near the cold wall where the gradients of the mean temperature distributions are close to zero. The distribution of the third and fourth order moments show in this region maximum values. Measurements also indicated the existence of strong temperature intermittencies in this flow regime that seem to characterize the interaction between the flow regions with significantly different intensities of temperature fluctuations, e.g. see [13, 15, 16].

The decrease of the turbulent heat transport under conditions of stable stratification leads, naturally, to decrease of the temperature boundary layer thickness. This corresponds to the displacement of the region of maximum values of the skewness and flatness factors, i.e. the values S_t, E_t. This region corresponds to the outer edge of the thermal boundary layer near the heated wall. For stable stratification, as Fig. 5 shows, there is a significant increase in the maxima for those skewness and flatness which can be taken as an indication of the increase in the temperature intermittency.

Opposite effects can be observed on the turbulent transport properties for conditions of unstable stratification. These conditions yield the thickness of the temperature boundary layer to increase until its thickness almost coincides with the channel height. As the data in Fig. 5 show, the skewness factor is constant over most part of the cross section. The skewness factor shows positive values whereas the flatness factor is negative. This is a clear indication of the fact that the probability of large positive temperature fluctuations is rather high. These correspond to ejection of hot fluid away from the wall.

The results of the amplitude analysis are confirmed by measurements of the frequency spectra of the temperature fluctuations. For comparison, Fig. 8 presents corresponding spectra $G(k_1)$ for the stable and unstable stratification. The figure clearly shows the difference between these spectra in the low frequency range. The energy in the low frequency fluctuation is much smaller for the stable stratification than for the unstable one. The figure shows that the high frequency end of the spectrum (eddies of small dimensions) is not affected by the buoyancy forces for stable and unstable stratification conditions. This fact is in good agreement with results of Obukhov's theoretical studies presented in [17]. *Obukhov* introduced a linear scale L_* which determines the minimum dimension of those turbulence eddies on which the buoyancy forces have influences. It is shown in [17] that effects of density stratifications are important for the motions of the larger eddy sizes whereas the turbulent motion of the smaller eddies is practically unaffected by buoyancy.

Following the suggestion by Prandtl, mixing length can be defined from momentum transport and heat transport:

mixing length for momentum transport: $$l_u = \sigma_u / \left|\frac{du}{dy}\right|$$

mixing length for heat transport: $$l_t = \sigma_t / \left|\frac{dt}{dy}\right|$$

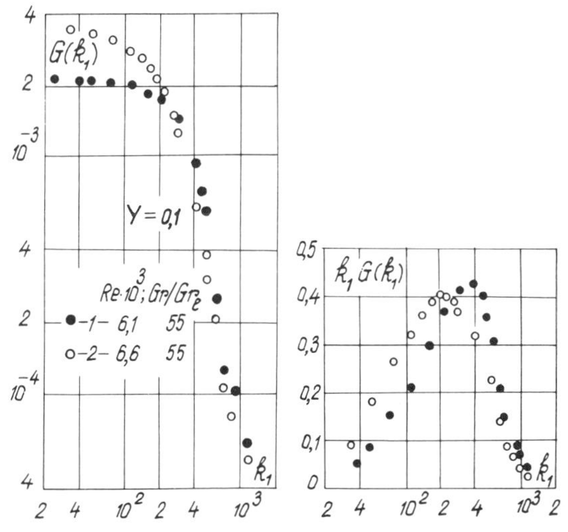

Fig. 8. One-dimensional power spectra of temperature fluctuations under the conditions of stable (*1*) and unstable (*2*) density stratification

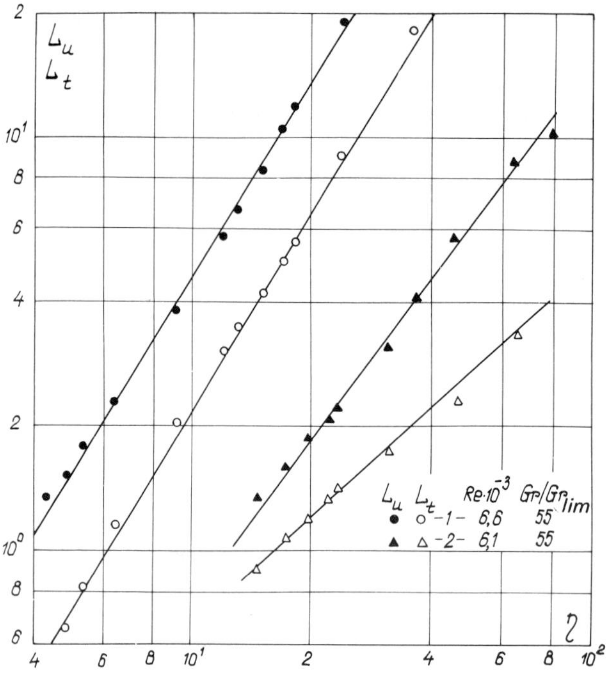

Fig. 9. Dependence of dimensionless scales for heat (L_t) and momentum transfer (L_u) on the dimensionless distance from the wall under unstable (*1*) and stable (*2*) stratifications

It is well known that these lengths can be interpreted as characteristic dimensions of local scales, i.e. as mean sizes of turbulent vortices carrying the main energy, e.g. see [18]. Figure 9 provides in a dimensionless form the aforementioned values as functions of the dimensionless coordinate η for both stable and unstable density stratification. These distributions show some interesting features. Firstly, the scales of the temperature field are smaller than the velocity scales. This is consistent with the Prandtl number being smaller than 1 for air. Secondly, it is clearly seen from Fig. 9 that L_u and L_t are linearly dependent on η within the near-wall region. This fact stresses what has been said above that the turbulent energy transport is carried out by large vortices with sizes that increase linearly with the distance from the wall, e.g. see [5]. Figure 9 also shows that for unstable stratification the ratio for L_u/L_t is nearly constant whereas for stable stratification L_t increases less with the distance from the wall than L_u. This is due to a stronger attenuation of the heat transport than the momentum transport in the near wall region. An explanation for this has been given in [19] where a wave-like transport of momentum has been postulated with a higher efficiency for momentum transport than for heat transport.

References

1. B. S. Petukhov, A. F. Polyakov, Yu. L. Shekhter: Turbulent flow and heat transfer in gravitational field. Teplofiz. Vys. Temp. (High Temperature) **16**(3), 624–639 (1978)
2. B. S. Petukhov, A. F. Polyakov, Yu. L. Shekhter, V. V. Troitsky: Gravitational field effect upon the structure of turbulent nonisothermal flow in horizontal cylindrical channels. Dokl. Akad. Nauk SSSR (Reports of the USSR Academy of Sciences) **236**(4), 820–823 (1977)
3. V. A. Kuleshov, A. F. Polyakov, Yu. V. Tsypulev: Experimental studies of temperature field for turbulent stratified air flow in a flat channel. Teplofiz. Vys. Temp. (High Temperature) **15**(6), 1316–1318 (1977)

4. B. S. Petukhov, A. F. Polyakov, Yu. L. Shekhter, Yu. V. Tsypulev: "Experimental Study of the Statistical Parameters of Temperature Fluctuations in a Viscous Air Sublayer", in *Trudy Moskovskogo Energeticheskogo Instituta:* Teploobmen i Gidrodinamika (Proc. of Moscow Energy Inst.), Vol. 235 (1975) pp. 3–13
5. A. S. Monin, A. M. Yaglom: *Statistical Hydromechanics,* Vol. I (Nauka, Moscow 1965) p. 640
6. A. F. Polyakov: "Limits of Archimed Forces Influence upon Flow and Heat Transfer in Horizontal Near-Wall Turbulent Flow", in *Teploobmen i fizicheskaya gazodinamika* (Heat Transfer and Physical Hydrodynamics) (Nauka, Moscow 1974) pp. 23–33
7. J. P. Hartnett, I. C. Y. Koh, S. T. McComas: Comparison of estimated and experimental friction factors for turbulent fluid flows in rectangular channels. J. Heat Transfer **84**(1), (1962)
8. B. S. Petukhov, A. F. Polyakov, V. A. Kuleshov: "Temperature Distribution in a Turbulent Air Flow with Constant Properties", in *Trudy Moskovskogo Energeticheskogo Instituta* (Proc. of Moscow Energy Inst.), Vol. 313 (1976) pp 111–117
9. A. M. Obukhov: "Turbulence in Temperature Nonuniform Atmosphere", in *Trudy Instituta Teoreticheskoi Geofiziki Akad. Nauk* (Proc. of Institute of Theoretical Geophysics of the USSR Academy of Sciences), Vol. 1 (1946) pp. 95–115
10. A. S. Monin: "Turbulent Regime in Near-Earth Air Layer". Informatsionnii Sbornik Glavnogo Upravleniya Gidrometeoslujbi (Informational collection of papers of the main department of weather service), Vol. 1 (1950) pp. 13–27
11. A. S. Monin, A. M. Obukhov: Nondimensional turbulence characteristics in near-earth atmospherical layer. Dokl. Akad. Nauk USSR (Reports of the USSR Academy of Sciences), **93**(2), 223–226 (1953)
12. A. S. Monin, A. M. Obukhov: "The Main Regularities of Turbulent Mixing in Near-earth Atmospherical Layer", in *Trudy Geofizicheskogo Instituta Acad. Nauk.* (Proc. of Geophysical Institute of the USSR Academy of Science), Vol. 24, (151) (1954) pp. 163–187
13. B. S. Petukhov, A. F. Polyakov, Yu. V. Tsypulev: Experimental study of velocity and temperature for a turbulent air flow in an asymmetrically heated flat channel. Teplofiz. Vys. Temp. (High Temperature). **16**(5), 987–996 (1978)
14. A. T. Popovich: Statistical lawsend of the viscous sublayer in turbulent flow. Ind. Eng. Chem. Fundam. 8(4), 609–614 (1969)
15. S. Tanimoto, T. J. Hanratty: Fluid temperature fluctuations accompanying turbulent heat transfer in a pipe. Chem. Eng. Sci. **18**(5), 307–311 (1963)
16. M. Hishida: "Structure of Turbulent Temperature and Velocity Fluctuations in the Thermal Entrance Region of a Pipe", in VI International Heat Transfer Conference, Toronto, Canada, Vol. 2, (1978) pp. 531–536
17. A. M. Obukhov: On Archimed forces effect on temperature field structure in a turbulent flow. Dokl. Akad. Nauk USSR (Reports of the USSR Academy of Sciences) **125**(6) 1246–1248 (1959)
18. V. M. Ievlev: *Turbulent Motion of High-Temperature Medium* (Nauka Moscow 1975) p. 256
19. G. S. Turner: *Buoyance Effects in Fluids* (Mir, Moscow 1977) p. 432

Part III

Complex Flows

Introduction. Recirculating and Complex Two- and Three-Dimensional Flows

H. McDonald

Complexity, like beauty, is in the eye of the beholder, and here we eschew defining precisely what we mean by "complex" other than to suggest that at this stage in the development of our understanding of shear flows "complex" implies three dimensionality and/or local or global breakdown of the thin shear layer approximations. The papers in this section comprise an experimental study of turbulent duct flow with one wall roughened, by Humphrey and Whitelaw, two papers on the prediction of three-dimensional laminar and turbulent boundary layers on three-dimensional bodies of revolution, one by Cebeci, Khattab and Stewartson and the other by Patel and Choi. These are followed by a combined theoretical-experimental study of the flow over a two-dimensional fence by Durst and Rastogi and a paper on the prediction of the flow resulting from a round jet discharging into a confined cross flow by Jones and McGuirk. Lastly an entirely experimental study on transitional flow around a sphere placed on a wall is reported by Okamoto. With the exception of the last paper all the studies reported are more or less directly concerned with current efforts to obtain a quantitative predictive ability (and presumably also understanding) for the flows in question. It is this common thread of "prediction" which provides the focal point for the present introductory observations.

First of all the current emphasis on prediction places very stringent demands on the experimentalist. In this case he is seeking not only to explain or understand mechanisms but also he must provide a carefully documented, dense, accurate data set, such that comparisons between prediction and measurement can reliably be used to calibrate predictions. This includes initial and boundary conditions, in view of the very natural human trait to assume the most favorable values for those items which are not measured but whose initial and boundary conditions are required in making the predictions. In the jargon of the day such detailed studies are termed "bench mark experiments" and the paper by Humphrey and Whitelaw gives some indication of the tremendous effort (and in this case, experimental measurement sophistication, now taken somewhat for granted) required to produce the quality and quantity of data necessary to qualify as benchmark study. Also the Humphrey-Whitelaw experiment illustrates three features of good benchmark studies. First, the geometry of the problem is simple (a straight rectangular duct) giving the predictor little added source of error and/or complexity from the geometry, other than its three dimensionality. Second, by fairly simple means (in this case a roughened wall) a complex internal flow is established, which with some selectivity can test certain features of the turbulence model approximations. Thirdly, the investigators have sought to explain their observations in terms of mechanisms, and this serves to guide predictors as to the type of approximation which may or may not be admissible, and in addition further substantiates the measurements.

In developing the benchmark philosophy it is not necessary to concentrate on turbulent flow or on measurement. There is indeed a great interest by predictors of turbulent flow in laminar experiments and in precise laminar flow calculations. The explanation for this interest in laminar flows rests in the dual nature of the problem faced by predictors of complex turbulent flows. First of all the predictor must (usually) numerically solve a nonlinear system of partial differential equations which contain two types of approximation, one based on physics (i.e., deleting certain terms from the averaged Navier-Stokes equations) and one based on numerics (the various forms of discretization and linearization approximations).

Next he must be concerned with the turbulence model per se and the degree to which he is able to represent either the statistical correlations or the sub-grid motion not resolved by the calculation. Turbulent benchmark experiments can guide the turbulence modeling effort. Laminar studies can aid in the resolution of the physical-numerical approximation problems. Thus it is not altogether surprising to find, for instance, Humphrey, Taylor, Whitelaw: J. Fluid Mech. *83* (3), (1977) making this same type of detailed benchmark study for laminar flow in curved rectangular ducts. The numerical problems faced by the predictors of these complex laminar flows can be very considerable. The resolution of the laminar problem can be very helpful when eventually the comparable turbulent problem is treated. Errors having their origin entirely in the numerical solution procedure can in some cases dominate the results and can lead (and has led) to the turbulence model being debited with the resulting poor agreement with experiment (or, quite the converse, in some cases the turbulence model has been incorrectly credited with the inaccurate but good agreement with experiments resulting from poor numerical resolution).

The interest and value of laminar flow studies for turbulent predictors is apparent in the three-dimensional boundary layer papers of both Cebeci, Khattab and Stewartson and of Patel and Choi. The importance of numerical error and the reasons for removing it are clearly evidenced in the recirculating flow paper by Durst and Rastogi, where it was found that the turbulence model was probably better than hitherto suspected (at least for this class of flows) when evaluated in circumstances where the numerical errors have been at least reduced, if not entirely eliminated. In the two three-dimensional boundary layer studies both were concerned with problems which arise equally with laminar and turbulent flow, the problem of numerical integration stability with negative cross flows and what is in fact a very closely related problem, that of axial flow reversal ("separation"). Both groups also chose the same body to treat, another example of achieving complexity with simplicity (an apparent contradiction), since use of the same body whose geometry can be defined analytically is very convenient in that it can reduce the geometric sources of error and also provides for easy comparison between different investigations. In the oral presentation of these two papers the authors gave very strong emphasis to their numerical problems. Although still discernable, the emphasis on numerical difficulty is more muted in the written versions given here. Cebeci et al. prefer to make an explicit spanwise march allowing for reverse flow in the cross flow profile by a zig-zag scheme, which may remove the numerical problem but still faces the problem of information flow against the sweep direction. The explicit spanwise sweep was selected by Cebeci et al. for convenience in defining the separation zone, which once determined is excluded from the computational region. On the other hand, Patel and Choi in one approach selected a spanwise implicit structure to overcome the reverse cross-flow velocity profile problem successfully, although then they faced the problem (not treated in the paper) of eliminating the separated flow region from the computational domain i.e., from the implicit spanwise sweep. Presumably the separated flow region could eventually be excluded from the implicit sweep in much the same way that people currently allow interior boundaries to appear in fully implicit schemes (see for instance, AIAA Paper 79-1546). Patel and Choi also raised the spectre of substantial interaction between the boundary layer and the external inviscid flow. All in all it is clear that while substantial progress is and has been made in predicting thin three-dimensional turbulent unseparated boundary layers, there remains much to be done numerically, and probably in laminar flow, before the boundary layer on the ellipsoid at incidence will be accurately predicted.

Durst and Rastogi were led to question their own and other numerical predictions for the turbulent flow over a fence partly as a result of the insensitivity of the solution to changes in the turbulence model and hence the turbulent viscosity (the inference being that the turbulent viscosity was being swamped by the numerical diffusion) and partly as a result of the disparity with other investigators' seemingly similar computations. The problem appears to have been traced to the coarse mesh computations performed earlier. When the mesh is refined the solution behavior now seems much more consistent, although mesh independence of the solution has not yet been demonstrated, presumably due to the high computational cost. The current status appears to be that the gross feature of this turbulent separated flow are remarkably well predicted and that future work will examine the detailed predictions of the mean flow. Durst and Rastogi however are now quite concerned with the computational cost of obtaining the fine mesh solutions, but presumably as the more efficient numerical techniques currently developed become more widely known and utilized the computational expense of obtaining the fine mesh solutions should be diminished.

The paper by Jones and McGuirk presents three-dimensional calculations for a round turbulent jet discharging into a confined cross flow. The gross features of the flow are fairly well predicted but the coarse mesh used (25 x 15 x 15) together with the recollection of the Durst-Rastogi findings warns against reading too much into the results. What is impressive with these results is that computations of this complexity can be performed by people in the turbulence modeling community. Indeed it has to be observed that unless the turbulence modeler/predictor develops a facility for using computational techniques, he is serverely restricted both in evaluating turbulence models and more generally in making flow predictions. Obviously, as one can glean from the Durst-Rastogi experience, this facility with computational technique must be more than simply a passing acquaintance. As Jones and McGuirk point out it is likely that there are significant truncation errors in their calculations and a true evaluation of the turbulence model must await a relatively error free computation. Obtaining these relatively error free calculations with the computational techniques currently in widespread use by the turbulent shear flow community will be computationally expensive and may not even be possible. It may well be time for the turbulence modeling community to carefully examine the computational advances of recent years with a view to implementation of some of these advances, as unwelcome and diversionary as this might appear. What is also now becoming clear is that the observations and evidence concerning the predictive capabilities of the turbulence models for complex flows based on the earlier coarse mesh results must now be questioned.

Turbulent Flow in a Duct with Roughness

J. A. C. Humphrey

Department of Mechanical Engineering, University of California,
Berkeley, CA 94720, USA

J. H. Whitelaw

Department of Mechanical Engineering, Imperial College of Science and Technology,
London, England

Abstract

Measurements are reported of the mean flow and turbulence characteristics of a flow developing in a straight duct of rectangular cross-section asymmetrically roughened by placing elements of square cross-section transverse to the main flow direction. The results, obtained by means of a laser-Doppler velocimeter, show that Reynolds stress induced cross-stream motion occurs earlier and is stronger than that arising in corresponding smooth duct flow. However, in response to strong pressure variations normal to the roughness elements, the transverse velocity component eventually overcomes the weaker stress-driven motion. Measurements of the stresses indicate high levels of turbulence intensity and anisotropy near the roughened and side walls of the duct, respectively, and suggest that generation and dissipation of kinetic energy of turbulence predominates at the roughened wall. Regions of negative contributions to the generation of kinetic energy, where they arise, are small. The measurements show clearly that no existing law of the wall relationship can adequately describe the wall velocity distributions which are strongly dependent on the transverse flow coordinates.

Introduction

Of special concern here are the respective roles of secondary flows and shear-layer interactions in straight duct flows with asymmetric roughness. For a useful discussion of these effects in a more general context, the reader is referred to the informative review by Johnson (see *Bradshaw* [1]).

Secondary Motions

Secondary motions are of special interest in relation to the flows in ducts with asymmetric roughness which arise where there is lateral curvature of the main flow and/or gradients in the Reynolds stresses. The former gives rise to pressure driven secondary flow, and the latter to normal-stress driven secondary flow. Both kinds of secondary flows have been investigated extensively. A recent review and experimental results pertaining to flows where lateral curvature is important are given by *Humphrey* [2]. Examples of calculations of 3-D curved flows in laminar and turbulent regimes are respectively given by *Humphrey* et al. [3] and *Patankar* et al. [4]. Secondary flows of the second kind have been measured experimentally by, among others, *Brundrett* and *Baines* [5], *Gessner* and *Jones* [6], and *Melling* and *Whitelaw* [7]. Corresponding predictions have been made by, for example, *Naot* et al. [8] and *Launder* and *Ying* [9], using higher order stress closure models of turbulence. It is worth noting that simpler models based on the notion of an isotropic turbulent viscosity cannot possibly predict Reynolds stress driven secondary motions.

Shear Layer Interactions

To a large extent, the manner in which asymmetrically roughened straight duct flows develop is a function of the nature and intensity of the interactions arising among the respective wall shear layers. In their experimental study of fully developed asymmetric 2-D channel flow, *Hanjalic* and *Launder* [10] have shown that this interaction results in displaced surfaces of zero shear stress and mean rate of strain. The effect is due to the strong turbulence diffusion of shear stress from the rough to the smooth channel wall which results in a negative production of turbulent kinetic energy. From experimental determination of the terms in the conservation equation for turbulent kinetic energy, *Hanjalic* and *Launder* [10] have also found that dissipation of kinetic energy is balanced primarily by the diffusional influx of this quantity from the vicinity of the rough wall into the shear layer interaction region.

In an effort to determine a general experimental equation to predict average resistance coefficients, *Fujita* [11] has recently measured longitudinal mean velocity and wall shear stress distributions in fully developed flow for all combinations of rough and smooth walls in a straight duct of square cross-section.

The present work, reported in part in Ref. [12], represents an attempt to quantify in more detail the mean flow and turbulence characteristics in an asymmetrically roughened duct of square cross-section. The measurements are of the longitudinal and transverse components and yield values of mean velocity and the corresponding normal and shear stresses.

Flow Apparatus, Experimental Technique and Measurement Procedure

The experimental equipment, flow configuration and measurement procedure used in this study have been explained in detail by *Humphrey* [2], but a brief discussion is in order here. Water at room temperature flowed at 0.58 m/s (Re = 2.6×10^4) from a constant head tank into a large plenum chamber and through a test section shown schematically in Fig. 1. The test section was made of 1 cm thick plexiglass with the dimensions shown in the figure. Rib-like elements of square cross-section cut from cold-drawn brass were placed on one wall, transverse to the main flow direction, at intervals of 0.99 hydraulic diameters. This resulted in a pitch-to-height ratio of approximately 10:1. Careful alignment of the duct with the inlet contraction and trip plate in the plenum chamber ensured symmetry in the measurements to within experimental error.

Bulk average velocities (U_B) were monitored by means of an orifice flowmeter designed in accordance with BS 1042 specifications. Measurements of turbulent flow properties at six longitudinal stations were obtained by means of a forward scatter, frequency-shifted, laser-Doppler velocimeter. The velocimeter and associated electronic components consisted of a 5 mW He-Ne Spectra Physics laser, an 18,000-lined bleached rotating diffraction grating with focussing lens and two additional lenses, one for crossing the first order beams, the second to focus the measuring volume onto the pinhole of an EMI 9558 B photomultiplier tube. The use of rotating gratings in laser-Doppler velocimetry has been discussed by, among others, *Durst* et al. [13]. In the present work, the grating provided a frequency shift of 0.243 MHz and facilitated measurement in highly turbulent flow regions by means of a tracker. The radial grating, its mounting and supporting platform were constructed such that measurements at ±45° to the main

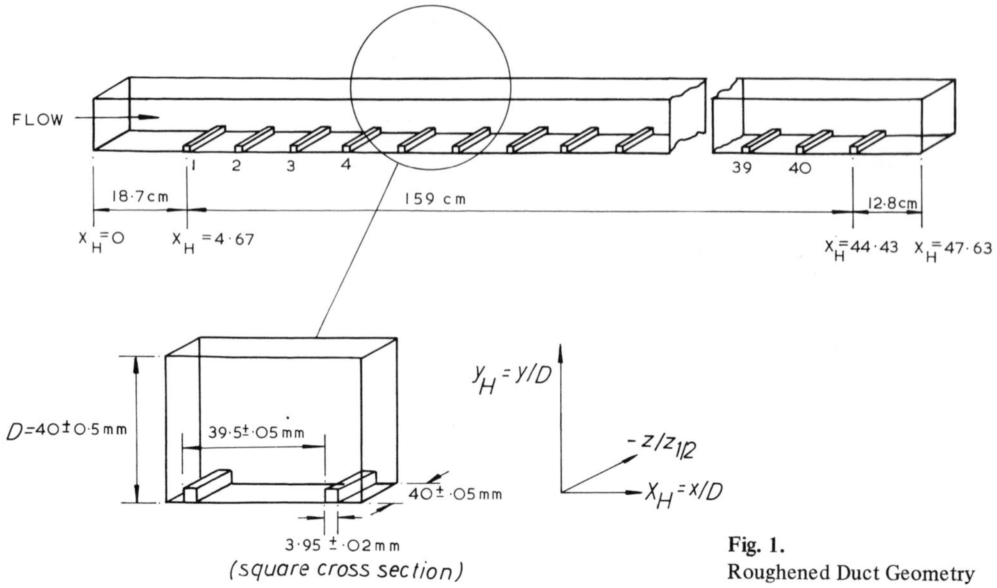

Fig. 1.
Roughened Duct Geometry

flow direction could be obtained. The first order beams diffracted by the rotating grating had a half angle of 1.55° and produced a calculated measuring volume of approximately 1.82 x 0.1 x 0.1 mm^3. The aligned optics were mounted on an *x, y, z* traversing mechanism which could be located relative to the test section with a precision of ±0.01 mm in the horizontal directions and ±0.1 mm in the vertical direction.

The electronic instrumentation consisted of a DISA 55L20 frequency tracking demodulator, a Tektronix 7603 oscilloscope, a DISA 55B30 true time integrator, and a DISA 55D35 rms meter. In addition, two Solartron LM 1420.2 digital voltmeters were used for displaying the integrator and rms meter results. Signals of high signal-to-noise ratio were monitored by the tracker at drop-out levels of less than 2%. The out-put from the tracker was presented to the true time integrator and rms meter from which mean velocities and turbulence intensities were obtained. It is estimated that approximately 10^5 individual Doppler bursts led to each single average and rms result. Between 3 to 5 mean velocity measurements and 10 to 15 rms measurements were taken to find the final averaged values.

Two components of velocity and three Reynolds stresses were measured in the flow of Fig. 1. On average, 12 traverses of 25 points each were made at each longitudinal station. These positions correspond to three main locations with measurements in the plane of the rib and between pairs of ribs, respectively. The measurements spanned the width of the duct and showed symmetry in the profiles to within experimental error. Therefore, the data over the whole cross-section was averaged with respect to the symmetry plane. Contour plots of the measurements were obtained by least squares polynomial regression followed by interpolation. The quality of the polynomial fits was ascertained by reference to the residual errors, and bias due to polynomial curvature was shown to be unimportant.

The error sources and accuracy of the measurements have been quantified in Ref. [2]. Turbulent and non-turbulent broadening effects and electronic noise were found negligible. Systematic errors in the longitudinal velocity and turbulence intensity were 0–1.5%, whereas maximum random errors were approximately ±2.5% and ±3%, respectively. These errors were due mainly to the signal processing instrumentation and orifice meter readings.

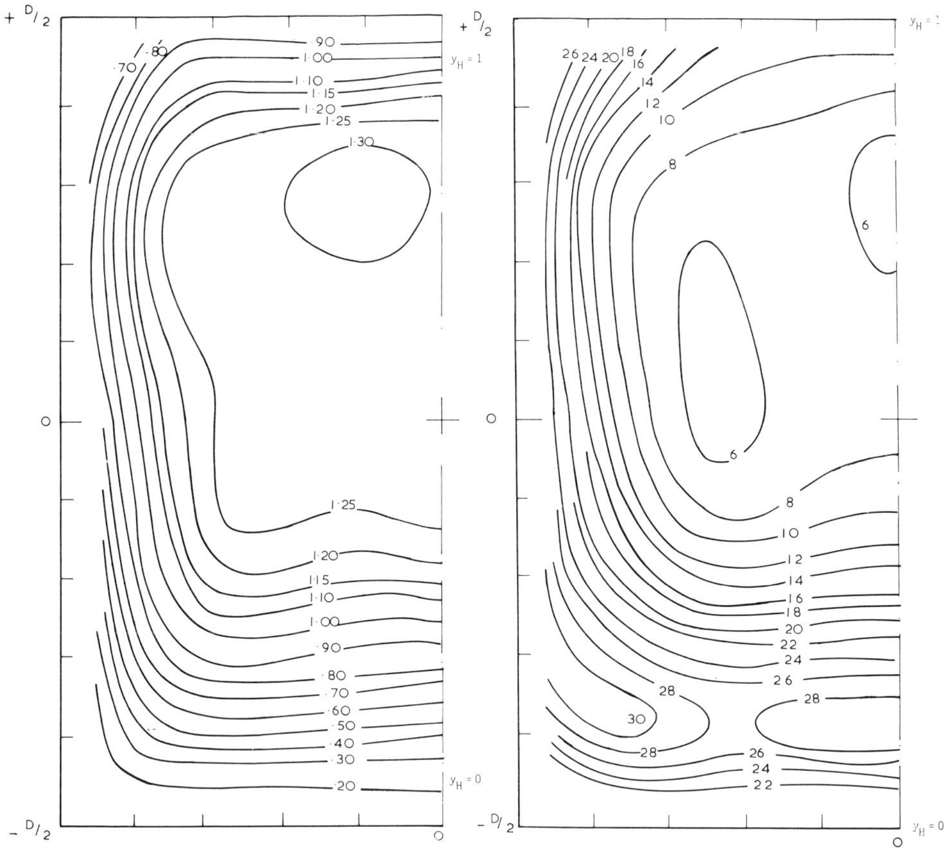

Fig. 2. U_x/U_B and $(\tilde{u}_x/U_B)$ 100 at x_H = 6.15

Discussion of Experimental Results

The discussion presented here centers on the results for the first and last pair of longitudinal locations, i.e. at x_H of 6.15 and 6.65, corresponding to the location midway between ribs 2 and 3 and over rib 3, respectively, and at x_H of 41.68 and 42.17, corresponding to the location midway between ribs 38 and 39 and over rib 39, respectively. Measured values of the mean velocities and Reynolds stresses are presented in contoured form in Fig. 2 to 13. Values are normalized by the bulk average velocity U_B as indicated, and the coordinate geometry corresponds to that shown in Fig. 1.

It should be noted that the entrance characteristics of the present flow correspond closely to those measured by *Melling* and *Whitelaw* [7] in a smooth duct investigation at a location corresponding to x_H = 5.6 in the present geometry.

Mean Velocities

Mean flow contours are shown in Figs. 2 and 3 for the stations halfway between ribs 2 and 3 and over rib 3, respectively. At both stations, the initial values for U_x show a plug-like distribution in the core region of the flow with very steep gradients towards the walls, except for the

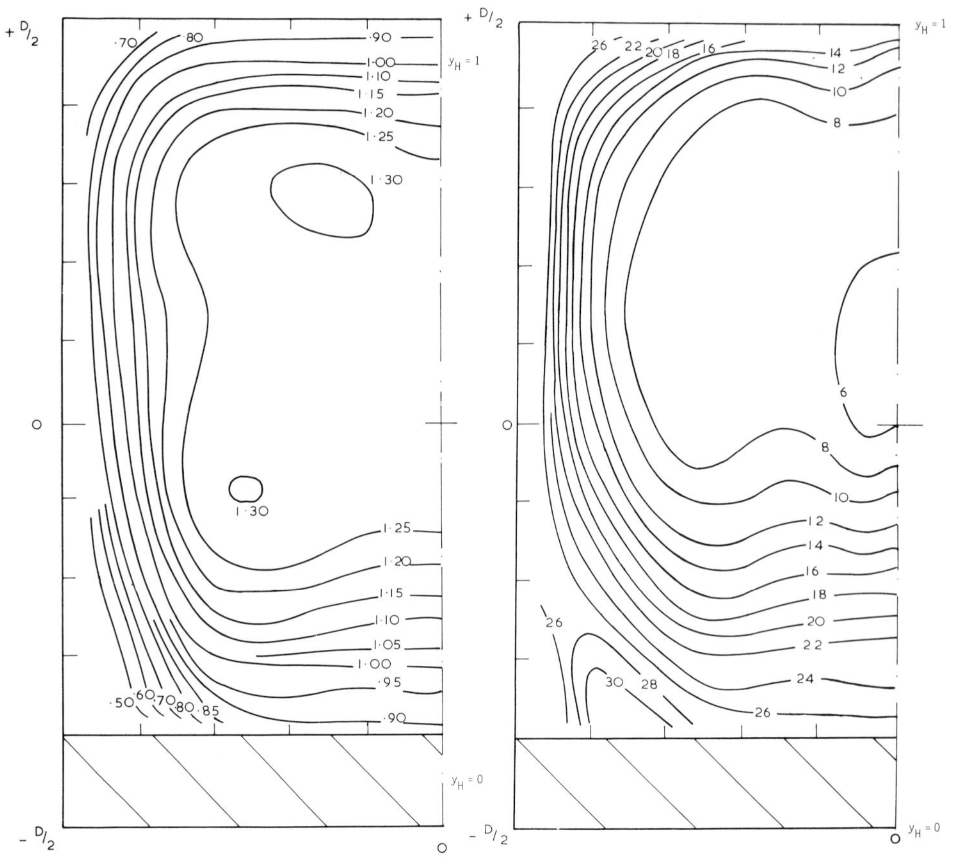

Fig. 3. U_x/U_B and $(\tilde{u}_x/U_B)$ 100 at $x_H = 6.65$

bottom wall region ($y_H = 0$) at the between-rib station where the flow moves considerably more slowly. Both plots show peak values of the U_x component slightly displaced from the symmetry plane ($z = 0$). The results, nevertheless, are reminiscent of developing flow in a square duct with smooth walls, particularly in the region removed from the ribs. Bulging of the U_x contours at these two stations suggests the early appearance of a normal stress-driven secondary motion, and the measurements of U_y at the duct corners between ribs 2 and 3 support this observation. Thus, in the top half of the duct at this station (see Fig. 4), the measurements of U_y show two regions of negative values (movement from the smooth to the rough wall) straddling a region of positive values. The signs for U_y imply the presence of two counterrotating vortices in this region of the flow similar to those measured by, for example, *Melling* and *Whitelaw* [7] in the corners of a turbulent smooth duct flow. In their case, however, the secondary flow reached values of up to $0.02 \times U_B$ only after a duct developing length $x_H > 35$. The bottom corner between ribs 2 and 3 shows a small region of positive U_y contained by relatively large and negative values which suggests a single vortex rotating in the sense of the signs shown by U_y. Bulging of U_x contours towards the duct corners also appears over rib 3. Here, however, the corresponding results for U_y given in Fig. 5 show an enhanced transverse velocity component which is mainly positive everywhere (directed away from the ribs) ex-

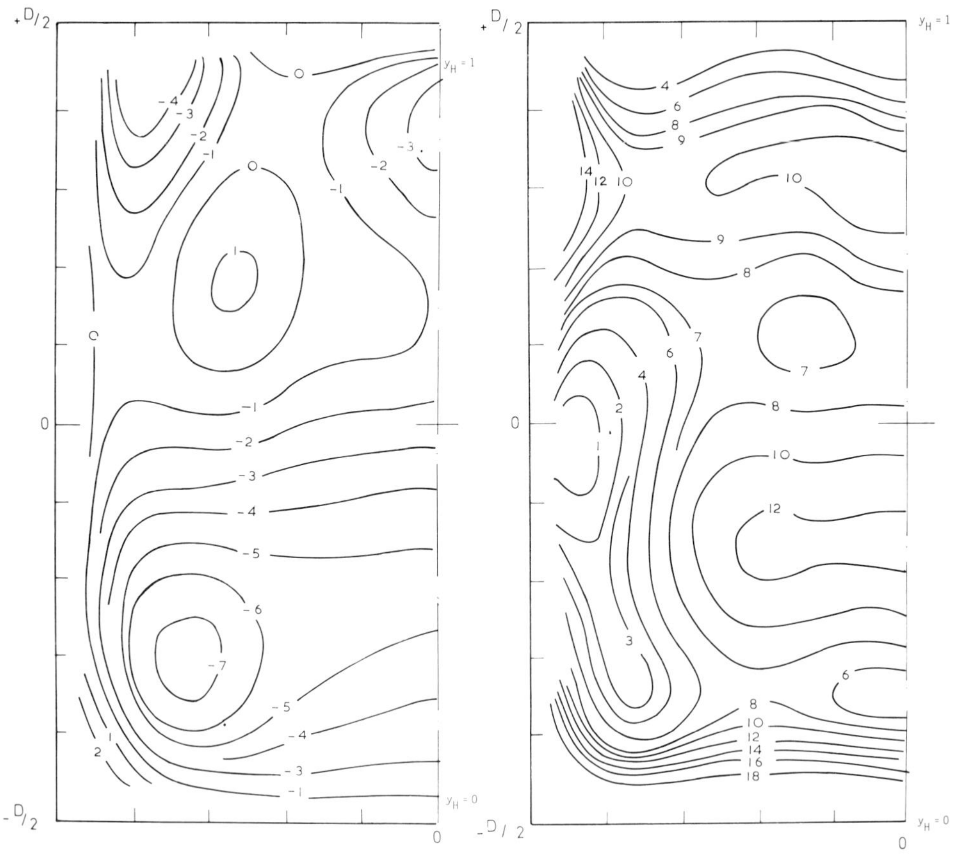

Fig. 4. U_y/U_B and $(\tilde{u}_y/U_B)$ 100 at x_H = 6.15

---

cept for a region near the center ($z/z_{1/2}$ = 0) of the duct top wall. The signs for U_y imply the presence of a cross-stream rotational motion in the top half of the duct directed mainly upwards along the side wall, along the top wall towards the duct symmetry plane and downwards between $z/z_{1/2}$ = 0 and $z/z_{1/2}$ = −0.6. Only a small region of negative U_y (and of small magnitude) remains near the side wall at y_H = 0.6. The peaks appearing at about y_H = 0.8 in the upstream plots for U_x are explained by the negative values for U_y near the duct symmetry plane. Slower moving fluid near the top wall of the duct is displaced by the transverse component into the core region of the flow, splitting it in two.

The measurements of U_y between ribs 2 and 3 and over rib 3 are qualitatively similar in the top wall region, but different in the region near the ribs. Thus, between ribs 2 and 3, the transverse component is mostly negative showing the effect of a "sudden expansion" on the developing flow. Over rib 3, a "sudden contraction" effect causes the reverse phenomenon to occur and U_y is positive, i.e., directed away from the rib. The result is for this pressure-driven motion to modify the normal stress-driven secondary motion as described above.

The U_x contours at the downstream positions displayed some bulging towards the top corner region of the duct (y_H = 1, $z = -z_{1/2}$). However, the presence there of a cross-stream rotational motion is not supported by the results for U_y which show that this component is

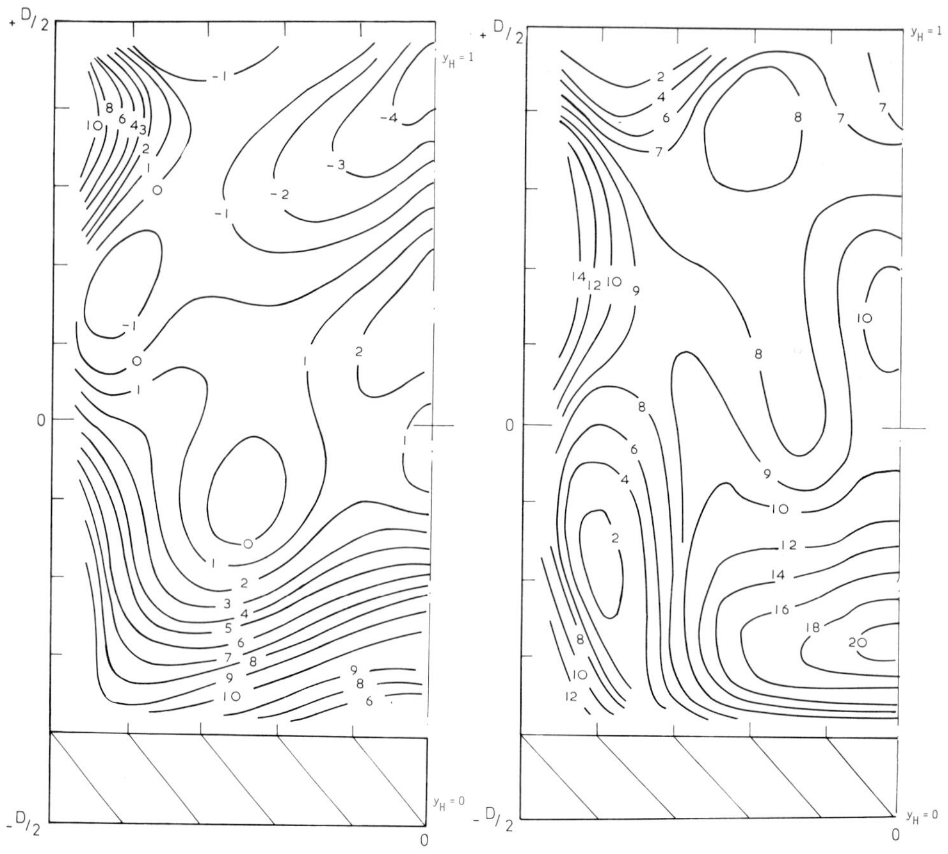

Fig. 5. U_y/U_B and $(\tilde{u}_y/U_B)\,100$ at $x_H = 6.65$

everywhere positive and large (except for a small region around $y_H = 0.1$ between ribs 38 and 39). In general, the high U_x velocity region shown in Fig. 6 and 7 is found displaced towards the top wall and away from the ribs where resistance to flow is higher. Downstream values of U_y, plotted in Figs. 8 and 9, show that the motion transverse to the main flow direction has increased considerably between the upstream and downstream stations. This is particularly evident in the side wall region of the duct where U_y reaches values of up to $0.24 \times U_B$. Large ratios in the values of $\overline{u_x^2}/\overline{u_y^2}$ (ranging between 3 and 140) indicate a high level of anisotropy at the side walls and suggest the possibility of strongly driven secondary motion of the second kind. However, values of $\overline{u_z^2}$ are not available to ascertain what the magnitude of the variation in $|\overline{u_y^2} - \overline{u_z^2}|$ is over the cross-stream plane and even though variations in $\overline{u_y^2}$ alone are certainly pronounced, it is unlikely that variations in cross-stream stresses are causing such a large transverse motion. In addition to a possible normal stress-driven motion is the effect of the transverse pressure gradient $(\partial P/\partial y)$ tending to deflect fluid approaching the ribs into the bulk of the flow. Because of its lower inertia, slow moving fluid near the side walls will be affected more strongly by the transverse pressure gradient than fluid in the core region of the flow. This can explain the large values for U_y observed at downstream stations and, also, the preferred

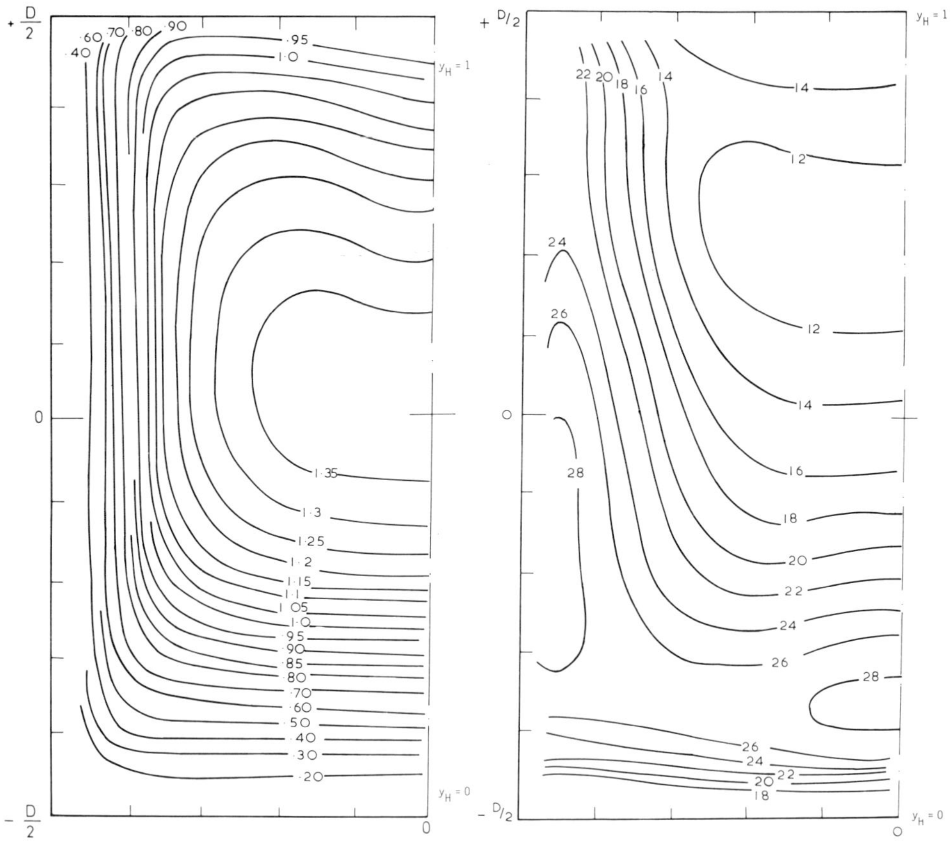

Fig. 6. U_x/U_B and $(\tilde{u}_x/U_B)$ 100 at $x_H = 41.68$

sense of rotation in the secondary motion displayed by the results for U_y over rib 3, which probably arises from a combination of both pressure gradient and stress-induced forces.

Reynolds Stresses

At the two upstream stations, plots of the longitudinal turbulence intensity show very large values of $\tilde{u}_x$ and its spatial gradient near the rough wall ($y_H = 0$) and near the side wall ($z \simeq -z_{1/2}$). The lowest values for $\tilde{u}_x$ appear in the upper half of the roughened duct ($0.5 < y_H < 0.9$) where, in addition, gradients of $\tilde{u}_x$ are relatively small. Distortion of the contours is in agreement with the sense of the transverse component, U_y, at these locations. At both upstream stations,

$$\frac{(\overline{u_x^2})_{\text{rough wall}}}{(\overline{u_x^2})_{\text{smooth wall}}} \cong 4,$$

which provides an indication of the degree of turbulence enhancement after the flow has inter-

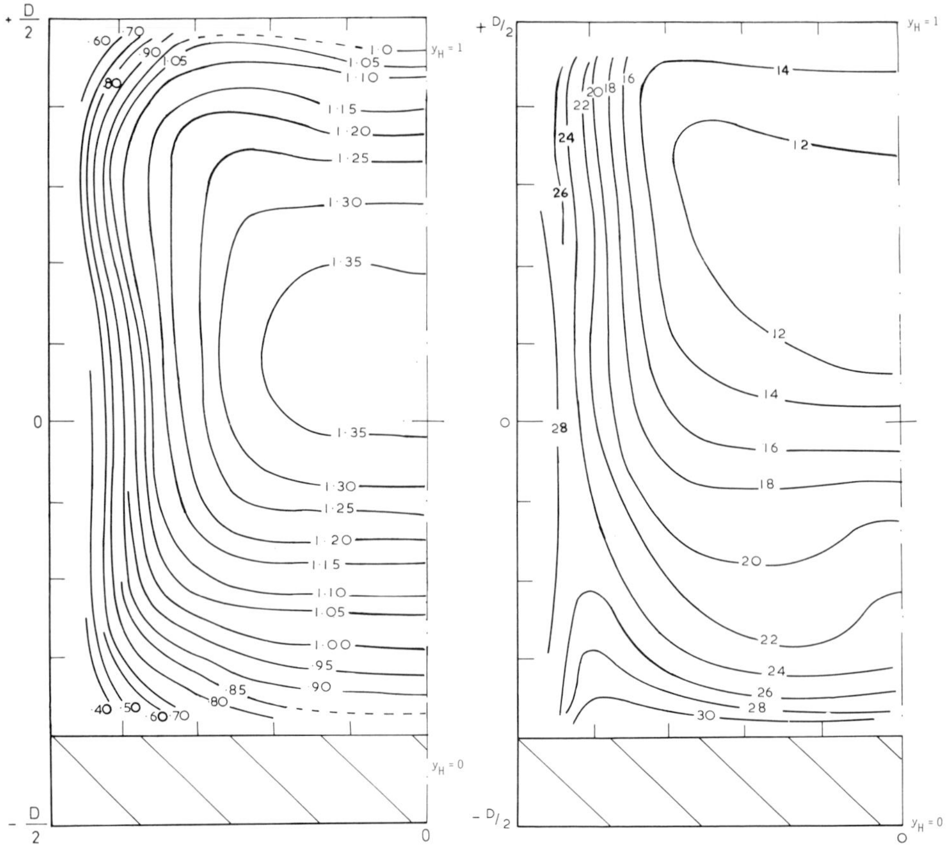

Fig. 7. U_x/U_B and $(\tilde{u}_x/U_B)$ 100 at $x_H = 42.17$

acted with the first three ribs. All plots for $\tilde{u}_x$ at the downstream stations show significant curvature in the contours at the side wall of the duct in the sense of the positive U_y component direction. Thus, it would appear that longitudinal stress, after being generated about the ribs, is transported by the transverse flow towards the top wall of the duct but more strongly at the side walls where U_y is larger. In the duct symmetry plane and at all between-rib locations, the $\tilde{u}_x$ plots showed maximum values of this turbulence component displaced, by approximately one rib height, from the bottom of the roughened wall. A similar displacement was not observed at any of the over-rib locations where the maximum in $\tilde{u}_x$ lies very close to the top of the rib (near $y_H = 0.1$).

The contoured results for $\tilde{u}_y$ at the two down-stream locations (Figs. 8 and 9) display some characteristics also found at the upstream stations, but several salient differences merit attention. A double hump, characterising the $\tilde{u}_y$ profile along the symmetry plane at the upstream stations, does not appear downstream and suggests that this is a characteristic of the flow in its developing stage. Along the side wall at the downstream location, a region of low $\tilde{u}_y$ has been extended towards the top (smooth) wall, presumably by the relatively large transverse velocity component there (see Fig. 9). The result is to isolate a small region of high $\tilde{u}_y$ (11% to 12%) in the top half of the figure which has no counterpart at the corresponding upstream location.

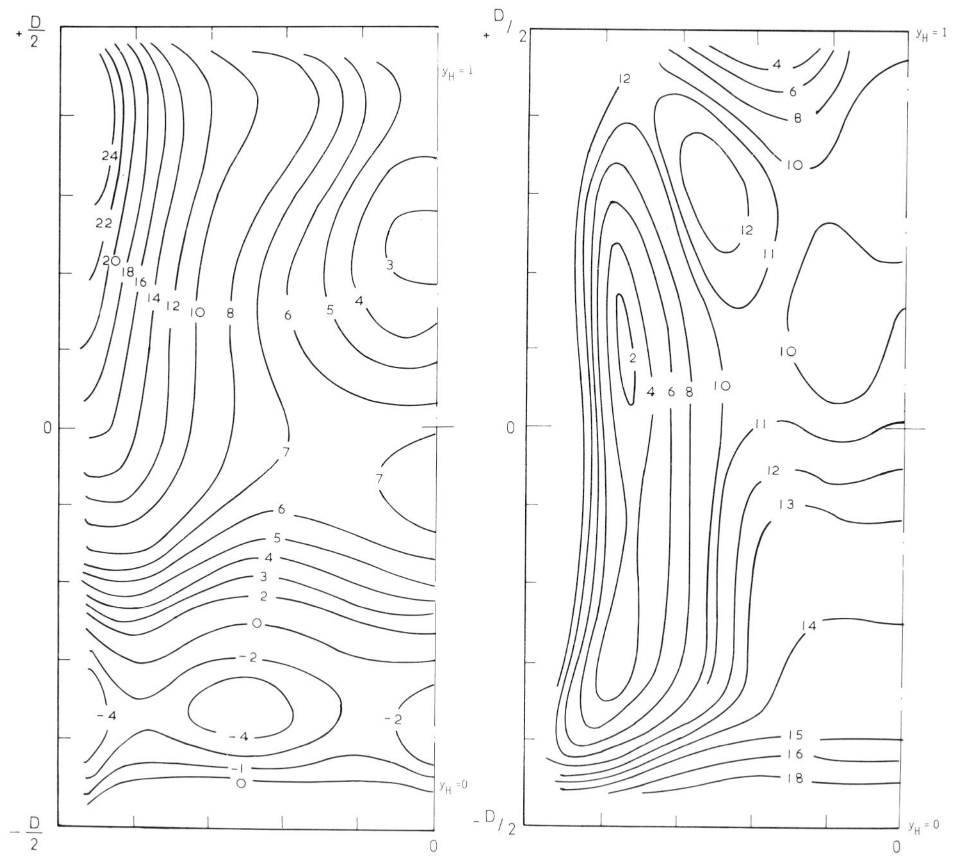

Fig. 8. U_y/U_B and $(\tilde{u}_y/U_B)$ 100 at $x_H = 41.68$

Curvature (in the sense of the positive transverse velocity component, U_y) is also evident in the contours located near the symmetry plane in the lower half of the plot. The pattern for $\tilde{u}_y$ over rib 39 is similar to that over rib 3, although absolute values of the turbulence intensity at respective locations are somewhat different. Particularly noticeable were the higher values for $\tilde{u}_y$ located near the side wall at $0.1 < y_H < 0.5$ and a more extended region of lower $\tilde{u}_y$ between $y_H = 0.7$ and $y_H = 1.0$.

A striking feature shown by the plots for $\tilde{u}_y$ is the relatively large difference, both in absolute value and spatial distribution, between measurements at between-rib and over-rib locations in the region near to the duct side wall. Between ribs, low values of $\tilde{u}_y$ are found near $y_H = 0.5$ straddled by high values near the rough and smooth walls, whereas at the over-rib locations, the reverse is observed. Thus, between ribs and near the side walls, it is possible to find $\overline{u_x^2}/\overline{u_y^2} > 100$, whereas at over-rib locations for the same cross-section coordinate position, $\overline{u_x^2}/\overline{u_y^2} \simeq 4$ is found. These numbers illustrate: (a) the very high degree of anisotropy which can arise in the present flows and (b) the periodic variation of the flow with x_H. In addition to depending on x_H, the degree to which the flow near the ribbed wall affects that at the opposite smooth wall is a function of z, with the largest interactions arising near the side walls ($z = \pm z_{1/2}$). At the over-

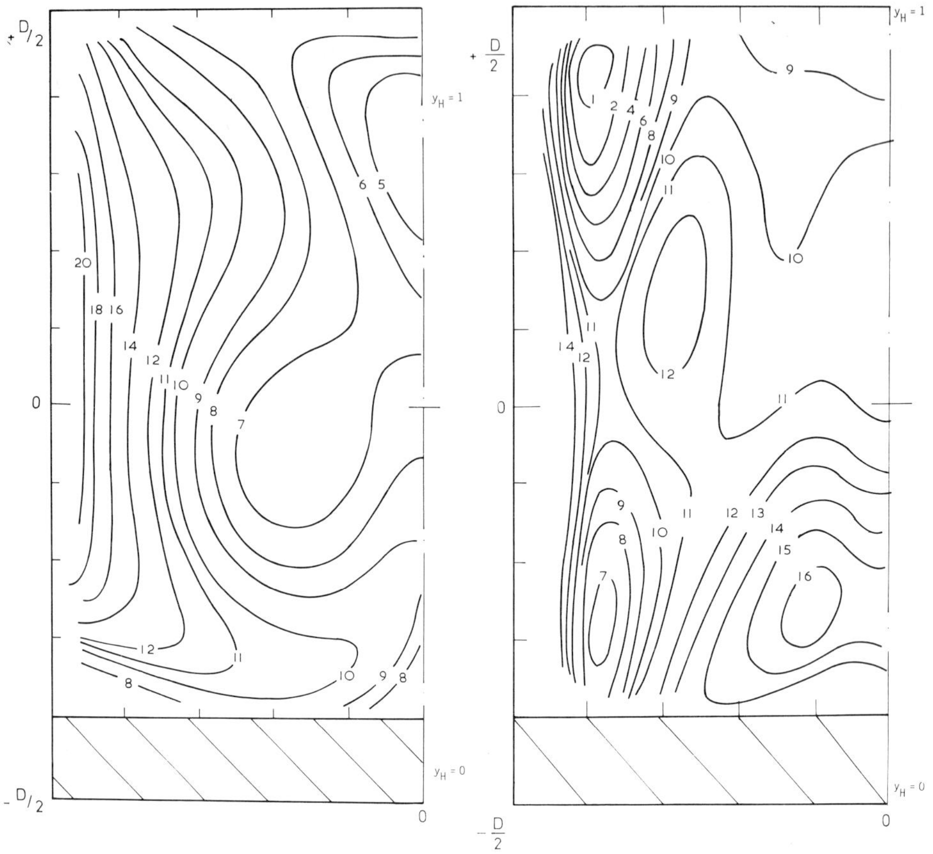

Fig. 9. U_y/U_B and $(\tilde{u}_y/u_B)\,100$ at $x_H = 42.17$

rib positions, absolute values for $\overline{u_x u_y}$ peak at about $y_H = 0.2$, whereas at the between-rib positions, maximum absolute values are found very near the bottom wall ($y_H < 0.1$). The values for $\overline{u_x u_y}$ immediately over rib 39 (shown in Fig. 13) are positive, whereas for all other stations, at the same y_H position, they are negative. Careful analysis of the unprocessed experimental data confirmed this observation and the possibility of distortion through curve fitting was discarded. The change in sign in $\overline{u_x u_y}$ suggests the possibility of a recirculation zone immediately over the rib, but measurements of the longitudinal velocity component were not made in sufficient detail and near enough to the rib to verify this possibility.

Generation terms for $\overline{u_x^2}/2$ and $\overline{u_y^2}/2$ for the present geometry are given by:

$$G\left(\frac{\overline{u_x^2}}{2}\right) = -\overline{u_x^2}\,\frac{\partial U_x}{\partial x} - \overline{u_x u_y}\,\frac{\partial U_x}{\partial y} - \overline{u_x u_z}\,\frac{\partial U_x}{\partial z}$$

$$G\left(\frac{\overline{u_y^2}}{2}\right) = -\overline{u_x u_y}\,\frac{\partial U_y}{\partial x} - \overline{u_y^2}\,\frac{\partial U_y}{\partial y} - \overline{u_y u_z}\,\frac{\partial U_y}{\partial z}$$

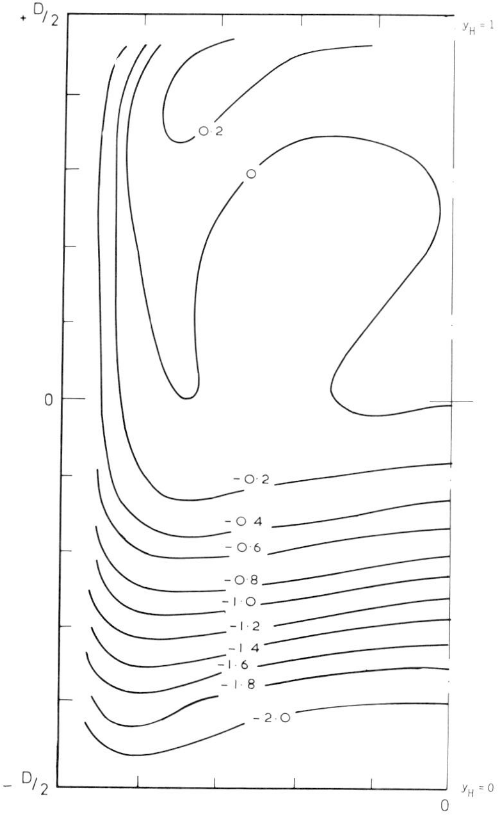

Fig. 10. $(\overline{u_x u_y}/U_B^2)\,100$ at $x_H = 6.15$

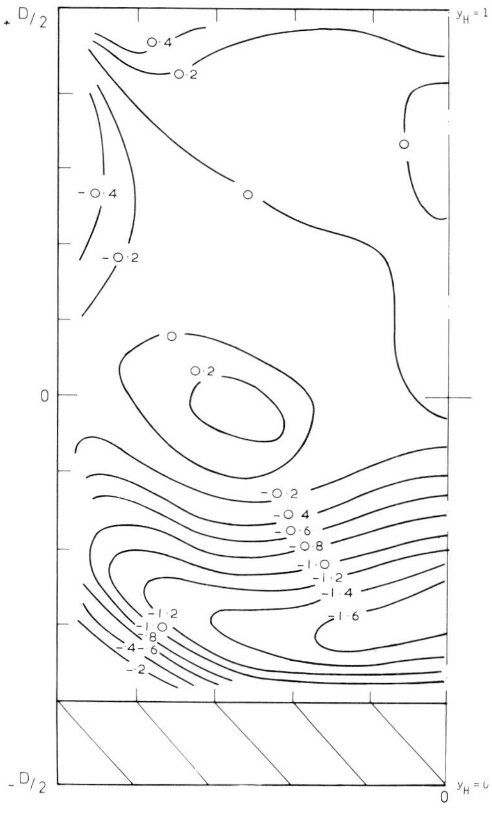

Fig. 11. $(\overline{u_x u_y}/U_B^2)\,100$ at $x_H = 6.65$

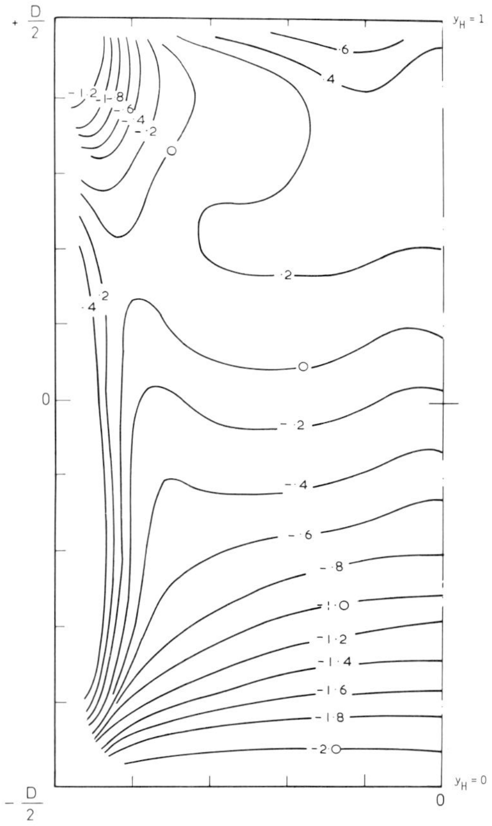

Fig. 12. $(\overline{u_x u_y}/U_B^2)\,100$ at $x_H = 41.68$

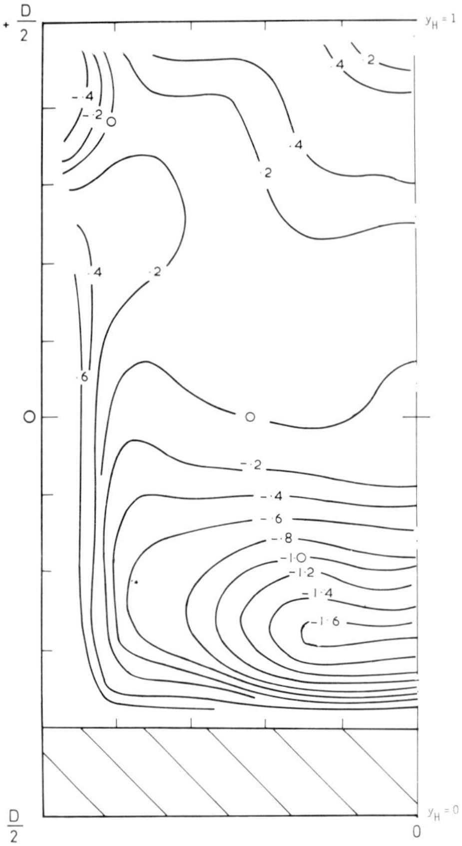

Fig. 13. $(\overline{u_x u_y}/U_B^2)$ 100 at $x_H = 42.17$

and are used here to correlate some of the observations presented for the normal stresses (ignoring pressure redistribution and turbulent diffusion). Positive values of $\partial U_x/\partial y$ and negative values for $\overline{u_x u_y}$ over most of the bottom half of the duct combine to produce fairly high levels of generation of $\overline{u_x^2}/2$ in this region of the flow (see figures for $\tilde{u}_x$). Along the symmetry plane at the downstream location, zero values for $\overline{u_x u_y}$ and $\partial U_x/\partial y$ do not coincide. The result is for a negative contribution to the generation of $\overline{u_x^2}/2$ to arise, but the effect is small. In the bottom wall region immediately preceding a rib, $\partial U_y/\partial x$ is positive and large enough so that generation of $\overline{u_y^2}/2$ by this gradient contributes significantly to the values of the transverse stress.

Law of the Wall Velocity Profiles

Plots of mean longitudinal velocity, U_x/U_B, in "law of the wall" coordinates, are shown in Fig. 14. The profiles are given for two z positions in the rough and smooth wall regions of the furthest pair of downstream stations. The data in these figures have been interpolated from the contoured results presented in Figs. 4 and 5. The profiles show that, although near the smooth and rough walls (except for near the ribs) a logarithmic behavior appears, large differences in profile constants arise and departure from linearity below the profile occurs earliest in the smooth wall region of the flow, especially near the side wall ($z/z_{1/2} = -0.6$). Similar conformity of near-wall profiles with logarithmic distributions has been found by *Hanjalic* and *Launder* [10] and *Fujita* [11] and has led the former to conclude that their flow was essentially

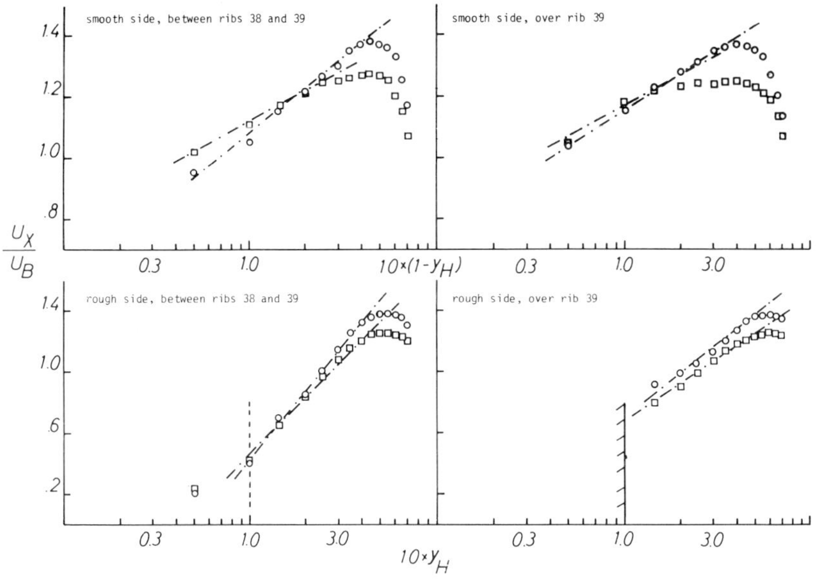

Fig. 14. Logarithmic plots of longitudinal velocity showing changing "law of the wall" behavior: (○) $z/z_{1/2} = 0$, (□) $z/z_{1/2} = -0.6$

in local equilibrium, i.e., unaware of the dissimilar structure prevailing at opposite walls. In the present case, however, near the side walls of the duct, the transverse velocity component U_y virtually trebles its value and, as has been discussed, significantly alters the characteristics of the turbulence in this region. The complexity of the measurements (particularly near the side wall) suggests especially strong interactions between dissimilar flow structures along the side walls in roughened duct flows as well as in the core region.

Conclusions

Measurements taken in an asymmetrically roughened straight duct geometry show the earlier development and higher intensity of a normal stress-driven secondary motion relative to that observed in a corresponding duct with smooth walls. Eventually, however, the stress-driven motion appears to be overcome by pressure gradient driven flow in the transverse direction, with values as high as $0.24 \times U_B$ appearing for U_y near the side walls. Results for the longitudinal velocity component show that the bulk of the flow is shifted towards the top half of the duct, and near the bottom wall region at between-rib positions is decelerated considerably. Although a logarithmic "law of the wall" is followed (albeit with varying values of the constants) in regions near the duct symmetry plane, departure from this universal behavior becomes significant for $|z/z_{1/2}| > 0.6$, where the side wall strongly affects the flow.

Reynolds stresses are of the order of four times larger at the roughened wall than at the facing smooth wall. These high values are associated with the strong generation of turbulent kinetic energy at the roughened wall. Interaction between the dissimilar flow structures prevailing at the rough and smooth walls is particularly intense along the duct side walls, where the transverse velocity component is strongest. Periodic variation of the flow near the ribs is shown by all the variables measured and is particularly evident in the plots for $\tilde{u}_y$, not only near the ribs but along the side wall of the duct as well. Regions where $\overline{u_x^2}/\overline{u_y^2} > 100$ have been observed which illustrates the high degree of anisotropy possible in this flow. Regions of generation (and suppression) of the normal stresses, and hence of the kinetic energy of turbulence, k, have been identified. In zones of the flow where the generation of k is large, dissipation can be expected to be large also.

Models of turbulence which incorporate a turbulent viscosity assumption, if used to predict the highly anisotropic flow presented here, must by necessity produce an incorrectly calculated mean flow field. Algebraic stress models and full stress transport models are free from a turbulent viscosity assumption and, therefore, are more likely to succeed in accurately predicting asymmetrically roughened duct flow. Algebraic stress modelling requires, however, that spatial variations in the turbulent stresses normalized by kinetic energy of turbulence $(\overline{u_i u_j}/k)$ be small everywhere, and this is a condition unlikely to be satisfied throughout the present flow. Stress transport models depend on the adequacy of the higher order closure approximations which arise. The experimental results provided here should help to establish the suitability of these approximations by a comparison with calculated results.

Acknowledgements. The authors would like to acknowledge discussions held with Professor B. E. Launder during the early stages of this research. We are indebted to Mr. R. Church for the precisely constructed test section in which the measurements were made. Financial support received from the Science Research Council and Imperial Chemical Industries during the course of this work is gratefully acknowledged.

References

1. P. Bradshaw: *Turbulence,* Topics in Applied Physics, Vol. 19 (Springer, Berlin, Heidelberg, New York 1978) p. 109
2. J. A. C. Humphrey: "Flow in Ducts with Curvature and Roughness". Ph. D. Thesis, University of London (1977)
3. J. A. C. Humphrey, A. M. K. Taylor, J. H. Whitelaw: Laminar flow in a square duct of strong curvature. J. Fluid Mech. **83**, 509 (1977)
4. S. V. Patankar, V. S. Pratap, D. B. Spalding: Prediction of turbulent flow in curved pipes. J. Fluid Mech. **67**, 583 (1975)
5. E. Brundrett, W. D. Baines: The production and diffusion of vorticity in duct flow. J. Fluid Mech. **19**, 375 (1964)
6. F. B. Gessner, J. B. Jones: On some aspects of fully-developed turbulent flow in rectangular channels. J. Fluid Mech. **23**, 689 (1965)
7. A. Melling, J. H. Whitelaw: Turbulent flow in a rectangular duct. J. Fluid Mech. **78**, 289 (1976)
8. D. Naot, A. Shavit, M. Wolfshtein: Numerical calculation of Reynolds stresses in a square duct with secondary flow. Waerme Stoffuebertrag. **7**, 151 (1974)
9. B. E. Launder, W. M. Ying: Prediction of flow and heat transfer in ducts of square cross-section. Heat Fluid Flow **3**, 115 (1973)
10. K. Hanjalic, B. E. Launder: Fully developed asymmetric flow in a plane channel. J. Fluid Mech. **51**, 301 (1972)
11. H. Fujita: "Turbulent Flows in Square Ducts Consisting of Smooth and Rough Planes". Research Reports of the Faculty of Engineering, MIE University, Vol. 3 (1978) p. 11
12. J. A. C. Humphrey: "Turbulent Flow in an Asymmetrically Roughened Duct of Square Cross Section", in *Laser Velocimetry and Particle Sizing,* ed. by H. D. Thompson, W. H. Stevenson (Hemisphere Publishing Corporation 1979) p. 220
13. F. Durst, A. Melling, J. H. Whitelaw: *Principles and Practice of Laser Doppler Anemometry* (Academic Press, London 1976)

Prediction of Three-Dimensional Laminar and Turbulent Boundary Layers on Bodies of Revolution at High Angles of Attack*

T. Cebeci

Aerodynamics Research Group, McDonnell Douglas Aircraft Company,
Long Beach, CA 90846, USA

A. A. Khattab

Aerodynamics Research Group, McDonnell Douglas Aircraft Company, Long Beach, CA 90846, USA

K. Stewartson

Department of Mathematics, University College, London, England

Introduction

This paper describes one phase of the work done towards the development of a general boundary-layer method for calculating three-dimensional boundary layers on bodies of revolution at incidence. In a previous paper [1] we addressed ourselves to the problem of computing boundary layers near the nose region and with the onset of leading-edge separation; this is important for the calculation of transition by stability theory and for the prediction of downstream flow properties. Taking the prolate spheroid as a paradigm, we performed laminar-flow calculations on the windward and leeward planes of symmetry for various angles of attack. The study revealed that, if the angle of incidence α is less than 41°, separation does not occur at the nose no matter how thin the body. In contrast, the value of α which provokes separation at the leading edge of a two-dimensional airfoil tends to zero with the thickness ratio of the airfoil.

In this paper we consider both laminar and turbulent flows and discuss the calculation of three-dimensional boundary layers with flow reversal in the circumferential velocity component as well as the calculation of three-dimensional separation lines. Again we take the prolate spheroid as a paradigm; this time we limit the calculation to $\alpha = 6^\circ$. At present, studies are also being performed for larger angles of incidence and we hope to report them at a later date.

Formulation for Prolate Spheroids

Governing Equations

For a prolate spheroid at incidence (see Fig. 1) and incompressible laminar or turbulent flow, the governing boundary-layer equations can be written as

$$\frac{\partial}{\partial x}(h_2 u) + \frac{\partial}{\partial \theta}(h_1 w) + \frac{\partial}{\partial y}(h_1 h_2 v) = 0 \tag{1}$$

* This work was supported by the Naval Surface Weapons Center under contract N60921-78-C-0158.

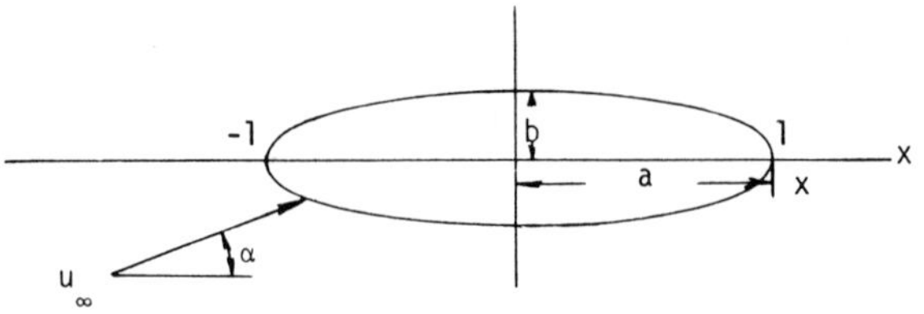

Fig. 1. Notation for prolate spheroid at incidence

$$\frac{u}{h_1}\frac{\partial u}{\partial x}+\frac{w}{h_2}\frac{\partial u}{\partial \theta}+v\frac{\partial u}{\partial y}+w^2K_2 = -\frac{1}{\rho h_1}\frac{\partial p}{\partial x}+\frac{\partial}{\partial y}\left(\nu\frac{\partial u}{\partial y}-\overline{u'v'}\right) \tag{2}$$

$$\frac{u}{h_1}\frac{\partial w}{\partial x}+\frac{w}{h_2}\frac{\partial w}{\partial \theta}+v\frac{\partial w}{\partial y}-uwK_2 = -\frac{1}{\rho h_2}\frac{\partial p}{\partial \theta}+\frac{\partial}{\partial y}\left(\nu\frac{\partial w}{\partial y}-\overline{w'v'}\right) \tag{3}$$

where h_1, h_2 are metric coefficients and K_2 is the geodesic curvature of the surface lines $\xi(\equiv x/a)$ const. The external velocity distribution is given by simple analytical formulas, and the pressure gradients in (2) and (3) can be easily computed from them.

The solution of the system of Eqs. (1) to (3) requires boundary and initial conditions. The boundary conditions are:

$$\begin{aligned} &y=0, \quad u=v=w=0; \\ &y\to\infty, \quad u\to u_e(x,\theta), \quad w\to w_e(x,\theta) \end{aligned} \tag{4}$$

The initial conditions in the longitudinal direction can easily be determined by taking advantage of the symmetry conditions (see, for example, *Cebeci* et al. [1]). The specification of the initial conditions in the circumferential direction is not quite so easy when body-oriented coordinates are used because of the singularity of h_1, h_2 and K_2 at the nose ($x=-1$). However, these difficulties can be avoided by using a suitable transformation near $x=-1$, as described by [1], and the initial conditions in the circumferential direction successfully generated.

Turbulence Model

The presence of Reynolds shear-stress terms $-\overline{u'v'}$ and $-\overline{w'v'}$ introduces additional unknowns to the system of equations given by (1) to (3). Here we use an algebraic eddy-viscosity (ϵ_m) formulation developed for three-dimensional flows by *Cebeci* [2]. We write the Reynolds shear-stress terms as

$$-\overline{u'v'}=\epsilon_m\frac{\partial u}{\partial y}, \qquad -\overline{w'v'}=\epsilon_m\frac{\partial w}{\partial y} \tag{5}$$

and define ϵ_m by two separate formulas. In the so-called inner region of the boundary layer, $(\epsilon_m)_i$ is defined by

$$(\epsilon_m)_i \equiv \{0.4\,y[1-\exp(-y/A)]\}^2\left[\left(\frac{\partial u}{\partial y}\right)^2+\left(\frac{\partial w}{\partial y}\right)^2\right]^{1/2} \tag{6}$$

where

$$A = 26 \frac{\nu}{N} u_\tau^{-1}, \quad N = (1 - 11.8\, p^+)^{1/2}, \quad p^+ = \frac{\nu u_e}{u_\tau^3} \frac{\partial u_e}{\partial x}$$

$$u_\tau = \left(\frac{\tau_{w_s}}{\rho}\right)^{1/2}, \quad \frac{\tau_{w_s}}{\rho} = \nu \left[\left(\frac{\partial u}{\partial y}\right)_w^2 + \left(\frac{\partial w}{\partial y}\right)_w^2\right]^{1/2} \tag{7}$$

In the outer region, ϵ_m is defined by

$$(\epsilon_m)_p = 0.0168 \left| \int_0^\infty [(u_e^2 + w_e^2)^{1/2} - (u^2 + w^2)^{1/2}]\, dy \right| \tag{8}$$

The inner and outer regions are established by the continuity of the eddy-viscosity formulas.

Numerical Method

We use Keller's two-point finite-difference method (called the Box method) to solve the governing equations of the previous section. In the nose region, we first solve the equations expressed in nose coordinates. At some x-station, we switch to x, θ coordinates and, with initial conditions in the circumferential direction given by nose-coordinate solutions and with initial conditions on the windward and leeward sides determined from the line of symmetry solutions, we solve Eqs. (1) to (3) subject to the boundary conditions given by (4). As in our previous studies in three-dimensional flows, we first transform the equations. We define the similarity variable η by

$$\eta = \left(\frac{u_0}{\nu s_1}\right)^{1/2} y \tag{9}$$

and introduce a two-component vector potential such that

$$h_2 u = \frac{\partial \psi}{\partial y}, \quad h_1 w = \frac{\partial \phi}{\partial y}, \quad h_1 h_2 v = -\left(\frac{\partial \psi}{\partial x} + \frac{\partial \phi}{\partial \theta}\right) \tag{10}$$

In addition, we define dimensionless parameters f and g related to ψ and ϕ by

$$\psi = (u_0 \nu s_1)^{1/2} h_2 f(x, \theta, \eta), \quad \theta = (u_0 \nu s_1)^{1/2} h_1 g(x, \theta, \eta) \tag{11}$$

Here s_1, which denotes the arc length along the x-coordinate, is defined by

$$s_1 = \int_{-1}^{x} h_1 dx \tag{12}$$

Introducing the relations (9) to (11) into (1) to (4), and using the definitions of eddy viscosity, we get the following equations:

$$(bf'')' + s_1 \beta_1 + s_1 f'' e - s_1 K_2 (g')^2 = s_1 \left(\frac{f'}{h_1} \frac{\partial f'}{\partial x} + \frac{g'}{h_2} \frac{\partial f'}{\partial \theta}\right) \tag{13}$$

$$(bg'')' + s_1 \beta_2 + s_1 g'' e + s_1 K_2 f' g' = s_1 \left(\frac{f'}{h_1} \frac{\partial g'}{\partial x} + \frac{g'}{h_2} \frac{\partial g'}{\partial \theta}\right) \tag{14}$$

$$e' = \frac{1}{h_1}\frac{\partial f'}{\partial x} + \frac{1}{h_2}\frac{\partial g'}{\partial \theta} + f'\left(\frac{1}{2\,s_1} - K_2\right) \tag{15}$$

Here primes denote differentiation with respect to η, β_1 and β_2 are dimensionless pressure gradient parameters and

$$f' = u/u_0, \quad g' = w/u_0, \quad b = 1 + \epsilon_m/\nu \tag{16}$$

The boundary conditions become

$$\begin{aligned} &\eta = 0, \quad f' = g' = e = 0; \\ &\eta = \eta_\infty \quad f' = u_e/u_0, \quad g' = w_e/u_0 \end{aligned} \tag{17}$$

The numerical solution procedure we use depends on the complexity of the flow. In regions where both u and w are positive, we usually use the standard Box method. We write (13) to (15) and (17) in terms of a system of five first order equations by introducing new dependent variables $v(x, \theta, \eta)$, $t(x, \theta, \eta)$. By setting $f' = u$, $g' = w$, we then can write the five first-order equations as

$$u' = v \tag{18a}$$

$$w' = t \tag{18b}$$

$$(bv)' + s_1\beta_1 + s_1 ve - s_1 K_2 w^2 = s_1\left(\frac{u}{h_1}\frac{\partial u}{\partial x} + \frac{w}{h_2}\frac{\partial u}{\partial \theta}\right) \tag{18c}$$

$$(bt)' + s_1\beta_2 + s_1 te + s_1 K_2 uw = s_1\left(\frac{u}{h_1}\frac{\partial w}{\partial x} + \frac{w}{h_2}\frac{\partial w}{\partial \theta}\right) \tag{18d}$$

$$e' = \frac{1}{h_1}\frac{\partial u}{\partial x} + \frac{1}{h_2}\frac{\partial w}{\partial \theta} + u\left(\frac{1}{2\,s_1} - K_2\right) \tag{18e}$$

$$\begin{aligned} &\eta = 0, \quad u = w = e = 0; \\ &\eta = \eta_\infty, \quad u = u_e/u_0, \quad w = w_e/u_0 \end{aligned} \tag{19}$$

We next consider the net cube shown in Fig. 2 and write difference approximations to (18) as discussed in several references (see, for example, *Cebeci* and *Bradshaw* [3]). Equations (18a) and (18b) are approximated using centered difference quotients and averages about the mid-

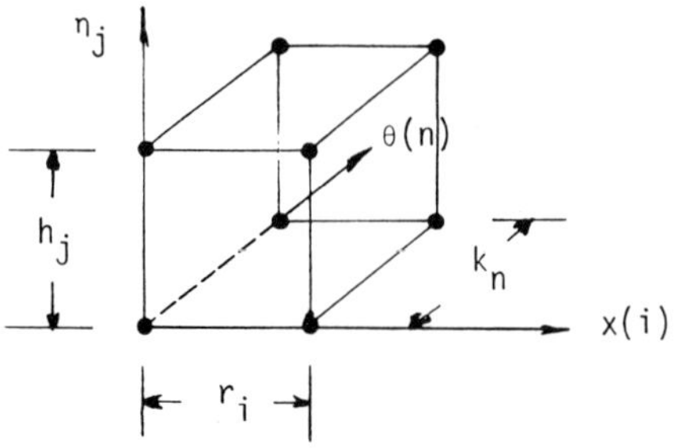

Fig. 2. Finite-difference pattern for the standard Box

point $(x_i, \theta_n, \eta_{j-1/2})$. The difference equations which are to approximate (18c), (18d) and (18e) are written about the midpoint $(x_{i-1/2}, \theta_{n-1/2}, \eta_{j-1/2})$ of the cube whose mesh widths are r_i, k_n and h_j. In centering the parameters u, v, w and t at the midpoint $(x_{i-1/2}, \theta_{n-1/2}, \eta_{j-1/2})$ we have written them in the form

$$\bar{q}_{j-1/2} \equiv \frac{\bar{q}_j + \bar{q}_{j-1}}{2}, \quad \bar{q}_j = \frac{1}{4}(q_j^{i,n} + q_j^{i-1,n} + q_j^{i,n-1} + q_j^{i-1,n-1}) \tag{20}$$

thus using four values of each parameter on the corners of the box at a given j-point. Here the three values of $\bar{q}_j$, namely $q_j^{i-1,n}$, $q_j^{i,n-1}$, $q_j^{i-1,n-1}$, are known and the value of $q_j^{i,n}$ is unknown. In centering the normal velocity component denoted by e_j, we do not make use of its values on the four corners of the section of the box but only its value at the center. We write e_j at $(x_{i-1/2}, \theta_{n-1/2}, \eta_{j-1/2})$ as

$$\bar{e}_{j-1/2} = \frac{1}{2}(\bar{e}_j + \bar{e}_{j-1}) \tag{21}$$

and treat $\bar{e}_j$ as the unknown. This was necessary in order to avoid unacceptable oscillations in the solutions.

The difference equations for (18) form a nonlinear algebraic system which we linearize by Newton's method and solve the resulting linear system by the block-elimination method discussed in *Cebeci* and *Bradshaw* [3].

When the transverse velocity component changes sign across the layer, the solution of the boundary-layer equations may break down if the region of flow reversal is sufficiently large. To avoid integration against the flow, we must make appropriate changes in our solution procedure. In our study we do this in two different ways. One way is to use a zig-zag scheme with the standard Box scheme described in [4] for unsteady laminar flows; in this scheme we still retain the procedure for (18a) and (18b) and center them at $(x_i, \theta_n, \eta_{j-1/2})$. To write the difference approximation for the Box centered at $(x_{i-1/2}, \theta_{n-1/2}, \eta_{j-1/2})$ we examine previously computed values of $w_{j-1/2}^{i,n}$. If $w_{j-1/2}^{i,n} \geqslant 0$, we use the standard Box method, and if $w_{j-1/2}^{i,n} < 0$, we write (18c), (18d) and (18e) for the Box centered at P using quantities centered at P, Q, and R, where

$$P \equiv (x_{i-1/2}, \theta_n, \eta_{j-1/2}), \quad Q \equiv (x_i, \theta_{n-1/2}, \eta_{j-1/2}), \quad R \equiv (x_{i-1}, \theta_{n+1/2}, \eta_{j-1/2}) \tag{22}$$

The pattern of the zig-zag scheme depends on whether the solution procedure, for a given x_i-station, starts from the leeward plane or from the windward plane. In either case we march in the θ-direction. The zig-zag pattern for each case is shown in Fig. 3.

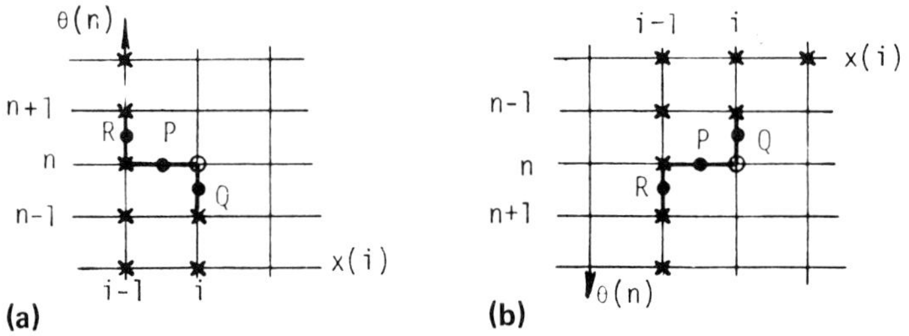

Fig. 3, a, b. Finite difference pattern for the zig-zag Box when the solution procedure begins from **(a)** the windward plane and **(b)** the leeward plane. In each case x denotes known values and ○ computed by the zig-zag Box

To illustrate the centering of the equations with this zig-zag scheme, let us consider a model equation for (18c),

$$(bv)' + s_1 ve = s_1 \left(\frac{u}{h_1} \frac{\partial u}{\partial x} + \frac{w}{h_2} \frac{\partial u}{\partial \theta} \right) \tag{23}$$

The finite-difference approximations to this equation for the case when the marching starts from the windward side and proceeds to the leeward side are:

$$(bv)'(P) + (s_1 ve)(P) = \left(\frac{s_1}{h_1} \right) (P) u(P) \frac{\partial u}{\partial x} (P)$$

$$+ \left(\frac{s_1}{h_2} \right) (P) \left[\tilde{\theta} w(Q) \frac{\partial u}{\partial \theta} (Q) + \tilde{\phi} w(R) \frac{\partial u}{\partial \theta} (R) \right] \tag{24}$$

Here

$$\tilde{\theta} = \frac{\theta_{n+1} - \theta_n}{\theta_{n+1} - \theta_{n-1}}, \quad \tilde{\phi} = \frac{\theta_n - \theta_{n-1}}{\theta_{n+1} - \theta_{n-1}} \tag{25}$$

The resulting algebraic system is again nonlinear and its solution is obtained by using the procedure followed in the standard Box method.

The second scheme uses a new method recently developed by *Cebeci* and *Stewartson* for three-dimensional flows. We shall refer to it as the characteristic Box method. It is based on the solution of the governing equations written along local streamlines. As in the zig-zag scheme we usually use this method when $w_{j-1/2}^{i,n} \leqslant 0$. In the characteristic Box method, we again retain the procedure for (18a) and (18b) and center them at $(x_i, \theta_n, \eta_{j-1/2})$. To obtain the difference approximations to (18c) and (18d), we first write them in streamline form, that is;

$$(bv)' + s_1\beta_1 + s_1 ve - s_1 K_2 w^2 = \lambda \frac{\partial u}{\partial s} \tag{26a}$$

$$(bt)' + s_1\beta_2 + s_1 te + s_1 K_2 uw = \lambda \frac{\partial w}{\partial s} \tag{26b}$$

where s denotes the local streamline direction (see Fig. 4) and

$$\lambda = s_1 [(u/h_1)^2 + (w/h_2)^2]^{1/2}. \tag{27}$$

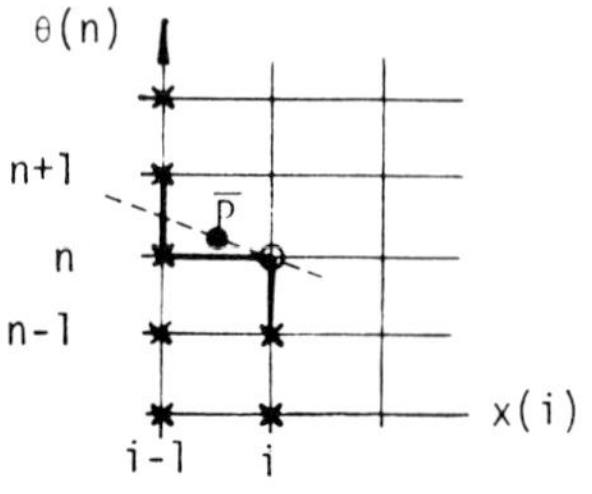

Fig. 4. Finite-difference pattern for the characteristic Box. The heavy lines show the pattern for the zig-zag scheme for continuity eq. × denotes known quantities and ○ those computed by the characteristic Box

With local streamline direction being known for an assumed u and w profiles, we then center (26) on $\bar{P}$ and write the difference approximations to them as usual. To solve the continuity equation (18e), we use the aforementioned zig-zag scheme and center (18e) as before. When $w_{j-1/2}^{i,n} \geqslant 0$, we then revert back to the standard Box scheme and center them as described before.

Results

The computation procedure near the forward stagnation point is the same for both laminar and turbulent flows. We use the nose coordinates and the line of symmetry solutions as described in [1]. Successive stations are then computed using the characteristic Box until $x = x_0$ when we switch to the general body coordinates (x, θ, η). We solve the governing equations with the standard Box method until the circumferential velocity component w changes sign which in the laminar case varies from $x = -0.225$ at $\theta = 180°$ through $x = 0.394$ at $\theta = 90°$ to $x = 0.761$ at $\theta = 0°$. We then switch the solution procedure from the standard Box to the zig-zag Box or to the characteristic Box. Comparison between these two methods are generally quite favorable but we feel that the zig-zag Box is less sensitive to the structural properties of the boundary layer and our final results are obtained using the characteristic Box.

For $x < 0.25$, the variation of u, v, w is sufficiently smooth that variable intervals of from 0.1 to 0.025 in x and of $10°$ in θ were sufficient to give reliable values of these quantities, but thereafter the step length in x is reduced to 0.01 and of θ to $2\frac{1}{2}°$ at $x = 0.25$.

In the laminar study the numerical solution begins to develop large gradients around $\theta = 115°$ as x approaches 0.30 and at $x = 0.31$ it separates at about this angle, the calculations being started from the windward line of symmetry up to this point. Continuing the calculations in this way, a separation line $\theta = \theta_s(x)$ is found on which θ_s decreases monotonically until it vanishes at $x = 0.840$. Careful examination of the solution shows, reasonably convincingly, that the separation line is an envelope of limiting streamlines (see *Brown* [5]). The calculations are repeated starting from the leeward line of symmetry, a procedure normally quite accurate but with less margin of safety near separation since the mainstream velocity has a positive component in the direction of θ. For $x \leqslant 0.30$, both methods agree well but for $0.31 \leqslant x \leqslant 0.34$ the leeward integration can be continued right through to $\theta = 0$ even though separation has occurred in the windward integration. For $x \geqslant 0.35$, the leeward integration

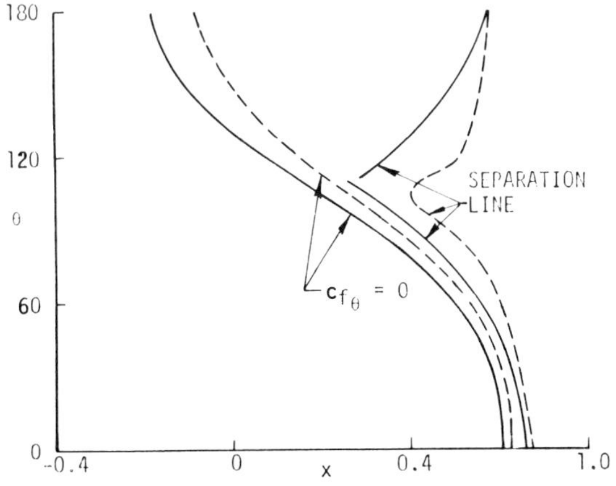

Fig. 5. Laminar zero-c_{f_θ} lines and separation lines for a prolate spheroid at $\alpha = 6°$. Solid lines denote present calculations and dashed lines those of *Wang* [6]

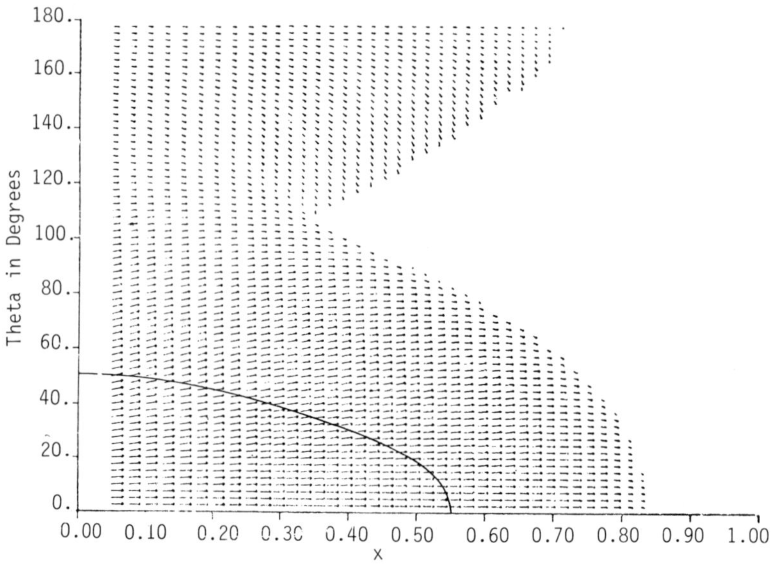

Fig. 6. Laminar skin-friction vectors for a prolate spheroid at $\alpha = 6°$. The arrows in the diagram generally give the magnitude and direction of the skin-friction vectors over the aft portion of the spheroid and upstream of the separation line. Below the solid curve, however, the magnitude of these vectors are so large that for clarity the length of the arrows has been kept constant. In the neighborhood of $x = 0.31$, the skin friction vectors computed from the windward line of symmetry, where available, have been allowed to override those computed from the leeward line of symmetry

breaks down also, at values of θ varying from $110°$ at $x = 0.35$ to $180°$ at $x = 0.72$. The manner of breakdown is, we believe, also associated with a Brown singularity but the final breakdown occurs so fast that we are unable completely to verify this conjecture. Figure 5 shows a comparison between the zero $c_{f\theta}$-line and the separation line as calculated by the present method and by *Wang* [6]. In Fig. 6 we show the local skin-friction near the separation line by arrows of appropriate length and direction. They bring out very clearly the way these lines are turning

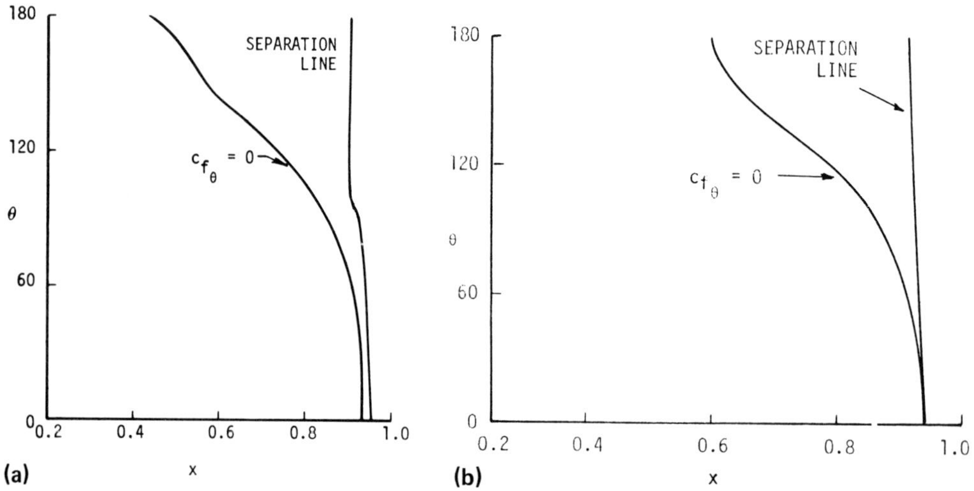

Fig. 7a, b. Turbulent zero-$c_{f\theta}$ line and separation line for a prolate spheroid at $\alpha = 6°$. (a) Transition at $x = -0.475$. (b) Transition at $x = 0.275$

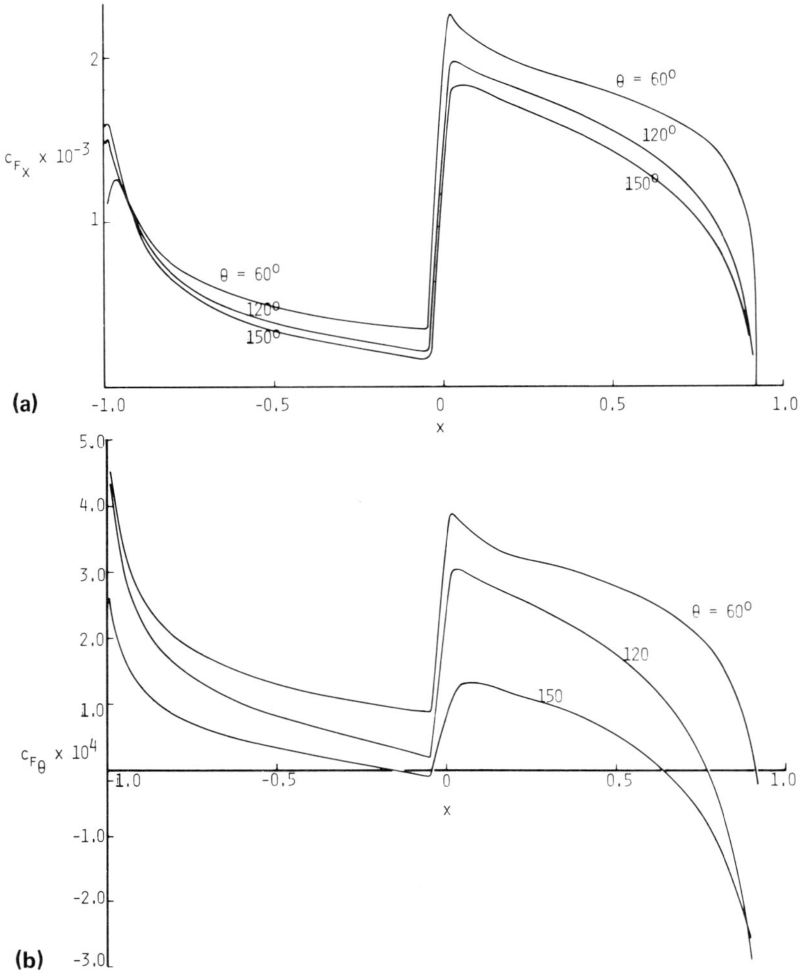

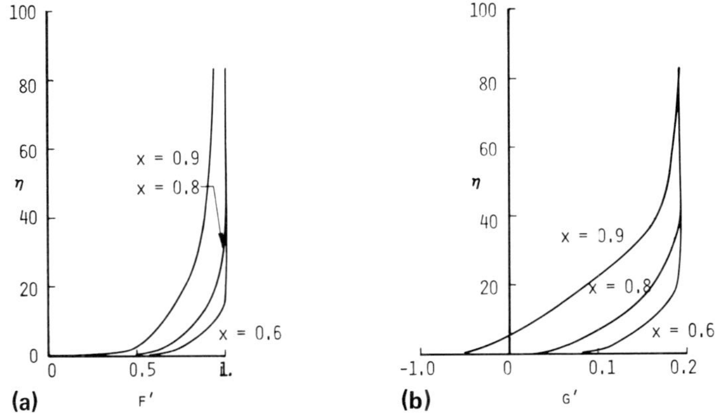

Fig. 9. The turbulent velocity profiles for three x-distances at $\theta = 100°$. (**a**) Longitudinal components. (**b**) Circumferential components

◀ **Fig. 8.** The variation of local skin-friction coefficients for three values of θ with transition at $x = 0$. (**a**) Longitudinal component. (**b**) Circumferential component against the direction of the main-stream velocity. As is seen, these results differ significantly from those of *Wang;* for example, his separation "ok" ("arrow" in Turkish) only extends upstream to $x \sim 0.50$, whereas our calculations indicate that separation is initiated near $x = 0.31$.

The turbulent flow calculations were performed for a unit Reynolds number of 3×10^6/ft, an angle of 6° and for transition at –0.475, 0, 0.2 and 0.275. The eddy-viscosity formulas given by Eqs. (6) to (8) are not corrected by an intermittency expression to account for the transition region which, according to the recent measurements of *Meier* and *Kreplin* [7], extends over more than 15% of the length. Thus transition, and the finite length over which it occurs, represent an important phenomenon in wind tunnel tests and information of flow profiles downstream of transition are required before detailed comparison with experiment can be made. Unfortunately, information of this type is not available at the present time and the relative merits of alternative turbulence models cannot be appraised. It is to be expected, however, and as suggested by *Cebeci* and *Meier* [8] that turbulence models more complicated than the algebraic formulation used here, cannot easily be justified especially in view of the uncertainties imposed by transition.

Figure 7 presents the separation lines for the two transition locations. The differences are comparatively small for the 6° angle of attack. As can be seen, negative crossflow occurs a considerable distance downstream of transition and is followed by flow separation of the turbulent boundary layer at around $x = 0.95$. The separation, which was determined by the breakdown of the calculations and preceeded by a precipitous fall in the skin friction, is almost independent of circumferential position and much less spectacular than the laminar separation of Fig. 5.

The longitudinal and circumferential distribution of the skin-friction coefficients are shown on Figs. 8 and 9 for three values of θ and transition at $x = 0$: transition and separation can readily be observed. Corresponding profiles of the longitudinal and circumferential velocity profiles are presented on figure 10 for three positions downstream of transition and at $\theta = 100°$. The negative crossflow velocity at $x = 0.9$ is shown together with the trend towards separation of the longitudinal velocity.

References

1. T. Cebeci, A. K. Khattab, K. Stewartson: "Studies on three-dimensional laminar and turbulent boundary layers on bodies of revolution at incidence I. Nose Separation". AIAA Paper 79-0138 (1979)
2. T. Cebeci: Calculation of three-dimensional boundary layers, I. Swept infinite cylinders and small cross flow. AIAA J. **12**, 779 (1974)
3. T. Cebeci, P. Bradshaw: *Momentum Transfer in Boundary Layers* (McGraw-Hill/Hemisphere, Washington, D. C. 1977)
4. T. Cebeci: The laminar boundary layer on a circular cylinder started impulsively from rest (To appear in J. of Comput. Phys.)
5. S. N. Brown: Singularities associated with separating boundary layers. Philos. Trans. R. Soc London **251**, 409–444 (1965)
6. K. C. Wang: Boundary layer over a blunt body at low incidence with circumferential reversal flow. J. Fluid Mech. **72**, 49–65 (1975)
7. H. U. Meier, H. P. Kreplin: "Experimental Investigation of the Transition and Separation Phenomena on a Body of Revolution" (Submitted to the 2nd Symposium on Turbulent Shear Flows, London 1979)
8. T. Cebeci, H. U. Meier: "Modelling Requirements for the Calculation of the Turbulent Flow Around Airfoils, Wings and Bodies of Revolution,". AGARD Conf. on Turbulent Boundary Layers; Experiments, Theory and Modelling, Den Haag, The Netherlands. (1979)

Calculation of Three-Dimensional Laminar and Turbulent Boundary Layers on Bodies of Revolution at Incidence

V. C. Patel and D. H. Choi*
Iowa Institute of Hydraulic Research, The University of Iowa
Iowa City, IA 52242, USA

Abstract

The capabilities and limitations of present-day three-dimensional boundary-layer calculation procedures are explored by using two different methods to predict laminar and turbulent boundary layers on bodies of revolution at incidence. The test cases are rather severe insofar as they include extreme flow features such as large and reversing crossflows and two types of separation. It is shown that boundary-layer calculation procedures have not yet reached a stage of development where they can be used with confidence to predict such complex flows.

1. Introduction

The prediction of the development of the boundary layer on a body of revolution at incidence represents a servere test of any calculation procedure since the flow is characterized by crossflow reversal, localized thickening of the boundary layer, formation of longitudinal vortices, occurrence of singular and free-vortex types of separation and strong viscous-inviscid interaction. Judging by the requirements of a successful solution procedure as well as the questions of the limitations of first-order boundary layer theory in such configurations, laminar flow is just as challenging as turbulent flow. In fact, laminar flow on a body at incidence offers the opportunity to study the phenomena peculiar to three-dimensional flows in general without the complications of turbulence models, and can be used to evaluate the strategies underlying the numerical procedures that are being developed for three-dimensional turbulent boundary layers.

As indicated by recent reviews of three-dimensional turbulent boundary-layer calculation procedures made by *Nash* and *Patel* (1977) and *Landweber* and *Patel* (1979), it is now popular to seek direct numerical solutions of the differential equations after making some closure assumptions for the two Reynolds stresses occurring in the first-order momentum equations. The methods differ from one another in one or more of three important respects: (1) techniques used to describe the body geometry and construct an appropriate coordinate system, (2) turbulence closure model and (3) finite-difference schemes used in the numerical solution procedure. A reliable assessment of the relative influence of these three quite distinct factors has not yet been attempted since the task is enormous. The present study was, however, designed to explore this and, at the same time, to investigate the performance of some of the available methods in the presence of the phenomena mentioned above. The geometrical aspect

* Present Address: Analytical Methods Inc., Bellevue, Washington, WA 98004, USA

is partly eliminated since bodies of revolution are easily described with precision and a convenient coordinate system can be readily identified. The problem of turbulence modeling is isolated by first considering laminar flow to evaluate the numerical content.

The method of *Chang* and *Patel* (1975), designed originally for ship hulls, and that of *Nash* and *Scruggs* (1976), developed for finite wings, have been selected as representative boundary-layer calculations procedures. Both methods were modified to perform calculations on aritarary axisymmetric bodies in surface-fitted cylindrical polar coordinates. These methods differ substantially in their numerical content, and, for turbulent flow, they employ quite different closure models. As indicated earlier, comparative calculations in laminar flow enable the evaluation of the relative merits of the numerical techniques. The two methods have therefore been used to calculate the boundary layer on a prolate spheroid at incidence. The results are compared with those obtained earlier by *Wang* (1970, 1974, 1975) by yet another numerical method and the experimental observations of *Han* and *Patel* (1979). The performance of the two methods in turbulent flow is then assessed by comparing their predictions with the recent experiments of *Ramaprian* et al. (1978) on a combination body.

2. Input Pressure Distributions

Before describing the boundary-layer calculations, it is instructive to examine the general features of the pressure distribution on bodies of revolution at incidence. Consider first the potential-flow pressure distribution, and the corresponding surface streamlines, on a spheroid of axis ratio 4.3 : 1 at an incidence of 30 degrees shown in Fig. 1. The boundary layer develops under the influence of this pressure distribution, at least in regions where there is little viscous-

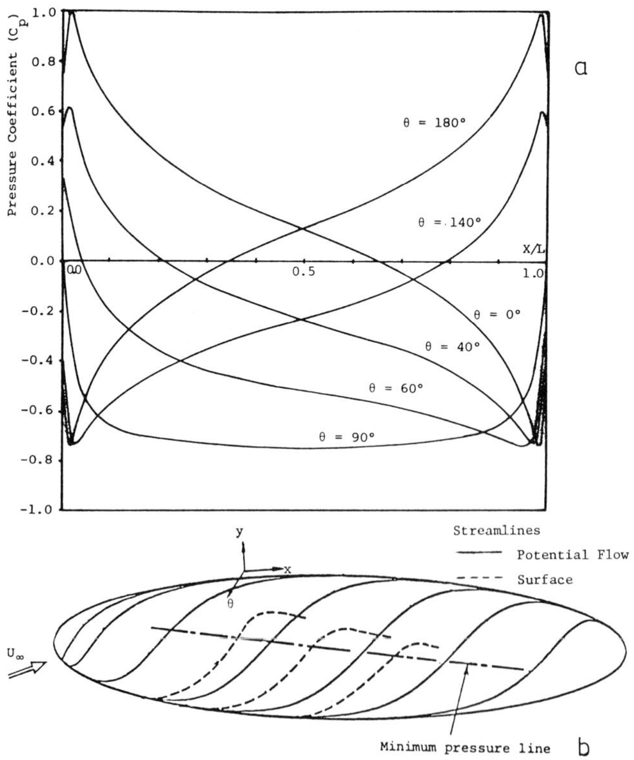

Fig. 1. Pressure distribution and streamlines on a spheroid at $\alpha = 30°$

inviscid flow interaction and no separation. In the circumferential direction, the pressure gradient is favorable from the windward to the leeward side over the windward part of the body and the locus of the points of minimum pressure occurs close to the leeward side over the front and gradually moves to the windward side over the tail region. The general direction of the limiting streamlines, close to the surface in the boundary layer, can be inferred from the pressure distribution and the curvature of the potential-flow streamlines, as shown by the dotted lines. Beyond the minimum-pressure line, the circumferential flow encounters an adverse pressure gradient and a point is reached beyond which it cannot proceed toward the leeward side. Thus, the limiting streamlines begin to turn back towards the windward side. Visual studies of the surface streamlines (see, for example, *Han* and *Patel*, 1979) indicate that, at sufficiently large angles of attack, the limiting streamlines coming from the windward and leeward sides merge along a fairly well defined line that may be identified with an open or free-vortex type of separation. The increasing adverse pressure gradient in the axial direction over the tail of the body, on the other hand, eventually leads to a closed or singular separation which is characterized by flow reversal (see *Han* and *Patel*, 1979). The primary objectives of the boundary layer solutions should therefore be the prediction of the line of circumferential flow reversal, beyond which lies open separation, and the line of closed or singular separation, which demarcates fluid originating from upstream from that appearing to come from the wake. The calculations presented here address these issues.

The reversal of the circumferential flow is accompanied by local thickening of the boundary layer and an interaction between the viscous flow and the inviscid flow outside the boundary layer. This results in a departure of the actual pressure distribution from that predicted by potential-flow theory. Figure 2 shows a comparison between the inviscid and actual pressure distribution in the case of the combination body tested by *Ramaprian* et al. (1978). The lack of agreement between the two is indicative of the magnitude and extent of the viscous-inviscid interaction.

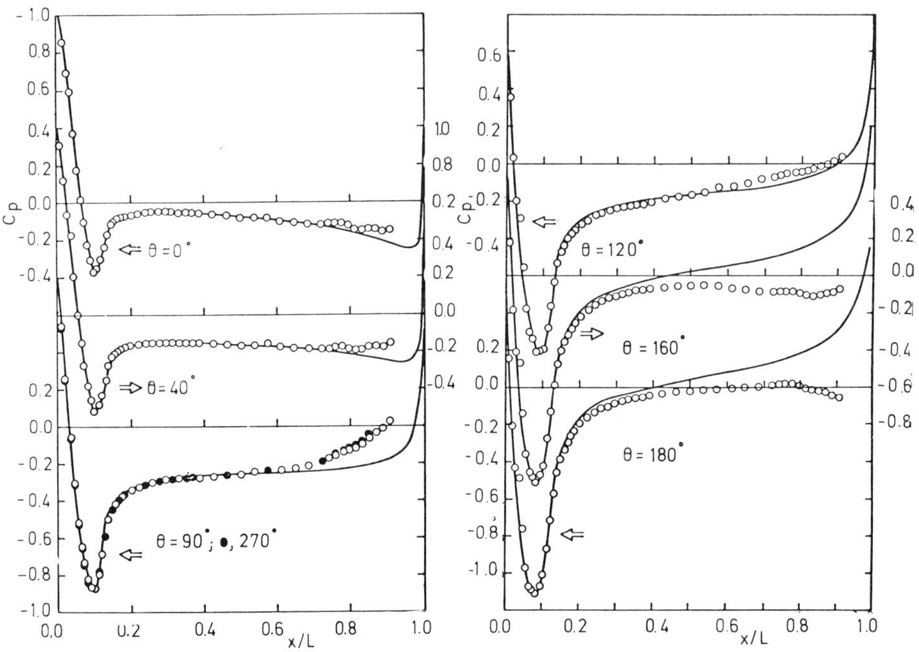

Fig. 2. Pressure distribution on the combination body, $\alpha = 15°$

In the subsequent laminar boundary-layer solutions for spheroids, use has been made of the potential-flow pressure distributions due largely to the availability of similar solutions obtained earlier by Wang. The calculations of the turbulent boundary layer on the combination body were performed, however, with the measured pressure distribution.

3. Boundary-Layer Calculation Methods

3.1. Equations and Closure

The body-fitted cylindrical polar coordinate system (x, y, θ) shown in Fig. 1 is the natural choice for the solution of the boundary layer equations on a body of revolution. It is perhaps useful to emphasize that, in general, x and y denote the directions of the coordinates but do not necessarily represent the physical distances measured along and normal to the body (hence the metric coefficients h_1 and h_2). In such a coordinate system, the first-order turbulent boundary-layer equations are:

continuity:

$$\frac{1}{h_1}\frac{\partial U}{\partial x} + \frac{1}{h_2}\frac{\partial V}{\partial y} + \frac{1}{h_3}\frac{\partial W}{\partial \theta} + (K_{21} + K_{31})\,U + K_{23}W = 0 \tag{1}$$

x-momentum:

$$\frac{U}{h_1}\frac{\partial U}{\partial x} + \frac{V}{h_2}\frac{\partial U}{\partial y} + \frac{W}{h_3}\frac{\partial U}{\partial \theta} - K_{31}W^2 + \frac{1}{h_1}\frac{\partial\left(\frac{p}{\rho}\right)}{\partial x} + \frac{1}{h_2}\frac{\partial(\overline{uv})}{\partial y} - \frac{\nu}{h_2}\frac{\partial}{\partial y}\left(\frac{1}{h_2}\frac{\partial U}{\partial y}\right) = 0 \tag{2}$$

θ-momentum:

$$\frac{U}{h_1}\frac{\partial W}{\partial x} + \frac{V}{h_2}\frac{\partial W}{\partial y} + \frac{W}{h_3}\frac{\partial W}{\partial \theta} + K_{31}UW + \frac{1}{h_3}\frac{\partial\left(\frac{p}{\rho}\right)}{\partial \theta} + \frac{1}{h_2}\frac{\partial(\overline{vw})}{\partial y} - \frac{\nu}{h_2}\frac{\partial}{\partial y}\left(\frac{1}{h_2}\frac{\partial W}{\partial y}\right) = 0 \tag{3}$$

where (U, V, W) are components of mean velocity in the (x, y, θ) directions and $\overline{uv}$ and $\overline{vw}$ are the Reynolds shear stresses.

In the present treatment, the coordinate x is identified with the *distance measured along the axis of the body* from the nose. If the shape of the body is given by $r(x)$, then the metric coefficients and the curvature parameter K_{31} are

$$h_1 = \left[1 + \left(\frac{dr}{dx}\right)^2\right]^{1/2}, \qquad h_3 = r(x)$$

$$K_{31} = \frac{1}{r\left[1 + \left(\frac{dr}{dx}\right)^2\right]^{1/2}}\frac{dr}{dx}$$

The metric coefficient h_2 and the associated curvatures K_{21} and K_{23} depend on the normal coordinate y. They are, in general,

$$K_{21} = \frac{1}{h_1 h_2} \frac{\partial h_2}{\partial x}, \quad K_{23} = \frac{1}{h_2 h_3} \frac{\partial h_2}{\partial \theta}$$

If the physical distance from the surface is used as a scale in the y-direction, as in the method of Chang and Patel, then we have the simple relations:

$$h_2 = 1, \quad K_{21} = K_{23} = 0$$

Equations (1), (2), and (3) are subject to the usual boundary conditions, namely

$$\begin{aligned} &U, V, W = 0 \text{ at } y = 0 \quad \text{and} \\ &U, W \to U_e, W_e \text{ as } y \to \delta \end{aligned} \tag{4}$$

where the subscript e refers to the edge of the boundary layer, $y = \delta$.

For turbulent flow, the method of Chang and Patel uses the eddy-viscosity model while the ADI method of Nash and Scruggs employs the turbulent kinetic-energy equation as the closure relationship. Details of the models are given in these original references.

3.2. Solution Procedure of Chang and Patel

The numerical scheme used by Chang and Patel is of the Crank-Nicolson type. It uses the central-difference formulation in the direction normal to the surface (y) so that the local truncation error is of the second order. The solution marches along the θ-(secondary flow) direction as well as the x-(main flow) direction. This restricts the method to a special class of problems, namely those in which the direction of the crossflow (W component of velocity) does not undergo a reversal in sign.

The computation domain in the direction normal to the surface extends from the wall, where the no-slip condition applies, to some distance beyond the edge of the boundary layer, where the velocity components U and W approach their potential-flow values, U_e and W_e. Typically, 25 to 35 grid points are distributed across this domain. Additional grid points are added as the boundary layer grows to insure that the whole boundary layer lies within the integration domain.

The step size in the cross-stream direction, $\Delta\theta$, may be chosen arbitrarily and need not be constant. The choice is dictated primarily by the rate of change of quantities, such as the boundary-layer thickness, in the θ-direction. Numerical tests performed earlier by Chang and Patel suggested that the step size $h_1 \Delta x$ in the marching direction of the order of two to three boundary-layer thicknesses is required to obtain acceptable accuracy and that even smaller values are needed if calculations are to proceed into regions of very rapid boundary-layer growth.

The finite-difference equivalents of Eqs. (2) and (3) are solved first with the boundary conditions of Eq. (4). The equation of continuity is then solved for the V component. For a body of revolution at incidence, the solution is first determined along the divergent plane of symmetry (i.e., the windward side, $\theta = 0^\circ$) at the new x station. The solution at other θ locations are then sought successively. Since the θ-momentum equation reduces to an identity on a plane of symmetry, leaving the system of equations indeterminate, the differential equation obtained by differentiating the θ-momentum equation is used in place of the θ-momentum equation. The associated computer program can therefore be used to perform independent solutions along a plane of symmetry. Further details of the solution procedure are given in *Chang* and *Patel* (1975).

3.3. Solution Procedure of Nash and Scruggs

The numerical scheme used in this procedure is an alternating-direction-implicit (ADI) method. The main difference between this method and the previous one is that here the equations are solved simultaneously in the plane normal to the streamwise direction. In other words, the scheme is capable of handling the reversal of the circumferential flow. The zone-of-influence principle associated with the hyperbolicity of the equations in planes parallel to the surface is taken into account by using "upwind" differences in representing the derivatives in the circumferential direction.

The number of grid points in the normal direction is kept constant by employing y/δ as the coordinate. The distribution of these grid points across the boundary layer is uniform for laminar flow and non-uniform for turbulent flow. In the latter case, the points are densely spaced in the region of large gradients close to the wall. The thickness of the integration domain is set equal to $\Omega\delta$, where Ω is constant (= 1.26).

The finite-difference forms of the momentum equations and the kinetic-energy equation can be expressed as

$$A_1 F_{j+1,k}^{(i+1)} + A_2 F_{j,k+1}^{(i+1)} + A_3 F_{j,k}^{(i+1)} + A_4 F_{j-1,k}^{(i+1)} + A_5 F_{j,k-1}^{(i+1)} = A_6 F_{j,k}^{(i)} \tag{5}$$

where F is a solution vector $(U, W, \overline{q^2})$ and i, j, k are node-point indices in the x, y, θ directions, respectively. The forward marching procedure advances in the positive x-direction, from a cross-plane i, where the solution is assumed to be known, to a cross-plane $i + 1$, at which a new solution is sought. The cross-plane is scanned alternately in the k-, and j-directions, converting Eq. (5), respectively, into the successive forms:

$$A_1 F_{j+1,k}^{(i+1)} + A_3 F_{j,k}^{(i+1)} + A_4 F_{j-1,k}^{(i+1)} = B_1 \tag{6}$$

$$A_2 F_{j,k+1}^{(i+1)} + A_3 F_{j,k}^{(i+1)} + A_5 F_{j,k-1}^{(i+1)} = B_2 \tag{7}$$

in which B_1 and B_2 contain the passive terms originating from the left-hand side of Eq. (5). The coefficients of Eqs. (6) and (7) form a block-tridiagonal matrix and is solved by a standard technique which is analogous to the factorization of an ordinary tridiagonal matrix. The normal component of velocity, V, is determined from a separate integration of the continuity equation after each k and j scan. A, B are updated in successive iterations and the solution process is continued until convergence is obtained with respect to the velocity components at each grid point within a specified tolerance.

The boundary conditions imposed in this procedure are

$$\begin{aligned} &U = V = W = \overline{q^2} = 0 \text{ at } y = 0 \\ &\frac{\partial U}{\partial y} = \frac{\partial W}{\partial y} = \overline{q^2} = 0 \text{ as } y \to \Omega\delta \end{aligned} \tag{8}$$

Furthermore, plane-of-symmetry conditions are applied as boundary conditions at the two lateral extremes of the domain of integration in the θ-direction, i.e., at the upper and lower θ-stations which lie on the plane-of-symmetry. These conditions are simply

$$\frac{\partial U}{\partial \theta} = \frac{\partial q^2}{\partial \theta} = W = 0 \tag{9}$$

4. Laminar Boundary Layer on a Spheroid

A prolate spheroid of axis ratio 4:1 was chosen as the test case, since, as noted earlier, the results can be compared with those obtained by previous investigators. Here it is assumed that the pressure distribution on the body is given by the exact potential-flow solution. Calculations have been performed using both methods for several incidences, namely, 0, 2, 6, 12 and 30 degrees. However, in the interest of brevity, we shall discuss only the results corresponding to an incidence of 6° in detail.

In general, it is not possible to start the solution from the nose of the body using the cylindrical polar coordinates (x, y, θ) since this would require the solution to march, over some region, in a direction opposing the flow. Proper treatment of the flow in the neighborhood of the stagnation point would require a special coordinate system that is aligned roughly with the local external streamlines and their orthogonals. The problems involved are considerable and are discussed by *Wang* (1974) and *Cebeci* et al. (1979). In the present work, this difficulty was avoided by starting the solutions from an axial station just downstream of the stagnation point. The initial velocity profiles were determined as follows. A separate calculation was performed along the leeward plane of symmetry, starting from the stagnation point and using the plane-of-symmetry option in the Chang and Patel method, to determine the boundary-layer thickness. The initial profiles were then generated by assuming a linear variation of δ between $\theta = 180°$ and 0°, the well known stagnation-point profile, and the assumption that the flow is collateral everywhere with the local freestream direction. Although this is an approximation, it was found that perturbations in the initial profiles ceased to influence the subsequent solutions within a short distance or a few axial steps.

4.1. Plane-of-Symmetry Boundary Layer

The method of Chang and Patel was first used to perform separate calculations of the boundary-layer development in the plane of symmetry. This required the simultaneous solution of the x-momentum equation, the differentiated θ-momentum equation and the equation of continuity using the specified distributions of U_e and $\partial W_e/\partial\theta$ in the inviscid flow outside the boundary layer. The primary quantities that are calculated are U, $\partial W/\partial\theta$ and V. Note that $\partial W/\partial\theta$ is a measure of the rate of flow convergence or divergence into the plane of symmetry.

Typical results of these calculations are illustrated in Figs. 3 and 4 for $\alpha = 6°$. Figure 3 shows that the plane of symmetry solutions can be used to identify the points of zero wall shear stress, $(x/L)_S$. These are the points of singular separation. The line joining the two singular points is the line of "singular" separation in the sense of Maskell, or the "closed" separation line defined by *Wang* (1972). The precise location of this line on the body is discussed later on.

Another special feature that can be identified from these solutions is the location $(x/L)_R$ where $\partial W/\partial\theta$ changes sign across the boundary layer, i.e., $(\partial^2 W/\partial y\partial\theta|_w = 0)$. Since the sign

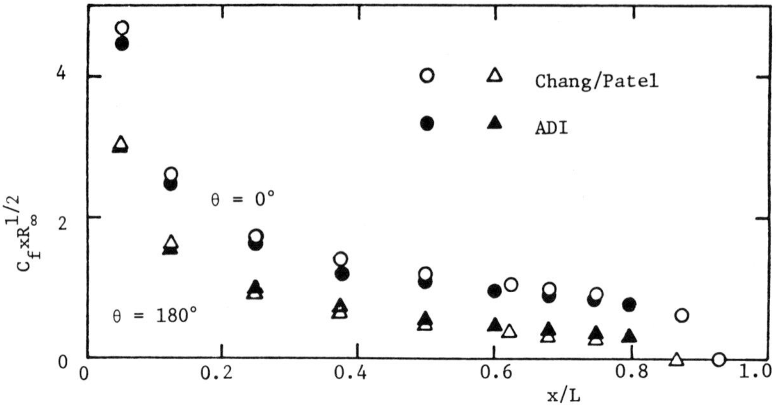

Fig. 3. Wall shear stress along the plane of symmetry. Spheroid, $\alpha = 6°$

of $\partial W/\partial\theta$ indicates flow convergence or divergence into the plane of symmetry, the point $(x/L)_R$ demarcates a major change in the flow pattern on either side of the symmetry plane. For example, reference to Fig. 4 indicates that, on the leeward side ($\theta = 180°$), $\partial W/\partial\theta$ is negative across the boundary layer at $x/L = 0.25$ and therefore the streamlines are converging into the plane of symmetry. At the three downstream stations ($x/L = 0.50$, 0.75 and 0.85), however, $\partial W/\partial\theta$ is positive near the surface and negative in the outer part of the boundary layer, indicating flow divergence near the surface and convergence in the outer part. A careful examination of the results gives $(x/L)_R = 0.38$.

Table 1. Circumferential flow reversal and separation points calculated on the plane of symmetry of a 4:1 spheroid at $\alpha = 6°$

Calculations	Windward		Leeward	
	$(x/L)_R$	$(x/L)_S$	$(x/L)_R$	$(x/L)_S$
Wang (1970)	0.895	0.928	0.45	0.862
Cebeci et al. (1980)	0.875	0.92	0.38	0.86
Present	0.885	0.923	0.38	0.865

Table 2. Movement of circumferential flow reversal and separation points on the symmetry plane with incidence. 4:1 Spheroid. Present calculations

Incidence	Windward		Leeward	
α, [deg]	$(x/L)_R$	$(x/L)_S$	$(x/L)_R$	$(x/L)_S$
2	0.785	0.875	0.60	0.80
6	0.885	0.923	0.38	0.865
12	0.955	0.965	0.25	0.925
30	None	0.998	0.035	0.98

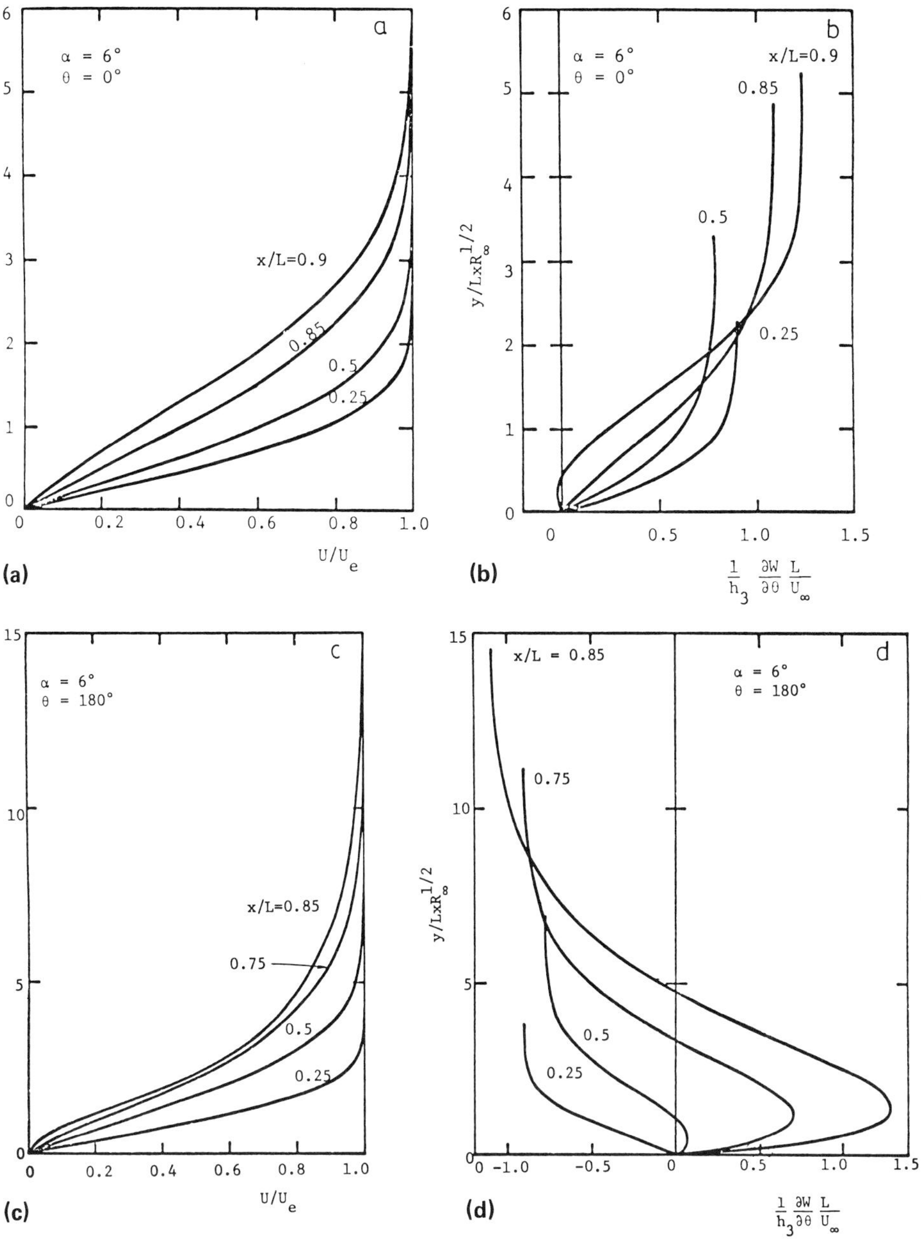

Fig. 4a–d. Velocity profiles on the windward plane of symmetry spheroid, $\alpha = 6^\circ$

The present results for $\alpha = 6^\circ$ are compared with those of *Wang* (1970) and *Cebeci* et al. (1980) in Table 1. It is seen that there is good quantitative agreement inspite of the fact that the solutions were obtained with three different numerical methods.

The locations of the points of singular separation, $(x/L)_S$, and circumferential flow reversal, $(x/L)_R$, calculated for various incidences upto $\alpha = 30^\circ$ are given in Table 2. It is seen that, on the windward side of the plane of symmetry both R and S move downstream and come closer together as the incidence is increased, while on the leeward side R moves upstream and S moves downstream. The downstream movement of S on both sides has been confirmed by the recent flow visualization experiments of *Han* and *Patel* (1979). (See Fig. 16 and its discussion in that reference.)

As demonstrated by *Wang* (1970, 1972, 1974), the plane of symmetry solutions can be used to infer the flow patterns elsewhere on the body. From the previous discussion it is clear that the flow at the separation point on the leeward side is locally divergent, making S a saddle point of separation. On the windward side, the flow is initially divergent and then convergent for a short distance (between R and S) just ahead of separation. Thus, the calculations suggest that the point of separation on the windward side is a nodal point. This led Wang to propose the flow pattern shown in Fig. 5a. This model of flow separation satisfies the topological law which requires that the number of nodal points exceed the number of saddle points by two.

While the agreement between the present calculations and those of Wang, and Cebeci et al., can be used to support the separation model shown in Fig. 5a, it should be noted from Table 2 that the distance between R and S on the windward side diminishes rapidly as the incidence increases. At $\alpha = 6°$, for example, this distance is of the order of three local boundary layer thicknesses, while at $\alpha = 12°$ it is less than one boundary layer thickness. Since the calculations were performed with the potential-flow pressure distributions, and do not take into account the higher-order effects and viscous-inviscid interaction which are undoubtedly present so close to separation, it is perhaps premature to infer that the model of Wang continues to apply in the real flow. Indeed, there are two reasons to suggest that the model should be discarded at least for $\alpha \sim 6°$. First, the experiments of Han and Patel on a 4.3 : 1 spheroid indicated that for incidences of 5° and larger, the flow at the point of separation on the windward side is divergent. This contradicts the calculations and suggests that the point of separation on the windward side is also a saddle point. The topological law then requires two additional nodal points on the singular separation line into which the divergent flow from the two saddle points on the plane of symmetry converges. Secondly, the complete ADI solutions discussed later, and those of *Cebeci* et al. (1980), have confirmed the occurrence of singularities on the side of the body. Since the singularities are predicted by the solutions of the first-order boundary-layer equations with potential-flow pressure distribution, and there is no information available on the flow further downstream, the type of singularities (i.e. nodal or saddle) existing in a real flow cannot be determined with certainty. Nevertheless, the experimental evidence, the numerical predictions of such singularities, and the requirements of the topological law indicate that they should be nodal points and the separation model should be revised as shown in Fig. 5b. It should be emphasized, however, that this model is proposed

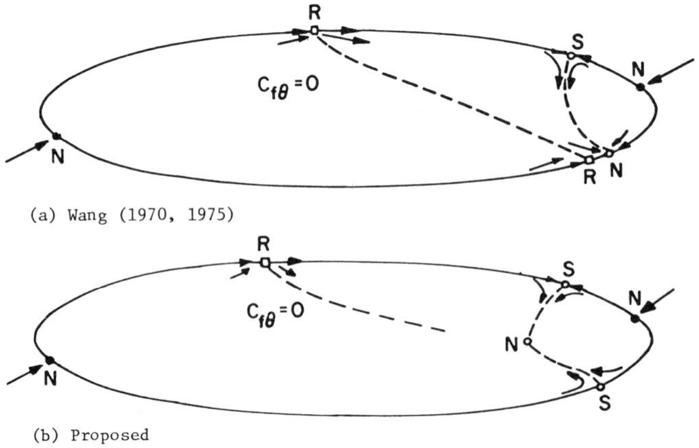

Fig. 5 a, b. Separation Patterns at $\alpha \sim 6°$

for moderate incidences of the order of 6°. It is known from the experiments of *Han* and *Patel* (1979) that the separation pattern undergoes further changes as the incidence increases with the occurrence of open or free-vortex separation and reattachment. Also, the applicability of Wang's model of Fig. 5a over a range of low incidences cannot be ruled out since no experimental evidence is available for smaller incidences.

The plane-of-symmetry solutions obtained by the ADI method were found to be generally similar to those discussed above. As seen from Fig. 3, however, these solutions could not be continued all the way to the upstream separation point on the plane of symmetry since they terminated at a singular point (zero wall shear stress) of separation occurring on the side of the body well upstream of the separation points on the plane of symmetry. From Fig. 3 and Table 1, it is evident that in the region where both methods apply, their results are in close agreement.

4.2. Off-Plane-of-Symmetry Boundary Layer

The amount of information that is required to describe a three-dimensional flow adequately is quite large. In the interest of brevity we shall present only that information which conveys the overall features and some of the peculiarities of three-dimensional flows.

Calculations by both methods were initiated at a cross-section a short distance downstream of the stagnation point using the initial conditions described earlier. In the method of Chang and Patel, calculations were performed along 12 θ = constant lines. As the calculations progressed downstream, the circumferential flow reversed at some value of θ close to the leeward plane of symmetry. According to the restriction of the method, the θ-stations where the circumferential flow reversal occurred were dropped before advancing the solution further in the axial direction. Thus, this method predicts the boundary-layer development in the region where the circumferential flow is unidirectional, and, in practice, gives the location of the line on the body along which the circumferential component of wall shear stress, $C_{f\theta}$, vanishes. The ADI method, on the other hand, provides the complete solution from the windward to the leeward plane of the symmetry since it can handle the reversal of the circumferential flow. In order to maintain second-order accuracy, the step size in the θ-direction was maintained constant at 20° and therefore the solution progressed along 10 lines of constant θ. As mentioned earlier, the ADI solution terminates when reversal of flow in the axial direction is indicated.

Figure 6 presents an overview of the results for $\alpha = 6°$. The lines of circumferential flow reversal ($C_{f\theta} = 0$) predicted by the two methods are compared with those calculated by *Wang*

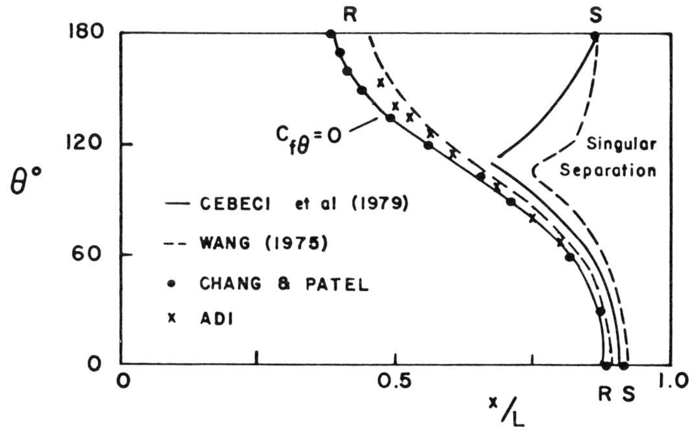

Fig. 6. Lines of Circumferential Flow Reversal and Singular Separation. Spheroid, $\alpha = 6°$

(1975) and more recently by *Cebeci* et al. (1980). It is seen that the predictions of the four methods are in reasonable agreement with respect to the location and shape of the line. In particular, close agreement is observed between the present solutions with the method of Chang and Patel and those of Cebeci et al. The figure also shows that the ADI solution continued upto $x/L = 0.8$, where it terminated due to the vanishing of both components of wall shear stress in the neighborhood of $\theta = 70°$. This location of the singularity is substantially different from those predicted by Wang ($x/L = 0.75, \theta = 106°$) and Cebeci et al. ($x/L = 0.68, \theta = 110°$). The reasons for these differences are not readily identified since the three methods differ not only in their numerical content but also in the manner in which the singularity is approached. For example, the ADI solution at each axial location is obtained by scanning the entire circumferential flow field while the other two solutions were determined by two separate calculations, marching towards the singularity from the leeward and windward sides of the plane of symmetry. Additionally, the circumferential step size of 20° in the present calculations is perhaps too coarse for an accurate determination of the flow in the region around the singularity. Note that Cebeci et al. employed $\Delta\theta$ as small as $2\frac{1}{2}°$ to obtain their detailed results.

Inspite of the aforementioned quantitative differences between the various solutions, it is instructive to examine some of the characteristic features of the boundary layer on the body as predicted by the ADI method. Figure 7 shows the circumferential variation of the two components of the wall shear stress, namely C_{fx} and $C_{f\theta}$, and the integral quantity δ_1, which is a measure of the physical displacement thickness of the boundary layer. It is interesting to note the gradual development of the reversal in $C_{f\theta}$, and the decrease in C_{fx}. We see that both components of C_f approach zero at $x/L = 0.8$ in the neighborhood of $\theta = 70°$. This is the singular point referred to earlier. It is seen that, at the axial station where this singularity is observed, the boundary layer on the plane of symmetry is still far from separation and that the value of δ_1 becomes large in the neighborhood of the point where C_f approaches zero. Figure 8 was prepared specifically to indicate the evolution of the circumferential flow reversal and demonstrate the qualitative similarity between the laminar-flow calculations and the recent turbulent-flow measurements of *Ramaprian* et al. (1978). Figure 9 shows the calculated velocity field projected onto the circumferential plane at two axial stations. Figure 9a, at $x/L = 0.5$, is similar to those of *Wang* (1975) and *Cebeci* et al. (1978). It was on the basis of such figures, and particularly on the basis of the sense of the normal component of velocity on the leeward plane of symmetry, that Wang and Cebeci et al. concluded that there was no evidence for the presence of a vortex within the boundary layer, although Wang added that this result was different from Nonweiler's (see *Thwaites*, 1960) which showed a vortical motion. The present result at $x/L = 0.68$ shown in Fig. 9b shows that the normal component of velocity undergoes further changes as the flow progresses downstream and indicates the development of a vortex whose center is moving gradually away from the leeward plane of symmetry. The changes in the normal component of velocity between $x/L = 0.5$ and 0.68 are obviously related to the increasing flow divergence from the leeward plane of symmetry. Thus, the present calculations indicate the tendency of the boundary layer to roll up into a vortex at some distance beyond the initiation of the flow divergence from the leeward plane of symmetry. The vortex is, however, buried within the boundary layer and consequently does not imply detachment of the flow from the body, as required in an open or free-vortex separation and observed at higher incidences.

The behavior of the boundary layer in the vicinity of the singularity is further illustrated in Figs. 10 and 11. From the former it is seen that the approach to zero C_f from the leeward as well as the windward side is quite rapid and that this is accompanied by a local thickening of the boundary layer. It is clear from this figure that these features are akin, in many ways,

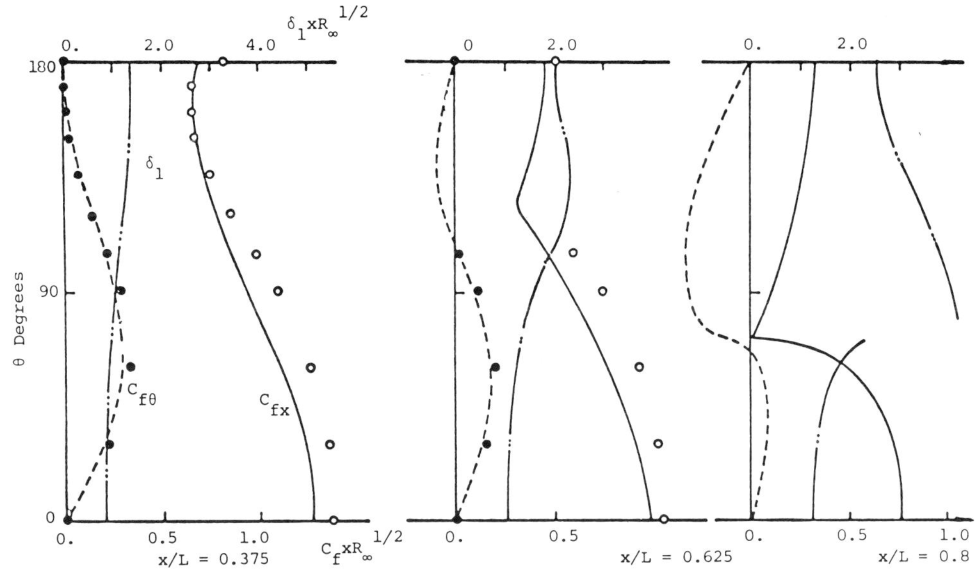

Fig. 7. Circumferential distribution of shear stress components and δ_1. Spheroid, $\alpha = 6°$. Lines: ADI, Symbols: *Chang* and *Patel*

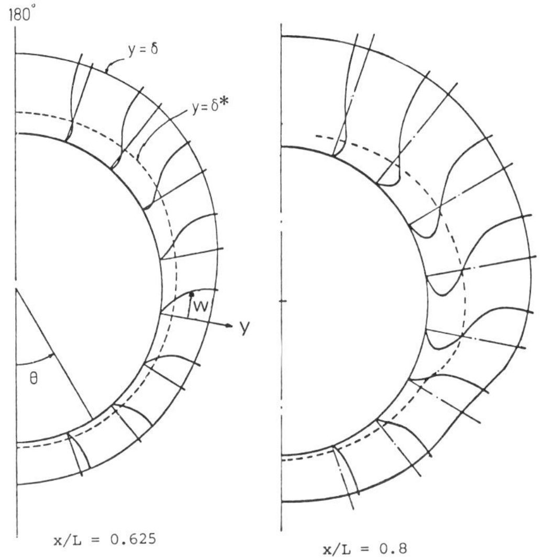

Fig. 8. Circumferential velocity component. Spheroid, $\alpha = 6°$

Fig. 9. Velocity field projected onto the cross-sectional plane, spheroid, $\alpha = 6°$

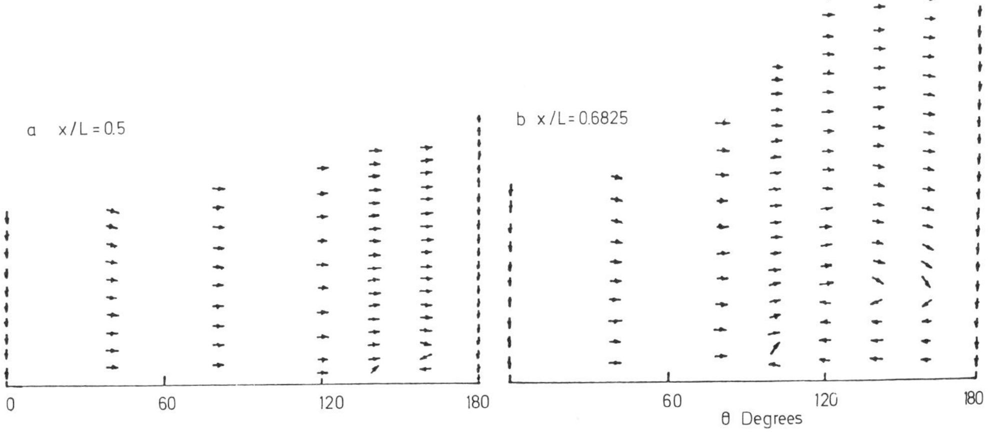

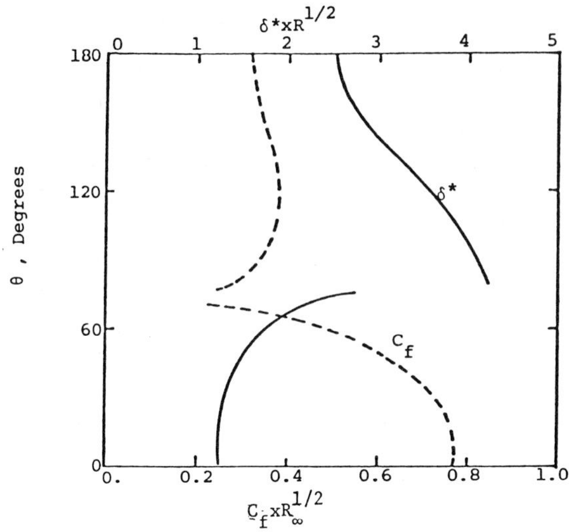

Fig. 10. Circumferential distribution of C_f and δ^* on spheroid at $x/L = 0.8$, $\alpha = 6°$

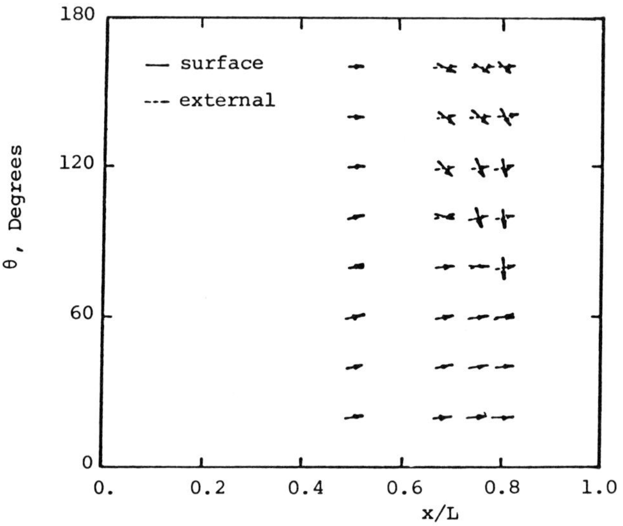

Fig. 11. Surface and external streamlines. Spheroid, $\alpha = 6°$

to those characterizing a two-dimensional separation. The surface and external-flow streamline directions shown in Fig. 11 appear to suggest that the singularity on the side of the body is a *spiral* nodal point. This is implied by the spiral nature of the streamlines converging into the singularity as well as the abrupt change in the direction of the surface streamlines in passing from the leeward to the windward side across the singularity. The evidence for this cannot, however, be considered conclusive until the details of the flow beyond the singularity are fully understood.

The final point of discussion is the concept of open separation introduced by *Wang* (1972). He postulated that this is a line into which streamlines, originating from upstream, coalesce, leading to a local thickening of the boundary layer and the possible formation of a vortex. Wang also indicated that such a line lies a short distance to the lee of the line of circumferential-flow reversal. This model of open separation was confirmed experimentally by *Han* and *Patel* (1979). Their results indicated that an open separation is absent at $\alpha = 5°$ but occurs at incidences of 10° and greater. While the calculations performed here indicate many of the features described by Wang and observed in the turbulent boundary-layer experiments of *Ramaprian* et al. (1978) the presence of an open separation at $\alpha = 6°$ is not apparent either from the present calculations or those of *Cebeci* et al. (1980). The identification of the precise location of open separation lines is difficult for two reasons: (a) the spacing of the grid in the θ-direction must be fine enough to enable an accurate tracing of the surface streamlines, and (b) the line of open separation, as defined, is the intersection with the surface of the body of a surface in the fluid across which there is a large velocity gradient (i.e., a surface of vorticity concentration). While the former can be overcome by further refinement in the grid, with a commensurate increase in computer time, the latter would require a considerable amount of analysis of the results. The problem of determining the open separation and its relation to the line of circumferential-flow reversal as well as the formation of a longitudinal vortex therefore deserves a separate study. In order to capture this phenomenon without ambiguity it is perhaps advisable to examine the solutions at higher incidences.

5. Calculations for Turbulent Flow

The originators of the two methods utilized here have already verified their performance relative to the numerical content and turbulence models against two-dimensional and some three-dimensional turbulent boundary-layer data. The general reliability of the methods for simple axisymmetric flows was further confirmed by demonstrating acceptable level of agreement between the predictions and the measurements of *Freeman* (1932) on an airship model as well as those of *Ramaprian* et al. (1978) on the combination body. For the present, attention is focussed on the applicability of these methods to the complex flow on a body of revolution at incidence. Both methods were used to calculate the boundary layer development on the combination body at an incidence of 15 degrees. The geometry of the body is indicated in Fig. 12. Measurements of velocity profiles were made at Stations 1 through 7 at several circumferential positions.

As in the case of the laminar boundary layer solutions, it is convenient to discuss the boundary layer along the plane of symmetry prior to considering the general three-dimensional flow.

5.1. Plane-of-Symmetry Flow

The method of Chang and Patel was again used to calculate the development of the boundary layer along the windward and leeward sides of the symmetry plane by starting at Station 1 with the measured velocity profile and the variation of $\partial W/\partial\theta$ across the boundary layer esti-

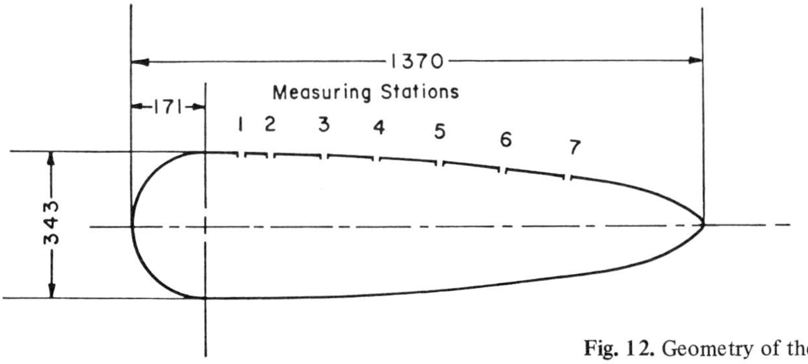

Fig. 12. Geometry of the combination body

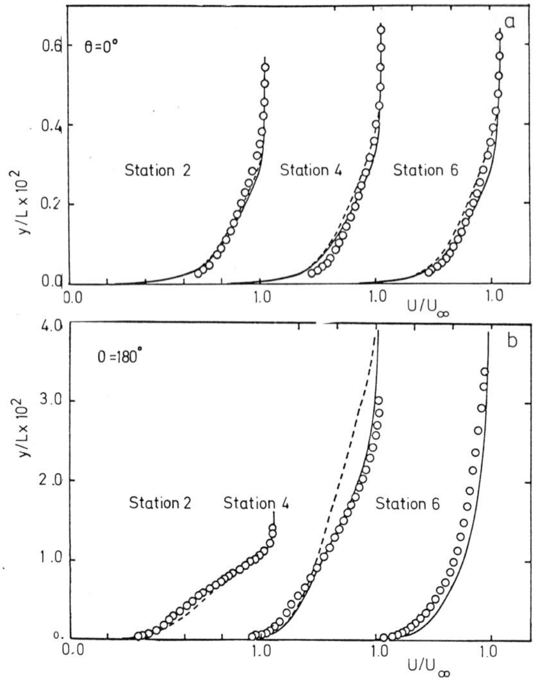

Fig. 13. Turbulent boundary layer on plane of symmetry of combination body, $\alpha = 15°$

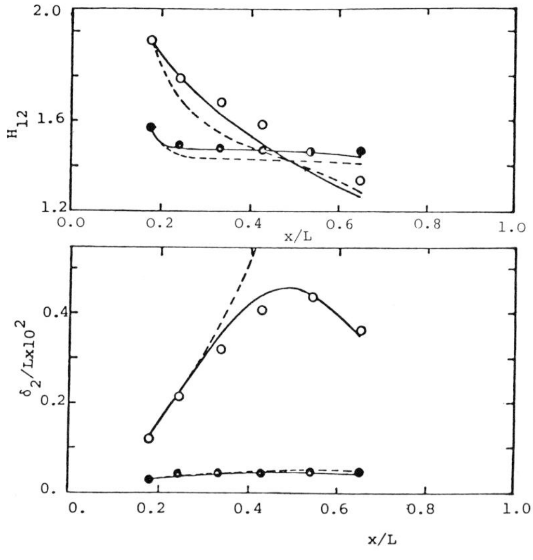

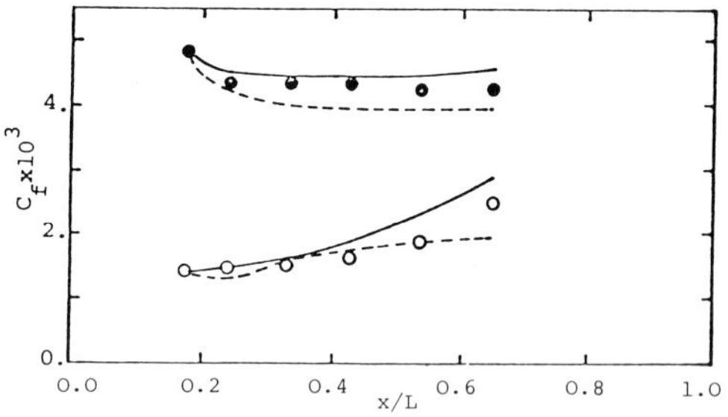

Fig. 14. Plane-of-symmetry solutions. Integral parameters. Symbols-Data, • $\theta = 0°$, ○ $\theta = 180°$, —— Chang and Patel --- ADI

mated from the measured W profile at the first θ-station away from the plane of symmetry. The results for both sides of the plane of symmetry ($\theta = 0°$ and $180°$), are shown in Figs. 13 and 14. Also included in these figures are the results obtained with the ADI method. Recall that the ADI method predicts the plane of symmetry flow as a part of the complete solution. It is clear that both methods give adequate agreement with the measurements all the way up to Station 6 along the windward plane of symmetry ($\theta = 0°$). Along the leeward plane of symmetry ($\theta = 180°$), the results of the method of Chang and Patel are in reasonable agreement with the data, but those of the ADI method show a significant departure from the data by Station 4. It is shown in the next section that the poor performance of the ADI method is not restricted to the leeward plane of symmetry. Possible reasons for this are discussed there. Finally, it is of interest to note that the Chang and Patel method predicted a reversal in the direction of $\partial W/\partial\theta$ (i.e., changeover from flow convergence into the plane to divergence out of the plane) on the leeward plane of symmetry at $x/L = 0.3$, which agrees with the origin of the circumferential flow reversal line indicated by the data (see *Ramaprian*, et al. 1978).

5.2. Off-Plane-of-Symmetry Flow

The strategy for the complete three-dimensional flow solutions by both methods was similar to that employed in laminar flow and discussed in Section 4. However, the present calculations were started with the measured velocity profiles at Station 1. The profiles of turbulent kinetic energy required in the ADI method were obtained from the Reynolds stresses calculated in the Chang-Patel method and, using structural similarity. The measured distribution of the velocity components outside the boundary layer, i.e., U_e and W_e, were smoothed and specified as boundary conditions in the Chang and Patel method. In the ADI method, the pressure gradients were also used and were obtained by differentiating the smoothed measured surface pressure distribution.

The solutions are shown in Fig. 15. In the interest of brevity, only the longitudinal and circumferential (U and W, respectively) velocity profiles at three axial and three circumferential positions are shown, although the measurements were made at many more stations. (Profiles at Station 6 at $\theta = 150°$ in Fig. 15 are not shown due to the failure of both methods, as discussed below.) Fig. 15, together with Fig. 13, suggests that a large part of the boundary-layer development on the windward side of the body is well predicted by both methods.

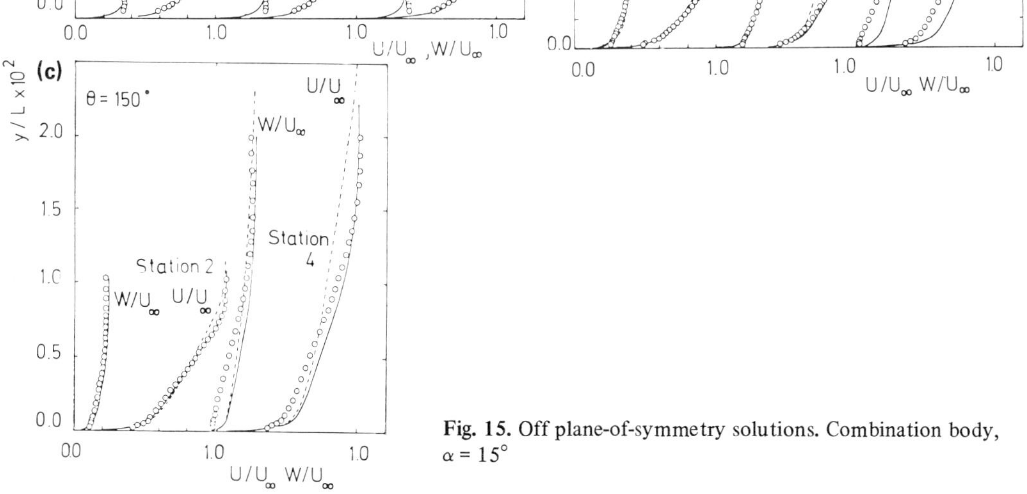

Fig. 15. Off plane-of-symmetry solutions. Combination body, $\alpha = 15°$

The results of the method of Chang and Patel indicate that the agreement with the data deteriorates as the solution progresses towards the line of circumferential flow reversal, and even when the agreement with respect to the longitudinal component of velocity is satisfactory, there is some disagreement in the circumferential component near the wall. The former difficulty is attributed to a loss of numerical accuracy in the θ-derivatives due to the rapid thickening of the boundary layer near the circumferential flow-reversal line predicted by the calculations and observed in the experiments. It is also known that most available procedures fail to predict the boundary layer development in regions where the crossflow becomes large. The lack of agreement in the wall region, on the other hand, may be attributed to either a deficiency in the isotropic eddy viscosity model, or more likely, to inaccuracies in the measurement of the flow angle by means of a three-hole yaw probe.

Figure 15 also shows that the performance of the ADI method is considerably worse than that of the Chang and Patel method. Although the method gives acceptable accuracy over a large part of the body on the windward side, where the results are similar to those of the Chang and Patel method, the boundary-layer thickness is grossly overestimated elsewhere. Since the ADI method solves for the entire circumferential flow simultaneously, it is believed that the failure of the method on the leeward side is closely related to the problem observed on the plane of symmetry. Although many numerical experiments were performed to study this aspect it was not possible to isolate the causes in a satisfactory manner. Among the likely candidates are the following: (a) The finite difference formulation in the θ-direction is of the second order everywhere on the body except at the plane of symmetry, where it is necessarily of the first-order. This leads to the same value of the θ-derivative at the plane of symmetry as that at the θ-station just removed from it. In view of the importance of accurately accounting for the rate of convergence or divergence, it is possible that the accuracy of the solution is affected everywhere. This problem could possibly be resolved by taking much smaller steps in the θ-direction with a commensurate increase in computer time and storage. (b) It is known (*Mitchell*, 1969) that the Neumann type of boundary condition ($\partial/\partial\theta = 0$) being imposed on the plane of symmetry leads to a persistent error. A possible solution to this problem is to perform a separate calculation for the plane-of-symmetry flow and prescribe the resulting solution as a boundary condition in the ADI method. (c) Yet another possibility is that the flow on the leeward side may be physically sensitive to small perturbations, in which case the level of numerical accuracy achieved in either method may not be sufficient to predict it. These issues are being examined in the ongoing studies.

6. Conclusions

The potentials and limitations of two representative methods for the solution of three-dimensional laminar and turbulent boundary-layer equations have been examined here through a series of calculations. It is apparent that the flow on a body of revolution at incidence presents a major challenge for such methods due to the occurrence of large and reversing crossflow and different types of separation. Although the calculations have been performed using either the potential flow or the measured pressure distribution, the complete solution cannot be obtained without due consideration of the viscous-inviscid interaction associated with the above phenomena. The limited number of calculations, and comparisons with previous work and experimental data, demonstrate that the existing methods of boundary-layer calculation have not reached a stage of development where they can be used with confidence. The methods give adequate description of the boundary layer development in regions where the crossflow is unidirectional. It is possible to extend the applicability of the simpler method of

Chang and Patel to regions where the crossflow reverses direction across the boundary layer (i.e., between the leeward plane of symmetry and the $C_{f\theta} = 0$ line) by modifying the finite-difference formulation to incorporate the so-called "zig-zag" scheme in which consideration is given to the local direction of the crossflow. This has been demonstrated by *Hoekstra* (1978) in his adaptation of this method to calculate the boundary layer over the bulbous bow of a ship from. The ADI method, on the other hand, is conceptually capable of describing crossflow reversals, but needs further development to resolve the difficulties associated with the specification of the leeside boundary conditions. Both methods would require considerable modification if they are to be used to calculate the shape of the line of closed or singular separation. The recent calculations of *Cebeci* et al. (1980) suggest that this involves two separate solutions, starting from the windward and leeward sides of the plane of symmetry. The method of Chang and Patel appears to be more suitable for this approach than the ADI method.

Acknowledgement. This paper is based upon research supported by the U.S. Army Research Office under Grant No. DAAG29-76-G-0036.

References

Chang, K. C., Patel, V. C. (1975): "Calculation of Three-Dimensional Boundary Layers on Ship Forms". Iowa Institute of Hydraulic Research, The University of Iowa, Report No. 178

Cebeci, T., Hirsch, R. S., Kaups, K. (1978): "Calculation of Three-Dimensional Boundary Layers on Bodies of Revolution at Incidence". Douglas Aircraft Co., Report No. MDC J7643

Cebeci, T., Khattab, A. K., Stewartson, K. (1979): Studies on three-dimensional laminar and turbulent boundary layers on bodies of revolution at incidence. I. Nose separation (1980). AIAA Paper 79-0138

Cebeci, T., Khattab, A. K., Stewartson, K. (1980): "Prediction of Three-Dimensional Laminar and Turbulent Boundary Layers on Bodies of Revolution at High Angles of Attack", in *Turbulent Shear Flows II,* Selected Papers from the 2nd Intern. Symp., London, ed. by L. J. S. Bradbury, F. Durst, B. E. Launder, F. W. Schmidt, J. H. Withelaw (Springer, Berlin, Heidelberg, New York) p. 189–198

Freeman, H. B. (1932): "Measurements of Flow in the Boundary Layer of a 1/40-Scale Model of the U.S. Airship AKRON". NACA Report No. 430

Han, T., Patel, V. C. (1978), Flow separation on a spheroid at incidence. J. Fluid Mech. **92**, Part 4, 643

Hoekstra, M. (1978): "Boundary-Layer Flow Past a Bulbous Ship Bow at Vanishing Froude Number". Proc. International Symposium on Ship Viscous Resistance, Göteborg (1978) pp. 2:1–2:38

Landweber, L., Patel, V. C. (1979): Ship boundary layers. Ann. Rev. Fluid Mech. **11**, 173–205

Mitchell, A. R. (1969): *Computational Methods in Partial Differential Equations* (John Wiley and Sons, New York)

Nash, J. F., Patel, V. C. (1977): "Advances in Turbulent Boundary Layer Calculation Methodology" (Submitted to the 1st International Symposium on Turbulent Shear Flows, Pennsylvania, USA)

Nash, J. F., Scruggs, R. M. (1976): "An Implicit Method for the Calculation of Three-Dimensional Boundary Layers on Fuselage Configurations". Report LG76ER0199, Sybucon Inc., Atlanta

Ramaprian, B. R., Patel, V. C., Choi, D. H.: Mean flow measurements in the three-dimensional boundary layer over a body of revolution at incidence. J. Fluid Mech. (to be published)

Thwaites, B. (1960): *Incompressible Aerodynamics* (Oxford University Press pp. 405–406

Wang, K. C. (1970): Three dimensional boundary layer near the Plane of Symmetry of a spheroid at incidence. J. Fluid Mech. **43**, 187–209

Wang, K. C. (1972): Separation patterns of boundary layer over an inclined body of revolution. AIAA J. **10**, 1044–1050

Wang, K. C. (1974): Laminar boundary layer near the symmetry-plane of a prolate spheroid. AIAA J. **12**, 949–958

Wang, K. C. (1974): Boundary layer over a blunt body at high incidence with an open-type of separation. Proc. R. Soc. London Ser. A: **340**, 33–55

Wang, K. C. (1975), Boundary layer over a blunt body at low incidence with circumferential reversed flow. J. Fluid Mech. **72**, 49–65

Turbulent Flow over Two-Dimensional Fences

F. Durst and A.K. Rastogi*

Institut für Hydromechanik, Universität Karlsruhe
D-7500 Karlsruhe, Fed. Rep. of Germany

Abstract

Turbulent flows over two-dimensional obstructions placed in a channel are studied for different blockage ratios within the range 0.0 to 0.5. The lengths of the recirculation zone behind obstructions with different geometrical parameters are reported. These were obtained by flow visualization in a wind tunnel. The governing partial differential equations for separated flows are solved by finite difference schemes. Two schemes employing upwind and skewed upwind differences are used and compared. Calculations employ the k-ϵ turbulence model which has been modified to take account of the effects of the mean streamline curvature. The sensitivity of the two finite difference schemes to turbulence model changes, such as the streamline curvature correction, is examined. Present calculations are compared with the available measurements and also with previous calculations for the same flow situations. The calculations presented in this paper provide definite improvements over previous calculations and compare satisfactorily with the experimental data on the size of the recirculation zone. General conclusions for computational studies of separated flows are presented in the paper that suggest the use of finer numerical grids than those reported in previous calculations of such flows. The flow computations show that accurate computations require fine grids and optimization of the grid distribution.

Nomenclature

A, B	Finite difference coefficients
A_1, A_2, A_3, C_s'	Empirical constants
B	Obstruction width
F	Curvature function
G	Kinetic energy production
H	Obstruction height
H_D	Channel height
k	Kinetic energy of turbulence
n	Number of grid points
P	Pressure, main grid location
R	Radius of curvature
Ri	Richardson number
S	Source
U	Mean longitudinal velocity
V	Mean transverse velocity
$\overline{uv}$	Turbulent shear stress
x	Longitudinal coordinates
y	Transverse coordinate
y^+	$\rho \cdot yU_\tau/\mu$
ϵ	Dissipation of turbulent kinetic energy
ρ	Density
μ	Viscosity
σ	Prandtl number
ϕ	General dependent variable
V_ϕ	Resultant velocity
τ	Wall shear stress

* Det Norske Veritas, D-2000 Hamburg, Fed. Rep. of Germany presently at Universität Karlsruhe

Subscripts

a, b Above, below
c Curvature
n, s, e, w North, south, east, west
o On the block
p Main grid point
r Recirculation
t Turbulent
u, d Upstream, downstream
y, x Coordinate direction

Introduction

Turbulent flows over obstructions are an important group of separated flows that occur in many practical flow situations. This practical importance and the fact that experimental and theoretical techniques have become available to study such flows, have focused increased attention to refined investigations of turbulent separated flows. The present paper only deals with turbulent flows over two-dimensional obstructions, one important group of separated flows. Flows of this kind occur in heat transfer applications, control of boundary layers, river flows, etc.

At the 1st Symposium on Turbulent Shear Flows, the authors presented detailed velocity measurements, see [1], obtained with a one component laser-Doppler anemometer in the flow around a square obstacle placed in a two-dimensional channel providing a blockage ratio of 0.5. Numerical computations of this flow, also reported in [1], employed the standard k-ϵ turbulence model and a solution procedure using upwind differences with special treatment of the corner points of the obstruction. Although these calculations did not provide accurate results for the velocity distribution inside the recirculation region, the length of the recirculation region was very well predicted leading the authors to believe that the general shape of the recirculation zone behind two-dimensional obstructions could be reliably predicted with the existing calculation procedures. The authors were concerned, however, to find in [2] (available only in 1978) that the calculations obtained for smaller blockage ratios ($\lesssim 0.1$), using the same calculation procedure and the same turbulence model as used in [1], grossly underpredicted the size of the recirculation zone behind a thin obstacle; for example, the length was underpredicted by 30% and the maximum height by 37% *. It was also found in [2] that changing the empirical constants in the k-ϵ turbulence model did not change the calculations appreciably. Although these changes in the constants were not based on physical reasoning, the fact that the calculation method was not responsive to them, leads to the belief that the grids used in the computation of [2] were too coarse and the calculations suffered from false diffusion associated with upwind finite difference schemes. The seemingly discouraging results of [2] motivated the authors to look into separated flows over obstructions with small blockage ratios and results of this investigation are reported in this paper.

The present investigation deals only with integral properties of turbulent flows over two-dimensional obstacles, such as the length of the recirculation zone and its shape. The paper does not consider details of the velocity field or the turbulence structure. The reasons for this are manyfold; one being the practical significance of the integral parameters of separated flows which should be correctly predicted before attempts are undertaken to calculate the details of the flow field correctly.

The reasons for large discrepancies between experimental results of separated flows and previous computations, even for gross flow features, are unclear. The turbulence models em-

* It should be mentioned here that these calculations employed coarser grids than those used by the authors in [1].

ployed in the computational procedures are often blamed. However, there is not enough evidence for such a blame, especially since previous predictions were claimed to be insensitive to turbulence model changes. It is the purpose of this paper to clarify the existing uncertainties about the capability of existing turbulence models on the basis of results of calculations which are sensitive to changes in turbulence modelling.

Some features, such as the treatment of the near-wall flow, in the calculation procedures for separated flows have been taken from the boundary layer calculation methods with some minor modifications. In future investigations, these have to be replaced by laws and functions that are more appropriate for separated flow regions. Such extensions are necessary for providing computations that represent the details of the separated flows correctly. However, more refined calculations must await detailed measurements carried out with reliable experimental techniques.

In addition to refinements of turbulence models, reliable and detailed measurements of mean velocity distributions and distributions of turbulence quantities are also needed to meaningfully test the employed numerical prediction schemes. Such detailed measurements are presently underway at the University of Karlsruhe and based on these measurements, the authors plan to develop the computational procedures and turbulence models for separated flows further in order to provide computer programs that correctly predict details of local, time averaged properties of the velocity field.

Initial investigations were carried out to yield information on the overall pattern of turbulent flows over two-dimensional obstacles. Furthermore, the measurements reported in the present work are also restricted to the size of the recirculation zone behind two-dimensional obstacles. These measurements extend the existing information on the length of the separation region over the complete range of blockage ratios below 0.5; this makes it possible to examine the calculations over a wider range of flows. Variations of important geometrical parameters and their influence on the flow were also studied.

Calculations were carried out in the present investigation using two upwind finite difference schemes; an upwind central hybrid finite difference and a skewed upwind hybrid finite difference scheme, in order to see the influence of the finite difference scheme on the solution of the governing partial differential equations. Calculations with both the schemes used the k-ϵ turbulence model. Various changes have also been incorportaed in the k-ϵ turbulence model, including two different corrections to simulate the influence of streamline curvature on turbulence. Calculations have been carried out with these changes in the turbulence model with both differencing schemes in order to examine how receptive they are to turbulence model changes. Although the two curvature corrections employed are not the most realistic ones available to simulate the influence of streamline curvature on turbulence, they have been used in the present work primarily to study the sensitivity of the finite difference schemes to the turbulence model modifications. They have, none the less, some physical basis and have resulted in the desired influences in other flow situations. Therefore, results obtained with these models will be of practical interest, even though the exact numbers produced may be questionable from a physical point of view. The reliability of a particular numerical scheme and/or a turbulence model is examinded by comparing the calculations with the present measurements and also other measurements available in the literature. Present calculations are also compared with the calculations reported in [2]. These comparisons lead to some general conclusions regarding computations of separated flows.

Experimental Investigation

Existing measurements on two-dimensional turbulent separated flows over obstructions are restricted either to large blockage ratios ($\gtrsim$ 0,5), e.g. [1] or were carried out for small blockage ratios ($\lesssim$ 0.2), e.g. [3, 4 and 5]. It was one of the purposes of the present measurements to fill the existing gap of information and to provide measurements covering the complete range of blockage ratios of interest. Another purpose of this investigation was to measure the influence of various geometrical parameters of the two-dimensional obstacles in order to identify the ones that have the largest influence on the size of the recirculation zone.

The main physical difference between the flows at low and high blockage ratios is given by the resulting pressure gradients, which at high blockage ratios influence the flow more than any turbulent transport processes. At small blockage ratios, the flow is entirely controlled by turbulent diffusion. This has been thought to be the main reason why the use of the eddy viscosity hypothesis with a two-equation turbulence model, e.g. the k-ϵ model in [1, 2], provides reasonable results for the pressure controlled flow at large blockage ratios but functions poorly at small blockage ratios at which the pressure influence on the flow is less dominant. Since it was the major aim of the present investigation to improve available calculation methods and to make them applicable also to the turbulent flows over two-dimensional obstacles at small blockage ratios, experimental data was needed to compare computational and experimental results at blockage ratios between 0.0 to 0.5.

Equipment, Techniques and Results

The size of the recirculation zone behind various two-dimensional obstructions was measured in the return flow wind tunnel with open test section shown in Fig. 1 which provided a low turbulence ($\lesssim$ 0.5%) jet between the elliptic (1,050 x 730 mm) contraction and the receiving diffusor. Most of the experiments were carried out at 15 m/s though the tunnel could be driven up to 60 m/s. A test section suited for the present investigations was built (see Fig. 2) in the wind tunnel. The test section consisted of a base plate with a well designed leading edge with a boundary layer trip, two side walls and a top plate. The side walls and the top plate also had leading edges to prevent flow separation. The side walls could be moved so that the aspect ratio of the test section could be adjusted to yield two-dimensionality in the central part of the flow. Most of the measurements were carried out with a 800 mm wide test section; flows with all the obstructions were found to be two-dimensional for this test section width. As shown in Fig. 2, obstructions could be fixed to the base plate about 300 mm downstream of the sand paper trip.

The blockage ratio for each obstacle could be easily changed with the help of the movable to plate. It was confirmed that at 150 mm upstream of the fence location, the wall boundary layer was fully turbulent and two-dimensional. Three rectangular fences and one sharp edged

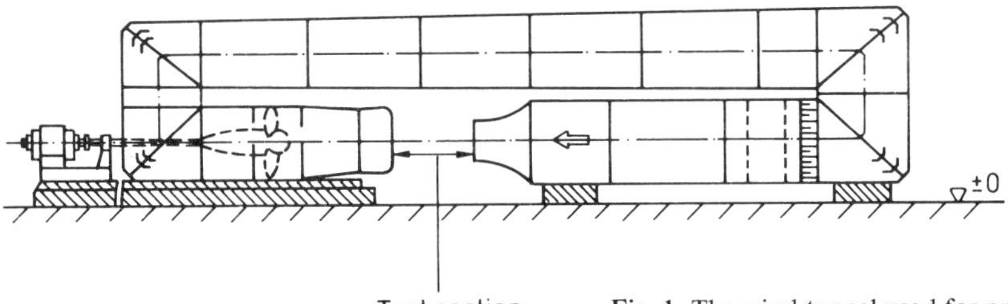

Fig. 1. The wind tunnel used for present measurements

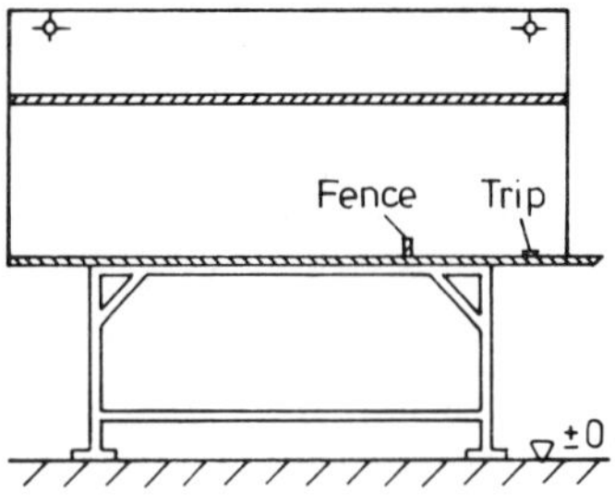

Fig. 2. The test section

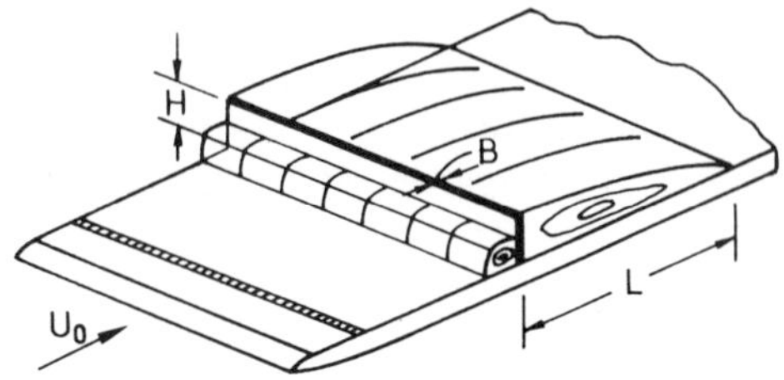

Fig. 3. Flow configuration

fence were used in the present measurements. Figure 3 shows qualitatively the flow situation along with the geometrical parameters.

Two flow visualization methods were used in order to determine the size of the recirculation zone behind the fences. In the first method, five parts of machine oil were mixed with 1 part petroleum and the mixture painted on the base plate. The oil mixture also contained fine particles of titaniumdioxide in order to mark the wall streamlines. Photographs were taken from above once the wall streamline pattern was stationary. The location of the reattachment point behind the obstructions could in this way be found to within an uncertainty of 5 mm, i.e. the length of the recirculation zone could be determined to better than 0.5 H. Figure 4 presents the measured lengths for the four fences used in this investigation. It is seen that the two thin fences give the same normalized recirculation lengths and the influence of making the separating edge sharp (45°) is to increase the length of recirculation by about one fence height, irrespective of the blockage ratio. The last points in Fig. 4 on the dotted lines have been measured without the top plate. The lengths of recirculation measured with the square fence are smaller than those measured with the thin fences but in both cases, the strong influence of the blockage ratio on the recirculation length is apparent. The boundary layer thickness, 150 mm upstream of the fence location was measured to be about 10 mm. The separation region upstream of the fence was not investigated in this work. The previous measurements [1] had shown that this region is limited to about one obstruction height upstream of the obstruction.

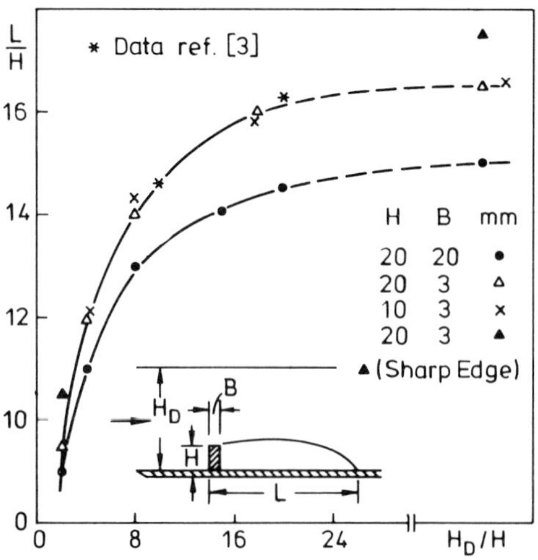

Fig. 4. Measured length of recirculation region

The general shape of the downstream recirculation zone was observed using the helium bubble technique. In this method, helium filled soap bubbles are injected under pressure upstream of the fence in the direction of the main flow. By adjusting the soap, helium and air-flow rates and pressures in the bubble generator, neutrally buoyant bubbles were produced which followed the streamlines. By illuminating these bubbles correctly, the streamline pattern beyond the fence was photographed. The bubble generator used in the present investigation was manufactured by Armfield Technical Co. Ltd. With the rectangular fences, the maximum height of the recirculation zone was measured to be about twice the fence height for a blockage ratio of 0.05. Detailed measurements of the mean velocities and the turbulent fluctuations were also made for one of the fences using hot-wire anemometry upsteam of the fence and downstream of the reattachment point. Because the aim of the present work was prediction of the size of the recirculation zone, these measurements are not reported here.

Calculations

As a result of their superior stability characteristics, it has been a common practice in numerical fluid mechanics to use upwind finite difference schemes or hybrid schemes (central for cell Peclet $\leqslant 2$ and upwind for Peclet numbers > 2) to solve the elliptic partial differential equations which govern flows with regions of separation. It is also known that upwind finite difference schemes suffer from numerical diffusion which can lead to large errors when the velocity vectors do not align with the coordinate lines. The authors suspect that the presence of numerical diffusion is indicated in previous predictions of separated flows by the fact that changes in turbulence models did not appreciably influence the obtained results, e.g. see [2, 6]. It is likely that too coarse numerical grids were employed for these studies although grid independence tests have been carried out. What is often not appreciated is the fact that upwind finite difference schemes are sluggish to grid refinement tests and this can mislead a user into believing that grid independent solutions have been obtained. This was indicated by the authors' calculation of the fence flows, details of which are given in this section.

Raithby [7] has proposed the skewed upwind finite difference scheme which has the advantages of the upwind schemes and has been shown [7, 8] to be numerically less diffusive. This scheme has also been employed in the present work for calculating the turbulent fence flows. The main features of the skewed scheme will be given and the results will be compared with the ones obtained with the upwind hybrid scheme. Both these schemes have been used with the standard k-ϵ turbulence model. The k-ϵ turbulence model has also been modified to take account of some specific features of the separated flows, e.g. influence of streamline curvature on turbulence. These modifications are briefly described and the sensitivity of the two schemes to these changes is investigated. Results of these calculations are compared with the present measurements over the complete range of blockage ratios and with some of the existing data for small blockage ratios.

Governing Equations

The governing equations for turbulent, two-dimensional separated flows can be written in the cartesian coordinate system in the following form:

$$\frac{\partial}{\partial x}(\rho u \phi) + \frac{\partial}{\partial y}(\rho v \phi) = S_\phi + \frac{\partial}{\partial x}\left(\frac{\mu_t}{\sigma_\phi}\frac{\partial \phi}{\partial x}\right) + \frac{\partial}{\partial y}\left(\frac{\mu_t}{\sigma_\phi}\frac{\partial \phi}{\partial y}\right) \tag{1}$$

The above equation reduces to the continuity equation for $\phi = 1$ and $S_\phi = 0$ and to the momentum equations in direction x and y for

$$\phi = U, S_\phi = -\frac{\partial P}{\partial x} + \frac{\partial}{\partial x}\left(\mu_t \frac{\partial U}{\partial x}\right) + \frac{\partial}{\partial y}\left(\mu_t \frac{\partial V}{\partial x}\right) \quad \text{and}$$

$$\phi = V, S_\phi = -\frac{\partial P}{\partial y} + \frac{\partial}{\partial x}\left(\mu_t \frac{\partial U}{\partial y}\right) + \frac{\partial}{\partial y}\left(\mu_t \frac{\partial V}{\partial y}\right)$$

respectively. The eddy viscosity hypothesis has been used for the turbulent diffusion of ϕ. The above equations have to be solved with the following boundary conditions

$U = V = 0$ along the walls

$\frac{\partial}{\partial x} = 0$ along the downstream plane of the calculation domain

$U = U_0 =$ constant in the free stream.

The velocities parallel to the walls at the grid points nearest the wall have been calculated from the law of the wall. The wall law should take account of the local pressure gradient in the direction considered as was done in [9] for boundary layer calculations. It may be mentioned here that eventually in separated flow calculations, the law of the wall has to be replaced by more appropriate laws or other wall treatments. No attempts have been undertaken in this paper to refine the predictions by improved wall treatments for separated flows.

Turbulence Model

The turbulent viscosity appearing in the momentum equations has been obtained using the $k - \epsilon$ turbulence model [10]:

$$\mu_t = \rho C_\mu k^2/\epsilon \tag{2}$$

k and ϵ in Eq. (2) are obtained by solving the partial differential equations for these quantities. These equations are obtained from Eq. (1) by substituting

$\phi = k, \quad S_\phi = G - \rho\epsilon, \quad \sigma_k =$ constant for k and

$\phi = \epsilon, \quad S_\phi = C_1 \frac{\epsilon}{k} G - C_2\rho\epsilon^2/k, \quad \sigma_\epsilon =$ constant for ϵ where

$$G = \mu_t \left[2\left(\frac{\partial V}{\partial y}\right)^2 + 2\left(\frac{\partial U}{\partial x}\right)^2 + \left(\frac{\partial V}{\partial x} + \frac{\partial U}{\partial y}\right)^2\right]$$

G represents the production of turbulent energy by the mean motion. At the points nearest to the walls the production of turbulent energy is evaluated using the local wall shear stress and account is taken of the fact that turbulent energy is produced only outside the wall sublayer ($y^+ \leqslant 11$). Similar considerations are also taken into account for the dissipation term of the k-equation at the points nearest to the walls. Such refined wall treatments are important around the obstruction where the numerical grid has to be fine. The ϵ-equation is not solved for the grid nodes nearest to the wall. The values of ϵ at these points are fixed using the condition of local equilibrium of turbulence. The five empirical constants appearing in the above closed set of differential equations have been given the usual values which have also been employed in [1] for separated flow calculations.

The standard $k - \epsilon$ model has been modified to take into account the influence of streamline curvature on turbulence. Recently some advanced turbulence models [11] have been developed for simulating the influence of streamline curvature on turbulence and models such as these should be eventually used for separated flow calculations. The emphasis of the present work was more on testing the sensitivity of the two finite difference schemes to changes in turbulence models. For this purpose, two curvature corrections have been chosen which are simple to apply in the framework of the $k - \epsilon$ turbulence model and also have some physical basis.

The MNA Curvature Correction. This correction has been used by Militzer et al. [12] for separated flow calculations and is based on the finding of *Irwin* and *Smith* [13] that the Reynolds stress $(-\overline{uv})$ is strongly correlated with the curvature parameter F in a curved wall jet. According to this correction, the turbulent energy production is modified in the following way:

$$G' = S_c G \quad \text{with} \quad S_c = A_1 \{1 - \exp [A_2(F - A_3)]\} \tag{3}$$

where $A_1 = 1.15, A_2 = 1.13$ and $A_3 = 0.18$.

F in Eq. (3) in a two-dimensional recirculating flow is defined as:

$$F = \frac{\sqrt{U^2 + V^2}}{R \, \partial V_\theta / \partial R} \tag{4}$$

where V_θ is the local resultant velocity and R is the local radius of curvature defined by:

$$\frac{1}{R} = \frac{UV \left(\dfrac{\partial V}{\partial y} - \dfrac{\partial U}{\partial x} \right) + U^2 \dfrac{\partial V}{\partial x} - V^2 \dfrac{\partial U}{\partial y}}{(U^2 + V^2)^{1,5}} \tag{5}$$

In [12] the influence of this correction on the calculations was not determined and moreover the expressions used for F and R were incorrect.

The LPS Curvature Correction. This correction has been used by *Launder* et al. [14] for boundary layer flows on curved surfaces and rotating bodies. Although semi-intuitive in nature, this modification of the $k - \epsilon$ turbulence model for flows with streamline curvature is based on considerations of the stability of a turbulent eddy on a curved surface.

This correction modifies the dissipation term of the ϵ-equation in the following way:

$$\rho C_2 \frac{\epsilon^2}{k} (1 - C_c \mathrm{Ri_t}) \quad \text{where } C_c = 0.2 \text{ and}$$

$$\mathrm{Ri_t} = \frac{k^2}{\epsilon^2} \frac{V_\theta}{R^2} \frac{\partial}{\partial R} (R V_\theta) \quad \text{is the turbulent Richarson number.}$$

The LPS correction differs from the MNA correction in that it modifies an already modelled term in the ϵ-equation whereas the MNA correction actually modifies an exact term in the k-equation.

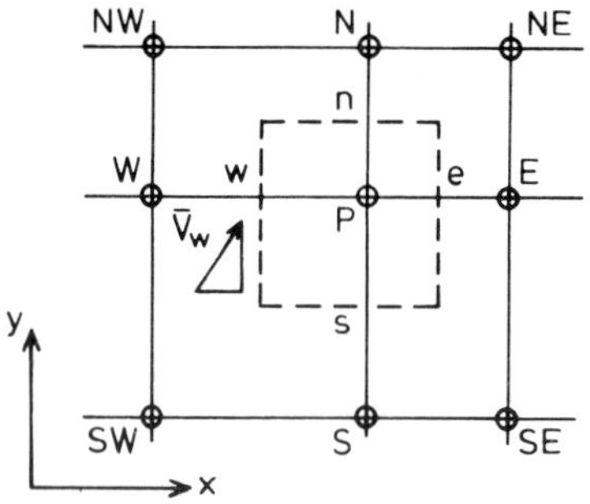

Fig. 5. The finite difference grid

Solution Procedure

The upwind hybrid finite difference scheme (TEACH) used in the present work is the one developed at Imperial College [15]; it has already been adequately described in the literature so that further descriptions are not required. However, specific features of the skewed upwind scheme developed by *Leschziner* [16] and used here will be briefly described. This scheme was proposed by *Raithby* [7] and is a 9 point finite difference scheme as opposed to a 5 point upwind scheme used in TEACH. As shown in Fig. 5, integration of the conservation equation (1) for ϕ over the control volume for a grid point P requires the convective and diffusive fluxes at the control volume faces represented by the points, w, e, n, and s. Considering only the west face, w, the convected value of ϕ is calculated in the skewed scheme taking into account the velocity vector at w, e.g. for U and V + *ve* at w, the convected ϕ at w, ϕ_w, is obtained by interpolating between ϕ_W and ϕ_{SW} or ϕ_{SW} and ϕ_S depending upon the relative magnitude of V and U at w. Similarly ϕ_W will be interpolated between ϕ_W and ϕ_{NW} or ϕ_{NW} and ϕ_N for + *ve* U and $-$*ve* V on the west face w. In the procedure used in the present work, linear interpolation is used. *Raithby* [7] has also suggested a weighted interpolation which involves the calculation of exponential functions and is computationally more expensive. In this way the ϕ difference equation for P contains ϕ values at the 9 grid points shown in Fig. 5. The terms containing the corner points NW, NE, SE and SW have been calculated within the known values from the previous iteration and the difference equations are solved in the following form using the tridiagonal matrix algorithm.

$$Ap\phi p = \sum_{i}^{EWNS} A_i\phi_i + B_i \tag{6}$$

Diffusive fluxes appearing in the above coefficients are calculated as usual by assuming a linear ϕ variation between the grid nodes. The skewed upwind finite difference scheme is more accurate for calculating convective fluxes than the usual upwind finite difference schemes for those parts of the flow which are convection dominated. When convection and diffusion are equally important, in cases of grid Peclet numbers of up to 2, the procedure employed use central differences and for large Peclet number, skewed upwind differences. Thus the scheme used is actually a skewed upwind hybrid scheme. In the calculations carried out in the present work, only the momentum equations were solved by the skewed scheme. The equations for k and ϵ were solved by the upwind finite difference scheme. Since the conservation equations for k and ϵ are dominated by production and dissipation terms, the finite difference scheme for their solution was not very critical. The numerical treatment for the obstruction corner described in [1] was retained in the present work in all the calculations carried out for thick obstacles. For thin fences, however, only one grid line was placed in the middle of the fence.

Results of Computations

Thick Fence

The authors began the present work by repeating the calculations of one of the flow cases studied in [2]. In this case, the flow was over a fence of square cross section with exactly the same boundary conditions used by *Good* and *Joubert* [4] in their measurements over a thin fence. Although there are no measurements available for this particular case of a thick fence, it was selected for the preliminary calculations because the computer program developed in [1] could be used with little additional effort to yield results comparable with those in [2]. The calculations were carried out for the blockage ratio of 0.11 with two numerical grids. The top boundary of the calculation domain was treated as a wall for most of these runs. One calculation was also performed with a free stream along the top boundary.

Figures 6 and 7 show the computed results for the above mentioned case obtained with various numerical grids. Some initial calculations had indicated significant influence of the grid distribution on the calculated values of the recirculation length and therefore, it was decided to carry out systematic grid-dependence tests before drawing any conclusions from the calculated results. Most of the results shown in Figs. 6 and 7 and the conclusions drawn here refer to the upwind finite difference scheme and the curvature correction employed referred to the LPS correction which showed more response than the MNA correction. Only a few calculations were performed with the skewed finite difference scheme because of the rather slow rate of convergence and large computing times. Figure 6 shows the influence of the number of grid points and their distribution in the y-direction. For these calculations, the number ($n_x = 28$) and the distribution of grid points in the x-direction was kept unchanged. It is seen from Fig. 6 that the calculated length of the recirculation zone changes only very slightly between $n_y = 26$ and $n_y = 34$. This means that for the geometry under consideration, e.g. square obstruction with a blockage ratio of 0.11, 26 grid points in the y-direction are sufficient to give reliable solutions provided they are distributed as shown in Fig. 8. Circles on Fig. 6 show the results of calculations carried out with a uniform grid distribution in the y-direction and prove the significance of problem-oriented grid distribution for numerical calculations of separated flows. As shown in Fig. 6, the use of non-uniform distributions in the y-direction with 26 points but without any curvature correction provides the same length of recirculation as obtained with 26 points equally distributed with curvature correction. Calculations presented in [2] were obtained by using a gradually expanding grid in the y-direction, the smallest Δy being next to the bottom wall rather than the top of the obstruction as shown in Fig. 8. This explains the significant difference in the calculations of [2] and the present calculations. The results of [2] are indicated by x_2 on Fig. 6 for the same flow calculations. Computations with equally distributed 19 points in the y-direction employing the skewed differencing scheme are also shown in Fig. 6 and provide the same values of L/H as obtained by the upwind difference scheme. Results presented in Fig. 6 show that with and without curvature correction, 26 or more points are sufficient in the y-direction if they are distributed in a manner similar to the distribution shown in Fig. 8. This was also found for a much finer ($n_x = 53$) grid distribution in the x-direction as shown in Fig. 7 and therefore, the above conclusion is independent of the grid number in the x-direction.

The sensitivity of the calculated results to the choice of grid in the x-direction was also examined and the results are shown in Fig. 7. For $n_x = 53$, the grid distribution employed is shown in Fig. 8. This figure shows that non-uniform grid distributions were employed and similar grids were used for all n_x except for $n_x = 61$ for which a few more points were provided in the region of reattachment. Calculations shown in Fig. 7 indicate that the computed recirculation length remained unchanged for $n_x \gtrsim 53$ provided the grid distribution was kept

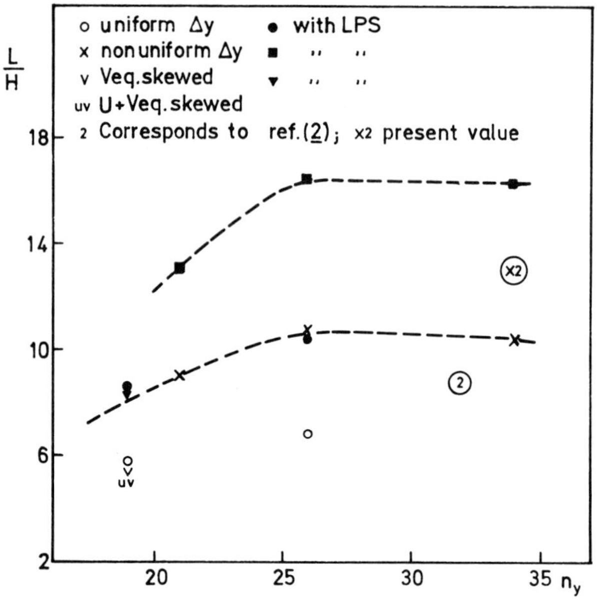

Fig. 6. Computed separation length as a function of grid points in the y-direction

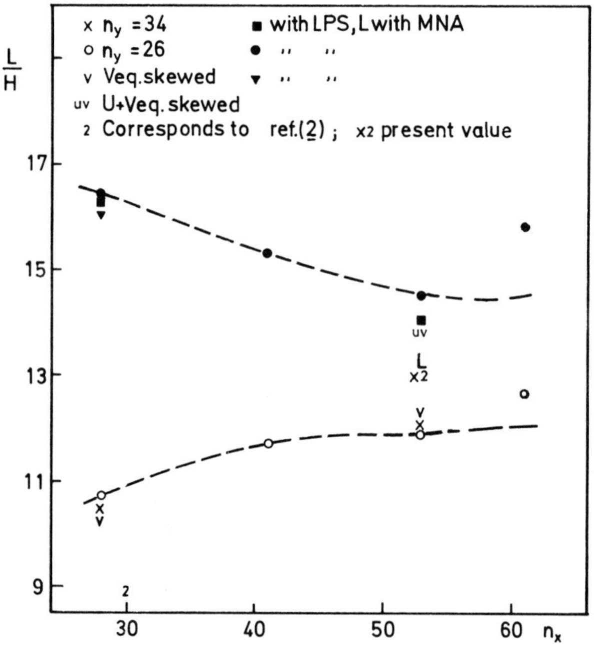

Fig. 7. Computed separation length as a function of grid points in the x-direction

unchanged. This intuitive convergence is shown by the dotted lines of Fig. 7. The last points ($n_x = 61$) on Fig. 7, however, indicate that changing the grid distribution in the reattachment point region still changes the calculated separation length by small amounts. This means that the grid distribution in the x-direction as indicated in Fig. 8, is not the optimum one. It was unfortunately not possible to use finer grids of different distributions in the time and with the computer storage available for this work. However, on the basis of sensitivity of the calculated results to the curvature corrections Fig. 7 shows that n_x should be larger than 50 for

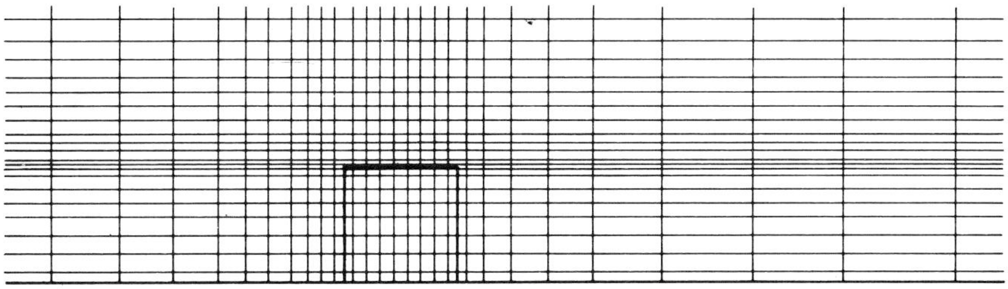

▲ **Fig. 8.** The fine grid used in the present calculations, $H_D/H = 9$, $n_x = 53$, $n_y = 34$, $n_u = 16$ upto $-13\,H$, $n_0 = 8$, $n_d = 29$ upto $35\,H$, $n_r = 16$, $\Delta_{min} = 0.12$ H, $\Delta x_{max} = 3\,H$, $n_a = 25$, $n_b = 9$, $\Delta y_{min} = 0.04\,H$, $\Delta y_{max} = 1.32\,H$

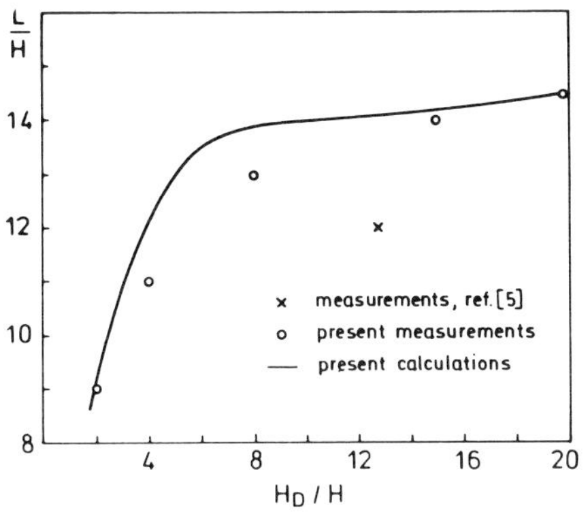

◀ **Fig. 9.** Calculated and measured values of the recirculation length, $H = B = 20$ mm

$n_y \gtrsim 26$. Figure 7 also shows clearly that too coarse a grid can provide too large an influence of changes in turbulence modelling resulting from insufficient resolution of the flow field.

The final computations were carred out using the LPS curvature correction with the upwind difference scheme. The upwind scheme converges faster than the skewed scheme and both the schemes produce similar results for the separation lengths. Figure 9 compares the computed values of the length of recirculation behind the square obstruction with the present measurements over the range of blockage ratios from 0.05 to 0.5. The computed values agree well with the measured values, the maximum discrepancy being 9% of the measured value at the blockage ratio of 0.25. Calculations were carried out for all blockage ratios using the same value for the boundary layer thickness (δ = 10 mm) on the base plate and the top plate and for the same boundary conditions at the inlet plane and the calculation domain. The initial profiles for other dependent variables such as k and ϵ were also kept unchanged for all the calculations and were taken to be those of boundary layer flows. Considering this uncertainty in the initial conditions used in the calculations, the computed values of L/H shown in Fig. 9 are very satisfactory. Figure 9 also shows the measured value of L/H from [5] for the same flow configuration. The 10% discrepancy between the present measurements and those of [5] is almost certainly due to different boundary conditions upstream of the obstruction. Present calculations also show separation on top of the obstruction as measured in [5]. Computations with exactly the same boundary conditions as in [5] were not carried out since the aim of the present calculations was to be able to calculate the shape of the separation region first.

The second flow situation for which calculations were carried out in the present work was the turbulent flow over a thin fence at a blockage ratio of 0.11 studied experimentally by *Good* and *Joubert* [4]. They used sharp edged fences of various heights and adjusted the tunnel roof to provide a zero pressure gradient in the free stream. Calculations were done with a uniform longitudinal velocity and constant pressure boundary conditions along the north face of the calculation domain and as mentioned earlier only one grid line was located passing through the centre of the fence. The baffled edge was not numerically simulated. The integral

Table 1. Integral parameters of flow over a thin fence

Parameter	Present calculations	Previous [2] calculations	Experiments [4]
Recirculation length, L	$14\,H$	$9.2\,H$	$13.3\,H$
Max. height, h_R, at x/H	$2.7\,H$ 6	$1.35\,H$ 4	$2.2\,H$ 5 to 7
Force coefficient	1.25	0.99	1.35
Max. forward velocity at x/H	$1.19\,U_0$ 4.9	$1.125\,U_0$ 2.25	$1.4\,U_0$ 2 to 5
Max. reverse velocity at x/H	$-0.86\,U_0$ 7.1	$-0.25\,U_0$ 2.25	$-0.49\,U_0$ 5 to 8

quantities obtained from the present calculations and from the previous calculations [2] are compared with the measured values in Table 1. The present calculations provide values for the integral parameters of this flow which compare rather well with the measured values. The largest discrepancy between the measurements in [4] and the present calculations is in the maximum reverse velocity. The detailed profile measurements in [4] were done only at discrete locations in the recirculation region, x/H = 2, 5, 8 and 10 and, therefore, the measured values given in Table 1 may not actually represent the values of the magnitude and the locations for the maximum reverse and forward velocities, really occuring in the flow. Moreover velocity profile measurements were carried out with a probe consisting of pitot static and yaw tubes which would definitely disturb the flow and give incorrect measurements in the reverse velocity region where the turbulence levels are high and the flow structure complicated. These possible errors in the measured reverse flow would also falsify the stream line pattern obtained from the measured velocity profiles as shown in Fig. 10 along with the present calculations. Reverse velocities, if measured too low, will clearly bring the measured streamlines towards the base plate. Considering these experimental problems, the measured streamlines compare fairly well with the calculated ones. The slope of the separation line at the fence edge and the length of the separation zone are well computed.

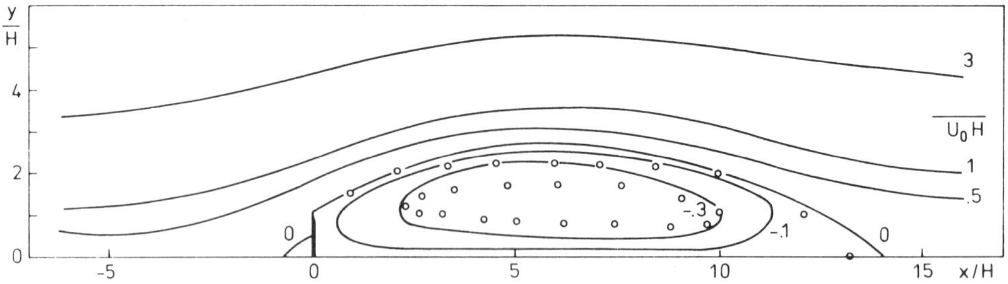

Fig. 10. Comparison of the calculated and the measured [4] streamline patterns

Concluding Remarks

Numerical calculations for turbulent flows over two-dimensional obstructions of different thicknesses and at different blockage ratios have been reported in this paper. Preliminary calculations were carried out for a flow over an obstruction with square cross section at a blockage ratio of 0.11. Upwind hybrid [15] and a skewed upwind hybrid [7 and 16] finite difference schemes were used with similar grids but varying grid numbers in order to investigate the sensitivity of the two numerical schemes to grid refinement. The above computations were carried out with three forms of the k-ϵ turbulence model – with the standard version [10], with the MNA curvature correction [12] and with the LPS curvature correction [14]. The preliminary calculations showed that the numerical schemes required fine grids to fully account for the changes employed in the turbulence model. Both the schemes were more sensitive to the LPS correction and therefore the final computations were carried out with the upwind hybrid finite difference scheme and the k-ϵ turbulence model with the LPS curvature correction. These calculations provided reasonable values for the integral parameters for turbulent flows over obstructions at all blockage ratios below 0.5. Results of the calculations show that for turbulent flow calculations over two-dimensional fences, the skewed upwind hybrid finite difference scheme does not offer any advantage with regard to grid requirements. This may not, however, be the case in separated flows, such as in free recirculating flows [8], where the length of recirculation is much smaller and hence the stream lines are much more skewed than in the fence flows.

The present calculations do provide improvements over the previous calculations carried out for turbulent fence flows. The authors are, however, somewhat disconcerted about the requirement of such fine grids for separated flows, and about the corresponding long computing times, e.g. the calculation for the thin fence (Fig. 10) required about 130 minutes of computing time on the UNIVAC 1108. This run needed 1,000 iterations with a 57 x 34 numerical grid covering the region between 30 H upstream and 33 H downstream of the fence. It may be worthwhile to test a higher order finite difference scheme such as the parabolic scheme given in [17] for separated flow predictions.

Another important conclusion of the present paper is that a two equation turbulence model, such as the k-ϵ model based on the time averaged equations of motion, is able to predict the gross features of turbulent separated flows over two-dimensional obstructions. The agreement between the present calculations and the available experimental data on the gross flow features, such as the shape of the recirculation zone, may be improved by using more realistic curvature corrections. It is clearly shown that, provided the numerical grid is fine enough and its distribution optimized, the upwind finite difference schemes are sensitive to the changes in turbulence models. It is also shown that the influence of the curvature correction in the turbulence equations on the time mean flow is much more significant than other effects, e.g. effects incorporated into the wall function treatment used in the present calculations.

The authors would like to stress that calculation procedures, such as the ones used in the present work, can be used to calculate the turbulent separated flow over an obstruction. These calculations should be extended to obtain details of the velocity field and these should be compared with measurements obtained by laser-Doppler-anemometry. LDA measurements in recirculating flows are becoming more and more available and are essential to improve calculation procedures for turbulent separated flows.

References

1. F. Durst, A. K. Rastogi: "Theoretical and Experimental Investigations of Turbulent Flows with Separation", in *Turbulent Shear Flows I,* ed. by F. Durst, B. E. Launder, F. W. Schmidt, J. H. Whitelaw (Springer, Berlin, Heidelberg, New York 1979) p. 208
2. D. Vasilic-Melling: "Three-Dimensional Turbulent Flow Past Rectangular Bluff Bodies". Ph. D. Thesis, University of London (1976)
3. I. P. Castro, J. E. Fackrell: "A Note on Two-Dimensional Fence Flows with Emphasis on Wall Constraint". CEGB (MEL) Report (Southampton, Hamshire, U. K.)
4. M. C. Good, P. N. Joubert: The form drag of two-dimensional bluff plates immersed in turbulent boundary layers. J. Fluid Mech. **31**, 547 (1968)
5. D. Crabb, D. F. G. Durao, J. H. Whitelaw: "Velocity Characteristics in the Vicinity of a Two-Dimensional Rib". FS/77/9, Department of Mechanical Engineering, Imperial College, London, England (1977)
6. S. B. Pope, J. H. Whitelaw: The calculations of near-wake flows. J. Fluid Mech. 73, 9 (1975)
7. G. D. Raithby: Skew upstream differencing schemes for problems involving fluid flow. Comput. Methods Appl. Mech. Eng. **9**, 153 (1976)
8. M. A. Leschziner: "On the Problem of Numerical Diffusion in First-Order Finite-Difference Schemes Applied to Free Recirculating Flows". Proc. of the 2nd GAMM Conference on Numerical Methods in Fluid Mechanics, Cologne, Fed. Rep. of Germany (1977)
9. A. K. Rastogi, W. Rodi: Calculation of general three-dimensional turbulent boundary layers. AIAA J. **15**, No. 2, 151–159 (1978)
10. B. E. Launder, D. B. Spalding: *Mathematical Models of Turbulence* (Academic Press, London and New York 1972)
11. M. M. Gibson: "Prediction of Curved Free Shear Layers with a Reynolds-Stress Model of Turbulence" (Submitted to the 2nd International Symposium on Turbulent Shear Flows, London 1979)
12. J. Militzer, W. B. Nicoll, S. A. Alpay: "Some Observations on the Numerical Calculation of the Recirculation Region of Twin Parallel Symmetric Jet Flow" (Submitted to the 1st International Symposium on Turbulent Shear Flows, Pennsylvania, USA 1977)
13. H. P. A. H. Irwin, P. Arnot Smith: Prediction of the effect of streamline curvature on turbulence. Phys. Fluids **18**, No. 6, 624 (1975)
14. B. E. Launder, C. H. Priddin, B. I. Sharma: The calculation of turbulent boundary layers on spinning and curved surfaces. J. Fluids Eng, 231–239 (1977)
15. A. D. Gosman, W. M. Pun: "Lecture Notes for Course Entitled – Calculation of Recirculating Flows". HTS/74/2, Heat Transfer Section, Imperial College, London (1973)
16. M. A. Leschziner: Private Communications
17. B. P. Leonard, M. A. Leschziner, J. J. McGuirk: "Third-Order Finite-Difference Method for Steady Two-Dimensional Convection". Proc. of the 1st International Conference on Numerical Methods in Laminar and Turbulent Flows, Swansea (1978) p. 807

Computation of a Round Turbulent Jet Discharging into a Confined Cross-Flow

W. P. Jones and J. J. McGuirk

Department of Chemical Engineering and Chemical Technology, Imperial College, London, S.W.7., England

Abstract

The paper presents an application of a finite-difference solution procedure for the three-dimensional elliptic equations of fluid flow to the problem of a round turbulent jet discharging at right angles into a confined cross-flow; the standard two-equation $k-\epsilon$ model is used to calculate the distribution of the Reynolds stresses. Results are presented describing the flow development for two different momentum flux ratios ($\rho_J V_J^2/\rho_0 U_0^2$) at the same blockage ratio (cross-flow channel height/jet diameter); the effect on the flow field is examined and found to be in qualitative agreement with the flow visualisation experiments of *Kamotani* and *Greber* [10]. Quantitative comparison is also made with data from reference [10] for jet trajectories, axial velocity, and temperature contours in planes parallel and perpendicular to the cross-flow. The trajectory measurements are well predicted for a range of momentum flux ratios from 8 to 72, but the comparison of predicted and measured contours shows only fair agreement. This discrepancy is certainly in part due to numerical error as the finest grid used contained only 20 x 15 x 15 nodes, which is insufficient to resolve all the high shear regions occurring in various parts of the solution domain. However, in addition, the use of a scalar eddy viscosity turbulence model is probably inadequate for this complex 3D flow.

Nomenclature

C_μ, C_1, C_2	Constants in turbulence model
D	Jet diameter at discharge
H	Cross-flow channel height
J	Momentum flux ratio $= \rho_J V_J^2/\rho_0 U_0^2$
k	Turbulence kinetic energy $= \frac{1}{2}(\overline{u'^2} + \overline{v'^2} + \overline{w'^2})$
P	Static pressure
Pr	Molecular Prandtl Number
R	Gas constant
T	Temperature
U, U_1	Velocity in x-direction
V, U_2	Velocity in y-direction
W, U_3	Velocity in z-direction
u', u_1'; v', u_2'; w', u_3'	Fluctuating velocity in x, y, z directions
x, x_1	Axial coordinate
y, x_2	Vertical coordinate
z, x_3	Transverse coordinate
U^*	Non-dimensional axial velocity $= U/U_0$
ΔT^*	Non-dimensional temperature $= (T-T_0)/(T_J-T_0)$
μ	Molecular viscosity
ρ	Density
ϵ	Dissipation rate of k
$\sigma_T, \sigma_k, \sigma_\epsilon$	Turbulent Prandtl Numbers for T, k, ϵ

δ_{ij} Kronecker delta

Subscripts

J Jet value
0 Cross-flow value

Introduction

Problem Considered

The flow situation consisting of a single, or multiple turbulent jets discharging into a cross-flow occurs in many branches of engineering practice. When the receiving medium is effectively infinite in extent the flow is relevant to such practical situations as smoke plumes from chimney stacks, cooling-tower plumes, and the jets of V/STOL aircraft in transition from vertical to forward flight. For rows of jets the cooling of the external surfaces of turbine blades or the discharge of warm water from pipes laid on the ocean bed provide further examples of practical applications, underlining the importance of this flow situation. A further complexity may be introduced if the cross-flow is confined, and here the practical relevance is to internal impingement cooling of turbine blades or, as in the present paper, the dilution air jets in gas-turbine engine combustion chambers. In this latter application air jets are discharged radially into the chamber through discrete circular holes in the chamber walls; it is of the utmost importance, therefore, for the designer of the combustion chamber to be able to predict both the trajectory and the rate of dilution of these jets. In the primary zone near the head of the chamber it is usual that the jets penetrate to the axis of the combustor, thereby locating the downstream end of the recirculation region which is used to anchor the flame. In this case some air is deflected upstream into the flame zone, and this can certainly influence the combustion processes taking place there; it is thus important to know what jet velocity is required for complete penetration and, when this happens, how the air divides as it flows upstream and downstream. In the so-called dilution zone further downstream, the jet trajectory and the rate of dilution of the hot combustion products by the incoming air influence greatly the outlet temperature profile at the end of the combustion chamber; this is, of course, of prime significance both for the performance and expected life of the turbine.

The flow just described is extremely complex, containing three-dimensionality, regions of reverse flow and simultaneous heat and mass transfer. For the purposes of the present paper, an idealised problem is considered which, however, still retains many of the complexities of the real flow. Figure 1 shows the schematics of the confined jet in a cross-flow problem to be considered here.

A single round jet is discharged through the lower wall of a rectangular wind tunnel (side-walls not shown) at right angles to the cross-flow direction. The position of the upper wall may be adjusted so that the effect of jet impingement on the tunnel roof may be studied. The flow may be non-isothermal, but if so, the temperature difference between jet and cross-flow is small enough so that density differences may be neglected. Measurements show that the cross-flow is deflected sideways by the penetration of the jet which, in turn, is bent over by the pressure difference between the upstream and downstream sides of the jet. The deflected jet is known to change cross-sectional shape from its original circular form to the kidney-shape sketched in Fig. 1, this being caused by the shear along the edges of the jet and the generation of a double vortex motion in the plane of the jet cross-section. The cross-flow is drawn

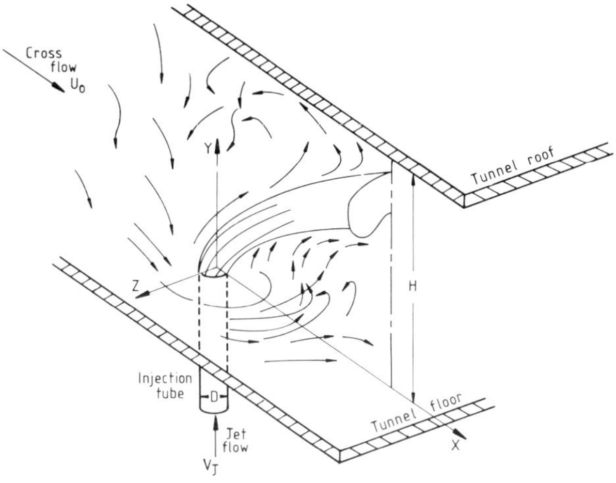

Fig. 1. Flow configuration

into the low pressure region behind the jet, and finally entrained into the jet proper, creating a region of negative streamwise (x-direction) velocity just downstream of the discharge. Finally, for high enough velocity ratios (jet/cross-flow velocity), or for situations where the tunnel roof is lowered close to the discharge plane, the jet impinges on the upper wall and the approaching cross-flow is decelerated sufficiently that a further recirculating flow region has been observed, this time upstream of the jet and on the upper wall.

Previous Work

Much of the published work on jets in a cross-flow has centred on the unconfined problem. For single jets the experiments of *Keffer* and *Baines* [1] and *Ramsey* and *Goldstein* [2] provided measurements of jet trajectories, velocity and temperature profiles; and recently *Moussa* et al. [3] presented a detailed investigation of the near-field ($-1.0 < x/D < 1.0$) for this flow. The theoretical work on this problem has mainly involved application of integral methods (see e.g. *Schatzmann* [4]) which suffer from the drawbacks that firstly an entrainment function must be specified, and secondly that the jet cross-section is usually asumed to be circular. This latter restriction has, however, been removed by *Adler* and *Baron* [5] who succeeded in predicting a kidney-shaped jet region with their integral method; the extension of these methods to predict the present problem is, however, unlikely to be profitable. A finite-difference scheme was used by *Bergeles* [6] to calculate the flow from a row of jets in unconfined cross-flow, but since he was interested primarily in the external surface blade cooling problem, he only considered very low velocity ratios (< 0.3), and indeed the numerical scheme was restricted to cases where no negative x-direction velocities occurred and as such is unsuitable for the present problem. *Patankar* et al. [7] have also used a finite-difference scheme together with a two-equation turbulence model to calculate the experiments of references [1] and [2] and obtained good agreement for velocity ratios from 2 to 10; a similar approach is used here for the present problem.

For confined cross-flow, the experimental data are much scarcer. *Stoy* and *Ben-Haim* [8] made measurements of jet trajectories for a single blockage ratio (H/D), but confined their attention to ranges of parameters applicable to impingement cooling of turbine blades. *Holdeman* and *Walker* [9] did investigate a geometry relevant to the combustor dilution jet problem, but made only measurements of the temperature field; these were then used to develop correlations to characterise the behaviour of rows of dilution jets in confined cross-flow. Only the experiments of *Kamotani* and *Greber* [10] represent both velocity and temperature measurements for the single jet problem sketched in Fig. 1, and even here the data are by no means plentiful; they do, however, provide some basis for comparison and assessment of the predictions to be presented later. Theoretical work on the confined problem is restricted to the calculations of *Tatchell* [*11*]. Again, however, he was concerned only with very low velocity ratios and used a numerical scheme even more restrictive than *Bergeles* [6].

Present Contribution

The preceding sections have indicated the practical importance of the confined jet in a cross-flow problem, and the lack of suitable computations with which to assess the ability of presently available numerical procedures and turbulence models to calculate this flow. This paper presents an application of a finite-difference scheme suitable for such complex flows, together with a two-equation turbulence model to the idealised flow problem discussed above, and examines the quality of the predictions on the basis of the measurements of *Kamotani* and *Greber* [10]. The details of the mathematical model and the calculation procedure used are given in the next section, which is followed by a discussion of the predicted flow pattern and its quantitative comparison with the available data.

Mathematical Model

Governing Equations

The flow in and around a jet discharging into a cross-flow is such that appreciable variations in transported quantities such as velocity and temperature arise in all three coordinate directions. In order to predict the flow and temperature fields in this configuration it is thus necessary to solve the full three-dimensional forms of the partial differential conservation equations for momentum, mass and thermal energy (temperature). In cartesian tensor notation the time-averaged forms of these equations may be written:

$$\bar{\rho}\,\bar{U}_j \frac{\partial \bar{U}_i}{\partial x_j} = -\frac{\partial \bar{P}}{\partial x_i} + \frac{\partial}{\partial x_j}\left(\mu \frac{\partial \bar{U}_i}{\partial x_j} - \bar{\rho}\,\overline{u_i' u_j'}\right) \quad (1)$$

$$\frac{\partial \bar{\rho}\bar{U}_i}{\partial x_i} = 0 \quad (2)$$

$$\bar{\rho}\bar{U}_j \frac{\partial \bar{T}}{\partial x_j} = \frac{\partial}{\partial x_j}\left(\frac{\mu}{\mathrm{Pr}} \frac{\partial \bar{T}}{\partial x_j} - \bar{\rho}\,\overline{u_j' T'}\right) \quad (3)$$

where overbars represent averaged quantities.

For the flows presently considered the temperature differences were sufficiently small for density fluctuations to be considered negligible. The density is obtained from the equation of state:

$$\rho = \bar{P}/R\bar{T}$$

Equations (1) and (3) contain turbulent momentum and heat fluxes which are here approximated via the high Reynolds number version of the two-equation $k-\epsilon$ turbulence model introduced by *Jones* [12] and *Jones* and *Launder* [13]. The Reynolds stress is linearly related to the mean rate of strain via a scalar turbulent viscosity:

$$\overline{\rho}\,\overline{u_i'u_j'} = 2/3\,\delta_{ij}k - \mu_T \left(\frac{\partial \overline{U}_i}{\partial x_j} + \frac{\partial \overline{U}_j}{\partial x_i} \right) \tag{4}$$

and the turbulent heat flux is obtained using the turbulent Prandtl number approach:

$$\overline{\rho}\,\overline{u_j'T'} = -\frac{\mu_T}{\sigma_T} \frac{\partial \overline{T}}{\partial x_j} \tag{5}$$

The turbulent viscosity is expressed as:

$$\mu_T = C_\mu \overline{\rho} k^2/\epsilon$$

where $k(\equiv \overline{u_i'u_i'}/2)$ and ϵ are the turbulence kinetic energy and dissipation rate of turbulence energy, the values of which are obtained from the solution of the transport equations:

$$\overline{\rho}\,\overline{U}_j \frac{\partial k}{\partial x_j} = \frac{\partial}{\partial x_j} \left(\frac{\mu_T}{\sigma_k} \frac{\partial k}{\partial x_j} \right) - \overline{\rho}\,\overline{u_i'u_j'} \frac{\partial \overline{U}_i}{\partial x_j} - \overline{\rho}\epsilon \tag{6}$$

$$\overline{\rho}\,\overline{U}_j \frac{\partial \epsilon}{\partial x_j} = \frac{\partial}{\partial x_j} \left(\frac{\mu_T}{\sigma_\epsilon} \frac{\partial \epsilon}{\partial x_j} \right) - C_1 \frac{\epsilon}{k} \overline{\rho}\,\overline{u_i'u_j'} \frac{\partial \overline{U}_i}{\partial x_j} - C_2 \overline{\rho} \frac{\epsilon^2}{k} \tag{7}$$

The model contains six empirical constants which are assigned the following values, found to give good agreement in a wide range of 2D shear flows (see, e.g. *Launder* and *Spalding* [14].

$C_\mu = 0.09$; $C_1 = 1.44$; $C_2 = 1.92$; $\sigma_k = 1.0$; $\sigma_\epsilon = 1.3$ and $\sigma_T = 0.9$.

Computational Procedure and Boundary Conditions

The computational results to be presented here required the solution of the steady flow 3-dimensional forms of seven coupled partial differential equations. These comprised the continuity equation, the *x, y* and *z* component momentum equations, the thermal energy equation and two turbulence equations for k and ϵ. The solutions were obtained using a finite-difference scheme embodied in a computer program called PACE [15] developed by one of the authors (WPJ) for Rolls-Royce Ltd. The finite-difference formulation comprises a linearised, implicit, conservative scheme using centred differencing except in regions where mean flow convective fluxes dominate diffusive processes in which case donor cell differencing is used. The method is formulated in terms of velocity and pressure using the staggered grid arrangement made popular by Harlow and co-workers, see e.g. [16]. In this formulation the determination of the pressure is not straightforward; the problem is that of obtaining a pressure field such that the solution of the momentum equations (in which the pressure gradient appears) satisfies the continuity equation. In PACE a guess and correct procedure similar to that of *Chorin* [17], *Amsden* and *Harlow* [16], and *Patankar* and *Spalding* [18] is utilised. The program is capable of providing both transient and steady-state solutions, the latter being obtained via iteration through some time-like iteration space. For steady flows the pressure and velocity fields are first initialised and the three algebraic finite-difference momentum

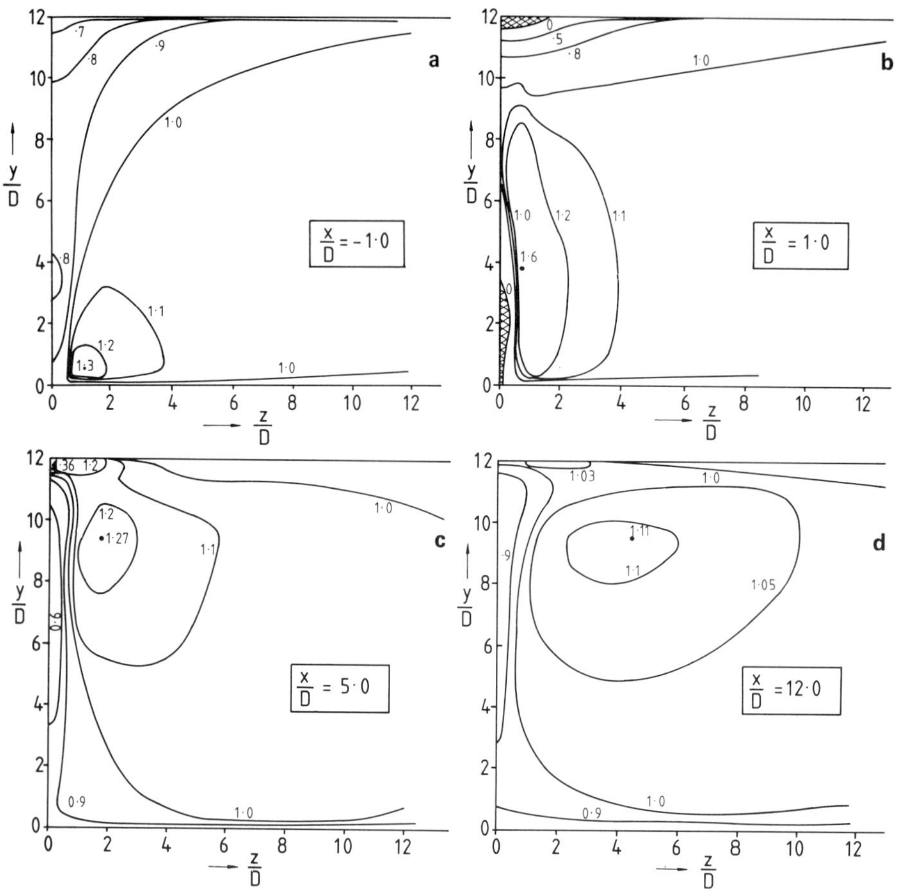

Fig. 2a–d. Axial velocity contours ($U^* = U/U_0$) in cross-stream (x = constant) planes. J = 72 (**a–d**), J = 18 (**e–h**)

equations then "solved" to yield an intermediate velocity field; this solution will, however, not in general satisfy the continuity equation. The intermediate velocity solution is then combined with the finite-difference formulation of the continuity equation to obtain an equation for the so-called "pressure correction". The solution of this latter equation then provides corrections to the velocity and pressure fields which drive the velocities towards simultaneous satisfaction of both the momentum and continuity equations. This process is repeated until a final converged solution is obtained. The computer code has been subject to extensive use and testing and no major problems with either convergence or stability have yet arisen.

The finite-difference solution domain used to obtain the results to be described below may be represented by a rectangular box whose upper and lower walls coincide with the walls of the cross-flow wind tunnel. The flow is symmetric about a vertical plane passing through the centre of the circular hole in the lower wall of the tunnel and one "side" of the solution domain thus corresponded to this. Under some circumstances, the position of the remaining "inflow" and "outflow" (upstream and downstream) boundary planes exerted a not insignificant influence on the solutions as did the location of the other "side" boundary. Before obtaining the results to be presented here, the sensitivity of the computed solutions to the location of these three last-mentioned solution domain boundaries was investigated

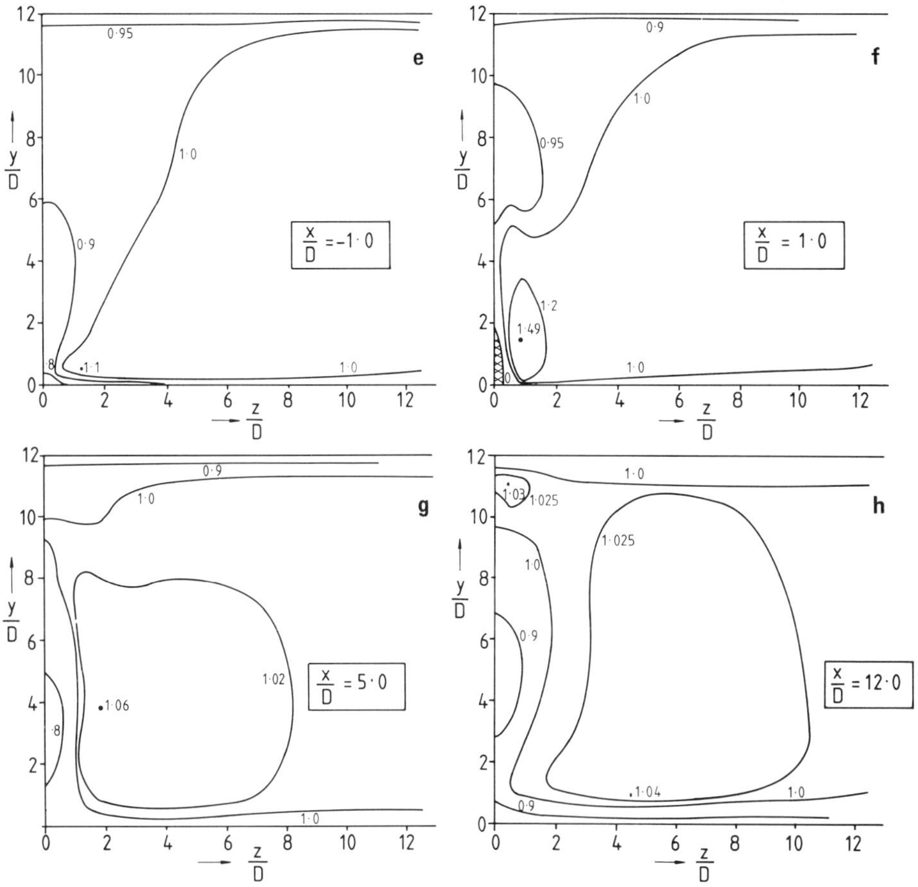

Fig. 2e–h. Figure caption see opposite page

and their position located sufficiently far away from the jet so that the details of the boundary conditions (particularly on the inflow and outflow) had a negligible influence on the computed flow in the vicinity of the jet. The inflow boundary was located ten diameters upstream of the hole whereas the outflow boundary was positioned twenty-four hole diameters downstream. The "side" plane boundary was eventually placed at a position coincident with the side wall of the tunnel in the experiments, i.e. approximately fifty diameters from the symmetry plane.

The elliptic equations appropriate to the present flow configuration require boundary conditions on all of the bounding surfaces of the solution domain. In the present work boundary conditions were supplied in terms of velocity and temperature which when taken together with the value of pressure at a *single* point serve to specify the problem completely. At the upstream (inflow) boundary uniform profiles of all the dependent variables were specified, whereas at the downstream (outflow) plane the gradients of dependent variables in the axial direction were set to zero. While these conditions may not have corresponded exactly to those prevailing in the wind tunnel, as was mentioned previously the location of the inflow and outflow boundaries was such that the imposed boundary conditions had negligible influence on the region of interest. For wall boundaries the finite-difference solutions were patched onto fully turbulent local equilibrium wall law profiles (see for example [14]). Finally, in the case of the lower wall from which the jet issued some additional special treatment was necessary in regions where the hole boundary bisected the surface area of the adjacent finite-difference control volume. Here the cell surface areas were modified so that the specified jet velocity produced the correct fluxes of mass and momentum through the hole.

In the next section results will be presented which were obtained using a grid consisting of 20 x 15 x 15 nodes (in the *x, y, z* directions respectively). Non-uniform spacing was used so that grid nodes could be grouped in regions where rapid variations were expected. This usually meant fine grid spacing near the discharge hole and increasingly larger grid spacing moving away from the hole in all three coordinate directions; an exception to this was the calculation made at the highest momentum flux ratio when fine grid spacing was also used near the upper wall where a recirculation region was expected. It was unfortunately not possible to use more grid nodes since these calculations required approximately 150 K words core store, which represented the maximum accessible storage on the University of London CDC 7600 machine. A typical calculation involved approximately 100 iterations to procure convergence, and required about 8 minutes execution time.

Results and Discussion

Figure 2 shows contours of the x-component of velocity on planes of constant x at various axial positions. Contours for two values of momentum flux ratio are shown, namely $J = 72$ and $J = 18$. The velocities are non-dimensionalised with the approach cross-flow velocity and the coordinate origin is chosen to coincide with the centre of the circular discharge hole in the tunnel floor. The contour plots allow some of the predicted general features of the flow to be inferred. For both momentum flux ratios the presence of velocity contours less than unity upstream of the discharge plane indicate the sideways deflection of the cross-flow by the jet which in turn is bent over by the induced pressure field. However, in the case of the higher momentum flux ratio the pressure difference is insufficient to prevent the jet from impinging on the top wall of the tunnel. The high shear on the lower wall and around the jet adjacent to the discharge leads to a horseshoe vortex around the jet inlet, and the axial flow is accelerated around the sides of the jet to produce velocity maxima. Examination of the predicted V and W velocity fields shows that these maxima are roughly coincident with the centre of the double vortex motions originating from the horseshoe vortex.

The velocity maxima predicted are quite large. At an axial distance of $X/D = 1.0$ for $J = 72$ a velocity maximum of 1.6 times the approach velocity is predicted and closer to the discharge even larger values are computed, e.g. although not shown a value of around 2 times the approach velocity is obtained at the axial position $X/D = 0$. These predicted velocity maxima which arise either side of the symmetry plane and well below the jet trajectory are in qualitative agreement with the measurements of *Moussa* et al. [3] albeit obtained in a slightly different flow configuration. For the lower momentum flux ratio the double vortex motions in the plane of the jet cross section deform the jet from its original circular shape to the well-known kidney shape; this can be seen most clearly from the contour at $X/D = 12.0$. However, in the case of $J = 72$ the flux ratio is such that the jet impinges on the top wall and the kid-

Fig. 3a–h. Temperature contours ($\Delta T^* = (T-T_0)/(T_J-T_0)$) in cross-stream planes. $J = 72$ (a–d), $J = 18$ (e–h) ▶

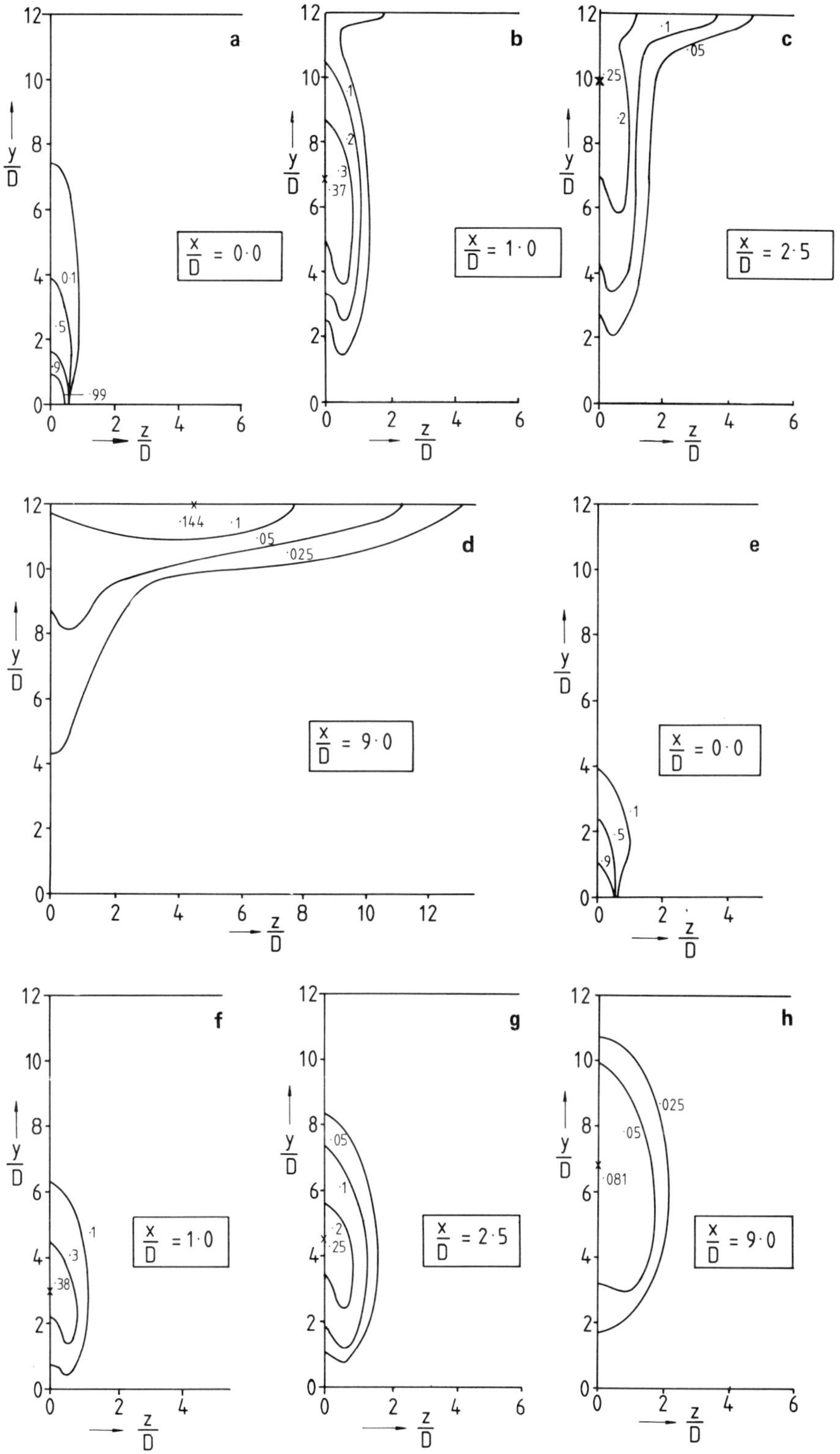

Fig. 3a–h. Figure caption see opposite page

ney shape does not therefore appear. Near the upper wall the approach flow is decelerated upstream of the impingement and a region with negative values of the x-component of velocity upstream of the jet is predicted. (Negative axial velocities are indicated in Fig. 2 by cross-hatching.) This result is in agreement with *Kamotani* and *Greber's* [10] smoke flow visualisation studies although the upstream extent of the negative velocity region seems to be underpredicted. Downstream of the impingement the jet flattens against the top wall and spreads in the transverse direction with velocity maxima either side of the plane of symmetry.

For both flows the low pressure region present immediately downstream of the discharge on the bottom wall induces a region with negative values of x-component of velocity. This region extends downstream to a value of X/D of approximately three but, although not shown, "lifts off" from the lower wall for values of X/D greater than about 1.5 (i.e. the velocity is positive immediately adjacent to the wall for $X/D > 1.5$). The "lifting" of the zone of negative velocity is due to the strong inflow towards the symmetry plane downstream of the discharge, and has also been noted in flow visualisation experiments; see, for example, reference [6].

The predicted contours of temperature on x-constant planes are shown for the same two momentum flux ratios in Fig. 3. For the lower flux ratio the jet is bent over rapidly by the influence of the cross-flow and correspondingly the jet temperature is convected downstream and diffuses out in both the y and z directions. Here, however, the familiar kidney shape of the jet cross-section is more clearly displayed than was the situation for the velocity contours. In the case of the flow with flux ratio 72 the impingement on the upper wall of the tunnel and the subsequent flattening of the jet as it develops downstream can be clearly seen as it produces temperature contours with maxima close to the wall and on either side of the plane of symmetry.

The calculated results presented so far are in qualitative agreement with the results of flow visualisation studies and we now turn to a more detailed examination of the accuracy of the predictions. In Fig. 4 the predicted and measured jet trajectories are plotted for a range of momentum flux ratios between 8 and 72. In the flows considered the jet temperature was 167 K above that of the approach cross-flow and the trajectory is thus here defined as the locus of the maximum temperature in the plane of symmetry. The calculated and measured trajectories are in good agreement though for the two larger flux ratios the predictions approach somewhat closer to the upper wall than the measured trajectories. Measured and predicted (non-dimensional) temperature contours on the symmetry plane, $z = 0$, are shown in

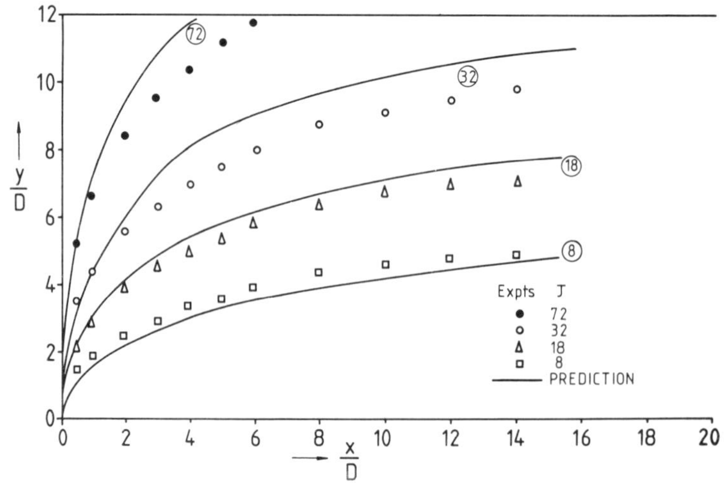

Fig. 4. Jet trajectories

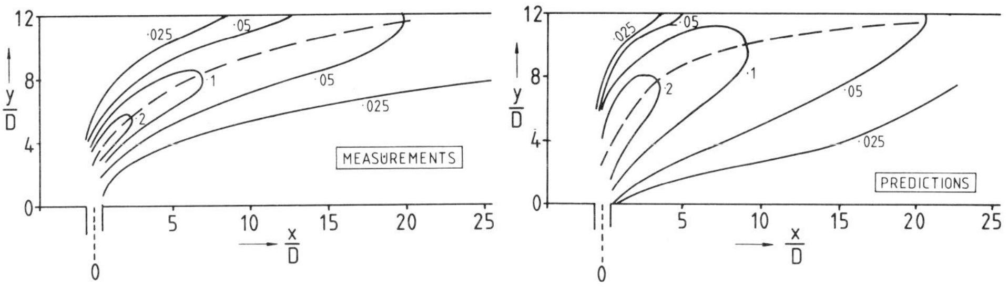

Fig. 5. Measured and predicted ΔT^* contours on central symmetry plane. $J = 32$, $H/D = 12$

Fig. 5 for a momentum flux ratio of 32. The contour shapes are in qualitative agreement although in the computations the spread of the temperature field in the direction normal to the jet trajectory is somewhat larger that the measurements indicate. Close to the jet hole this difference is accentuated by the fact that, in contrast to the measurements, heat is predicted to diffuse downwards into the region of negative axial velocity behind the jet and is then convected back towards the jet discharge. Displayed in Fig. 6 and 7 are the predicted and measured velocity and temperature contours for $J = 72$ and a blockage ratio (H/D) of 12 on a plane at $X/D = 12.0$. Here again the gross features of the flow are shown to be correctly predicted. However, it is evident that the high velocity fluid which arises from the upstream deflection and acceleration of the cross-flow is predicted to have spread over too great a region. Also it seems that the calculated secondary flow (V and W velocities) is too large – there are no measurements – with the result that low velocity fluid originating from the wake which forms behind the jet is convected upwards towards the tunnel roof more rapidly than is measured. This influences the predicted shape of the jet proper which lies immediately adjacent to the upper wall and which has a maximum velocity somewhat less than measured. Insofar as the temperature contours are concerned heat is predicted to spread at too great a rate in the z-direction; for example, the 0.02 temperature contour extends out to a position corresponding to a z/D value of about fourteen compared with a measured value of around ten. In addition, near the symmetry plane the temperature field

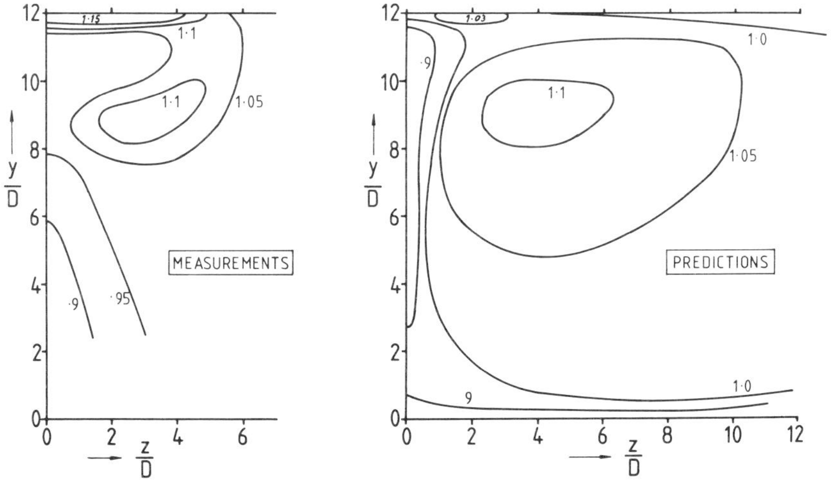

Fig. 6. Measured and predicted U^* contours on cross-stream plane $X/D = 12$. $J = 72$, $H/D = 12$

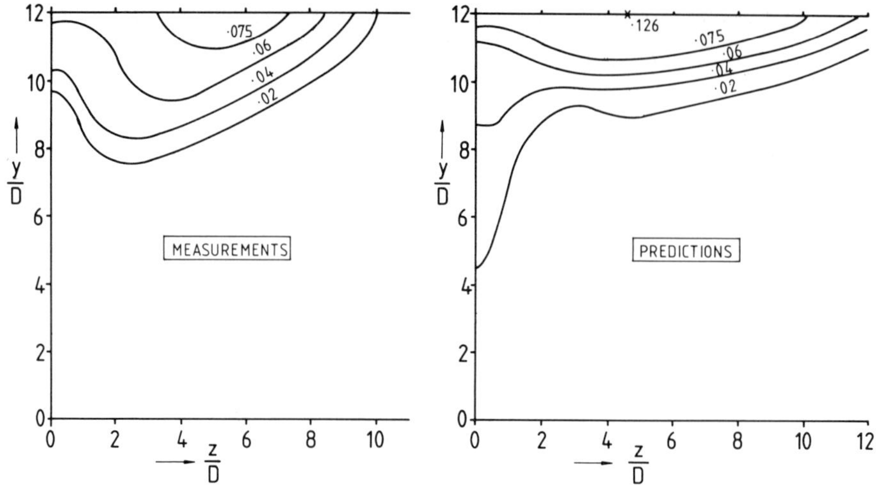

Fig. 7. Measured and predicted ΔT^* contours on cross-stream plane $X/D = 12$. $J = 72$, $H/D = 12$

is spread over too large a region in the *y*-direction; in all probability, this arises as a consequence of the previously mentioned (incorrectly) predicted diffusion of heat into the region of negative axial velocity immediately behind the jet discharge.

The precise reasons for the above discrepancies are difficult to identify. The nature of the flow under consideration is such that, with the computer storage available at the time the calculations were made, it was impossible to refine the finite-difference grid further. It was thus not possible to establish whether or not the solutions would have changed with more grid refinement. There are, however, some indications that they were not grid independent, for example runs with a *coarser* grid (15 x 10 x 10) produced similar trajectory predictions, but noticeable changes in contour shapes – further grid refinement is therefore desirable. Of course, another potential source of error is the k-ϵ turbulence model and, in fact, it is perhaps improbable that *any* turbulent scalar viscosity model can provide accurate calculation of the six components of the turbulent stress tensor, all of which are significant in three-dimensonal flows with recirculation. Overall the predictions appear to exhibit diffusion rates larger than those consistent with measured profiles. Since these can be attributed equally well either to numerical (diffusive) or "turbulence model" errors it is not possible to identify the precise source of disagreement between measurement and predictions. Finally, it should also be mentioned that the measurements are themselves not without error. At the plane considered in Fig. 6 and 7, the values of axial velocity lie between –10% and +15% of the approach cross-flow velocity. The comparisons shown here thus represent a stringent test of the prediction method and, for that matter, the measurement technique.

Notwithstanding the above comments the qualitative agreement is good and the gross features of the flow are well predicted for all flux ratios; for example, both the area of mixed fluid (in *x*-constant planes) and the rate of jet dilution are both roughly correct, as are the jet trajectories. It seems worthwhile to consider further extensions of the present work, and these will include the calculation of flows with rows of interacting jets such as are found in gas-turbine combustion chambers. In this situation the chamber walls and geometrical features of the configuration will serve to constrain the flow to a much greater extent that is the case for a single jet in a crossflow. It is therefore not unreasonable to anticipate that the grid refinement problems experienced here will be more readily overcome, and the calculation of such multiple jet flows may be a somewhat easier task than that of the present single jet flow.

Acknowledgements. The work reported in this paper was sponsored by Rolls-Royce Ltd., and one of the authors (JJM) gratefully acknowledges their financial support. The numerical calculations were performed on the University of London CDC 7600 computer.

References

1. J. F. Keffer, W. D. Baines: "The round turbulent jet in a cross-wind". J. Fluid Mech. **15**, Part 4 (1963)
2. J. W. Ramsey, R. J. Goldstein: "Interaction of a heated jet with a deflecting stream". J. Heat Transfer (1971)
3. Z. M. Moussa, J. W. Trischka, S. Eskinazi: "The near field in the mixing of a round jet with a cross-stream". J. Fluid Mech. **80**, Part 1 (1977)
4. M. Schatzmann: "A Mathematical Model for the Prediction of Plume Rise in Stratified Flows" (Submitted to the 1st International Symposium on Turbulent Shear Flows, Pennsylvania USA 1977)
5. D. Adler, A. Baron: "Prediction of a three-dimensional circular turbulent jet in crossflow". AIAA J. **17** (1979)
6. G. Bergeles: "Three-Dimensional Discrete-Hole Cooling Processes, an Experimental and Theoretical Study". Ph. D. Thesis, University of London (1976)
7. S. V. Patankar, D. K. Basu, S. A. Alpay: "Prediction of the three-dimensional velocity field of a deflected jet". J. Fluids Eng. (1977)
8. R. L. Stoy, Y. Ben-Haim: "Turbulent jets in a confined cross-flow". J. Fluids Eng. (1973)
9. J. D. Holdeman, R. E. Walker: "Mixing of a row of jets with a confined cross-flow". AIAA J. **15** (1977)
10. Y. Kamotani, I. Greber: "Experiments on Confined Turbulent Jets in a Cross-Flow". NASA CR-2392 (1974)
11. D. G. Tatchell: "Convection Processes in Confined, Three-Dimensional Boundary Layers". Ph. D. Thesis, University of London (1975)
12. W. P. Jones: "Laminarisation in Strongly Accelerated Boundary Layers", Ph. D. Thesis, University of London (1971)
13. W. P. Jones, B. E. Launder: "The prediction of laminarisation with a two-equation model of turbulence". Int. J. Heat Mass Transfer **15** (1972)
14. B. E. Launder, D. B. Spalding: "The numerical computation of turbulent flows". Comput Methods Appl. Mech. Eng. **3** (1974)
15. W. P. Jones: "PACE: A Computer Program for Solving Three-Dimensional Flow Problems". Rolls-Royce Ltd. Internal Report, 1975 (not for general circulation)
16. A. A. Amsden, F. H. Harlow: "The SMAC Method". Los Alamos Scientific Lab. Report LA-4370 (1970)
17. A. J. Chorin: "Numerical solution of the navier-stokes equations". Math. Comp. 22 (1968)
18. S. V. Patankar, D. B. Spalding: "A calculation procedure for heat, mass and momentum transfer in three-dimensional parabolic flows". Int. J. Heat Mass Transfer" **15** (1972)

Turbulent Shear Flow Behind a Sphere Placed on a Plane Boundary

Shiki Okamoto

Department of Mechanical Engineering, Shibaura Institute of Technology, Tokyo, Japan

Abstract

This paper describes an experimental study of the turbulent shear flow behind a sphere placed on the plane boundary. The experiment was carried out in an N. P. L. blow-down type wind-tunnel having a working section of 400 x 400 x 2,000 mm in size at the Reynolds number 4.74×10^4.

The surface-pressure distribution on a sphere was measured and the drag and lift coefficients were determined by integration of the surface pressure. The velocity and pressure distributions and turbulence in the shear layer behind a sphere were measured.

It was found that the drag coefficient is larger than that of a sphere alone and the wall wake behind a sphere becomes low and spreads transversely with an increase in the downstream distance.

1. Nomenclature

b	Position of peak velocity defect [mm]
C_D	Drag coefficient
C_L	Lift coefficient
C_P	Pressure coefficient = $(P - P_\infty)/(\rho U_\infty^2/2)$
D	Diameter of sphere [mm]
P	Static pressure [Pa]
P_∞	Free-stream pressure [Pa]
Re	Reynolds number = $U_\infty D/\nu$
U	Time-mean velocity [m/s]
U_∞	Free-stream velocity [m/s]
u', v', w'	Velocity fluctuation in X, Y, Z direction [m/s]
X, Y, Z	Coordinates axes [mm]
α	Latitude angle [deg]
δ	Thickness of shear layer [mm]
θ	Longitude angle [deg]
ν	Kinematic viscosity [m^2/s]
ρ	Density of air [kg/m^3]

2. Introduction

The flow past a blunt three-dimensional obstacle placed on the plane boundary was first studied by *Schlichting* [1] in connection with the effect of surface roughness, and since Schlichting, it has been studied for the most part in connection with the wind flow past buildings, cooling towers, chimney stacks, spherical gas tanks and other structures. *Klemin* et al. [2] studied the drag of a sphere placed on a ground plate, and *Sforza* and *Mons* [3] investigated the wall wake behind a rectangular body placed on the leading edge of a flat plate. *Hunt* [4] studied the flow past a building, and *Peterka* and *Cermak* [5] measured the mean velocity defect and turbulence in the wake behind a building.

Recently, the flow past a circular cylinder of finite length and a cone placed on the plane boundary was investigated by *Okamoto* and *Yagita* [6] and *Okamoto* et al. [7] respectively. The present paper is concerned with an experimental study of the flow field around a sphere and the turbulent shear layer behind a sphere placed on a plane boundary.

3. Experimental Apparatus and Procedure of Measurement

The experiment was carried out in an N. P. L. blowdown type wind-tunnel having a 400 x 400 mm working section of 2,000 mm length. The ground plate, an aluminium plate of 4 mm thickness, was set with a spacing of 25 mm from the lower wall of the tunnel exit in order to avoid the boundary layer developed on the tunnel wall.

The origin of the coordinates was taken at the bottom of the sphere with the X-axis facing downstream, the Y-axis to the right of downstream, and the Z-axis vertically.

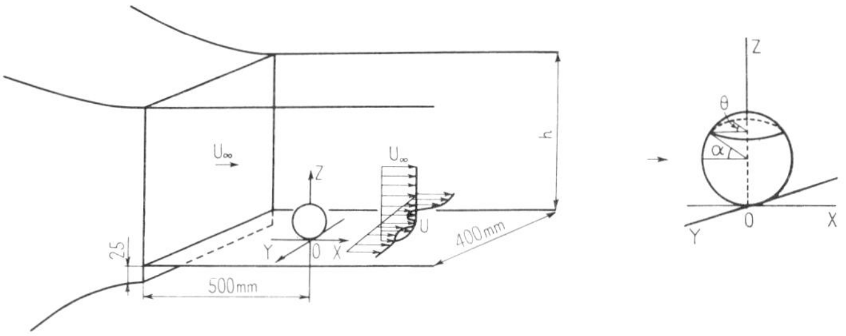

Fig. 1. Schema of apparatus and nomenclature

Pressure holes of 0.5 mm diameter were bored with an angular spacing of $\alpha = 10$ degrees on the surface of the sphere in the vertical plane $\theta = 0$ (Fig. 1). Measurements of the surface-pressures for any other values of θ were made by rotating the sphere about the vertical Z-axis. In order to examine the blockage effect on the surface-pressure distribution, spheres of 40, 50, 60 and 70 mm diameter were tested, and it was confirmed that there is no blockage effect for the spheres tested.

The velocity vectors were obtained by use of a cylindrical yawmeter of 6 mm diameter. The time-mean velocity and the static pressure were measured by Pitot and static-pressure tubes of 1 mm diameter and the turbulence by a hot-wire anemometer of the constant temperature type.

Measurements of the velocity and pressure distribution and turbulence in the flowfield behind a sphere were recorded at the section $X/D = 0 \sim 28$ with the Reynolds number of $Re = 4.74 \times 10^4$, which is less than the critical value.

4. Experimental Results

4.1. Surface-Pressure, Drag and Lift

The surface-pressures on the 60 mm sphere were measured. Figures 2 and 3 show the surface-pressure distribution on the upper hemisphere ($\alpha > 0$) and the lower hemisphere ($\alpha < 0$), where the broken line represents the theoretical value $C_P = 1 - (9/4) \sin^2 \theta$ of a single sphere

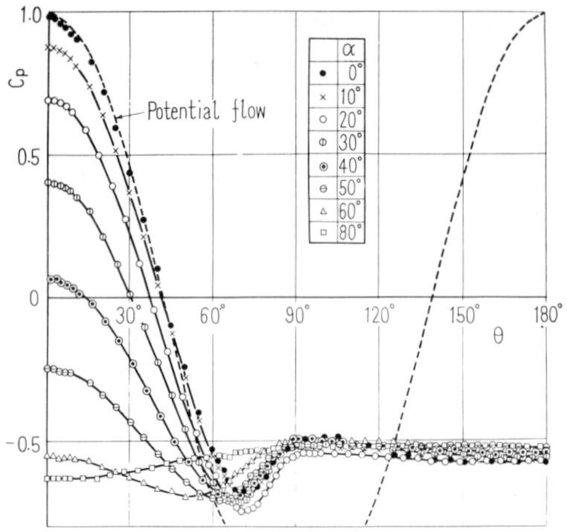

Fig. 2. Surface-pressure distribution on the upper hemisphere

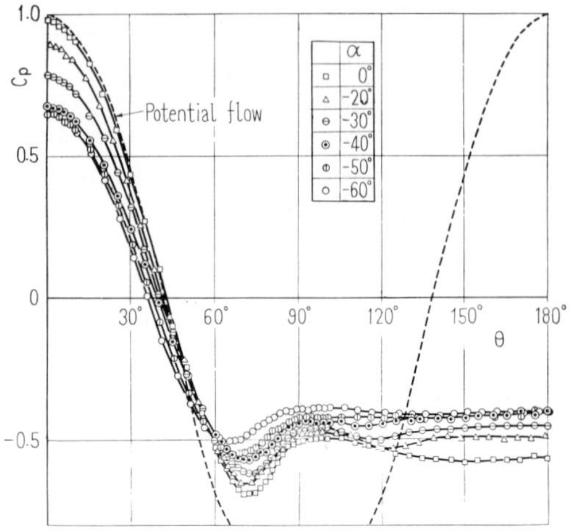

Fig. 3. Surface-pressure distribution on the lower hemisphere

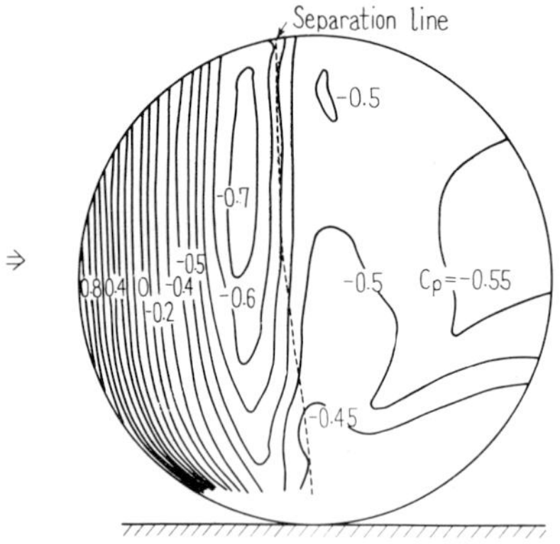

Fig. 4. Iso-pressure lines

Table 1. Coordinates of the separation line on a sphere

α	θ
77°	0°
70°	53.8°
60°	68.3°
40°	77.7°
20°	83.0°
0°	86.0°
−20°	87.4°
−40°	88.9°
−60°	90.0°
−80°	90.0°

based on the potential-flow theory. The surface-pressure distribution is different between the upper and lower hemisphere, and the stagnation point is at the position $\alpha = -5$ degree, $\theta = 0$ degree.

Figure 4 shows the iso-pressure lines in a side view of the sphere. It can be seen from this figure that the minimum pressure occurs at the position $\alpha = 30$ deg., $\theta = 70$ deg. The broken line in this figure shows the separation line obtained using the oil film method. The separation line bends backward as it approaches the ground plate. The coordinates of the separation line are shown in Table 1. It may be seen from this table that the separation is a laminar one.

The drag and lift coefficients of the sphere are determined from the surface-pressure distribution by the following equations:

$$C_D = \frac{2}{\pi} \int_{-\frac{\pi}{2}}^{\frac{\pi}{2}} \int_{0}^{\pi} C_P \cos^2 \alpha \cos \theta \, d\alpha \, d\theta$$

$$C_L = -\frac{1}{\pi} \int_{-\frac{\pi}{2}}^{\frac{\pi}{2}} \int_{0}^{\pi} C_P \sin 2\alpha \, d\alpha d\theta$$

The above-mentioned drag is, strictly speaking, the pressure drag, but it may be viewed to be the total drag, because the frictional drag of a sphere is negligible in comparison with the pressure drag.

As a result we have $C_D = 0.627$ and $C_L = 0.242$. The generation of the lift means the existence of a strong downwash behind a sphere. The value of C_D for a sphere on the ground plate is very large compared with the value of $C_D = 0.48$ of a sphere in the flow. The increase of C_D due to the presence of the ground plate was pointed out by *Klemin* et al. [2], whose experiment with a sphere of 609.6 mm diameter at the supercritical Reynolds number $Re = 1.39 \times 10^6$, gave $C_D = 0.200$ for the sphere in the flow and $C_D = 0.573$ for the presence of the ground plate and the collar.

Okamoto et al. [7] showed that if the curve for C_D or C_L is plotted against $D/(2H)$, where D is the maximum diameter of a body and H the height of a body, the results of a circular cylinder and a cone lie on the same curve. The results for the sphere also lie on the same curve as that for a circular cylinder and a cone, as shown in Fig. 5.

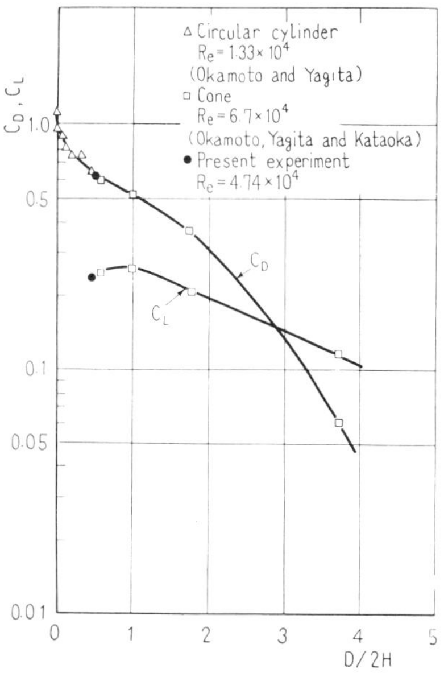

Fig. 5. Drag and lift coefficients

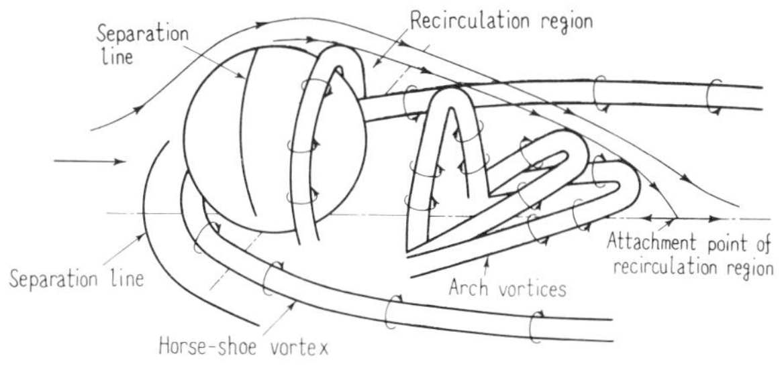

Fig. 6. Vortices

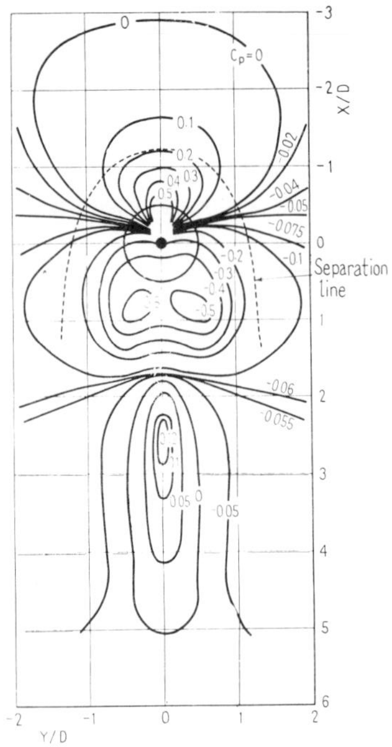

◀ **Fig. 7.** Static pressure distribution on the ground plate

Fig. 8. Flowfield behind the sphere observed by the tuft-grid method

▼

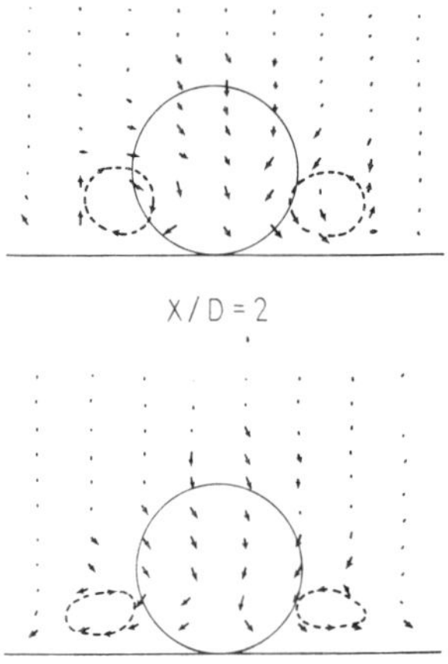

4.2. Vortices

In the flowfield around a sphere there exist vortices of two kinds;

(1) The horse-shoe vortex generated at the corner between the frontal face of the sphere and the ground plate.

(2) The arch vortices generated behind the separation line on the sphere.

These vortices are illustrated in Fig. 6, and similar vortices have also been observed in the case of a cone placed on the ground plate [7]. The arch vortices are shed downstream, raise and fall down to the ground plate as a result of the strong downwash behind the sphere. The movement of the vortices can be seen from the pressure distribution on the ground plate shown in Fig. 7. There are two areas of low pressure $C_P = -0.5$ behind the sphere at which the arch vortices are seen to be attached. It is supposed that the arch vortices rotate about those attachment points and fall down.

The recirculation region is formed behind a sphere, involving arch vortices. Because of a strong downwash behind a sphere, the recirculation region is reduced rapidly. The position $X/D = 2.5$ at a high pressure of $C_P = 0.12$ is seen to be the point of reattachment of the outer streamline of the recirculation region. Hence the length of the recirculation region is considered to be 2.5 D.

The broken line in this figure shows the separation line formed in front of the horse-shoe vortex. The horse-shoe vortex is shed downstream, passing by the sides of the sphere. The horse-shoe vortex was observed by the tuft grid method. Figure 8. illustrates the movements of the tuft seen from a downstream position at the tuft grid position of $X/D = 2$ and 4. It was found that the horse-shoe vortex becomes flat at $X/D = 4$ and disappears at about $X/D = 5$.

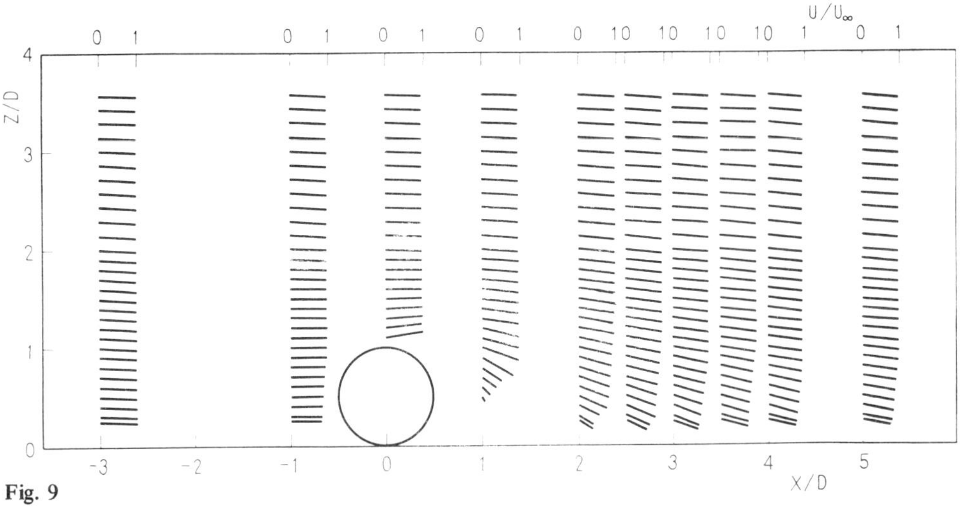

Fig. 9

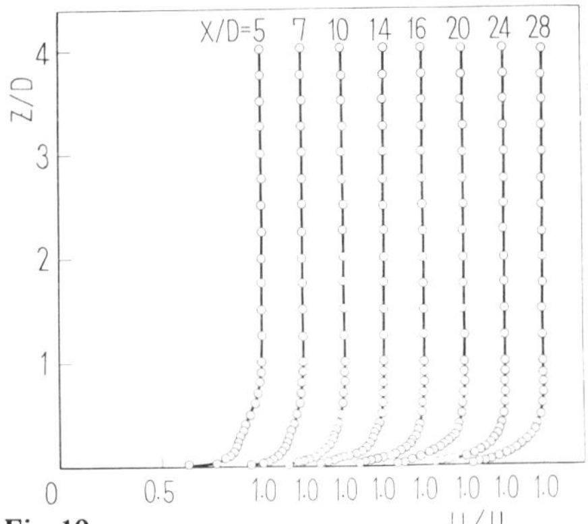

Fig. 10

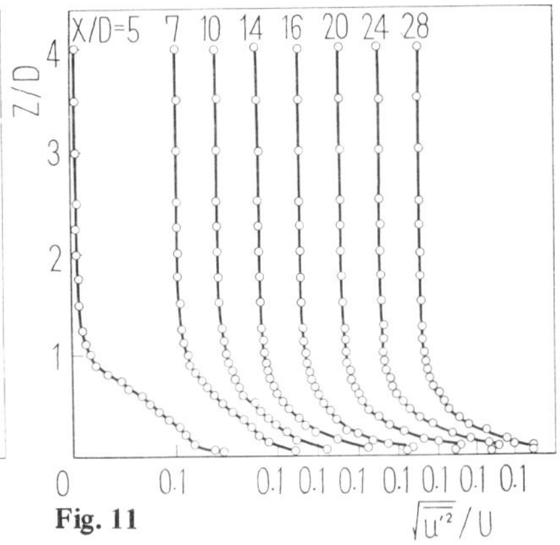

Fig. 11

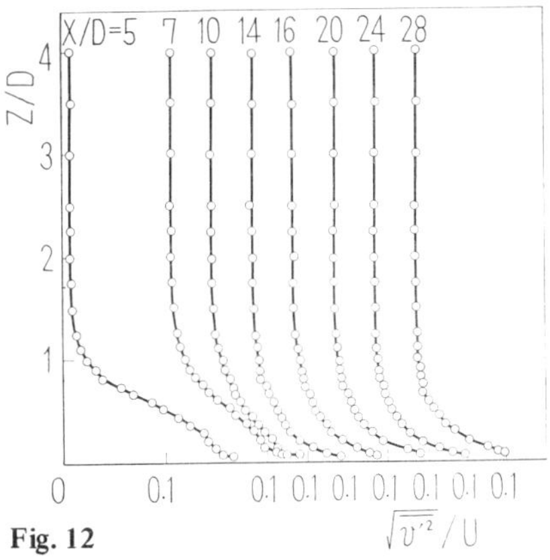

Fig. 12

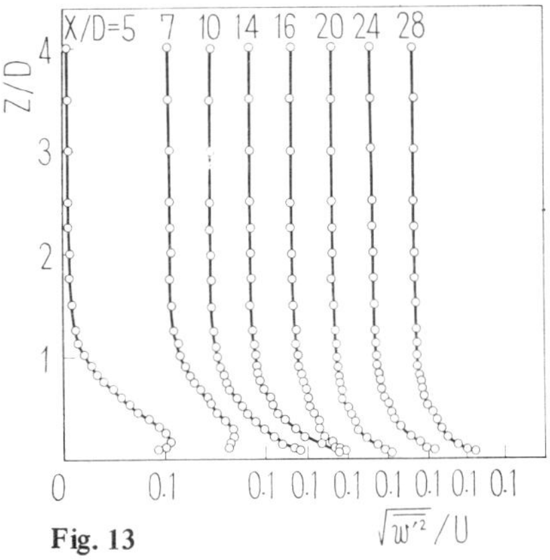

Fig. 13

Fig. 9. Distribution of velocity vectors in the center section ($Y = 0$)

Fig. 10. Profiles of mean velocity in the center section ($Y = 0$)

Fig. 11. Profiles of turbulence intensity in X-direction in center section ($Y = 0$)

Fig. 12. Profiles of turbulence intensity in Y-direction in center section ($Y = 0$)

Fig. 13. Profiles of turbulence intensity in Z-direction in center section ($Y = 0$)

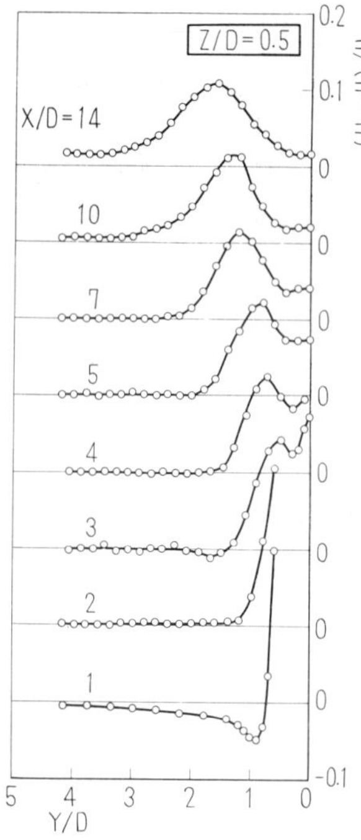

Fig. 14

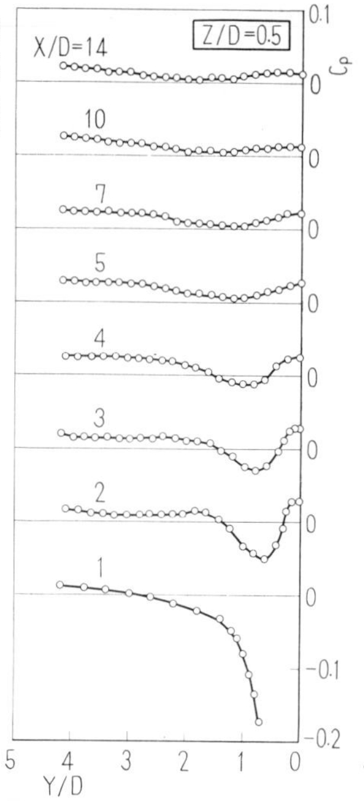

Fig. 15

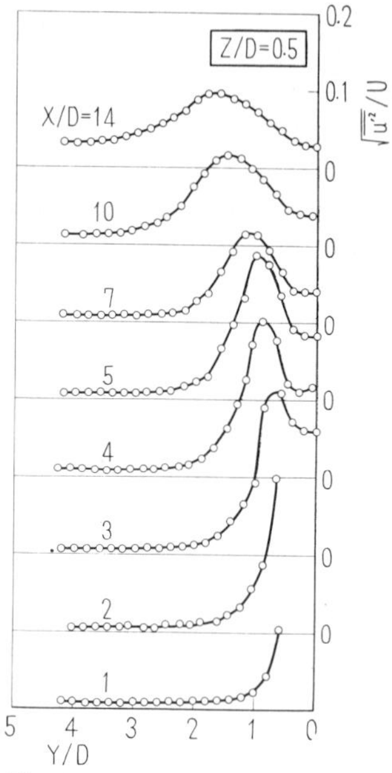

Fig. 16

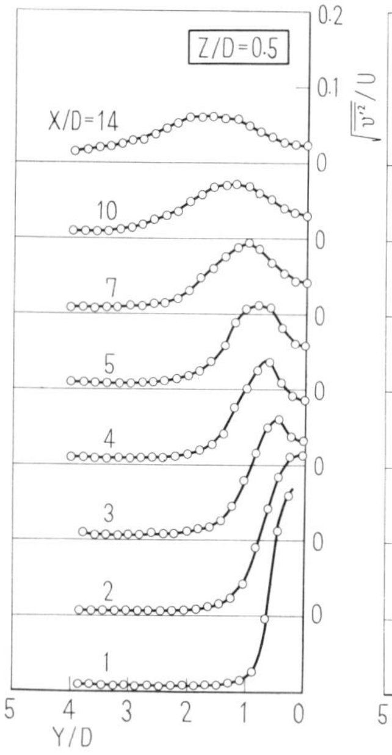

Fig. 17

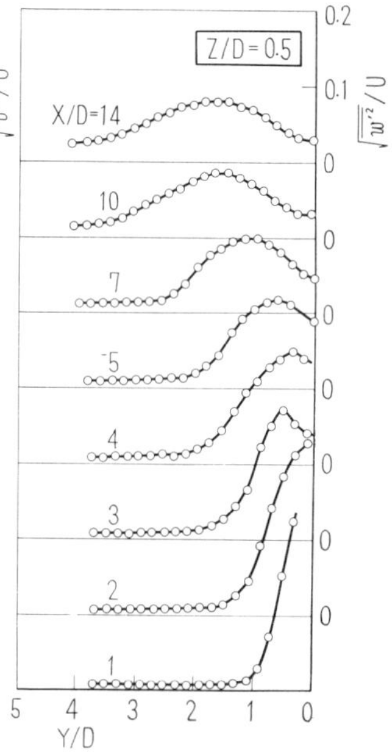

Fig. 18

Fig. 14. Profiles of mean velocity defects in the horizontal center section (Z/D = 0.5)

Fig. 15. Profiles of static pressure in the horizontal center section (Z/D = 0.5)

Fig. 16. Profiles of turbulence intensity in X-direction in the horizontal center section (Z/D = 0.5)

Fig. 17. Profiles of turbulence intensity in Y-direction in the horizontal center section (Z/D = 0.5)

Fig. 18. Profiles of turbulence intensity in Z-direction in the horizontal center section (Z/D = 0.5)

4.3. Velocity, and Turbulence Intensity in Center Section ($Y = 0$)

Figure 9 shows the distribution of velocity vectors in the center section. It is found from this figure that the recirculation region ends at about $X/D = 2.5$, as forecasted from the pressure distribution on the ground plate, and the turbulent shear layer is formed along the ground plate at $X/D > 5$. Figure 10 shows the time-mean velocity profiles at $X/D = 5 \sim 28$. The thickness of shear layer is nearly equal to $0.86\,D$ at $X/D = 5$, and it decreases with increasing downstream distance. The static pressure increases slightly in the shear layer.

Figures 11–13 shows the X-component, Y-component and Z-component of the turbulence intensity. The turbulence intensities of the free stream are 0.005, and they increase remarkably in the shear layer.

4.4. Velocity Defect, Static Pressure and Turbulence Intensity in Horizontal Center Section ($Z/D = 0.5$)

Figures 14–18 shows respectively the distribution of the mean velocity defects, the static pressure, the X-component, the Y-component and the Z-component of turbulence intensities in the horizontal center section ($Z/D = 0.5$). The position of both peak defect and peak turbulence intensities are shifted in the Y-direction when increasing the downstream distance. The peak

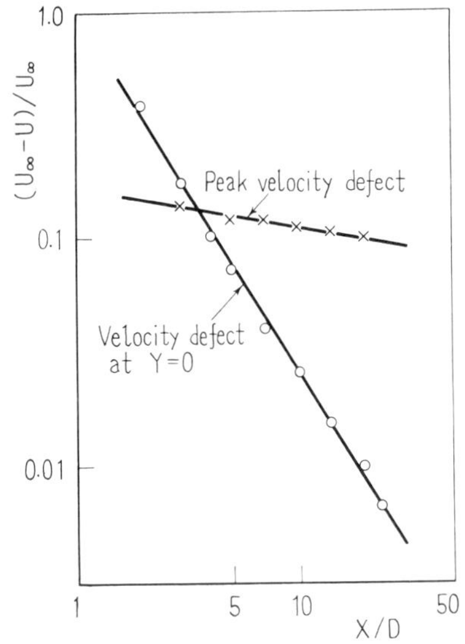

◀ **Fig. 19.** Peak velocity defect and velocity defect in the center section ($Y = 0$)

Fig. 20. Thickness of the shear layer and position of peak velocity defect
▼

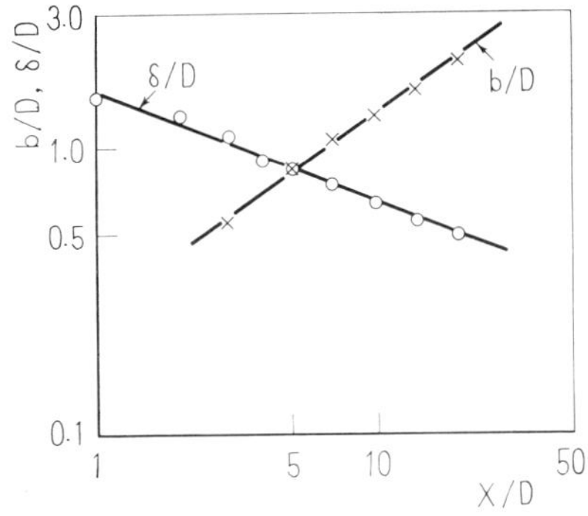

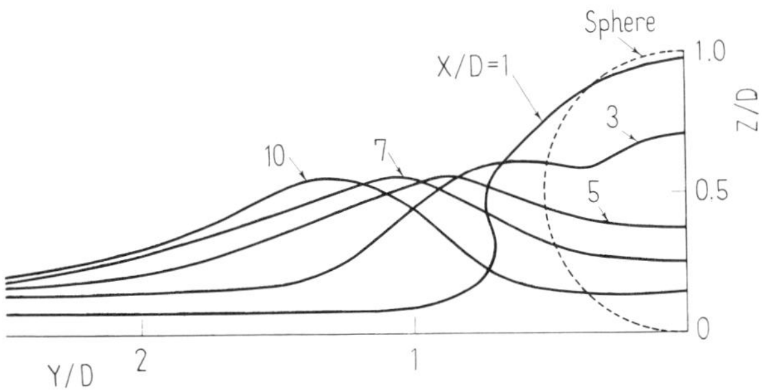

Fig. 21. Lines of $(U_\infty - U)/U_\infty = 0.10$

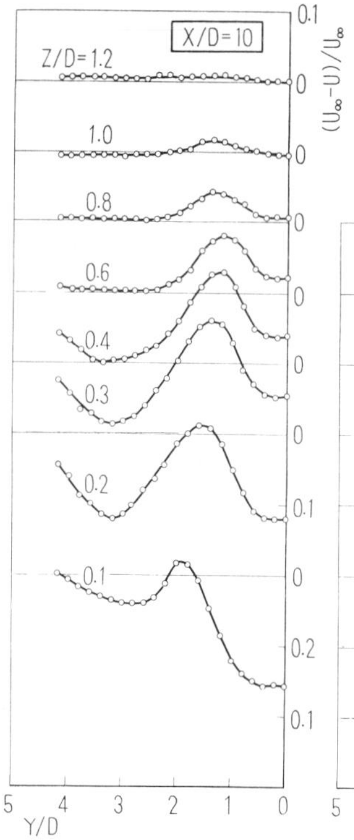

◀ **Fig. 22**

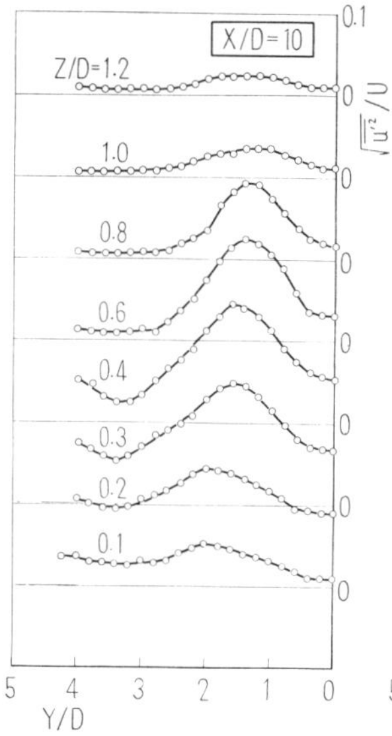

Fig. 23

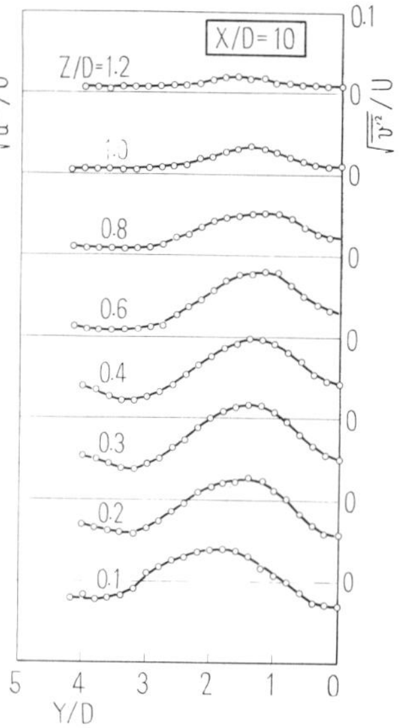

Fig. 24

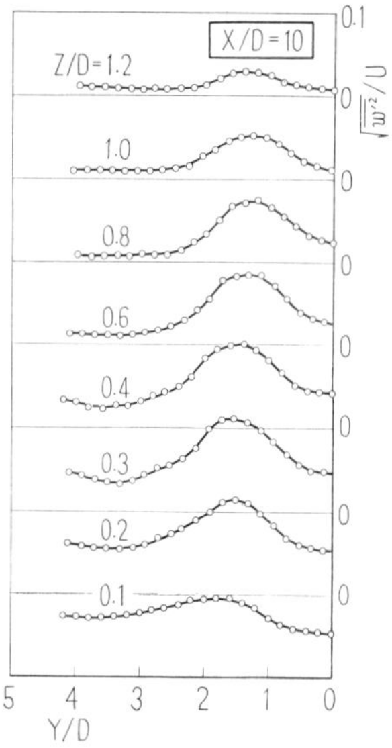

Fig. 25

Fig. 22. Profiles of velocity defect at $X/D = 10$

Fig. 23. Profiles of turbulence intensity in X-direction at $X/D = 10$

Fig. 24. Profiles of turbulence intensity in Y-direction at $X/D = 10$

Fig. 25. Profiles of turbulence intensity in Z-direction at $X/D = 10$

defect occurs at $Y/D = 0.8$ for $X/D = 5$ and at $Y/D = 1.3$ for $X/D = 10$, while the peak turbulence intensity in the X-direction occurs at $Y/D = 1.0$ for $X/D = 5$ and $Y/D = 1.5$ for $X/D = 10$.

The static-pressure coefficient becomes negative in the range of $0.3 < Y/D < 1.5$ for $2 < X/D < 5$ under the influence of the horse-shoe vortex while it becomes positive at $X/D = 5$, where the horse-shoe vortex disappears.

Figure 19 shows the velocity defect in the center section ($Y = 0$) and the peak velocity defect obtained from Fig. 14. The peak velocity defect decreases slowly in proportion to $(X/D)^{-0.17}$, while the velocity defect in the center section ($Y = 0$) decreases rapidly in proportion to $(X/D)^{-1.62}$, this is close to *Hunt*'s result [4] $(U_\infty - U)/U_\infty \propto X^{-1.5}$ for the wake behind a building.

Figure 20 shows the thickness of the shear layer obtained from Fig. 9 and the position ($Y = b$) of the peak velocity defect obtained from Fig. 14. The thickness of the shear layer decreases in proportion to $(X/D)^{-0.40}$ and the position of the peak velocity defect changes in proportion to $(X/D)^{0.69}$. This means that the wall wake behind a sphere becomes low and spreads transversely with increasing downstream distance. This change in the wall wake can obviously be seen from the lines of $(U_\infty - U)/U_\infty = 0.10$ shown in Fig. 21.

4.5. Velocity Defect and Turbulence Intensity at $X/D = 10$

Figure 22 shows the profiles of the velocity defect at $X/D = 10$. The peak velocity defect decreases with increasing vertical distance, and it almost vanishes at $Z/D = 1.2$. As the upward distance increases, the position of the peak velocity defect approaches the center until $Z/D = 0.5$ and then moves away from the center.

Figures 23–25 shows the distribution of the turbulence intensity in the *X*-, *Y*-, and *Z*-directions at $X/D = 10$, respectively. As the downstream distance increases, the position of the peak turbulence intensity moves in a manner similar to that of the peak velocity defect. The turbulence becomes almost isotropic at $Z/D = 1.2$. At $Z/D < 1.0$, the streamwise turbulence intensity is larger than the lateral and vertical turbulence intensities.

5. Conclusion

The experimental results are summarized as follows:

(1) The drag and lift coefficients of a sphere placed on a plane boundary at $Re = 4.74 \times 10^4$ are $C_D = 0.627$ and $C_L = 0.242$.

(2) Because of a strong downwash behind a sphere, the recirculation region behind a sphere is shortened. The length of the recirculation region is 2.5 *D*.

(3) The separation line on a sphere is shifted backward when approaching the ground plate.

(4) The arch vortices generated behind the sphere disappear at $X = 2.5\ D$. The horse-shoe vortex generated at the corner between the front face of the sphere and the ground plate is shed, passes by the sides of the sphere, and disappears at $X = 5.0\ D$.

(5) The velocity defect in the center section ($Y = 0$) decreases in proportion to $X^{-1.62}$, and the peak velocity defect decreases in proportion to $X^{-0.17}$.

(6) The thickness of the shear layer in the center section ($Y = 0$) decreases in proportion to $X^{-0.40}$, and the position of the peak velocity defect in the horizontal center section ($Z/D = 0.5$) changes in proportion to $X^{0.69}$.

(7) In the section $X/D = 10$, turbulence becomes isotropic at $Z/D = 1.2$, and at $Z/D < 1.0$, the streamwise turbulence intensity is larger than the lateral and vertical turbulence intensities.

Acknowledgement. The author expresses his thanks to Dr. Tetsushi Okamoto, professor emeritus of Tokyo Institute of Technology, for his cordial guidance.

References

1. H. Schlichting: Experimentelle Untersuchungen zum Rauhigkeits-Problem. Ing. Arch. **7**, 1 – 34 (1936)
2. A. Klemin, E. B. Schaefer, J. G. Beerer: Aerodynamics of the perisphere and trylon at world's fair. Trans. Am. Soc. Civ. Eng. No. 2042, 1449–1472 (1939)
3. P. M. Sforza, R. F. Mons: Wall-wake: Flow behind a leading edge obstacle. AIAA J. **8**, 2162–2167 (1970)
4. J. C. R. Hunt: The effect of single buildings and structures. Philos. Trans. R. Soc. London Ser. A: **269**, 457–467 (1971)
5. J. A. Peterka, J. E. Cermak: "Turbulence in Building Wakes". Proc. of the 4th International Conference on Wind Effects on Building and Structures 447–463 (1977)
6. T. Okamoto, M. Yagita: The experimental investigation on the flow past a circular cylinder of finite length. Bull. Jpn. Soc. Mech. Eng. **16**, No. 95, 805–814 (1973)
7. T. Okamoto, M. Yagita, S. Kataoka: Flow past cone placed on flat plate. Bull. Jpn. Soc. Mech. Eng. **20**, No. 141, 329–336 (1977)

Part IV

Coherent Structures

Introduction to the Section on "Coherent Structures"

P. Bradshaw

Department of Aeronautics, Imperial College of Science and Technology
London SW7 2BY, Great Britain

The papers in this section of the Proceedings, and the rather larger number of papers actually presented at the Symposium, are concerned mainly with free shear layers in general and mixing layers in particular. People with an interest in boundary layers should consult the outcome of the second research specialist workshop on Coherent Structures in Turbulent Boundary layers held at Michigan State University July 30 to August 1, 1979, organised by Dr. R. E. Falco, and also the Lehigh Workshop Proceedings (Smith and Abbott 1978). This division is an example of, although probably not a consequence of, the fact that research on coherent structures in boundary layers and in mixing layers has not yet been unified.

In boundary layers, there have been two interacting lines of work. The first is the study of the "large eddies" in the outer layer, which *Townsend* (1956) supposed to be weak but which were shown by *Grant* (1958) and others to be strong, in the sense of contributing significantly to the turbulent energy. The second line of work in boundary layers is the study of events in the viscous sub-layer and wall region, inspired by the flow visualization work of Kline's group at Stanford (e.g. *Kim* et al., 1971). The picture of boundary layer structure that has emerged from this research is of identifiable types of characteristic motion, strong, fully three-dimensional, and with a "time constant" of the order of the turbulent kinetic energy divided by the turbulent kinetic energy production rate (very roughly $3/(\partial U/\partial y)$). In this writer's view, there has been some failure to acknowledge the importance of low-Reynolds-number effects in the outer layer; most of the flow-visualization studies in boundary layers have been done at low Reynolds number. More generally, there has sometimes been a failure to ensure that the structures which are hypothesized to produce most of the turbulent energy and shear stress should fit logarithmic-law or defect-law scaling, at least to a first approximation. However, an impressive amount of detailed knowledge about the behaviour of boundary layer turbulence has been accumulated, and it should not be long before a consensus on the qualitative aspects, at least, of the structure is achieved.

In mixing layers, matters are more confused. Among the first to point out that turbulent mixing layers might, at least in certain circumstances, have a highly coherent and effectively two-dimensional (or axisymmetric) structure were *Crow* and *Champagne* (1971). Previously, the Berlin group led by the late Rudolf Wille had explored the later stages of transition in axisymmetric mixing layers, and shown that the "vortex ring" disturbances which are the instability eigenmodes could pair at intervals, so that the typical distance between vortex rings increased roughly proportionally to the distance downstream (e.g. *Wille*, 1963). However, the consensus among the rather large number of workers investigating jet turbulence at that time was that the flow eventually became fully three-dimensional, and that the three-dimensional large eddies, although exceptionally strong, were proper to the turbulent mixing layer and not a relic of transition (e.g. *Bradshaw* et al., 1964). Circular jets are known to be prone to various forms of instability, and it was not until the work of *Brown* and *Roshko* in a plane mixing layer (reported briefly in 1971 and more fully in 1974) that the possibility of persistent two-dimensonal "large eddies" attracted wide attention. Since most of the papers in this section of the Proceedings, with the exception of the excellent discussion on methodology by Yule, are concerned in some way with the presence of this two-dimensional (or axisymmetric) structure, a brief general discussion seems to be called for. A more detailed review of coherent structures is given by *Kovasznay* (see Fiedler 1978) who gives them the less emotive name "large scale structures".

It must be stated at once that the present writer believes (with probability about 0.9) that the two-dimensional structure observed in mixing layers is a relic of the transitional disturbances which it so closely resembles, and that it will eventually be broken up, by the small scale motion that inevitably accompanies it, into fully-three-dimensional, conventional large eddies. It is possible (probability nearly 0.1) that there are two asymptotic forms of the two-dimensional mixing layer, one with conventional three-dimensional large eddies and one with two-dimensional structure. There is a further possibility, which the present author regards as remote, that the two-dimensional structure so clear in Brown and Roshko's flow visualization pictures is the same as the apparently highly three-dimensional structure which produces the correlation measurements of *Bradshaw* et al. (which have been satisfactorily repeated by *Weber* 1974 and more briefly cross checked by *Chandrsuda* et al.1978). Controversy still ranges, although it remains entirely amicable; the present author's view is probably a minority one. The following remarks are intended to illuminate matters of controversy rather than continue the controversy itself.

As mentioned above, low Reynolds number effects complicate our picture of turbulent boundary layers at momentum-thickness Reynolds numbers less than about 5,000. The most relevant Reynolds number is $u_\tau \delta/\nu \equiv \sqrt{(\tau_{max}/\rho)}\delta/\nu$; in a mixing layer this latter Reynolds number is about 0.02 Ux/ν or about 0.1 $U\delta/\nu$ so that if low Reynolds number effects disappear at a universal Reynolds number $\sqrt{(\tau_{max}/\rho)}\delta/\nu = 2{,}000$, this would correspond to $U\delta/\nu = 2 \times 10^4$ or $Ux/\nu = 10^5$ in a mixing layer. Unless it is disturbed by an excessively large transition trip, the low-Reynolds-number turbulent boundary layer appears to be universal. A plausible explanation for the low-Reynolds-number effects on the outer layer is that the viscous superlayer, whose thickness is probably of the same order as that of the viscous sublayer, is rather thick at low Reynolds number. The mixing layer, on the other hand, seems to be extraordinarily sensitive to initial conditions. Transition in a mixing layer is due to the "inviscid" inflection point instability, so it is wrong to seek a universal Reynolds number marking the end of transition; *Bradshaw* (1966) offered the alternatives of a Reynolds number $Ux/\nu = 10^6$, or a downstream distance $x = 1{,}000\ \theta_0$, recognising that although the initial disturbance grew by an inviscid mechanism, the breakdown to finer scales and to fully developed turbulence might involve viscosity. The paper by Bernal et al., page 305, shows, by a subtle chemical technique that has important practical implications in the field of combustion and mixing, that small scale motion does not develop fully in a mixing layer until $U\delta/\nu = 10^4$ (it could be argued that the appearance of the small scale structure is by no means the same thing as the attainment of final equilibrium between the large scale and small scale structure, but this is perhaps the testimony of a hostile witness). In boundary layers, the initially two-dimensional Tollmien-Schlichting disturbances rapidly become three-dimensional, perhaps even before they interact significantly with each other; in the mixing layer, on the other hand, two-dimensional transitional disturbances can interact (pair), remaining two-dimensional unless there are rather large disturbances in the flow (*Chandrsuda* et al., 1978). The main argument for supposing that the large structure in the fully turbulent mixing layer may be closely related to the transitional eigenmode is that the instability Reynolds number for the range of wavelengths found in practice is of order 10; the Reynolds number of a fully developed mixing layer based on the *eddy* viscosity is of order 100 – that is, well above the critical! The argument is crude, but the presence of a factor of safety of 10 allows a certain crudity. It is therefore easy to see that the transitional eigenmode could persist for a very long time in the "turbulent" flow and conceivable that the two-dimensional mode could establish itself even if the initial boundary layer were turbulent (and therefore fully three-dimensional!). The question is whether the inflection point instability processes tending to maintain a two-dimensional disturbance will overcome the natural tendency of three-dimensionality to arise by the random vortex stretching precipitated by the smaller scale three-dimensional motion. Wygnanski's experiment described below shows that an *artificial* two-dimensional disturbance in the mixing layer is very slow to decay; and it is well known that two-dimensional or axisymmetric disturbances excited by upstream going pressure pulses can cause a strong feedback loop to appear (Ho and Nosseir, page 297). Of course, even an organized three-dimensional disturbance can persist for a long time, as shown by the work of Fuchs et al. (page 282) and the paper and film presented by Perry et al. at the Symposium: see also *Zilberman* et al. (1977). This writer believes that a distinction should be drawn between large structures with conventional lifetimes about three times the "time constant" $3/(\partial U/\partial y)$ mentioned above, and unusually-coherent structures with a much longer lifetime than this. The Brown-Roshko structures have a lifetime, between pairings, roughly equal to $3/(\partial U/\partial y)$; regarding this as a "half-life" gives a total lifetime about four times $3/(\partial U/\partial y)$ which is *not* an unusually large figure.

Two very important general points are raised by the controversy:

(I) is a turbulent flow unique, or can different structures exist for the same boundary conditions? This is a more serious matter than the possible existence of bistable flows, such as a separated flow in a diffuser. The principle of self-preservation (*Townsend,* 1956) is based on the assumption that the flow will eventually forget the effect of its initial conditions except infosar as they determine global quantities like momentum flux.

(II) if the largest eddies in a mixing layer are two-dimensional, then they cannot take part in the usual vortex-stretching energy "cascade" which, for this author at least, is the distinguishing feature of turbulence. This may imply that a field consisting of two-dimensional large structures and three-dimensional fine structure will be more difficult to predict than conventional turbulence, contrary to the hopes of those who seek simple coherent structures in turbulent flows.

An important question of experimental technique is whether space correlations can be relied on to show up the important large scale structure in a turbulent flow; Yule says that large structures, if important, should contribute to the turbulent energy or Reynolds stresses, or at least to turbulent entrainment (that is, to triple velocity products; it is difficult to imagine a structure that could contribute significantly to the triple products without contributing to the Reynolds stresses as well). This was the weak point in Townsend's original large eddy equilibrium hypothesis, which was based on the assumption that the smaller scale eddies contained virtually all of the turbulent energy. Yule's guidelines seem to imply that, except in pathological cases, a reasonably complete set of spatial correlations should show up the large scale structure and, according to the decomposition principles outlined by Townsend and discussed in more detail by Lumley and by *Payne* (1966), permit the ensemble-average shape of the large eddies to be determined. This is *not* a guarantee that one or two correlation measurements with separations along the coordinate axes will infallibly show up the large structure, but it is almost inconceivable that the strong two-dimensional structure seen in Brown and Roshko's flow visualization photographs should not give large values of correlation coefficient with spanwise separation, at least in some parts of the shear layer.

A fine detail of the present controversy concerns the behaviour of the spanwise correlations outside the shear layers; here, the scales are undoubtedly large, several times the shear layer thickness; however some recent work by *Wood* (1980) at Imperial College suggests that these large scales are attributable to the irrotational motion as such, which, being driven by pressure fluctuations, can easily correlate over very large distances.

The concentration on the mixing layer and the possibility of two-dimensional structures therein reflects the balance of interest in the papers presented at the Symposium; whether or not the orderly structure found by *Crow* and *Champagne* and by *Brown* and *Roshko* is universal, there is no doubt that their papers have provided a new impetus to research on turbulent shear flow and caused us at least to rethink, if not to abandon, the conventional ideas of turbulent structure. It seems unlikely that the large scale structure in boundary layers is closely related to that found in mixing layers, and the implication for calculation methods is that one should not necessarily expect to be able to calculate boundary layers and mixing layers (say) with the same set of empirical constants. However, it is to be hoped that, at least at the qualitative level, the schools of research on boundary layers and on mixing layers will have greater contact in the future than in the past.

References

Bradshaw, P. (1966): J. Fluid Mech. **26**, 225
Bradshaw, P., Ferriss, D. H., Johnston, R. F. (1964): J. Fluid Mech. **19**, 591
Brown, G. L., Roshko, A. (1974): J. Fluid Mech. **64**, 775
Chandrsuda, C., Mehta, R. D., Weir, A. D., Bradshaw, P. (1978): J. Fluid Mech. **85**, 693
Crow, S. C., Champagne, F. H. (1971): J. Fluid Mech. **48**, 547
Fuchs, H. V., Mercker, E., Michel, U. (1979): J. Fluid Mech. **93**, 185
Grant, H. L. (1958): J. Fluid Mech. **4**, 149
Kim, H. T., Kline, S. J., Reynolds, W. C (1971): J. Fluid Mech. **50**, 133
Kovasznay L. S. G. (1978): "Large Scale Structures in Turbulence – A Question or an Answer? ", in *Structure and Mechanisms of Turbulence I*, ed. by H. Fiedler, Lecture Notes in Physics, Vol. 75 (Springer, Berlin, Heidelberg, New York) p. 1
Payne, F. R. (1966): Ph. D. Thesis, Dept. of Aerospace Eng., Pennsylvania State University
Townsend, A. A. (1956): *The Structure of Turbulent Shear Flow,* 1st ed. (University Press, Cambridge)
Wille, R. (1963): Z. Flugwiss, **11**, 222
Wood, D. H. (1980): Ph. D. Thesis, Dept. of Aeronautics, Imperial College, London (available on microfiche)
Zilberman, M., Wygnanski, I., Kaplan, R. E. (1977): Phys. Fluids **20**, S 258

Phase Scrambling Effects and Turbulence Data Analysis

A. J. Yule

Department of Chemical Engineering and Fuel Technology
University of Sheffield, Sheffield, S1, 3JD, England

Abstract

It is described how investigations of coherent structures should study the various contributions to the apparent loss of eddy coherence, indicated by multipoint measurements, as eddies are convected downstream. These include the phase scrambling effects which result in low multipoint covariances, in spite of the presence of repetitive eddies, due to superimposed variations in the details of eddies and their movements. The effects on signals of various contributions to scrambling are considered and concepts such as individual eddy coherence, jitter and the efficiency of conditional sampling techniques are discussed. It is proposed that conditional sampling experiments must be specifically designed to measure, and differentiate between, the flow patterns of classes of individually similar coherent structures as a function of time, the local probability of occurrence of different types of coherent structure, and the probability distributions of interacting flow patterns, i.e. the biographies of coherent structures. The design of experimental and data analysis techniques, to achieve these aims, is discussed.

1. Introduction

There is a need for a more rationalised and systematic approach to the experimental study of coherent structures in turbulent shear flow. An aim here, is to encourage the establishment of a clearer framework for the diverse coherent structure/conditional sampling field of experimental turbulence investigation. Concepts are discussed which, although recognised at least implicitly by many researchers, are worthy of a fuller and more open discussion. With this objective much of the following has a discursive nature. It is argued that improvement in the investigation of coherent structure lies in recognition of the difference between "true" eddy coherence and apparent losses in coherence caused by phase scrambling effects. It is also noted that the concept of eddy coherence requires a clearer formulation when it is applied to the physical structures of individual eddies.

Phase Scrambling, is caused by a superposition of variations of the details, movements and histories of structures which results in low multipoint covariances, in spite of the presence of structures which may be individually, quite repetitive and highly coherent. These variations in structures, causing phase scrambling, are frequently referred to as eddy "jitter". The attenuation of recovered signals produced by these effects is often referred to, colloquially, as the "smearing" of signals.

Explicit recognition of the various phenomena contributing to measured coherence requires a more thorough consideration of the meanings of the conditioned forms of variables which occur when applying conditional sampling techniques. A description is first given of the factors which can influence fixed point signals when coherent structures are present. On this basis a description is then given of the experimental and data processing techniques which can be applied to measure the important parameters specifying the coherent structures and their histories. This is not a review of coherent structures and conditional sampling investigations but the discussion of the research field is intended to encourage a more effective application of the new techniques. Reviews, such as those of *Davies* and *Yule* [1] and *Laufer* [2] give an introduction to coherent structure orientated investigations.

In the following Coherent Structures (CS) are large eddies which may be repetitive in structure, remain coherent for significant downstream distances and which may be quasi-deterministic with repetitive "birth-life-death" cycles. To be worthy of the label, coherent structures should also contribute significantly to important phenomena such as turbulent energy and shear stress, entrainment, mixing and aerodynamic noise. However other connected features of turbulent flow can reasonably be termed coherent structures because of their repetiveness and coherence when viewed in a quasi-Lagrangian reference frame, e.g. wall layer features in boundary layers, laminar/turbulent interfaces and coherent "flamelets" in turbulent flames. All of these structures require basically the same investigative techniques for processing fixed point measurements, derived in the Eulerian reference frame, to provide information on the coherence and histories of structures in a quasi-Lagrangian frame of reference moving with the structures.

An interesting introduction to the possible roles of coherent large eddies, in time averaged correlations and spectra, is given by *Townsend* [3]. *Townsend* described how cross-correlations and wave number spectra can be deduced for assumed arrays of "model" eddies. For example a distribution of essentially similar eddies, with randomness in their separations, can result in a failure of the scales and essential periodicity of these eddies to be recognisable in cross-correlations. Thus "inference of velocity patterns from observed spectra and correlations in an uncertain process, usually involving preconceptions whose validity is always open to doubt". Further, *Townsend* [3] showed that the space-time structure function for a given eddy scale could, in a simplified analysis, by divided into variations in the actual eddy structures with time (as characterised by the velocity "amplitude" of the eddy) and variations in the positions of eddy centres. Once again *Townsend* commented that it is difficult to distinguish between the effects on space-time correlations of variations in structures (coherence) with time, and variations in the movements of structures from one structure to the next. It is proposed here that achievement of this ability, to distinguish between these effects and also between other phase scrambling effects, is a major objective and incentive for the development of conditional sampling methods of data processing. "Conditional Sampling" is here used in a wide sense to include all techniques of measurement and data analysis (including flow visualisations) which examine certain features of interest in a flow, rather than use classical time averaging.

Large eddy structure and coherence, and the interactions and breakdown of large eddies, are of more than purely academic interest. Improvement in the prediction and modelling of turbulent shear flow requires precisely this type of information and these eddies are known to be directly important in the fields of mixing, combustion and aerodynamic noise.

2. Areas to be Resolved in Coherent Structure Investigations

It is useful at the outset to specify the particular problem areas arising when considering flows in terms of coherent structures:

1) Flow fields of individual CS's: can they be measured?
2) Measurement of the variation in the structures of eddies found locally.

3) The biographies and interactions of the CS's: Do they break down or coalesce?
4) How do the CS's relate to the smaller scales of turbulence?
5) The "three-dimensionality" of eddies: How coherent are they in the direction perpendicular to the directions of the main flow and the highest velocity gradients?
6) What are the roles of CS's in the phenomena of, entrainment, macro and micromixing, combustion, noise, pressure fields etc.
7) How do CS's depend upon initial and boundary conditions?
8) Are CS's found only in certain "pathological" laboratory flows?
9) How can experimental observations of CS's be reconciled with and incorporated in, analytical and modelling treatments of turbulent flow?
10) Are coherent structures relevant when considering practical, complex engineering flows?

In order that we can have the measurement tools to answer these questions, it is essential that the phase scrambling effects can be understood and taken into account. The final two questions reflect the ultimate goals in this type of turbulence research. To "believers" in coherent structures they may appear rhetorical, however it must be said that it is not yet proven beyond reasonable doubt that CS's are essential features of most turbulent flows nor that they should necessarily be included in the "physics" of turbulence models. One could thus add a further aim: To elevate investigations of coherent structures from situations in which experiments are open to different interpretations according to one's "beliefs", to the situation where measurements give objective information on structures and their coherence or otherwise.

3. Eddy Coherence, Phase Scrambling and Conditional Sampling

3.1. Introduction and Notation

In the general case, as shown in Fig. 1 a coherent structure is recognised as it passes a conveniently chosen position P, origin of the x', y', z', coordinate system, where it has a distribution $a_s(x'', 0)$. The field of interest is either a scalar or vector quantity, such as temperature, specie concentration, vorticity or a velocity component. The distribution of a_s associated with an eddy should generally be divided into a coherent component a_c and an incoherent,

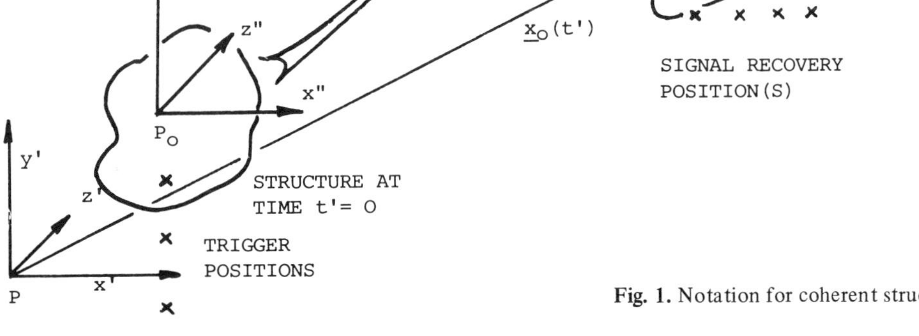

Fig. 1. Notation for coherent structures

presumably random, component a_R, so that $a_s = a_c + a_R$. As will be discussed, how this division is made is open to debate, because of the many possible random contributions to the eddies. However typically one expects a_R to consist largely of contributions from the smaller scales of turbulence.

The coordinate system x'', y'', z'' is "fixed" in the CS and is thus a quasi-Lagrangian system. Its origin P_0, the eddy "centre", is arbitrary in definition, but the definition is consistent for all structures. In the most general situation P_0 need not be identified with the same fluid element throughout the lifetime of the structure.

The motion of P_0 corresponds to an eddy convection velocity $u_0(t')$. If a is velocity the Eulerian velocity u is the sum of the eddy field and eddy convection components $u = u_s + u_0$. For a particular eddy the centre has a position $x'_0(t')$ relative to P.

In a typical conditional sampling situation a trigger is activated when each structure of interest is in the vicinity of P and, after a time delay t', the a distribution is measured at a fixed point (or points) known as the signal recovery position. The position at which the event giving $t' = 0$ is recognised is the trigger position. In the most general case the trigger position need not be at P, the structures need not pass through the trigger position and the trigger signal need not be derived from a fluctuations. For example, irrotational fluctuations outside the turbulent flow are often convenient trigger signals. Because of the nature of turbulent flow, in general one cannot expect each CS to be identical to its predecessor, the trajectories of each CS will not be the same, the distortions of structures during the time interval t' will not be exactly repetitive nor will the position of P_0 relative to P necessarily be the same, at $t' = 0$, for each CS. Figure 2 indicates how "jitter" of basically similar eddies, measured at time delays t' after trigger events, can cause quite drastic variations in fixed point signals. This type of information can be derived by simultaneously filming a flow whilst recording signals from probes.

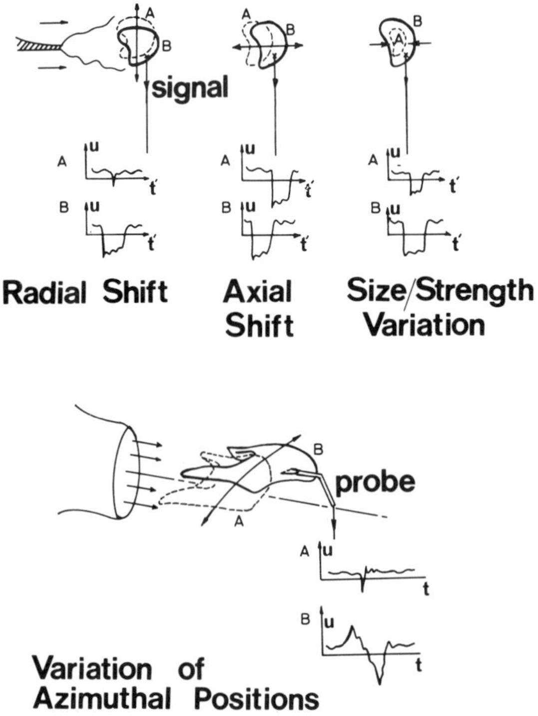

Fig. 2. Different modes of jitter and their effects on signals

It is readily seen that these various effects will combine to produce "smearing" of signals recovered at a time delay t', by ensemble averaging the recovered signals from many CS's. As this smearing is predominantly caused by phase shifting of $\boldsymbol{a}$ distributions from one structure to the next, it can be conveniently described as a phase scrambling effect. Other forms of conditional sampling to study the effect of CS's can be considered as particular forms of the above general case, and they must all suffer from phase scrambling. It should be noted that the frequently used intermittency conditioning of local signals is a particular case for which the trigger and recovery positions coincide and t' is zero.

One can conveniently consider the problems of signal interpretation in a CS – containing flow in two stages; firstly the problem is discussed when the observer moves with individual eddies and then consideration is given to variation in eddy trajectory and interaction.

3.2. Coherent Structures in Moving Reference Frames

Consider the situation where the observer moves with the x'', y'', z'' coordinate system for each individual eddy and variations in eddy trajectory are ignored. It is necessary to interpret experimental data to give information on the coherence of an individual eddy as a function of time and to differentiate between this and other phenomena which must be quantified such as eddy distortion, interaction and breakdown. It is first necessary to be explicit concerning the physical meaning of the "coherence" of an individual eddy. It is possible to interpret ensemble averaged fields for *many eddies*, only when one clearly defines how coherence, breakdown, etc. could be deduced from signals produced by *one eddy* (although in practice measurement of the instantaneous structure of a single eddy may not be feasible).

It is emphasized that discussion of the coherence of eddies should be differentiated from information on flow structure given by the classical coherence function. The coherence function, the correlation between signals for a given frequency component, is susceptible to the phase scrambling effects which conditional sampling should be capable of quantifying.

Coherence and the Choice of $\boldsymbol{a}_c$. With increasing t' each CS can vary in many ways, including growth by entrainment and an orderly deformation. Where structures move with no significant changes in their flow fields, it is clear that they are coherent. Difficulties arise as soon as any change at all occurs in a structure during a time interval and the differentiation between "orderly" change, and a loss of coherence, is then an arbitrary one. The clearest requirement for an eddy to be considered coherent in the time interval t' is that the eddy strength, as indicated by the scale or amplitude, $\Delta \boldsymbol{a}_c$, of fluctuations produced by the eddy, should not change appreciably, so that,

$$\Delta \boldsymbol{a}_c(t' = 0) \approx \Delta \boldsymbol{a}_c(t') \quad (1)$$

Fluctuations produced by the eddy should also be comparable with local fixed point fluctuations of $\boldsymbol{a}$ so that,

$$\Delta \boldsymbol{a}_c(x'_0, t') \approx [\overline{(a - \overline{a})^2}]^{1/2} \quad (2)$$

This is the requirement that the eddy is a significant feature of the local turbulence. For naturally decaying turbulence, such as free jets, (2) could supercede (1), i.e. the eddies can decay in strength, but be considered coherent, provided that they remain an important part of the local turbulence. Thus coherence is essentially the retention of identity and amplitude. Obviously care should be taken with such a simplistic viewpoint, depending upon the flow variable and flow type under consideration. For example the concentration of a marker chemical within eddies in a chemically reacting flow could decay due to molecular scale mixing and reaction, even though the eddies themselves retain coherence, as indicated by velocity, density or passive scalar fields.

Variations of condition (2) occur according to which aspects of the turbulent flow are of interest. For example for individual eddies to be significant contributors to the turbulent shear stress, $\overline{uv}$, requires the condition that $\Delta u_c v_c \approx \overline{uv}$.

The precise definition of the coherent component of an eddy a_c, is arbitrary, but it is clear that small scales within the large eddy should be an incoherent component of the structure. These small scales are swept around by the convective action of the main large eddy motion, they are essentially independent of the main motion and molecular forces ensure that they are relatively short lived. This does not apply in transitional flow where orderly vortices exist nor, perhaps, in the wall layers of confined flows. The small scales of these vortices are the viscous cores wich can remain coherent for significant distances downstream. This special case apart, it is reasonable to propose that investigation of the coherence of the larger scales requires, at some stage, a filtering out of those smaller scales which are not directly of interest.

The attribution to large eddies, of low wave numbers or frequencies, is not considered to be the most productive approach. From the experimental viewpoint, wave number spectra are ambiguous and are not convenient. Low pass (frequency) filtering of signals is straightforward but does not give the most meaningful results. For example discrimination is not made between fast moving "wide interfaces" and slow moving "narrow interfaces". Conditional sampling generally requires digital signal processing so that digital filtering procedures can be used to characterise $\boldsymbol{a}_c$. Digital spatial filtering was used by *Reynolds* [4] to derive large scale variables for the prediction of turbulence. Spatial filtering is not practical experimentally, with the possible exception of sophisticated quantitative flow visualisation techniques. However smaller scales are convected at nearly the local instantaneous velocity so that an approximate spatially filtered signal can be derived from single point time histories to give a formulation for $\boldsymbol{a}_c$:

$$\boldsymbol{a}_c(x'', t') = \int_{-\infty}^{+\infty} G(t' - \tau)\, \boldsymbol{a}_s(x'', \tau)\, d\tau \tag{3}$$

where the filter function G is (following Reynolds)

$$G(t' - \tau) = \left[\left(\frac{6}{\pi}\right)^{1/2} \frac{|u_s|}{l_R}\right] \exp\left[-6(t' - \tau)^2\, |u_s|^2 / l_R\right]$$

Here, the random component of the motion is characterised by a length scale l_R. For the concept of local strong coherent structures with length scale l_c, to be valid two further conditions seem reasonable:

$$l_c \gg l_R \quad \text{and} \quad \Delta a_c \gg \Delta a_R \tag{4}$$

Thus the scales and intensities of the coherent and random components of the eddies are widely separated. It is considered that spatial filtering processes quantitative point data in much the same way as the eye extracts the coherent large eddy motions from the rest of the flow in flow visualization films. It appears to be inevitable that any consideration of flows, either experimentally or theoretically, which contain coherent large eddies, must apply this type of filtering to separate the coherent and incoherent components.

Apart from such general guidelines, the concept of eddy coherence remains essentially ill defined and open to individual interpretation according to the context considered. For example if two eddies coalesce, to form a larger eddy, it is a matter of viewpoint as to whether the original eddies break down or not. Furthermore one must ask to what degree large eddies can deform randomly, in the large scale, before they should be considered to have lost their coherence, or whether coherence is only destroyed when energy has been transferred to the small, random, scales.

Repetitiveness of Individual Coherent Structures and Ensemble Averaging. For the concept of coherent structures, which introduce a deterministic component to the flow, to be valid it is essential that there is an important repetitive element in the local flow. It is neccessary to formulate the expected relations for this case, and then search for agreement in the experimental data. There should be not only an important repetitive local eddy motion, but also repetition in the pattern of events associated with the "birth", interaction and decay of the eddies.

Given the range of conceivable variations in an essentially repetitive (not periodic) basic structure, it is not possible to say exactly which parameters characterise an individual eddy, without preliminary knowledge of the flow structure. However, each coherent eddy will have characteristic length and amplitude scales $l_c(t)$ and $\Delta a_c(t)$. In addition, parameters are required to characterise the shape of an eddy, thus $b_1(t)$ is the first order description of eddy shape. One can envisage situations when it is useful to attribute and "age" to each eddy passing the trigger position, t_A say, where $t_A = 0$ at for example, a coalescence forming the eddy. Further parameters, $b_2, \ldots, b_n$, may be required for an adequate description of the local eddies, for example to differentiate between leading and trailing eddies which coalesce. In general the field of an eddy at time t' is characterised by

$$a_c = f\,(x'', t', \Delta a_c, l_c, t_A, b_1, \ldots, b_n) \tag{5}$$

If coherent structures exist, an objective of conditional sampling must be to determine f. Additionally, for the concept of coherent structures introducing a deterministic nature to the flow to be valid, simplifications of (5) should be possible.

A *deterministic* flow structure requires that if the flow pattern is known near P at $t' = 0$, it can be predicted at a later time $t' > l_c/u_0$. This implies that coherent structures which are identical when sampled at $t' = 0$ should behave similarly as they move downstream. Thus if $f^*(x'', t')$ refers to all structures which have certain specified values of the parameters $a_c, \ldots,$ b_n at $t' = 0$, then the ensemble average of many of these structures at time t', should be similar to the field for only one of these structures so that, for N eddies

$$\langle f^*(x'', t') \rangle \equiv \frac{1}{N} \sum_1^N f^*(x'', t') \approx [f^*(x'', t')]_{\text{One Eddy}}$$

or more usefully,

$$\left[\frac{1}{N} \sum_1^N f^{*2}(x'', t') - \langle f^*(x'', t') \rangle^2 \right] / \langle f^*(x'', t') \rangle^2 \ll 1 \tag{6}$$

Thus an objective of conditional sampling is to apply triggering criteria which will educe, from the complete flow, classes of structure which are repetitive spatially and temporally. If this precise selection is not achieved, so that a wide range of eddy types passing P are sampled, smearing, or phase scrambling of the ensemble averaged signals is introduced.

It is noted that the ensemble averaging of the f^* patterns automatically removes incoherent features in the structures. Thus, in practice, it is not necessary to process signals to derive a_c before ensemble averaging and the actual signal a_s can be used. However, it would be instructive to compare the averaged structures with those derived by spatially filtering individual signals.

If the joint probability distribution for the parameters characterizing the sampled eddies is $P(\Delta a_c, l_c, \ldots, b_n)$ the ensemble average of sampled eddies at the time delay t' is (in the moving reference frame),

$$\langle a_c(x'', t') \rangle = \int\int P(\Delta a_c, \ldots, b_n) f(x'', t', \Delta a_c, \ldots, b_n) dV \qquad (7)$$

where the integral is taken over all values of the parameters $\Delta a_c, \ldots, b_n$. Conditional sampling experiments should thus also be designed to determine $P(\Delta a_c, \ldots, b_n)$.

Consideration should be given to the interaction between adjacent eddies. When this interaction (including possible eddy coalescence) is of importance, it appears essential to consider *groups* of adjacent eddies as coherent structures, rather than single eddies. When an eddy group is considered as a CS, there is no loss of generality in the above arguments. The parameters specifying an individual CS must provide descriptions of the relative positions and strengths of the individual eddy components. The usual received meaning of an eddy, as a coherent region of rotational flow, can also differ from what could be considered a CS. Thus a CS can include not only a turbulent large eddy (or eddies), but also irrotational regions of entrainment associated with the eddy, and a section of the near field which experiences fluctuations caused by the eddy.

3.3. Coherent Structures in a Fixed Reference Frame

The coherence, distortions and, perhaps, interactions of structures should ideally be studied in a moving frame of reference, however this is impractical experimentally. Fixed point measurements are required and variations in the trajectories of individual eddies appear to be major contributors to the phase scrambling of multipoint signals. Returning to Fig. 1, and considering a CS with centre at $x_0'(0)$ at $t' = 0$, the eddy centre has moved, at time t' to

$$x_0'(t') = \int_0^{t'} u_0 dt' + x_0'(0)$$

In practice the limited available evidence indicates that $u_0(t')$ can differ considerably from one eddy to the next and also, in general, there will be variation in the value of $x_0'(0)$ from eddy to eddy due to the impossibility of deriving a perfect triggering criterion. As with considerations in the moving reference frame, it is advantageous to design the conditional sampling technique to select a narrow class of similar patterns which have a narrow range of $x_0'(0)$. In this way one should minimise the smearing of the conditionally sampled signals and obtain an ensemble averaged pattern representative of the instantaneous flow structure. In a situation where such a class of similar eddies, $f^*(x'', t')$ are sampled at time t', and these eddies have a probability distribution of their positions $P[x_0'(t')]$, the recovered ensemble averaged signal for a scalar quantity is,

$$\langle a_c(x', t')\rangle = \int\limits_{\substack{\text{All} \\ \text{Values}}} P[x_0'(t')] f^*[(x' - x_0'), t']\, dx_0' \tag{8}$$

For standard deviations of $P(x_0')$ comparable with l_c, recovered signals will be negligible, in spite of the existence of coherent eddies. When the conditional sampling technique is sampling a range of eddy types, Eq. (8). requires the inclusion of the parameters describing these types and the probability of occurrence of these different types. The integral must then be taken over all values of these variables in addition to the values of x_0'. It is easy to see that small variations in the size, strengths etc. of sampled eddies can combine with position variation to give severe attenuation of recovered signals even though the eddies are strong and completely coherent in the sampling time interval. Phase scrambling equations, such as Eqs. (7) and (8), can be taken further by making simplifying assumptions using experimental data.

It is more convenient to sample lengths of time history a different positions across a flow, at times relative to t' rather than to measure at a matrix of many closely spaced points, provided that a simple Taylor approximation of a convected unchanging structure can be made. In this situation the averaged structure at time t' in the fixed reference frame $\langle a(x', t')\rangle$, is related to the ensemble average time histories $\langle a(x_p', \tau)\rangle$, where x_p' is a recovery probe location and it is assumed that, at time $t', x' = x_p' - \langle u_0\rangle\tau$. This assumption is only valid if (a) sampled structures do not change significantly in a time interval $l_c/|u_0|$ and (b) convection velocities do not vary from structure to structure and do not change in this time interval. Available data indicates that neither of these assumptions is strictly valid so that the eduction of structures from ensemble averaged time histories, requires caution.

Additional contributions to phase scrambling occur when velocity, rather than scalar, signals are sampled as then individual eddy convection velocities are detected and these are additive to the velocity patterns of the eddies. Thus if a repetitive class of eddies is examined with a velocity pattern, in the moving frame, $u_c(x'', t') \equiv f^*(x'', t')$, the velocity measured in the fixed frame is $u[(x_0' + x''), t'] = u_0(t') + u_c(x'', t')$. The phase scrambling probability distribution is $P[u_0(t'), x_0(t')]$ and the ensemble averaged velocity pattern is

$$\langle u(x', t')\rangle = \iint\limits_{\substack{\text{All} \\ \text{Values}}} P[u_0(t'), x_0'(t')]\ \{u_0(t') + f^*[(x' - x_0'), t']\}\, du_0\, dx_0 \tag{9}$$

Available evidence suggests that eddy convection velocities are dependent on the lateral position of the eddy in the shear flow so that a simple separation of variables may not be possible.

4. Simple Example of Attenuation by Phase Scrambling

It is useful to investigate how an assumed signal is attenuated when ensemble averaged with a varying degree of phase shift or jitter. This analysis gives an insight into the order of magnitude of phase scrambling effects in conditionally sampled data. As shown in Fig. 3, the eddy signature of interest is assumed to be a sine wave with period L and peak $\hat{u}_c$ so that the distribution of the coherent component of the velocity in the x direction is $u_c = \hat{u}_c \cos [2\pi(x' - x_0')/L]$. The notation in Fig. 1 is followed. Let us consider the attenuation of this basic pattern which is produced by ensemble averaging many identical realizations allowing for phase shifting only in the x' direction. Thus at the time t', relative to the trigger event at $x' = 0$, the position x_0' of the sampled pattern, relative to the average position $\langle x_0'\rangle$ has a probability distribution $P(x_0' - \langle x_0'\rangle)$. An obvious first approximation to P would be a Gaussian

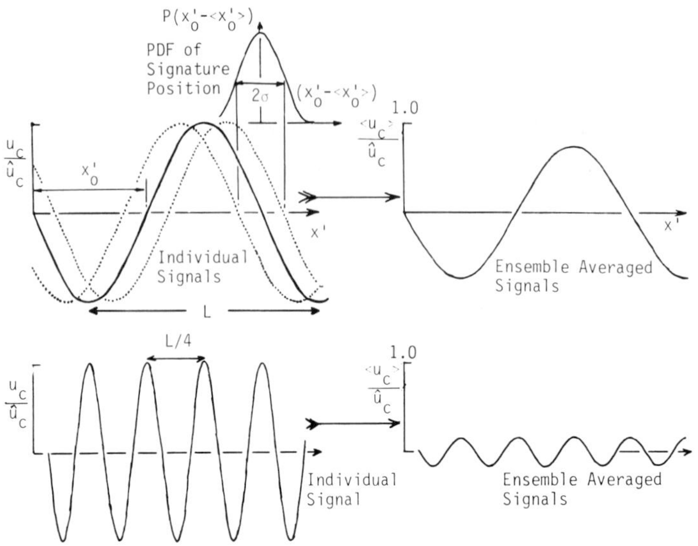

Fig. 3. Attenuation of cosine signals by phase scrambling

distribution, however, on reflection the extended tails of such a distribution are not realistic in this situation. Bearing in mind that the exact shape of the P distribution is of less importance than its standard deviation σ, a cosine "bell" probability distribution is a reasonable first assumption so that $P(x'_0 - \langle x'_0 \rangle) = \dfrac{1}{2\beta}\,[1 + \cos(\pi x/\beta)]$ and the half width of P, β, is related to the standard deviation by $\beta = \sigma/(1/3 - 1/\pi^2)^{1/2}$. Substituting for P and U_c in equation 8 and solving yields the ensemble averaged velocity pattern at time delay t'

$$\langle u_c(x', t') \rangle = \frac{L\hat{u}_c}{2\pi\beta} \sin\left(\frac{2\pi\beta}{L}\right) \cos[2\pi(x' - \langle x'_0 \rangle)/L] \tag{10}$$

Thus the basic cosine shape of the pattern in conserved but the amplitude of the recovered pattern is attenuated by the factor $L/2\pi\beta \sin(2\pi\beta/L)$. The upper part of Fig. 3 shows the ensemble averaged signal when $\sigma/L = 0.1$, i.e. the standard deviation of eddy positions is 10% of the basic eddy length scale. A reduction of 27% in the true amplitude of the eddy velocity pattern is produced. The situation becomes more serious when smaller scales of velocity variation within the eddy are considered. The lower part of Fig. 3 shows the effect of the same 10% variation, on a length scale $\frac{1}{4}$ of the main length scale. Here an amplitude reduction of 83% is produced.

It is interesting to note, from Eq. (10), that for $\beta > L/2$, i.e. for standard deviations σ/L greater than 24%, there is a phase reversal in the ensemble averaged pattern so that the measured velocity pattern is an inverted form of the true pattern. As this results from the simplifying assumption of a strictly periodic array of similar eddies, such as phase reversal is unlikely to happen in practice where eddy separations can vary. In addition, the various other contributions to phase scrambling are likely to smear recovered signals to negligibly small amplitudes before such a stage can be reached.

Most conditional sampling experiments measure time histories at a restricted number of points, rather than obtain conditionally averaged spatial distributions by using a matrix of

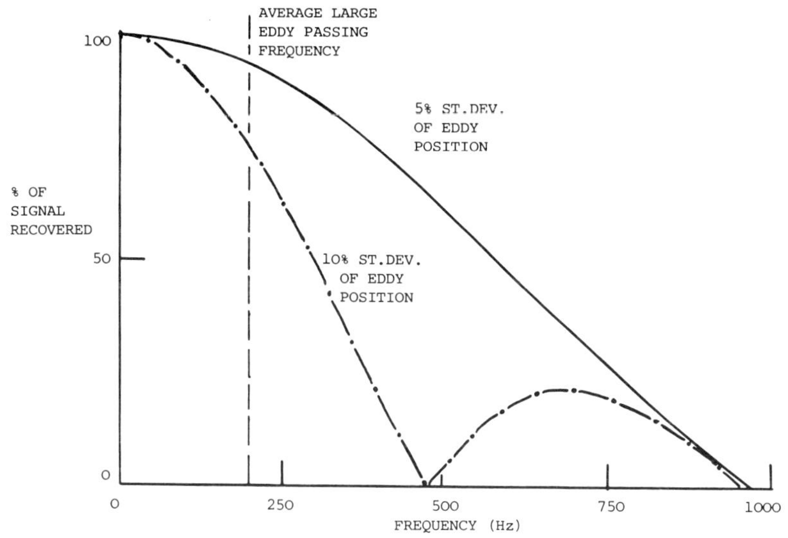

Fig. 4. Attenuation of frequency components of ensemble averaged signals, due to phase scrambling

many closely spaced points. The above discussion can be extended to the time domain so that variations in fixed point signals, due to eddy jitter, are considered. If one approximates by assuming that all eddies have the same convection velocity, Eq. (10) can be written in terms of ensemble averaged time histories, rather than spatial velocity distributions. Then one can consider how various frequency components in eddies are attenuated after ensemble averaging. A result is shown in Fig. 4. This can be considered to represent how frequency components of the Fourier transform of fixed point signals produced by passing eddies, are attenuated in the ensemble averaged time histories for many eddies. Thus, for example, for an average large eddy passing frequency of 200 Hz frequency components near 500 Hz are completely absent in ensemble averaged signals for 10% standard deviation of eddy position. The increase in the amplitude of recovered signal for frequencies greater than 500 Hz is an effect of the physically unreal assumption of a strictly periodic array of eddies. In a real experimental situation a variation of 10% in the positions of sampled eddies does not appear unlikely, although experimental data is lacking. Furthermore the contributions from "jitter" in other directions, coupled with variations in the actual structures of the sampled eddies, seem likely to make the results of these "one-dimensionsl jitter" analyses conservative estimates of the true signal attenuation. Obviously future experiments must attempt to quantify these various contributions to the smearing of ensemble averaged signals before such data can be reliably interpreted in terms of eddy structure and coherence. This smearing of multipoint signals also applies, to a worse extent, in cross-correlation measurements. If phase scrambling significantly attenuates low wave number components of cross-correlations, corresponding to the scales of the large coherent eddies, then higher wave number components will be very severely attenuated. One cannot thus simply use such averaged multipoint data to derive either relative or quantitive measures of the coherence or decay of small scale structure unless some knowledge exists of the phase scrambling effects. Thus for example, experiments [5] in flames show that thin reacting interfaces remain identifiably coherent for long downstream distances. These correspond to small scales in the turbulent flow which cross-correlations indicate to decay very rapidly.

5. Investigative Techniques

5.1. The Need for Caution in Coherent Structure Experiments

The design of experiments, flows and data analysis techniques for investigating coherent structures, require considerable care and planning if observations are to be useful and unambiguous. Because the concept of repetitive, coherent large eddy motions is an attractive one from the viewpoint of understanding turbulent flow, there is often the tendency to seek orderliness where perhaps none exists. One can specify several dangers which should be faced when embarking on this type of investigation:

Dangers of Coherent Structure Investigations

1) Confusion between turbulent and transitional flows: Thus invalid conclusions can be drawn Careful consideration should be made of flow regions in which the stabilising influences of viscosity, wall effects etc. are important.
2) Modification of flow to facilitate measurements: Thence assuming that the data also applies to "natural" flow. Before utlising sound fields, vibration, sparks, etc., to periodically force a flow or artificially instigate the growth of diturbances, it appears to be necessary to consider how relevant the results will be with regard to unforced turbulent flow.
3) Conditioning data too highly, or incorrectly so that a physically non-existent structure is educed. This is a very real danger when using sophisticated multiprobe-triggering and processing techniques, such as the phase-matched ensemble averaging noted below. However, provided that the experimenter is aware of the danger, simple statistical considerations could be used to check the physical realism of ensemble averaged data.
4) Disregarding the three-dimensionality of turbulent flow: Fully developed flow is three-dimensional in its time dependent structure, with fluctuations being of the same order in all directions. This must be taken into account when analysing data.

One can imagine many examples of item 3 above. For example Fig. 5 illustrates a flow visualization in a shear flow with important large eddy motions, in which the shapes of the

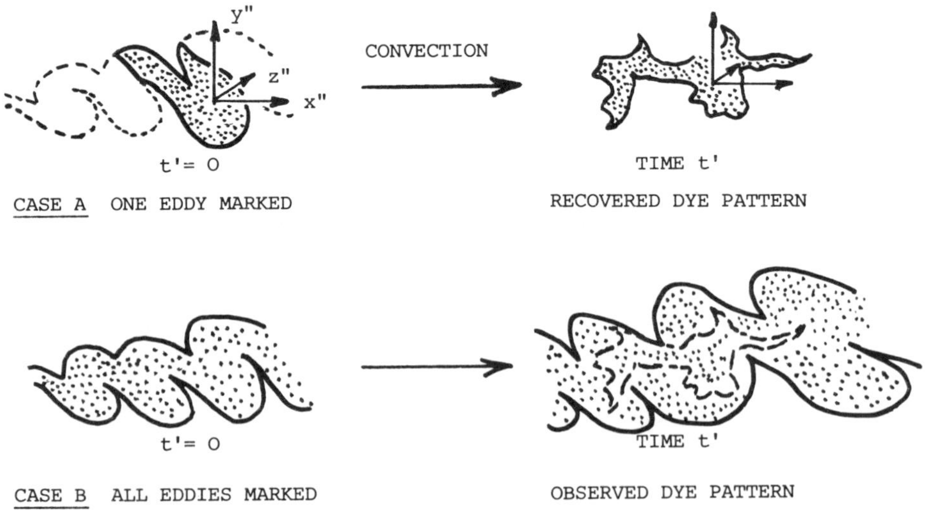

Fig. 5. Study of eddies by dye tagging

boundaries of structures are examined by injecting dye to mark the rotational flow region. If, as in the upper part of Fig. 5 one eddy *only* is tagged with dye at $t' = 0$, a pattern will be recognised at a later time t'. However if *all* eddies passing P are tagged with dye, the eddy boundaries near the "remains" of the eddy of interest at time t', can be quite different from the pattern for that eddy alone, Thus the boundaries of the dye at t', for a single eddy marked at $t' = 0$, do not coincide with the large eddy boundaries. This is indicative of a loss in coherence of the marked eddy. If many eddies are individually marked and the dye patterns at time delay t' are ensemble averaged one arrives at an average "eddy shape" at delay t' and an average position, and thus "convection velocity", for the eddies. However, individually, each eddy may have completely disintegrated, and thus lost its coherence at this time so that these conditionally sampled structures do not represent physical coherent structures in the flow.

This is an illustration of one of the pitfalls of conditional sampling techniques: the creation of apparent or "False Structure" by an unsuitable combination of measurement, signal processing and interpretive techniques. Similar interpretation problems can arise when using conditional sampling with point measurements of velocity. For example in a situation where signals at two points are completely uncorrelated there is still a finite joint probability for two desired criteria to occur simultaneously in each signal. It is thus important to differentiate between chance agreements between multipoint signals and a true relation between signals produced by structured flow.

5.2. Measurement Techniques

Experimental measurements in CS – containing flows can be divided into three stages (I) preliminary search for evidence of CS's, (II) development of discrimination/trigger techniques, and (III) application to measure the structures and distributions of CS's.

In general it is proposed that coherent structure investigations should include two aspects; flow visualizations and point measurements with good spatial resolution and frequency response. In the, presently unattainable, ideal experiment, fixed point measurements would be unnecessary, as a quantitative optical technique would be used to simultaneously examine large volumes of the flow as a function of time. Approaches to this ideal may eventually be made by using, for example, holography of marker particles together with an image analysis computer. However, the case is considered here only of fixed probes, as the principal quantitative tool, together with flow visualisation in a essential interpretive role.

Preliminary Studies

Flow visualisation cine-photography is an essential first technique to apply for the study of coherent structures in turbulent shear flows. The technique must, however, be carefully designed to suit the flow, and the aspect of the flow, of interest. For instance, eddies can best be tracked by marking them at their origins (injection of dye at a jet nozzle lip or hydrogen bubbles at a boundary layer wall). Instantaneous flow three-dimensionality creates flow visualization difficulties. Slit lighting of flow cross-sections can help overcome this. However, the line of sight integration properties of schlieren-type methods can be useful for indicating large scale structure in flows which appear highly three-dimensional. Water models can be useful as high Reynolds numbers are achieved with low velocities and quantitative marker techniques, such as hydrogen bubbles, can be exploited. Successful flow visualisation can not only provide indications of the most fruitful regions for investigation, but it can also give quantitative data on eddy passing frequencies, etc.

Development of Discrimination and Triggering Techniques

The earliest applications of conditional sampling in turbulent shear flow concentrated on the intermittency conditioning of data. Conditioned averages were derived for "turbulent" and "laminar" portions of signals or point signals were ensemble averaged at times relative to an assumed "viscous superlayer". In a coherent structure containing flow, this type of conditioning yields only very restricted information which can be misleading. For example, a turbulence intensity conditionally averaged for "turbulent" regions, will have contributions from signal variations due to the "phase scrambling" effects, i.e. variations in eddy structures and behaviours. On the other hand, conditional sampling experiments suitably designed to measure the structures, movements and statistics of coherent eddies, should automatically be capable of providing information on intermittency conditioned quantities, if this is considered to be of interest. The eddies are responsible for movements of the laminar/turbulent interface, where such an interface exists. However, the discrimination methods derived for intermittency sampling have an important place in current and future, fuller conditional sampling investigations.

If indications of structures are found in flow visualisations it is useful to film the flow at the same time as point measurements are made (with superposition of signals on the films). Observed flow patterns are correlated with the signals and suitable triggering criteria for sampling the structures are deduced. If structures are not visible, triggering or discrimination criteria must be developed, solely by the examination of signal records (pattern recognition). Particular care must then be taken that recovered signals are large enough to be truly representative of coherence and that a physically unrealistic false structure is not being created by the technique.

The triggering criteria, as has been discussed, should not only be sensitive to the occurrence or passage of an event or structure, but it should also be capable of discriminating between classes of structure as specified by variables such as size, strength and position. For the case of large eddies in shear flows, there are a range of useable techniques including for the velocity field: (I) *Signal Peaks,* associated with eddies as they pass the probe. Triggering is facilitated by first filtering the signals to remove the incoherent small scales. (II) *Potential Flow Peaks* in the near field associated with the passage of eddies. (III) *Interface Detection,* using turbulent/laminar criteria developed for intermittency sampling. Interfaces can also be detected optically, or by sampling probes, when flows are seeded with a passive scalar. Other variables can be used for devising triggering criteria when considering flows with other than isothermal, homogeneous and incompressible conditions. Thus near, or far, field sound amplitude fluctuations can be utilised for higher velocity flows, temperature or flame ionisation fluctuations have been used as triggers in burning flows and density fluctuations, measured by beam deflection techniques, have been studied for heterogeneous mixing.

It seems unlikely that triggering criteria, with the required levels of discrimination, can be derived from single point signals. However it is an essential step to make detailed studies of signals, at different positions, to attempt to identify signal features with flow patterns. Such signal characteristics are; the duration of turbulent "bursts", and their relation with eddy scale and position, the relation between signal peaks and eddy amplitude and position, and for example, the relative phase and amplitude characteristics of local u and v fluctuations. The possible complex variations in eddy structures require complex criteria for recognition of local individual eddy characteristics. This, in turn, implies the simultaneous use of several probes to supply the data from which the characteristics of each eddy (at $t' = 0$) can be deduced. An ideal is the use of an array of closely spaced probes across the flow. Characteristics of the simultaneous signal time histories can be examined to recognise repetitive patterns and measures of eddy scale, shape and position. Three-dimensionality requires that the array should

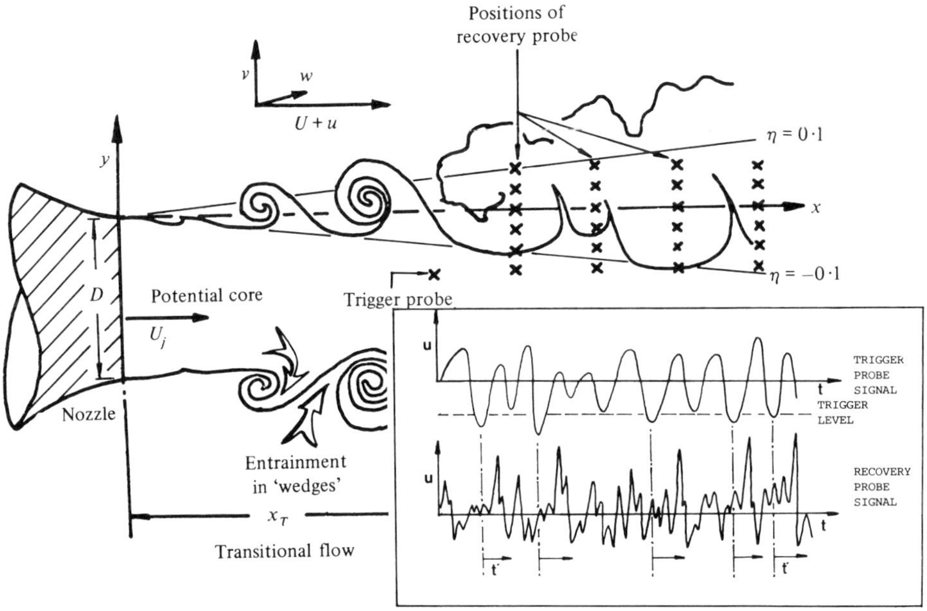

Fig. 6. Conditional sampling in the mixing layer of a round jet

be, at least, two-dimensional, in the plane $x' = 0$. The use of such an array to look for certain flow patterns is of course, in itself, a powerful from of condition sampling, even though it is introduced here as a means of discriminating between different classes of structure which are sampled again further downstream. Practical considerations may necessitate only a small number of probes being used for eddy discrimination. One can envisage, for example, signals from two interface detecting probes used to characterise eddy shape and size, plus a velocity measuring probe to characterise eddy strength.

Fixed probe techniques for coherent structure measurement can be roughly divided into two approaches which use either (I) Arrays or matrices, of many measurement points, or (II) Small numbers of trigger and recovery probes so that the recovery probe position is varied and ensemble averages are taken at each position. Figure 6 illustrates the second approach in which eddies are detected and measured in the mixing layer of a round jet [6]. The trigger events are large velocity peaks in the jet potential core and these are assumed to correspond to large eddies which move particularly close to the centre line. Thus phase scrambling effects are reduced by selecting a fairly narrow class of eddies. The second probe is moved to a matrix of points where time histories are ensemble averaged at time delays relative to the trigger events.

An array of simultaneous measurement points is preferable if it can be physically achieved. Signals simultaneously acquired from a three-dimensional matrix of probes should, in theory, be processable to derive the flow fields of the structures, their variations as a function of time and position, and the probability distributions of the important parameters describing them. Probe intrusion is a major problem, although multiple-measurement volume Laser Doppler Anemometry appears to be a technique which should overcome this. The capabilities of modern computers are now sufficient to acquire significant lengths of time history at acceptable data rates from, say, 50 probes. However, remembering that turbulence structure is strongly three-dimensional, one would need of the order 1,000 measurement positions, at separations $l_c/10$, to reasonably characterise the instantaneous flow structure. Even more would be re-

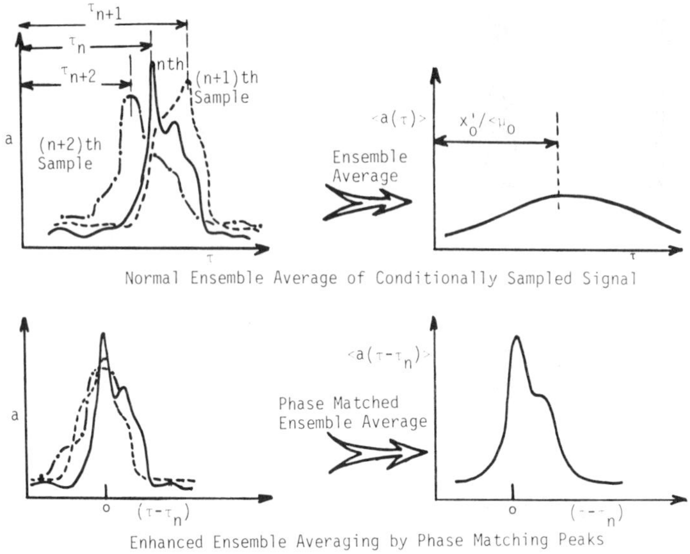

Fig. 7. Ensemble averaging using phase matching

quired to follow each structure downstream. For this reason it is likely that conditional sampling using small numbers of probes will continue to be used extensively. Eduction of the structures and variations of eddies is then achieved by careful design of the conditioning procedures.

Figure 7 illustrates a data processing technique for both enhancing recovered signals and also for quantifying phase scrambling effects. Essentially repetitive patterns, which arrive at different times at the recovery probe, produce a severely attenuated ensemble average. The Phase Matched Ensemble Average is obtained either by phase shifting each pattern so that a certain representative feature always coincides, or by shifting each individual pattern to give the best fit with the normal ensemble average. Phase shifting to give best fit with the normal cross-correlation is a possibility, but this assumes that the cross-correlation is dominated by the sampled structures. Variations of this phase matching technique, using shifting of patterns in both space and time and multiple recovery probes, can be designed for quantifying all of the phase scrambling variables which have been discussed. A difficult, but not intractable, problem in this type of measurement is understanding and quantifying three-dimensional effects. Techniques must be designed to differentiate between "spanwise" (z' direction) length scales of structures and phase scrambling effects which can reduce the apparent scales. In addition, constributions to losses in recovered signals come from movements of structures in all three dimensions (not simply in the $x'y'$ plane).

5.3. Signal Enhancement

It has been described how "phase scrambling" attenuates ensemble averaged data and how conditional sampling, and other techniques, should be used to quantify this effect and extract the true eddy structure by, for example, selecting narrow classes of repetitive eddy. An alternative approach, which has not received attention in the past, is to accept that ensemble averaged data have suffered attenuation by the various scrambling phenomena, and then attempt to correct the data by making use of independent information on "jitter" etc. The problems

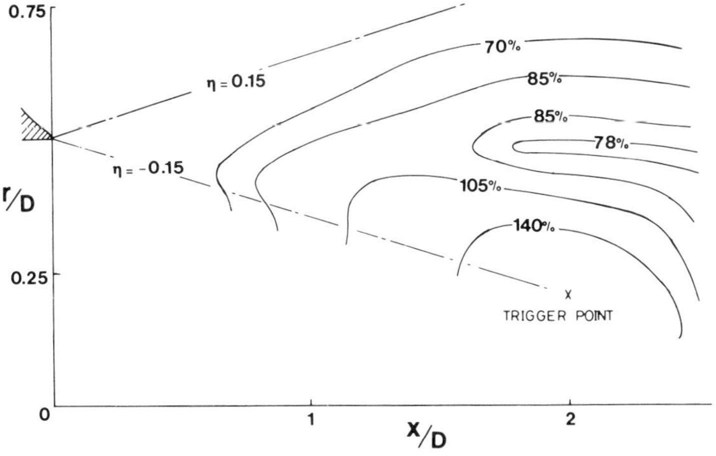

Fig. 8. Percentage of local rms signal recovered by ensemble averaging in round jet, Re = 10,000

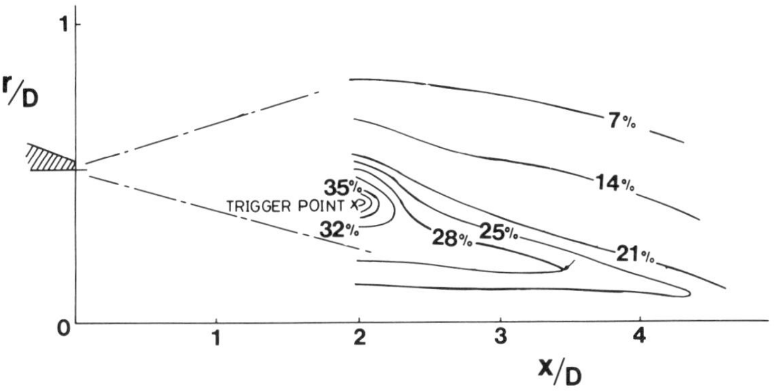

Fig. 9. Percentage of local rms signal recovered by ensemble averaging in a turbulent jet, Re = 2×10^5

of signal attenuation currently found in conditional sampling experiments are illustrated by Fig. 8 and 9. The figures also demonstrate the relative ease with which conditional-sampling techniques can be used in transitional flows. Figure 8 shows the contours of the approximate percentage of the local velocity signal recovered for a matrix of recovery positions in a transitional round jet (Re = 10^4) with triggering using a hot wire in the potential core, as shown in Fig. 6. The percentage of local signal amplitude recovered by the ensemble averaging is expressed as $100\langle \hat{u} \rangle / \sqrt{2}\, (\overline{u^2})^{1/2}$. The contours exceeding 100% near the trigger position indicate that the sampled eddies give unusually large contributions to the turbulence fluctuations in this region. Significant signals are recovered for most of this transitional mixing layer region, where the flow is dominated by regular and orderly vortex rings.

Figure 9 shows the signal recovered in the same jet at a higher Re (2×10^5) and it is seen that the peak amplitude of the ensemble averaged signal now decreases very rapidly with increasing distance from the trigger position. A fundamental problem when interpreting data such as these, is to differentiate between effects, on ensemble averaged signal amplitude, of actual losses in coherence of eddies and the smearing effects of phase scrambling. For example can one interpret the contour peaks, on the inner side of the mixing layer in Fig. 9 as indicative of maintenance coherence in this region, do they result from a more repetitive flow in this region or do the eddies move, on average, towards the centre of the jet?

In theory, if the phase scrambling variables and their probability distributions are known, one could correct measured ensemble averages by inserting a measured ensemble average velocity field into equations, such as equation 7, and solving them numerically. In practice this is likely to have difficulties and involve the inversion of large matrices. This enhancement of smeared signals is akin to the computer enhancement of blurred or distorted images in modern digital image analysis techniques.

Rather than obtaining full solutions to the phase scrambling equations, in general the first order effects of phase scrambling can be described by simpler equations which compensate for the major effects of smearing. For example, for the one-dimensional x' direction jitter of identical eddies $a_c(x' - x_0')$ assumed in Sect. 4 above, one can show that a first approximation to the attenuating effects of jitter can be described by a relation of the form:

$$a_c(x', t') = \langle a_c(x')\rangle - \theta^2 \partial^2 \langle a_c(x')\rangle / \partial x'^2 \tag{11}$$

where θ is related to the standard deviation σ of the jitter by an equation which is dependent on the assumed p.d.f. of the eddy position. For a cosine bell or Gaussian p.d.f. θ is within 5% of σ.

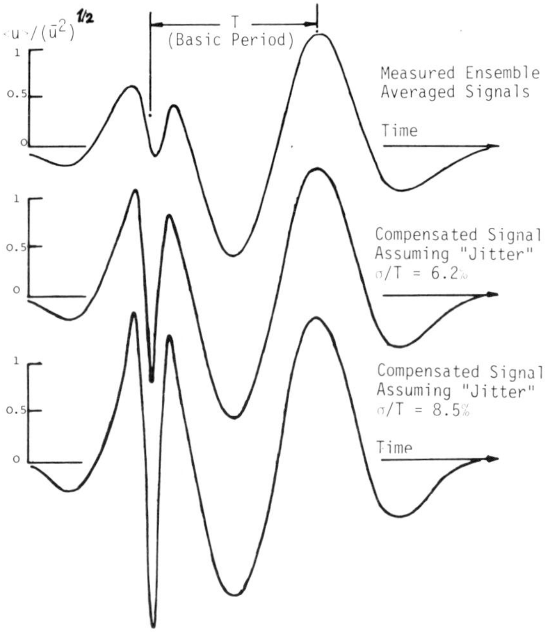

Fig. 10. Calculated effect of jitter on an ensemble averaged signal measured in the mixing layer of a round jet, Re = 10,000

Figure 10 shows a typical ensemble averaged signal [7] in the transitional region in the jet corresponding to Fig. 8, at $x = 2D$, $\eta = -0.025$. Also shown are the effects of two assumed degrees of eddy jitter when this signal is "compensated" by using an equation of the form of Eq. (11). There is no evidence of the actual degree of jitter in this flow but it can be seen that quite small standard deviations produce a significant change in the signal in the vicinity of the first small dip in the signal. This region is likely to correspond to the viscous cores of the sampled vortex eddies. These cores provide the high wave number component of the flow and are thus most likely to be smeared by the effects of jitter.

Acknowledgments. Research in the area of coherent structures in turbulent flames is supported at Sheffield by the Air Force Office of Scientific Research/AFSC, United States Air Force, under Grant AFOSR-77-3414 and the U.S. Office of Naval Research (Project SQUID), subcontract No. 8960-30.

References

1. P. O. A. L. Davies, A. J. Yule: "Coherent structures in turbulence". J. Fluid Mech. **69**, Part 3, 513–537 (1975)
2. J. Laufer: Annu. Rev. Fluid Mech. **7**, 307 (1975)
3. A. A. Townsend: *The Structure of Turbulent Shear Flow,* 2nd ed. (Cambridge University Press, England 1976)
4. W. C. Reynolds: "Computation of turbulent flows". Annu. Rev. Fluid Mech. **8**, 183–207 (1976)
5. N. A. Chigier, A. J. Yule: "The Structure of Eddies in Turbulent Flames – 1". Project Squid Tech. Report (March 1979)
6. A. J. Yule: "Large scale structure in the mixing layer of a round jet". J. Fluid Mech. **89**, Part 3, 413–432 (1978)
7. H. H. Bruun: "A time-domain analysis of the large-scale flow structure in a circular jet. Part 1. Moderate Reynolds number". J. Fluid Mech. **83**, Part 4, 641–671 (1977)

Mode Expansion of Coherent Structures in the Wake of a Circular Disk

H. V. Fuchs
Fraunhofer-Institut für Bauphysik, D-7000 Stuttgart 70, Fed. Rep. of Germany

E. Mercker[1]
University of Adelaide, South Australia 5001, Australia

U. Michel[2]
DFVLR-Abteilung für Turbulenzforschung, D-1000 Berlin 12, Fed. Rep. of Germany

Abstract

The apparently turbulent shear flow behind a circular disk in a cross flow is shown to be governed by quasi-periodic coherent motions. Their scale and structure bear an amazing similarity with those observed in Karman vortex streets behind cylinders in a cross flow. The main difference lies in a random oscillation of the orderly flow patterns around the axis of symmetry. This spatial (and the remaining temporal) randomness in, e.g., the axial velocity fluctuations calls for a cross-spectral analysis of signals detected at azimuthally displaced points in the wake. The specific symmetry imposed by the axisymmetric boundary conditions at the disk also enables a very illustrative description of the large scale flow phenomena by relatively simple, elementary modes. The resulting azimuthal constituents are quantified, by experiment, in terms of their respective magnitudes and individual power spectra. At what may be termed the "vortex shedding" frequency in analogy to the plane wake case, the most dominant mode is that for which the axial velocity and the aerodynamic pressure fluctuations on one side of the wake are in an exact anti-phase relationship with the respective signals on the other side.

1. Introduction

A remarkable agreement on the existence of large scale coherent structures in turbulent shear flows could be registered among those contributing to the present symposium. Numerous observations of quasi-deterministic phenomena in apparently random turbulence have already created widely differing techniques of identifying and analysing them. This may not surprise after certain guide-lines set by "classical" statistical turbulence models had to be abandoned.

It would seem premature to call for a unified approach towards all the different forms of appearance of coherent structures in shear flows, bounded and unbounded. Yet, after so many successful attempts of investigators to find their own way to these phenomena with a minimum of alterations in their conceptual models, a need for a more systematic and quantitative description of these structures becomes evident. The method applied herein is based on the following assumptions which were verified by experimental cross correlations of directly measured fluctuating flow properties:

a) The structures to be described are continuously present over the whole of the flow region.
b) At any given location in the flow the fluctuations measured are dominated by the large-scale motions.

[1] Present address: Technische Universität, D-1000 Berlin 12
[2] Present address: NASA Langley Research Center, Hampton, Va. 23665, U.S.A.

c) The time history of each turbulent signal is basically random.
d) No external forcing is applied to the flow to enhance the order or amplify the magnitude of the structures.

In particular, the turbulent signals need not be conditionally sampled in any way to obtain a clear-cut picture of the large-scale unsteadiness of the flow. That means, the resultant representation of coherent structures is by straightforward statistical averages of their strength, coherence and phase variations rather than by the history, life-time, and shape of pre-selected typical "eddies" or isolated "events" in the flow.

The structures are analysed over the whole range of relevant frequencies because discrimination against small-scale turbulence by low-pass filtering the turbulent signal is felt to introduce an unnecessary arbitrariness.

Table 1. Coherent structures in free turbulent shear flows

	Jets	Wakes
Plane (2 d)		
Round (3 d)	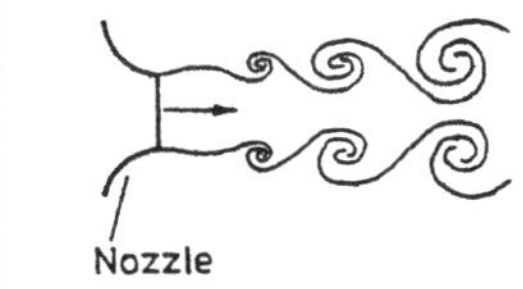	

Coherent structures are of particular interest in four classes of free shear flows a representative example of each class having been sketched in Table 1. Only for simplicity have the structures been drawn schematically as if they always resembled vortices following the overcome view of turbulent structures or eddies representing coherent regions of rotational flow. In reality, however, coherent structures in, e.g., a 3 *d* jet may look much more complicated than the axisymmetric toroidal ring vortices indicated. In fact, the mode expansion scheme applied below to a 3 *d* wake will be seen to yield a whole set of different modes of fluid motion.

The mode expansion scheme has already been used to describe the relative efficiency in radiating *noise* of the different modes of motion a circular jet can perform [1].

In the present context of wake flows it may be noted that the lowest-order modes can be directly related to the fluctuating drag and lift *forces* exerted on the body.

2. Historical Background

Orderly structures have been observed a long time ago in connection with the instability of thin shear layers.

Figure 1 shows how ring vortices are formed on a laminar shear layer shed from a so-called vortex-filament nozzle. The fixed-point frequency induced by these naturally excited instabil-

Fig. 1. Smoke photograph of ring vortices in the initial region of a circular jet at Re = 70,000 after *Wille* [2]

ities was found in [2] to scale on the boundary layer displacement thickness $\delta+$ and the velocity U at the nozzle exit,

$$St_\delta^+ = \frac{f\delta^+}{U} = 0.023 \qquad (1)$$

At that time, the observed structures were solely attributed to the so-called laminar/turbulent transition process. It was considered as the first part of a continuous transition from a relatively ordered to a fully turbulent state of the flow. Tripping the boundary layer was shown to destroy much of this order (compare Figs. 2a, b).

It was then not anticipated that another instability mechanism of the steadily broadening turbulent shear layer gives rise to a new set of large structures further downstream. The quasi-deterministic nature and an amazingly strong coherence in the fluctuating pressure and velocity fields induced by these structures in a high Reynold number round jet were first emphasized in [3].

At a fixed point within the jet or its near field the dominant frequency of these self-generated structures scales on the jet exit diameter D, e.g. at $x = 3\ D$ downstream

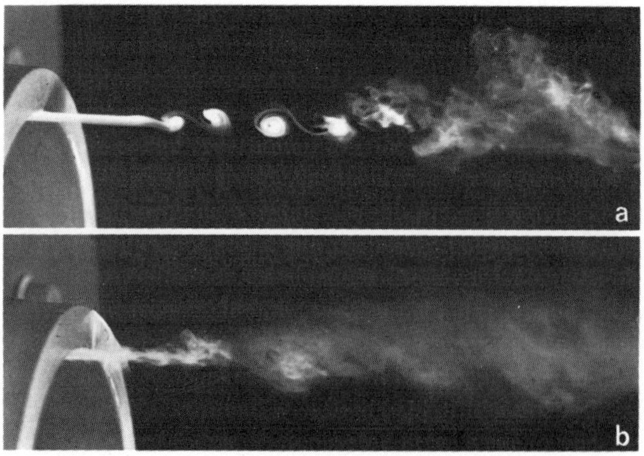

Fig. 2a, b. Effect of tripping the boundary layer at the nozzle exit on the rolling-up of the shear layer after *Wille* [2]. **(a)** Separated laminar shear layer. **(b)** Separated turbulent shear layer

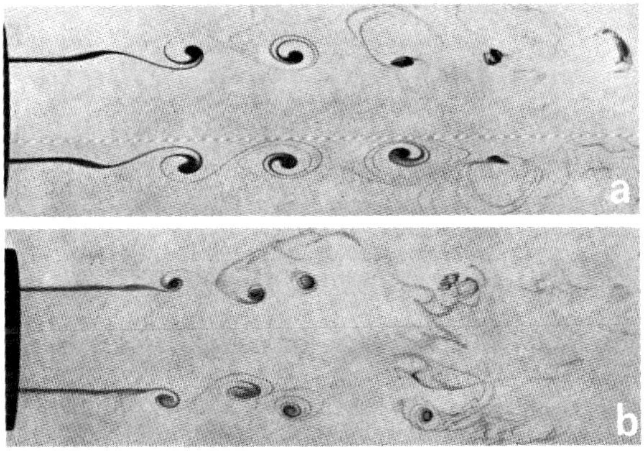

Fig. 3a, b. Dye photographs of the formation of vortices at different Re after *Wille* [2].
(**a**) Matching of initial and developing structures, Re = 1,780. (**b**) Pairing process, Re = 3,300

$$St = \frac{fD}{U} = 0.45. \tag{2}$$

Since δ^+ in Eq. (1) depends on Re according to

$$\delta^+/D \sim (\mathrm{Re})^{-1/2} \tag{3}$$

there is a certain range of lower Re where the frequency of the initial shear layer instability is of the same order as that due to the structures further downstream. This may be the reason why at low Re a very regular train of vortices is observed travelling with little distortion along the jet (Fig. 3a). At somewhat higher Re (Fig. 3b) pairing of neighbouring structures occurs thus enabling an adjustment of the then smaller initial scales to the larger scales at several D downstream. This adjustment by pairing is no longer possible at very high Re where the scales of both phenomena differ by orders of magnitude.

We may thus differentiate between different regions of flow development in free turbulent shear flows:

1. Separation of laminar shear layer
2. Rolling-up of initial shear layer
3. Pairing of vortices under certain conditions
4. Domain of large-scale coherent structures
5. Fully developed turbulent flow

Instead of what is usually termed "laminar/turbulent transition" we have thus defined a whole series of transitions between different states of the flow in which coherent structures play an important part. Three types of flow instability are discernible: that of the laminar shear layer, that of the developing shear layer and that of the large-scale structures themselves.

To indicate that the described phenomena are not a peculiarity of jet flows only, Fig. 4 shows the initial phase of regular flow disturbances on the shear layer separating from a sphere at low Re. The large structures developing downstream in the wake are shown in Fig. 5. In his excellent review paper *Roshko* [5] showed large-scale structures in a two-stream (helium/nitrogen) plane mixing layer (Fig. 6). Evidence for organized structure in the far region of a turbulent round jet is reproduced here in Fig. 7.

The illustrations in this paragraph were to show, in a qualitative manner, that with every shear flow are associated identifiable, characteristic structures. We have reason to believe that the

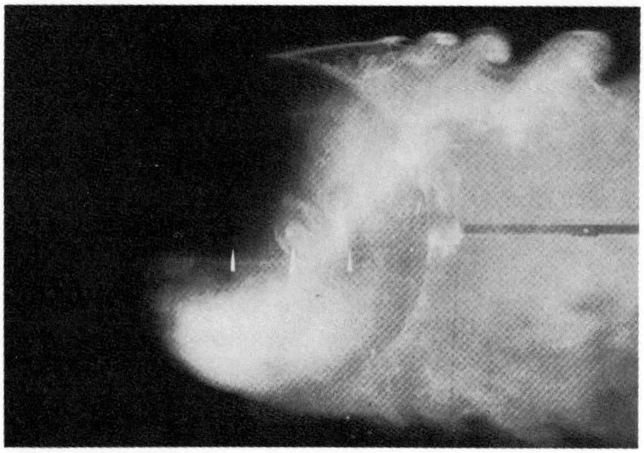

◀ **Fig. 4.** Smoke photograph showing roll-up of initial shear layer shed from a sphere at low Re = 23,000 after *Taneda* [4]

Fig. 5a, b. Simultaneous smoke photographs of large-scale coherent structures in the wake of a sphere at Re = 23,000 after *Taneda* [4]. (**a**) Top view. (**b**) Side view

▼

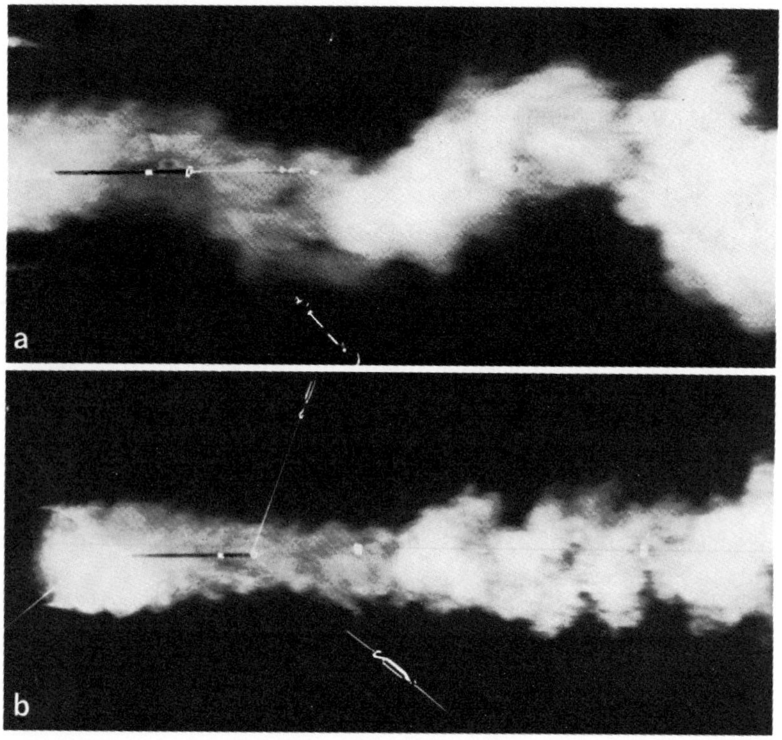

flow development is controlled by interactions of these structures with each other. When the free shear flows described interact with solid bodies a variety of additional structures are likely to be generated. Some of the structured flow phenomena may be enhanced (or paralysed) by external forcing or triggering, see e.g. Refs. [6] and [7].

A quantitative description of the large-scale phenomena in turbulent flows may be based on two-point correlations or their corresponding Fourier transforms rather than on conventional local intensity and spectrum analyses. The method described below is particularly suited for flows with an axisymmetric mean flow geometry.

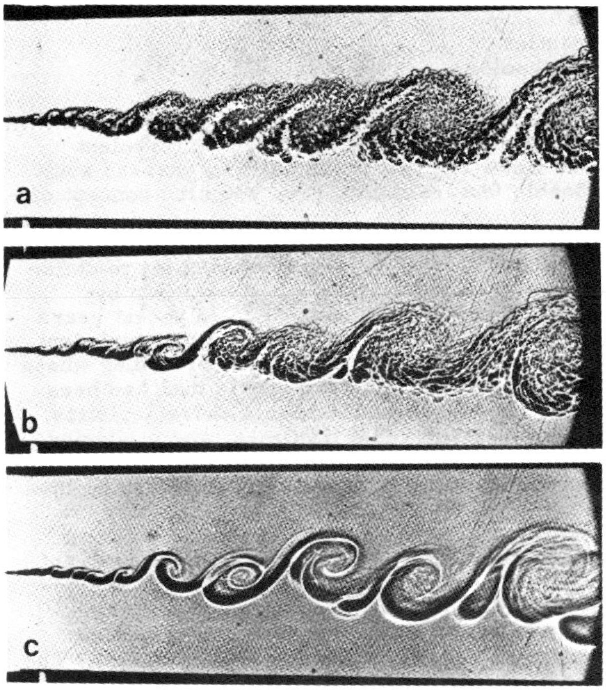

◀ Fig. 6a–c

▼ Fig. 7

Fig. 6 a–c. Organized structures as visualized in a two-stream plane mixing layer after *Roshko* [5]. Velocity ratio 0.38. (**a**) Re = 120,000. (**b**) Re = 60,000. (**c**) Re = 30,000

Fig. 7. View upstream in a wind tunnel of a coflowing jet with plane of symmetry illuminated by a vertical sheet of light after *Roshko* [5]. Velocity ratio 1.5, 80 jet dia. downstream, jet Reynolds number Re = 3,500

3. A Quantitative Description of Coherent Turbulence Structures

Ever since *Townsend* [8] put forward his "big eddies" idea attempts were made to condense more than just vague qualitative information on typical eddy structures from the shape of measured two-point correlation functions of the turbulent velocity field. Most of the experimental techniques applied, however, did not allow a strictly quantitative analysis of coherent structures in turbulent shear flows. Our own approach is fairly fundamental and almost universally applicable, although its specific virtue could, up till now, be demonstrated for axisymmetric free shear flows only. It is based on observations that certain flow fields are continuously dominated by large-scale coherent motions which are self-generated by the flow. It does not make use of any kind of conditional averaging either.

Under these fairly general conditions a maximum of useful information about the space-time structure of a turbulent field can be extracted from the complex two-point cross power spectral density function

$$R_\omega = C_\omega + iQ_\omega \tag{4}$$

where both the coincident and the quadrature spectra, C_ω and Q_ω, may be functions of the moving as well as of the fixed probe positions, indices 2 and 1. For a most convenient description of the coherent phenomena in turbulence the total information is suitably subdivided into three categories according to the following expression,

$$R_\omega = \tilde{p}_{\omega,1}\tilde{p}_{\omega,2} \frac{|R_\omega|}{\tilde{p}_{\omega,1}\tilde{p}_{\omega,2}} \exp(\mathrm{i}\psi_\omega), \tag{5}$$

with

A) $\tilde{p}_\omega = (\overline{p_\omega^2})^{1/2}$

denoting the local power spectral density, or strength, of the turbulent property measured,

B) $\gamma = |R_\omega|/(\tilde{p}_{\omega,1}\tilde{p}_{\omega,2}) = (C_\omega^2 + Q_\omega^2)^{1/2}/(\tilde{p}_{\omega,1}\tilde{p}_{\omega,2})$

denoting the degree of statistical correlation, or coherence, of signals at two distant points in the flow,

C) $\psi_\omega = \overline{\psi_{\omega,2} - \psi_{\omega,1}}$

denoting an averaged phase relationship between both signals.

Category A) is normally dealt with in conventional turbulence studies with $\tilde{p}_\omega$ or $\tilde{p}$ suitably normalized with the appropriate mean flow properties. Category C) contains all the information concerning the convection of disturbances by the flow. In this paper we shall focus our attention on mainly the coherence properties (B) and restrict our analysis to one specific class of flow configurations.

Strictly axisymmetric flow conditions allow three very decisive assumptions to be made. If, to start with, both probes are located on a constant radius r in a plane x = const normal to the axis of symmetry one may easily convince oneself that for circumferential displacements $\Delta\phi$

1) Q_ω is identically zero within experimental error provided that no mean swirl is superimposed on the flow, i.e. there is no preferred helical propagation for the disturbances while travelling downstream so that $\psi_\omega = \arctan Q_\omega/C_\omega = 0$. Hence we are dealing with the real function $R_\omega = C_\omega = R_\omega^*$.
2) For similar symmetry reasons, C_ω is symmetric in $\Delta\phi$. i.e. $C_\omega(-\Delta\phi) = C_\omega(\Delta\phi)$.
3) The most important feature of axisymmetric flows, however, is the spatial periodicity inherent in circumferential correlation functions at any arbitrary radius, $C_\omega(\Delta\phi + n\,2\,\pi) = C_\omega(\Delta\phi)$.

Under these very special conditions we may expand C_ω

$$C_\omega(\Delta\phi) = \sum_{m=0}^{\infty} C_{\omega,m} \cos(m\Delta\phi). \tag{6}$$

With its value at $\Delta\phi = 0$,

$$C_\omega(0) = \sum_{m=0}^{\infty} C_{\omega,m} = \overline{p_\omega^2} = \sum_{m=0}^{\infty} \overline{p_{\omega,m}^2} \tag{7}$$

it is made possible to define the fraction of fluctuating energy contained in each individual azimuthal constituent of order m by

$$\frac{\overline{p_{\omega,m}^2}}{\overline{p_\omega^2}} = \frac{C_{\omega,m}}{C_\omega(0)} . \tag{8}$$

Co-spectra C_ω are readily obtained with the aid of a digital computer by using a Fast Fourier Transform routine thus enabling several different kinds of representation of turbulence structure which were discussed in more detail in [9].

4. Experimental Results

The normalized velocity power spectrum,

$$\frac{\tilde{u}_\omega^2}{\tilde{u}_{\omega_{max}}^2} = f(St), \tag{9}$$

is seen in Fig. 8 to peak around $St_0 = 0.135$ in accordance with *Roberts'* [10] data. This characteristic Strouhal number does hardly vary in the flow region between $x = 3\,D$ and $9\,D$. The bulk of the fluctuating energy (about 40%) is contained in a rather narrow range of Strouhal numbers, $0.1 < St < 0.17$.

The periodicity around $St_0 = 0.135$ is also visible in the fluctuating pressure at $x = 3\,D$ (Fig. 9). These spectral analyses suggest that underlying the apparent chaos in axisymmetric wakes, there may be a well-ordered turbulence structure of a similar nature as in a Karman vortex street. A possible reason for their not having been observed in the past may be that the anti-symmetric structure, which so obviously dominates the vortex street in a plane normal to the cylinder does not find a similarly pre-determined orientation in the wake of an axisymmetric body as it does in the cylinder wake. While the resulting randomness will presumably render visualizations of the structures as difficult as in a round jet, it leaves our quantitative approach entirely unaffected since this is based on straight-forward long time averaged covariance measurements.

We commenced our analysis of axisymmetric wake turbulence by determining its coherence γ according to Eq. (5) as a function of St. For measuring positions facing each other across the wake ($\Delta\phi = 180°$) we found γ values above 0.85 at the vortex shedding frequency f_0.

In Fig. 9 the magnitude of the pressure cross spectrum, $S_\omega = |R_\omega|$, normalized with its maximum value at St_0, is plotted as

$$\frac{S_\omega}{S_{\omega_{max}}} = f(St), \quad \Delta\phi = 180° \tag{10}$$

along with the corresponding power spectrum of the pressure. One may clearly distinguish three easily separable frequency regimes:

(I) near f_0 the cross spectrum exhibits a peak which is exceptionally narrow indicating a particularly strong periodicity and coherence of the vortex shedding mode.

(II) at twice the vortex shedding frequency, $2f_0$, a second though less clear-cut peak is discernible. This may be taken as a first harmonic of the vortex shedding mode which could

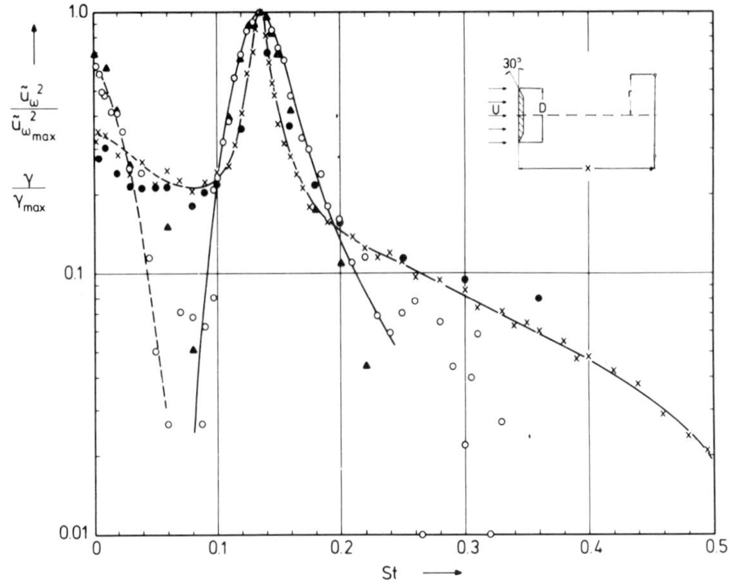

▲

Fig. 8. Power spectra and coherence spectra of streamwise velocity fluctuations in the wake of a disk. $D = 50$ mm, $U = 15$ m/s; x Power spectrum at $x = 9\,D$, $r = 0.83\,D$; ○ Coherence spectrum, $\Delta\phi = 180°$; (●, ▲ *Roberts'* [10] data)

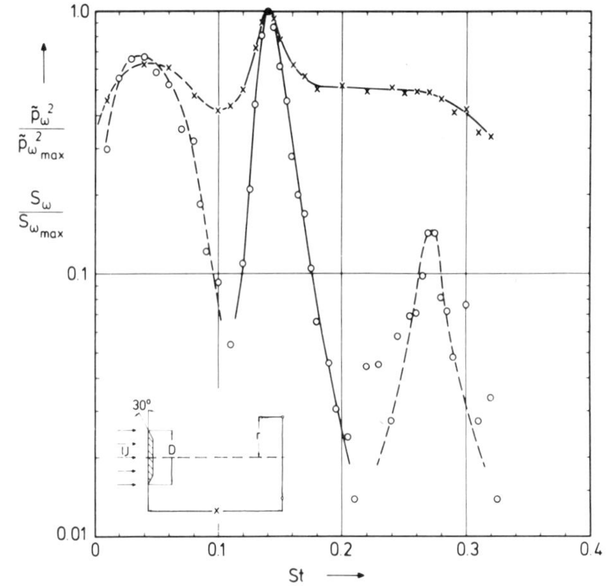

Fig. 9. Shape of power spectrum and cross spectrum of pressure fluctuations. x Power spectrum at $x = 3\,D$, $r = 0.75\,D$; ○ Cross spectrum, $\Delta\phi = 180°$

not be identified in the power spectrum of the pressure. Between these two peaks the cross spectrum drops to very small values indicating that outside these characteristic frequency regimes the underlying "base" cross spectrum is very small indeed.

(III) a third broad hum is seen in Fig. 9 to arise at frequencies well below f_0, which also manifests itself in the corresponding power spectrum. Its origin is up till now less obvious than that of the two other peaks. It may well be that this low frequency-coherence in the wake is somehow tied up with large-scale turbulence structures of the wind tunnel flow.

In the corresponding cross spectra of the axial velocity fluctuations the first regime is even more pronounced than that in the pressure with $S_\omega/S_{\omega_{max}}$ falling below 0.004 and

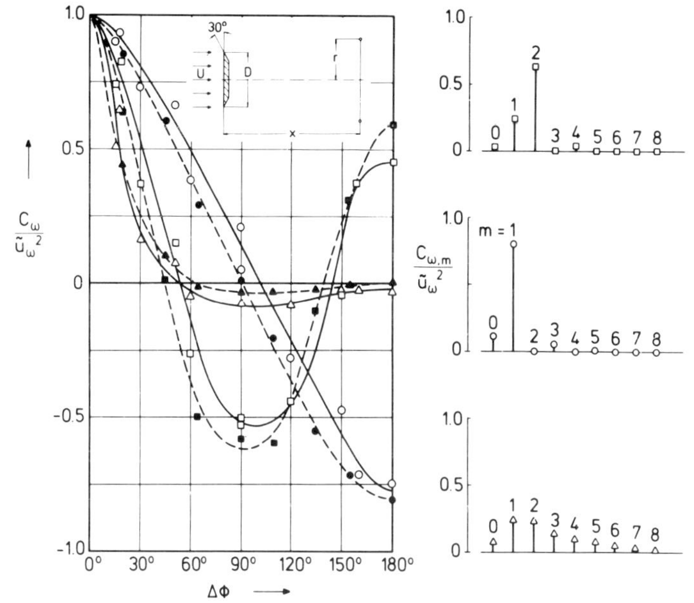

Fig. 10. Circumferential correlation and azimuthal constituents of narrowband streamwise velocity fluctuations at $x = 9\,D$, $r = 0.83\,D$; □ $St = 0.005$, ○ $St = 0.135$, △ $St = 0.270$; (■●▲, corresponding data of *Roberts* [10])

0.008 for $St = 0.065$ and 0.185, respectively. This prompted us to plot the related quantity

$$\frac{\gamma}{\gamma_{max}} = \frac{S_\omega}{S_{\omega max}} \; \frac{\tilde{u}^2_{\omega max}}{\tilde{u}^2_\omega} = f(St); \quad \Delta\phi = 180^\circ \tag{11}$$

in Fig. 8. Again *Roberts*' [10] data at $x = 9\,D$ fit in quite well. The scatter at higher frequencies renders the identification of a peak at $2f_0$ difficult.

In order to resolve the intrinsic structure of fluctuations in the three frequency regimes separately we will now analyse circumferential correlations for St components at 0.005, 0.135, and 0.27 of both velocity and pressure,

$$\frac{C_\omega}{\tilde{u}^2_\omega}; \quad \frac{C_\omega}{\tilde{p}^2_\omega} = f(\Delta\phi), \quad St = \text{const} \tag{12}$$

as depicted in Figs. 10 and 11. Again *Roberts*' [10] correlation data at $x = 9\,D$ are confirmed.

The numerical results of a Fourier analysis of azimuthal velocity and pressure correlations were also depicted in Figs. 10 and 11. The coefficients were plotted as

$$\frac{C_{\omega,m}}{\tilde{u}^2_\omega}; \quad \frac{C_{\omega,m}}{\tilde{p}^2_\omega} = f(m); \quad St = \text{const} \tag{13}$$

for the three frequency regimes defined above.

Not very surprisingly, at $St_0 = 0.135$ it is the $m = 1$ constituent which is, by far, the strongest at any axial position in the velocity as well as in the pressure field. Hardly more than 10% of the fluctuations is contained in any of the other modes. We concludethat the alternating vortex shedding off the upper and lower parts of a cylinder is parallelled by an anti-phase relationship of signals in the 3 *d* wake case. The latter proves the existence of a regular vortex formation in an axisymmetric wake which is consistent with the notion of a

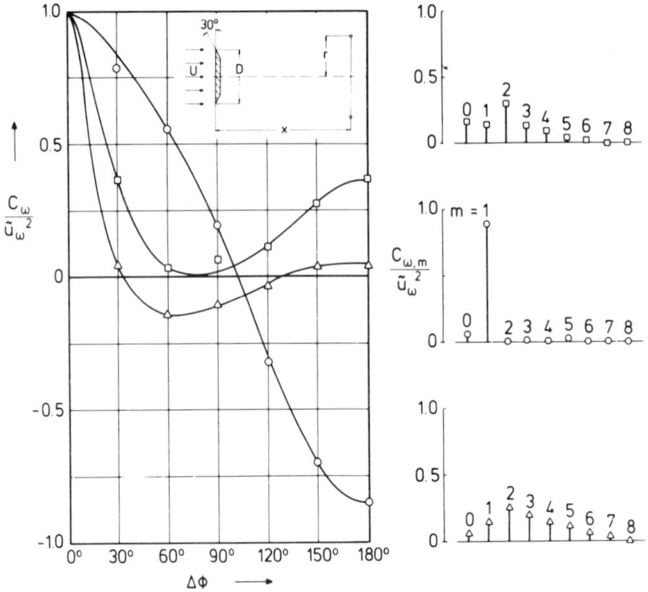

Fig. 11. Circumferential correlation and azimuthal constituents of narrowband pressure fluctuations at $x = 3\,D$, $r = 0.75\,D$.
□ $St = 0.005$,
○ $St = 0.135$,
△ $St = 0.270$

flapping or fish-tail whiggling type of motion of the whole wake. Since, contrary to the situation in the 2 *d* wake, this flapping motion continuously changes its orientation in a random fashion, the correlation technique favoured in our approach is particularly suited for detecting and quantifying it. Approximately 35% of the total fluctuating energy is thus found to be concentrated in the $m = 1$ mode of the wake turbulence.

In the frequency regime far below St_0 the $m = 1$ constituent is no longer a dominant feature of the axisymmetric wake. Instead, we find the $m = 2$ constituent gaining energy from the others as the wake develops further downstream. At $x = 9\,D$ (Fig. 10) more than 60% of the mean square velocities $\tilde{u}_\omega^2$ at $St = 0.005$ contribute to this particular mode which is characterized by an antiphase relationship at points displaced by $\Delta\phi = \pm 90°$ and by an inphase relationship for $\Delta\phi = 180°$. The corresponding correlation curve in Fig. 10 shows this very clearly. It is certainly not "consistent with a slow, random variation in the orientation of a flapping motion" as was erroneously suggested by *Roberts* [10].

At high frequencies the correlation curves in Figs. 10 and 11 show a more conventional behaviour in that they fall almost monotonously with minor negative lobes. These trends reflect the presence of a large number of constituents simultaneously. The corresponding plots of $C_{\omega,m}/\tilde{u}_\omega^2$ and $C_{\omega,m}/\tilde{p}_\omega^2$ indicate only a marginal preference for the $m = 1$ and 2 modes with contributions from $m \geqslant 8$ constituents negligible as in the other frequency regimes.

Our complete set of cross spectral analyses in, e.g., the plane $x = 9$ D enables correlation curves plotted and Fourier analyses performed on them for any arbitrary St. As an interesting variant we may now plot

$$\frac{C_{\omega,m}}{\tilde{u}_\omega^2} = f(St); \quad m = \text{const} \tag{14}$$

with mode number m as a parameter. Figure 12 clearly shows that the $m = 1$ contribution is strongest at $St = St_0$ where the power spectrum $\tilde{u}_\omega^2$ itself reaches its maximum (compare Fig. 8). This contribution gradually decays on both sides of $St = St_0$ whereas the influence of the $m = 2$ mode increases continuously with decreasing frequency.

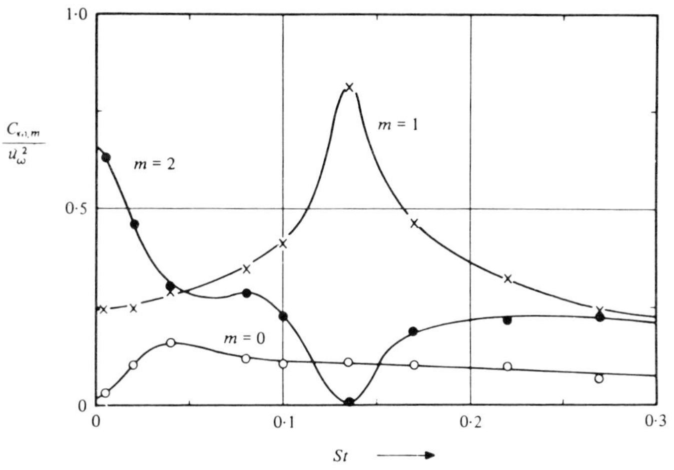

Fig. 12. Relative strength of individual constituents of wake structures. Correlations and Fourier analyses at $x = 9\,D$, $r = 0.83\,D$

Thus the two peaks in the power spectrum of Fig. 8 do not only correspond to frequency regimes where the coherence of signals is exceptionally high (compare the γ/γ_{max} plot in Fig. 8) but they are also seen to be associated with completely different geometrical structures of the wake. If we assume that all coherent structures in a wake presumably travel downstream at roughly the same speed, we might as well conclude that the two dominant structures, $m = 1$ and 2, must have longitudinal scales or wave lengths which differ by, at least, a factor of 10! Obviously, the higher mode $m = 2$ has a much larger axial scale just as if structures of a lower mode are merged in a larger structure of a higher mode on their way downstream.

To conclude this paragraph, we plot normalized power spectra

$$\frac{\tilde{u}^2_{\omega,m}}{\tilde{u}^2_{\omega_{max}}} = \frac{C_{\omega,m}}{\tilde{u}^2_\omega}\,\frac{\tilde{u}^2_\omega}{\tilde{u}^2_{\omega_{max}}} = f(St) \tag{15}$$

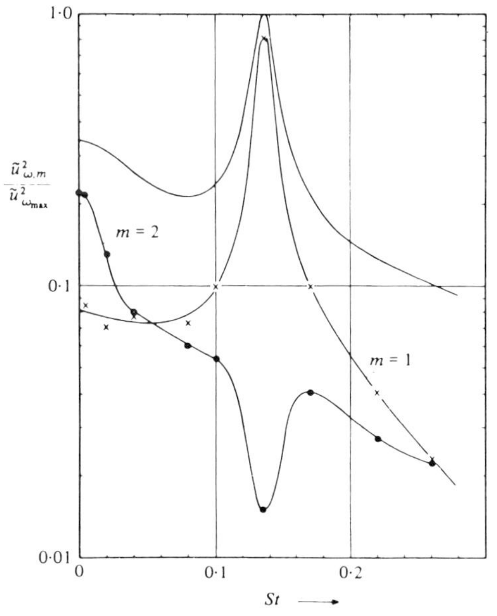

Fig. 13. Shape of power spectra of dominant constituents of wake structures at $x = 9\,D$, $r = 0.83\,D$. *Upper curve* shows overall (unresolved) velocity power spectrum

for the individual constituents m. In Fig. 13 the $\tilde{u}^2_{\omega,1}$ spectrum is seen to be even more narrowband in character than the overall, i.e. unresolved, spectrum $\tilde{u}^2_\omega$. On a logarithmic decibel scale, its peak lies only 1 dB below $\tilde{u}^2_{\omega_{max}}$ while one octave below or above the vortex shedding frequency $\tilde{u}^2_{\omega,1}$ lies 5 or 6 dB below the respective values of $\tilde{u}^2_\omega$. The $m = 2$ spectrum, on the other hand, peaks about 2 dB below the value of $\tilde{u}^2_\omega$ at very low *St*.

The $m = 0$ contribution is always smaller than that of the $m = 1$ and 2 constituents and remains relatively constant over the whole range of frequencies considered here. This is in marked contrast to our earlier findings in a circular jet with and without a cylindrical centre-body (Fig. 14). The modal structure of the jet which is characterized by a strong axisymmetric constituent is hardly affected by the presence of the body or its wake.

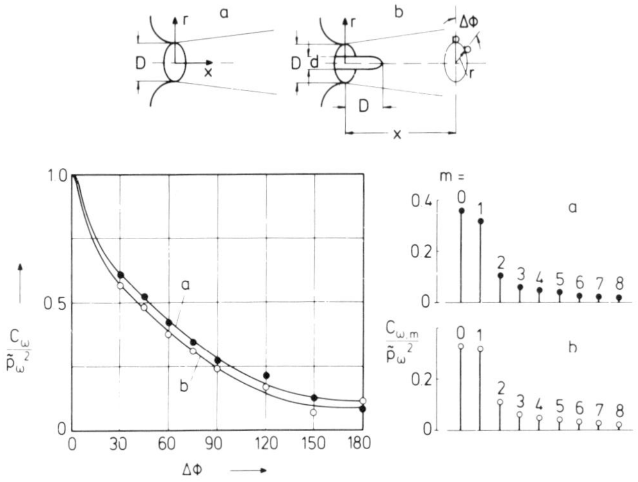

Fig. 14. Circumferential correlation and azimuthal constituents of a narrowband component of the pressure field inside the mixing region of a model jet (*a*) without and (*b*) with an axisymmetric centre-body.
$d/D = 0.4$, $x = 3\,D$, $r = 0.5\,D$, $Ma = 0.15$, $St = 0.45$

After having separated the predominant vortex shedding ($m = 1$) structure from the rest of the turbulence, we may now return to the periodic phenomena in wake flows. Previous investigators like *Hwang* and *Baldwin* [11] were hardly able to identify the main peak in their velocity spectra. *Calvert* [12] says that "it has been known for some time that there are periodic phenomena associated with such flows, but that these are not nearly so predominant as in two-dimensional configurations". Nowhere in the wake was this periodicity sufficiently strong to be identified by either inspection of an oscilloscope trace or smoke observations except at very low Reynolds number. The $m = 1$ spectrum in Fig. 13 reveals the periodicity inherent in the wake by removing, very efficiently, the apparent randomness of its orientation in space which has blurred the picture.

5. Conclusions

The specific space-time structure of the "turbulent" fluctuations in the wake of a disk has been discussed in a more general context in [9].

Our interpretation of the results is, as yet, based on a schematic or symbolic rather than on a physical description of the large scale structures. The elementary constituents of these structures could be determined in a systematic and quantitative manner. No attempt will be made, however, to speculate about the nature and the origin of the geometrically different structures identified in axisymmetric wakes. It is not at all clear at the moment how these could be associated with "horse-shoe type vortices" or "double-helix vortex loops". Nor do we see a possibility to relate our coherent flow structures to any kind of statistical eddy model of the turbulence. We are not even sure whether turbulence researchers would term the large scale phenomena turbulent as we did.

The proposed technique for describing the large scale coherent structures in axisymmetric flows may, with some modifications, be applied to other flow configurations, notably those which allow a Fourier analysis in one (or more) distinguished spatial coordinate like, e.g., in the spanwise or chordwise directions of 2 *d* wakes.

The expansion method is not restricted to ordinary long-time averaged two-point cross correlations. It may equally well be applied to turbulent signals which have been conditioned beforehand such as to reveal details of the structures where this is felt to be essential. This may be necessary when studying, e.g., certain intermittent phenomena in free or bounded shear layers.

Our approach, though generally applicable, will lose its special virtues when dealing with a turbulence field that is something like small scale homogeneous and isotropic in which case it would yield not just a few lower-order constituents but a very large number m of constituents of comparable strength and relevance.

Fortunately enough, an amazingly small number of azimuthal constituents representing relatively simple elementary structures or modes suffice to describe the unsteady flow in axisymmetric jets and wakes. The $m = 0$ and $m = 1$ constituents clearly dominate the initial region of jets at least down to $x = 10$ D. The power spectra of both modes are basically similar with their Strouhal frequency peak varying inversely proportional to the downstream distance x. In the wake behind a disk or a cone, on the other hand, it is the $m = 1$ constituent alone that governs the coherent structures at the vortex shedding Strouhal number St_0 whereas the $m = 2$ constituent dominates at very low frequencies.

Whether one can artificially excite the missing $m = 0$ mode in axisymmetric wakes by means of, e.g., an externally applied sound field is an open question. Destruction of all the periodic wake phenomena seems to be easily achievable; *Fail* et al. [13] report that "perforating the plate can eliminate the regular shedding of eddies and reduce the random low-frequency velocity fluctuations".

Coherent structures in jets and wakes as described above can be imagined to play an important role in

I) Jet exit feed-back mechanisms

The strong $m = 0$ pressure mode may induce corresponding pressure pulsations in phase over the whole of the exit plane. These can be estimated to cause effective mean thrust variations of the order of 0.25–0.5%.

II) Jet impingement forces

Similarly strong or even stronger large scale coherent pressure fields are likely to be built up when a jet impinges on a plane rigid wall. Corresponding surface pressure correlations by *Ho* and *Nosseir* [14] suggest powerful alternating forces to be exerted on the plate as a result of the $m = 0$ fluid loading.

III) Oscillating lift and drag forces

The large structures in a wake can likewise be associated with net forces. An oscillating drag force would, e.g., be exerted on the disk by the $m = 0$ structures. These, however, were not found to be strong in the experiments reported. The $m = 1$ modes which dominate the 3 *d* wake, on the other hand, can be associated with corresponding lift forces exerted on the axisymmetric body. Oscillating moments may also be generated in phase with this vortex shedding mode. The direction of both fluctuating lift and moment is to vary randomly in a plane normal to the mean flow direction. All the higher-order modes, $m > 1$, however, cannot produce equivalent resulting forces on the body.

6. Final Remarks

We hope that we could show that the large structures leave some well identifiable traces on conventional two-point correlations in jets and wakes as P. Bradshaw suggested in his introduction to this chapter. The expansion scheme which was herein applied to a set of azimuthal correlations may be developed further to become "a more rationalised and systematic approach to the experimental study of coherent structures in turbulent shear flow" that *Yule* [15] called for. To achieve this, at least as far as axisymmetric jets and wakes are concerned, one would like to compare more quantitative data from correlations with more qualitative information obtainable from flow visualisations like those by *Taneda* [4], *Pao* and *Kao* [16], and by *Perry* and *Lim* [17].

References

1. A. Michalke, H. V. Fuchs: "On turbulence and noise of an axisymmetric shear flow". J. Fluid Mech. **70**, 179-205 (1975)
2. R. Wille: "Beiträge zur Phänomenologie der Freistrahlen". Z. Flugwiss. **11**, 222-233 (1963)
3. J. C. Lau, H. V. Fuchs, M. J. Fisher: "A Study of Pressure and Velocity Fluctuations Associated with Jet Flows". I. S. V. R. Rep. 28 (1970)
4. S. Taneda: "Visual observations of the flow past a sphere at Reynolds numbers between 10^4 and 10^6". J. Fluid Mech. **85**, 187-192 (1978)
5. A. Roshko: "Structure of turbulent shear flows: a new look". AIAA J. **14**, 1349-1357 (1976)
6. A. Michalke, O. Wehrmann: "Akustische Beeinflussung von Freistrahlgrenzschichten". Proc. 3rd ICAS Congr., Stockholm (1962) pp. 773-785
7. S. C. Crow, F. H. Champagne: "Orderly structure in jet turbulence". J. Fluid Mech. **48**, 547-591(1971)
8. A. A. Townsend: *The Structure of Turbulent Shear Flow.* (University Press, Cambridge 1956)
9. H. V. Fuchs, E. Mercker, U. Michel: "Large scale coherent structures in the wake of axisymmetric bodies". J. Fluid Mech. **93**, 189-211 (1979)
10. J. B. Roberts: "Coherence measurements in an axisymmetric wake". AIAA **11**, 1569-1571 (1973)
11. N. H. C. Hwang, L. V. Baldwin: "Decay of turbulence in axisymmetric wakes". J. Basic Eng. **88**, 261-268 (1966)
12. J. R. Calvert: "Experiments on the low-speed flow past cones". J. Fluid Mech. **27**, 273-289 (1967)
13. R. Fail, J. A. Lawford, R. C. Eyre: "Low-Speed Experiments on the Wake Characteristics of Flat Plates Normal to an Air Stream". Aero. Res. Counc., London, R & M 3120 (1959)
14. C.-M. Ho, N. S. Nosseir: "Large Coherent Structures in an Impinging Jet" (this book)
15. A. J. Yule: "Phase Scrambling Effects and Turbulence Data Analysis". (this book)
16. H.-P. Pao, T. W. Kao: "Vortex structure in the wake of a sphere". Phys. Fluids **20**, 187-191 (1977)
17. A. E. Perry, T. T. Lim: "Coherent structures in coflowing jets and wakes". J. Fluid Mech. **88**, 451-463 (1978)

Large Coherent Structures in an Impinging Jet

Chih-Ming Ho and N. S. Nosseir
Department of Aerospace Engineering, University of Southern California,
Los Angeles, CA 90007, USA

Abstract

In a free jet the large coherent structures play an important role in the entrainment and mixing for the first few diameters from the nozzle. When a flat plate is placed normal to the jet, the influence of the structures is enhanced by a feedback phenomenon. Upstream propagating waves are produced by coherent structures as they impinge on the downstream plate. These waves can significantly change the initial stability process and alter the flow properties of a jet. In this paper, measurements of surface pressure and near-field pressure were made to reveal the importance of coherent structures in an impinging jet.

Nomenclature

a	Speed of sound
d	Diameter of the jet
M	Mach number
$p(t)$	Fluctuating pressure
p'	rms value of pressure fluctuations
p_0	Pressure in the stagnation chamber
$R(r)$	Correlation coefficient with transducers separated in radial direction
$R(\theta, \tau)$	Correlation coefficient with transducers placed at different azimuthal angles
Re	Reynolds number, $\mathrm{Re} = U_j d/\nu$
r	Radial coordinate
T	Passage period of large coherent structures
U_j	Jet exit speed
U_c	Convection speed of coherent structures
x	Streamwise coordinate
x_0	Position of the flat plate
ν	Kinematic viscosity of air
θ	Azimuthal angle
τ	Time delay of correlation

Introduction

A turbulent impinging jet is an interesting fluid mechanics research subject because the rapid divergence of the mean flow near the solid boundary results in strong augmentation of vorticies. The turbulent energy transfer due to vortex stretching in the flow field has been related to the high heat convection rate [1] around the stagnation point. On the other hand, the high fluctuating pressure level on the plate is a serious concern in the development of blown flaps utilized in STOL vehicles.

These topics have been investigated by many researchers. *Strong* et al. [2] studied the oblique impingement of a low speed jet. They observed that low frequency components of the surface pressure fluctuations increase with increasing angle of incidence. *Donaldson* et al. [3] carried out a detailed study of the impinging jet. Flow properties have been carefully documented for the free jet and for the impinging jet so that some characteristics of the two types of flow can be correlated. *Gutmark* et al. [4] studied the turbulent structures on the center plane of a plane jet impinging on a plate located 100 slot widths downstream. A neutral turbulence scale is found in the flow. Eddies larger than the neutral scale are augmented. Eddies smaller than the neutral scale are attentuated. *Foss* and *Kleis* [5] studied the case for shallow angle ($<12^\circ$) impingment. They found that the stagnation point does not occur at the location of the maximum surface pressure.

Lau et al. [6] studied the large coherent structures in a free jet. They examined the pressure and velocity traces in the potential core and entrainment region of a free jet. Their measurements agree with the model of vortex-like coherent structures. *Brunn* [7] can reconstruct the flow field of the structures by using multiple hot wire probes and an ensemble averaging technique. *Browand* and *Laufer* [8] observed the coalescence of coherent structures in a low Reynolds number water jet. Their data also indicate a constant Strouhal number at the end of the potential core irrespective of Reynolds number.

For the case of a flat plate placed near the nozzle ($x_0/d<5$), the large coherent structures are still well organized as they impinge on the plate. *Ho* and *Kovasznay* [9] have reported high spatial coherence in the surface pressure fluctuations produced by isolated impinging vortex rings. Therefore, the surface loading on a plate located close to a nozzle is expected to be quite different from that on a solid surface placed far downstream. Furthermore, when the jet is operated at high speed ($M>0.8$), a strong resonance [10, 11] appears, possibly caused by the organized structures.

The purpose of this paper is to provide a basic understanding of the role of the large coherent structures in the impinging flow. Most of the measurements are made in the pressure fields. Each near field microphone provides a measurement of the pressure signatures produced by the passing structures and the instability waves, while an array of these microphones demonstrates their development. The surface pressure measurements offer the information about surface loading caused by coherent structures. Two point statistics on the surface present the mode of stability of the coherent structures.

Facilities and Data Processing

The jet used in the experiment has a one-inch nozzle which is connected to a seventeen-inch diameter stagnation chamber through a smooth contraction. The turbulence level in the stagnation chamber is about 0.4%. The stagnation pressure is regulated by a low noise control valve. The variation of the stagnation pressure during the measurements can be maintained to within 2% of the mean guage pressure. Five large storage tanks (51 cubic ft each) allow the jet to be operated from 3 min. up to 10 min., depending on the Mach number. The Mach number can be varied from $M = 2.5$ to $M = 0.3$ but the velocity of the jet is limited to subsonic speeds in the present experiment. The jet is located in an anechoic chamber with dimensions of 12 ft. in height, 16 ft. in width and 20 ft. in length. The low cut-off frequency of the anechoic chamber is 150 Hz.

A flat steel plate (30″ x 30″ x 1/4″) is mounted on a traverse with four degrees of freedom. The streamwise and traverse movement can be remotely controlled. The amplitude of vibration of the plate was measured by an accelerator (Bruel & Kjaer 4324). Even under the test of high subsonic jet speed, $M = 0.9$, the amplitude of vibration was only 1.12×10^{-5} in.

The surface pressure fluctuations are measured by 1/8 inch Kulite piezo-resistive pressure transducers (type XTEL-1-190-25 g). There are 45 transducer sites on the plate. The unused sites are filled by flush mounted plugs. The static pressure distributions are measured from static pressure ports. The near field pressure fluctuations are measured by 1/8 inch Bruel & Kjaer microphones (type 4138). The measured signals are recorded on a Hewlett-Packard 14 channel tape recorder (type 3525a). The signals can be either recorded in FM mode or in direct mode. The frequency response in FM mode is from DC to 20 kHz. The direct mode has a frequency response from 300 Hz to 300 kHz. The analog data are digitized and processed by a minicomputer (type PDP-11/55).

Experimental Results

For a normally impinging jet and small nozzle-to-plate distance ($x_0/d<5$), the pressure measurements have two distinct types of signatures. The signal is random when the jet speed is low subsonic. At high subsonic speed ($M>0.8$), the pressure signal becomes sinusoidal. The sinusoidal wave patterns is caused by resonance. In order to obtain a complete understanding of the role of coherent structures, the flow fields are investigated in both the resonant and the non-resonant regions.

Collective Interaction

After the coherent structures in turbulent shear flow had become recognized as an essential feature rather than an atypical incident [12]. *Winant* and *Browand* [13] further pointed out that the growth of turbulent shear flow is due to the pairing of these structures. The passage frequency of the structures reduces to half the original frequency after two vortex structures pair. The passage frequency decreases linearly with the distance from the nozzle exit [14].

For an impinging jet in resonance, the acoustic waves produced by impinging coherent structures propagate upstream. These waves drive the flow at the nozzle of the jet so that the coherent structures shed at constant intervals and form the resonance. At the nozzle, the upstream propagating acoustic waves are expected to be phase locked with the downstream propagating hydrodynamic waves. Due to the pairing process, the frequency of the initial instability waves should always be higher than the frequency of waves produced at downstream locations. The phase lock between the high frequency instability waves at the nozzle and the low frequency acoustic waves is then impossible unless the acoustic wave is a subharmonic of the instability wave. However, the experimental evidence does not support the "subharmonic" speculation. Furthermore, the passage frequency drops about one order of magnitude over a short distance from the nozzle rather than gradually decreasing. Therefore the conventional pairing mechanism cannot explain the situation and a new model is needed for understanding the phenomenon.

Ho and *Huang* [15] observed multiple vortices merging into a single structure when a low frequency driving force is applied to a plane shear layer. The low frequency driving displaces the vortices from their original locations and forms an array of vortices in a wavy shear layer. Vortices located in the region convex toward the low speed side are unstable and interact together to form a large vortex. We call the phenomenon of simultaneous merging of several coherent structures as *collective interaction*. Whenever a collective interaction appears the passage frequency drops sharply and the shear layer experiences a large growth rate.

Though the collective interaction is observed in a plane shear layer, the same phenomenon can appear in an axisymmetric impinging jet. The resonance of an impinging jet can then be ex-

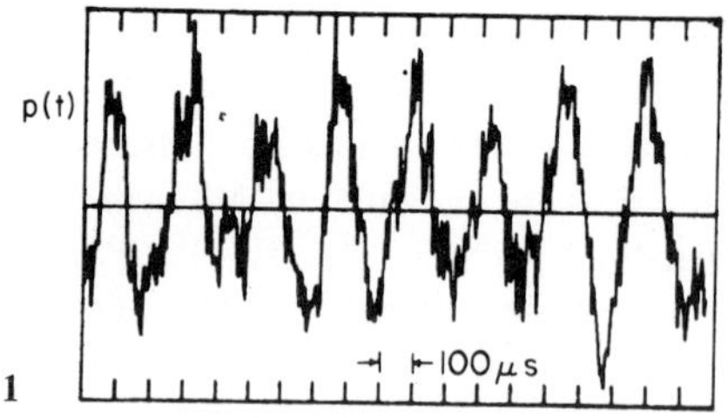

Fig. 1

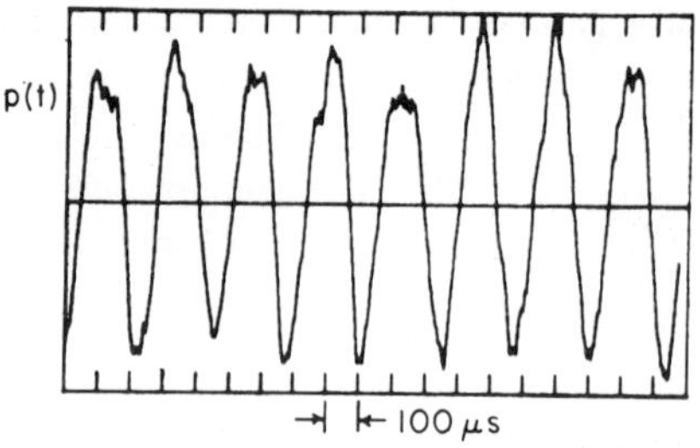

Fig. 2

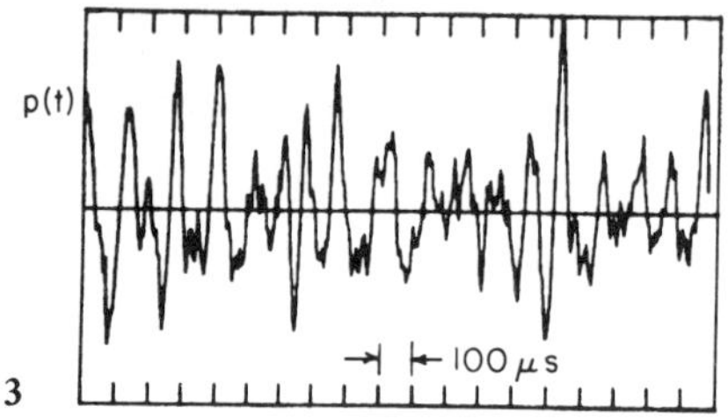

Fig. 3

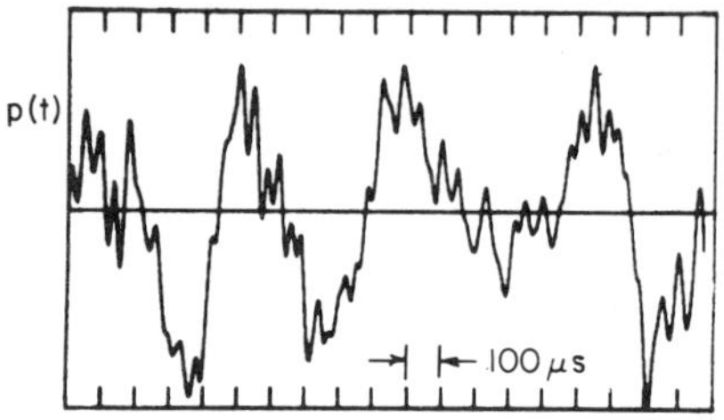

Fig. 4

Fig. 1. The pressure trace near the nozzle of a jet in Resonance ($M = 0.9$, $x = 0.13''$, $r = 0.5''$, $x_0/d = 4.5$)

Fig. 2. The pressure trace at a location downstream from the nozzle of a jet in resonance ($M = 0.9$, $x = 0.92''$, $r = 1.125''$, $x_0/d = 4.5$).

Fig. 3. The pressure trace near the nozzle of a non-resonant jet ($M = 0.5$, $x = 0.13''$, $r = 0.5''$, $x_0/d = 4.5$)

Fig. 4. The pressure trace at a location downstream from the nozzle of a non-resonant jet ($M = 0.5$, $x = 0.92''$, $r = 1.125''$, $x_0/d = 4.5$)

plained. The low frequency acoustic wave deforms the thin shear layer near the nozzle. Consequently several high frequency instability waves merge together under the influence of the wavy shear layer. A direct visualization in the high speed jet would be difficult, but the near field pressure traces fully reveal the collective interaction. From the output of a microphone placed close to the nozzle ($x = 0.13''$, $r = 0.5''$) (Fig. 1), the pressure signature shows a high frequency wave corresponding to the instability frequency superimposed on a low frequency pressure wave induced by the wavy shear layer. A short distance downstream the high frequency wave diminishes (Fig. 2) due to intense mixing of the collective interaction. *Lau* et al. [6] has related the positive peak of the pressure signal to the passage of the valley between two structures, and the negative peak to the passage of the bulge of the structure. In Fig. 2, the amplitude of the high frequency wave on the positive peaks of the low frequency waves is higher than those on the negative peaks. These characteristics further confirm the collective interaction mechanism, because the convex part of the shear layer toward the low speed side (i.e. the bulge) is less stable. Therefore, the high frequency waves on the bulge diminish faster than those on the valley. *Ho* and *Nosseir* [11] measured the frequencies of the instability wave and the coherent structure. The ratio of these frequencies can be as high as 10:1.

In the non-resonant case the collective interaction also appears, but the upstream propagating wave is not as strong as in the high subsonic case. The pressure trace measured in the first location (Fig. 3. $x = 0.13''$, $r = 0.5''$) has a rather low content of the low frequency wave. The signature at a location downstream from the first microphone (Fig. 4. $x = 0.92''$, $r = 1.125''$) shows the low-frequency upstream-propagating wave become dominant in the measured signal. In a free jet, *Laufer* and *Monkewitz* [16] also observe the influence of the large coherent structures on the initial instability.

The Probability Density Distribution of the Passage Period

The collective interaction mechanism forms a coherent structure of low passage frequency very close to the nozzle rather than farther downstream as with pairing. After formation of the structures, the shear layer will still grow as these structures convect downstream. What is the mechanism of this growth, collective interaction or pairing? The probability density distribution analysis of the passage period offers a definite answer to this question.

The signal was first lowpass filtered to suppress the signal generated by small scale turbulence. Then the time interval between zero-crossings of the filtered signal is chosen to represent passage periods of the large coherent structures. An example of the probability density distribution of the measured passage periods is shown in Fig. 5. Pairing is indicated by the multiple peaks.

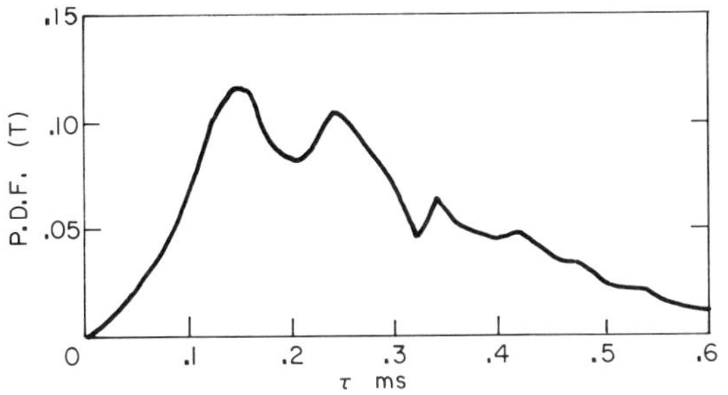

Fig. 5. The probability density distribution of the passage period ($M = 0.4$, $x = 0.92''$, $r = 1.125''$, $x_0/d = 4.5$)

The passage period at the first peak corresponds to the period of formation of the coherent structure. The passage period at the second peak is twice that at the first peak. This certainly indicates that the large coherent structure formed by the collective interaction do pair again downstream. Furthermore, the coherent structures have a high chance of pairing as shown by the high value of probability density at the second peak. There is another minor third peak. Its passage period is three times that of the non-paired structures. So, sometimes three structures merge.

Stability Modes of the Coherent Structure

At low Reynolds number, the coherent structure has a ring vortex shape [8]. When the Reynolds number increases, higher mode instability, a helical mode, appears. The pairing process also becomes more complicated as reported by *Browand* and *Laufer* [8]. *Fuchs* [17] measured the pressure field in a plane normal to a jet at $\mathrm{Re} = 4.2 \times 10^4$. The data indicate the existence of the high order instability modes.

In the present experiment, the Reynolds number varies from 1.8×10^5 to 5.4×10^5. The modes of stability are examined by placing five pressure transducers at equal distance from the stagnation point on the plate. Cross-correlations are made between two microphones at different azimuthal angles. The minimum angle difference is 45°. When the impinging jet is not resonant, the cross-correlations between outputs from these microphones have their maximum correlation at $\tau = 0$, which suggests the coherent structures are dominated by the axisymmetric mode. At high Mach number, where the jet becomes resonant, the same measurements are made. The cross-correlations are sinusoidal in shape, because of the resonance. An example is shown in Fig. 6. An important characteristic of all these cross-correlations is that

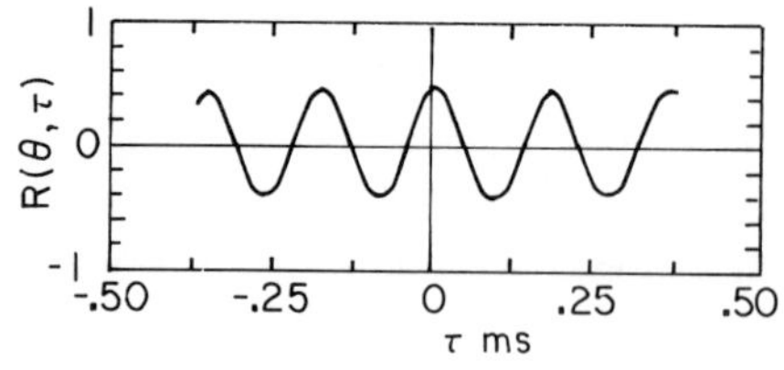

Fig. 6. Azimuthal correlation before prewhitening ($M = 0.8, x_0/d = 4, r/d = 1.5, \Delta\theta = 135°$)

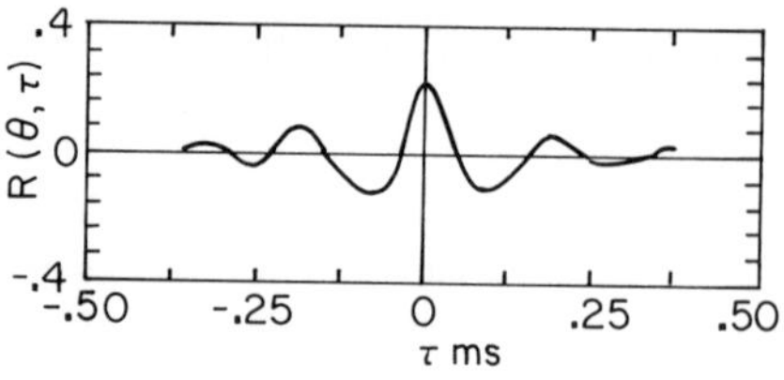

Figure 7. Azimuthal correlation after prewhitening ($M = 0.8, x_0/d = 4, r/d = 1.5, \Delta\theta = 135°$)

the correlation coefficient at $\tau = 0$ has its maximum value. This fact supports the contention that coherent structures are axisymmetric in a resonant jet.

When the coherent structures impinge on the plate, turbulence is generated. Surface pressure traces are distorted sine waves. A data processing method called prewhitening is quite useful for analyzing small random signals superimposed on a high amplitude pure tone. Conventional prewhitening technique [18] separates the random signal from the pure tone by subtracting an in phase sine wave from the original signal by analogue methods. A digital prewhitening technique used here makes the operation much simpler. The original signal is first Fourier transfored into frequency space. Then the pure tone is removed by replacing the high amplitude Fourier components at the pure tone with the averaged value of the two neighboring Fourier components. The random signal is recovered by the inverse Fou-

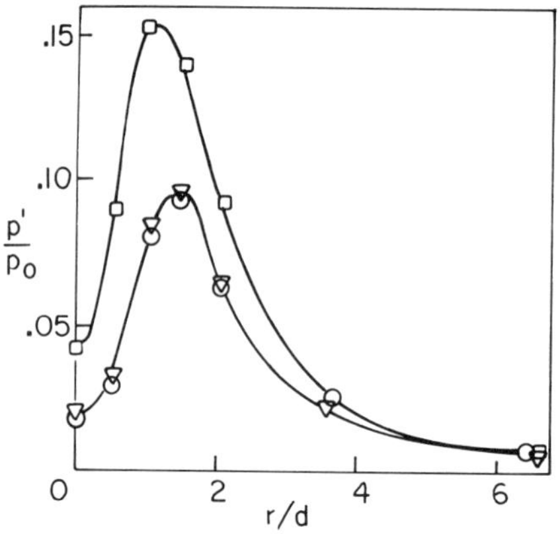

◀ **Fig. 8.** Pressure fluctuation levels ($x_0/d = 2$, ○ $M = 0.3$, ▽ $M = 0.5$, □ $M = 8$)

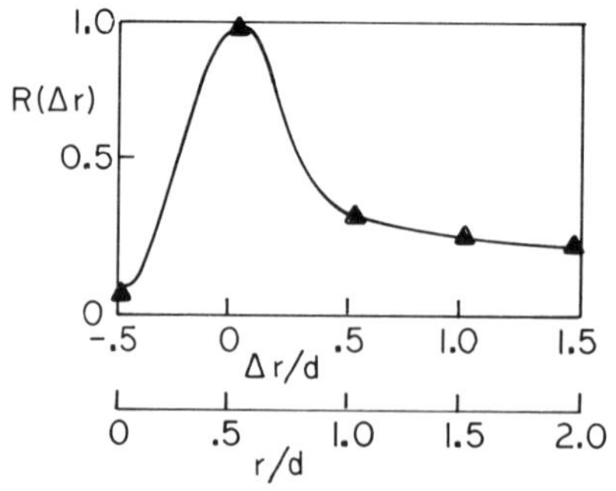

Fig. 9. Spatial cross-correlation of a non-resonant jet ($x_0/d = 4.5, M = 0.5$)

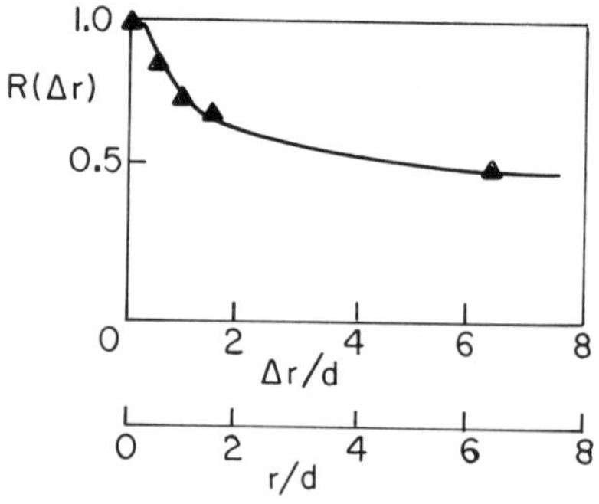

Fig. 10. Spatial cross-correlation of a resonant jet ($x_0/d = 2, M = 0.9$)

rier transform. A cross-correlation of the prewhitened signals of two transducers separated by $\Delta\theta = 135^\circ$ is shown in Fig. 7. The maximum correlation coefficient still appears at $\tau = 0$, it indicates that the turbulent portion of surface pressure fluctuations is also axisymmetric.

Surface Loading

The isolated impinging vortex ring experiment [9] and a water jet experiment [18] show that the coherent structures reach the plate near $r = 1.5''$. The rms value of surface pressure fluctuations has been measured at several radial positions (Fig. 8). The level of fluctuating pressure has a peak at $r = 1.5''$ where large coherent structures impinge. For a jet at $M = 0.3$ the peak fluctuating level is about 10% of the stagnation pressure. The surface pressure profile at $M = 0.5$ is the same as that at $M = 0.3$. This fact indicates that the distributions of normalized fluctuating pressure can collapse into a single curve for a non-resonant jet. However, when the jet is in resonance ($M = 0.8$), the peak fluctuating pressure is 15% of the stagnation pressure. The surface loading is apparently highly intensified during resonance. The intensification also depends on the position of the plate, x_0/d.

The cross-correlation of outputs from surface pressure transducers in the radial direction reveals many interesting flow phenomena. At $M = 0.5$ the jet is not resonant. The correlation coefficients between the stagnation point and any other point on the plate are extremely low (< 0.05). From oil trace visualization, *Donaldson* et al. [3] reported that a flow region with recirculation existed around the stagnation point. A point half an inch away from the stagnation point has high correlations with other points (Fig. 9). The optimum time delay of these correlations indicates a convected flow structures on the plate.

When the jet is in resonance, the spatial correlation coefficient (Fig. 10) has very high values, $R(\Delta r) > 0.5$, over a large area ($6 > r/d > 0$). Therefore the surface loading in a resonant jet not only has high levels but also is highly coherent.

Conclusion

At moderate Mach number ($0.9 > M > 0.3$) and in an impinging flow, it is difficult to make velocity measurements. However, the pressure signature can reveal important information about the coherent structures in the flow.

A collective interaction mechanism generates large coherent structures from the high frequency instability waves (or small vortices) due to the deformation of the shear layer by a low frequency feedback. The frequency reduction can be one order of magnitude or more. After the formation of the coherent structures, the passage period measurements indicate that these structures pair as the shear layer evolves downstream. The cross-correlation measurements on the flat plate show that the coherent structures are dominated by an axisymmetric mode. Furthermore, when the jet is in resonance, the plate experiences high level coherent dynamic loading on the surface.

Acknowledgements. This work is sponsored by AFOSR under contract F40620-78-C-0060. The authors would like to thank Prof. R. E. Kaplan for assistance in typing the paper.

References

1. M. C. Smith, A. M. Kuethe: Effects of turbulence on laminar skin friction and heat transfer. Phys. Fluids **9,** 2337 (1966)
2. D. R. Strong, T. E. Siddon, W. T. Chu: "Pressure Fluctuations on a Flat Plate with Oblique Jet Impingement". NASA, CR-839 (1967)
3. C. D. Donaldson, R. S. Snedeker, D. P. Margolis: A study of free jet impingment, Part 1. J. Fluid Mech. **45,** 381; Part 2 **45,** 477 (1971)
4. Z. Gutmark, M. Wolfshtein, I. Wygnanski: "The Plane Turbulent Impinging Jet". Technical Report, Tel-Aviv University, Israel (1976)
5. F. J. Foss, S. J. Kleis: "The Oblique Impingement of an Axisymmetric Jet". Annual Report to NASA Lewis Research Center (1972)
6. J. C. Lau, M. J. Fisher, H. V. Fuchs: The intrinsic structures of turbulent jets. J. Sound Vib. **22,** 379 (1972)
7. H. H. Brunn: A time-domain analysis of the large-scale flow structures in a circular jet. Part 1: Moderate Reynolds number. J. Fluid Mech. **83,** 641 (1977)
8. F. K. Browand, J. Laufer: "The Role of Large Scale Structure in the Initial Development of Circular Jet". Proc. of Fourth Biennial Symposium on Turbulence in Liquids, University of Missouri – Rolla (1975)
9. C. M. Ho, L. S. G. Kovasznay: "Unsteady Pressure Field of an Impinging Turbulent Vortex Ring". 27th Annual Meeting, American Physical Society, Division of Fluid Dynamics (1974)
10. G. Neuwerth: "Acoustic Feedback Phenomenon in Subsonic and Supersonic Free Jets That Strike a Perturbing Body". Ph. D. Thesis, Rhein-Westfalia University, Aachen, West Germany (1973)
11. C. M. Ho, N. S. M. Nosseir: The role of coherent structure in an impinging jet. Bull. Am. Phys. Soc. Sec. II, **23,** 1013 (1978)
12. G. Brown, A. Roshko: "The Effect of Density Differences on the Turbulent Mixing Layer". Turbulent Shear Flows AGARD Current Paper No. 93 (1971) pp. 23.1–23.12
13. C. D. Winant, F. K. Browand: Vortex pairing: The mechanism of turbulent mixing layer growth at moderate Reynolds number. J. Fluid Mech. **63,** 237 (1976)
14. R. A. Peterson: Influence of wave dispersion on vortex paring in a jet. J. Fluid Mech. **89,** 469 (1978)
15. C. M. Ho, L. S. Huang: Subharmonics and vortex merging in an unsteady shear layer. Bull. Am. Phys. Soc. Sec. II, **23,** 1007 (1978)
16. J. Laufer, P. A. Monkewitz: The influence of the dominant large scale structures on the initial instability of a low-speed circular jet. Bull. Am. Phys. Soc. Sec. II, **23,** 1004 (1978)
17. H. V. Fuchs: "Resolution of Turbulent Jet Pressures into Azimuthal Components". AGARD-CP-131 Paper 27 (1973)
18. K. C. Williams, K. R. Purdy: A prewhitening technique for recording acoustic turbulent flow data. Rev. Sci. Instrum. **41,** 1897 (1970)
19. C. M. Ho, N. S. M. Nosseir: The hydrodynamic feedback of an impinging water jet. Bull. Am. Phys. Soc. Sec II, **22,** 1280 (1977)

On the Development of Three-Dimensional Small Scales in Turbulent Mixing Layers

L. P. Bernal, R. E. Breidenthal, G. L. Brown, J. H. Konrad[1], and A. Roshko

California Institute of Technology
Pasadena, CA 91109, USA

Abstract

This paper summarizes the results of several experiments addressed to the question of how small-scale, three dimensional structure develops in plane mixing layers and how it affects the properties of the turbulent mixing layer and the large-scale, quasi two dimensional, organized primary structures of which the mixing layer is composed. The cases studied include mixing layers in water and in gases, the latter at uniform density and also at a density ratio of 7, using nitrogen and helium respectively in the two streams between which the mixing layer develops. The effects of Reynolds number and of initial conditions on these developments are determined from several flow visualization techniques and from measurements of mixedness, employing concentration probes for the experiments in gases and chemical-reaction indicators for those in water. It is found that the effect of increasing Reynolds number is mainly to introduce small-scale, three dimensional structure into the flow. This has the effect of increasing the internal mixedness of the shear layer but it does not break down the large primary structures, which continue to develop and interact.

Introduction

There is at present much interest in the properties and effects of the organized large structures that are found in various turbulent shear flows. One of the problems of interest concerns the effect of increasing Reynolds number and indeed the question has been raised [1] as to whether such structures maintain their organization or coherence at very large Reynolds number. There is of course no argument that increasing Reynolds number tends to increase disorganization in that more modes of instability are introduced to the flow. But the new modes that appear with increasing Reynolds number are of decreasing scale [2] and it is possible that they are only weakly coupled to the large structures, as secondary flows, and do not drastically alter those large, primary structures. Perhaps the more pertinent question is whether inviscid instability will "break down" the large structure with time, i.e., with increasing distance from its place of origin. Of course the superimposed smaller scale motions at higher Reynolds number might contribute to such breakdown as might disturbances in the flow external to the shear region in question.

These problems are most clearly exhibited in the case of the plane mixing layer, where the basic large organized structures are particularly simple, namely two dimensional vortices

---

1 Present Address: Hughes Aircraft Co., El Segundo, California.

or rollers which develop from the Kelvin-Helmholtz instability waves if the shear layer is initially laminar. They subsequently amalgamate with each other to form larger rollers [3, 4] and the process keeps on repeating itself with increasing distance downstream, apparently indefinitely, but this last assertion is in some dispute. In fact, the question is repeatedly raised whether the organized motion and the phenomenon of shear layer growth by amalgamation will persist at higher Reynolds numbers and at greater distances downstream than those at which they have been observed in laboratory experiments so far. To what extent will the large structures maintain a two dimensional character and whether spanwise organization will break down completely are other aspects of this question. Some of the doubts thus implied are founded on the fact that many years of detailed measurements with pressure probes and, especially, hot-wire anemometers failed to reveal the coherent structures which were more recently found in flow pictures. In fact, by now, the evidence from the pictures has been extended to Reynolds numbers comparable to or higher than those explored in the hot-wire era. What has become clear is that the information available in timeseries signals from probes at one or two points in a turbulent shear flow is quite insufficient to describe the instantaneous *spatial* structure which a flow picture reveals. When this spatial structure does not remain perfectly two dimensional and/or periodic it becomes very difficult to recognize its organization from time signals.

In this paper we discuss one aspect of these problems, that is the origins of small-scale three dimensional structure, its effect on some properties of the mixing layer and on the large structures. It is based on some of the results obtained in several investigations in progress or completed [5, 6] in our laboratory.

Appearance of Small-Scale Turbulence

An important step in the development with respect to Reynolds number can be seen on edge-view shadowgraphs [4] of a mixing layer between a stream of helium and a lower-speed stream of nitrogen shown in Fig. 1. These show the similarities and differences in structure that are produced by changing the pressure in the apparatus by a factor of four. At the lower Reynolds number the Kelvin-Helmholtz waves, which form after the flow has left the splitter plate, roll up to form vortices which grow as they convect downstream. The two pictures differ mainly

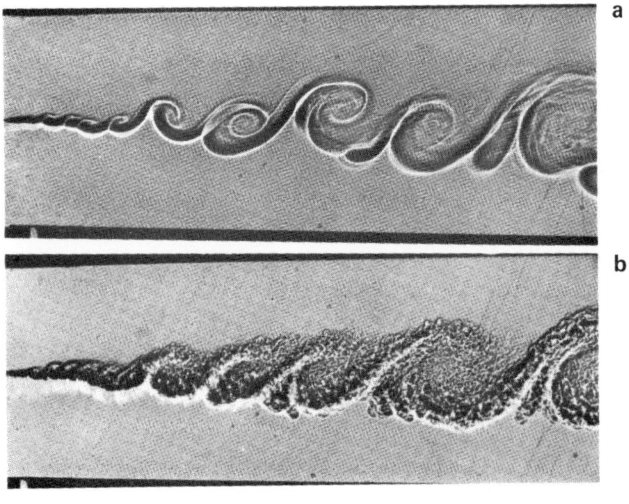

Fig. 1a, b. Spark shadowgraphs of a mixing layer between helium flowing at velocity U_1 = 1,000 cm/s and nitrogen at 380 cm/s. Channel height is 5 cm. Upper picture (**a**) taken at a pressure of 2 atm., lower picture (**b**) at 8 atm

in their small-scale structure, which has become quite well developed at higher Reynolds number. The large-scale vortical structure is similar in both cases and the overall development of the shear layer, as indicated by a visual measure of its growth rate and by mean flow measurements is approximately the same.

At a value of unit Reynolds number intermediate to those in Fig. 1 the appearance of the flow is as shown in Fig. 2a. Here a plan view taken simultaneously with and in correct geometrical relation to the edge view is also displayed. Only about half the total span (10 cm) is visible in the plan view. The spanwise organization of the large vortices may be observed in that

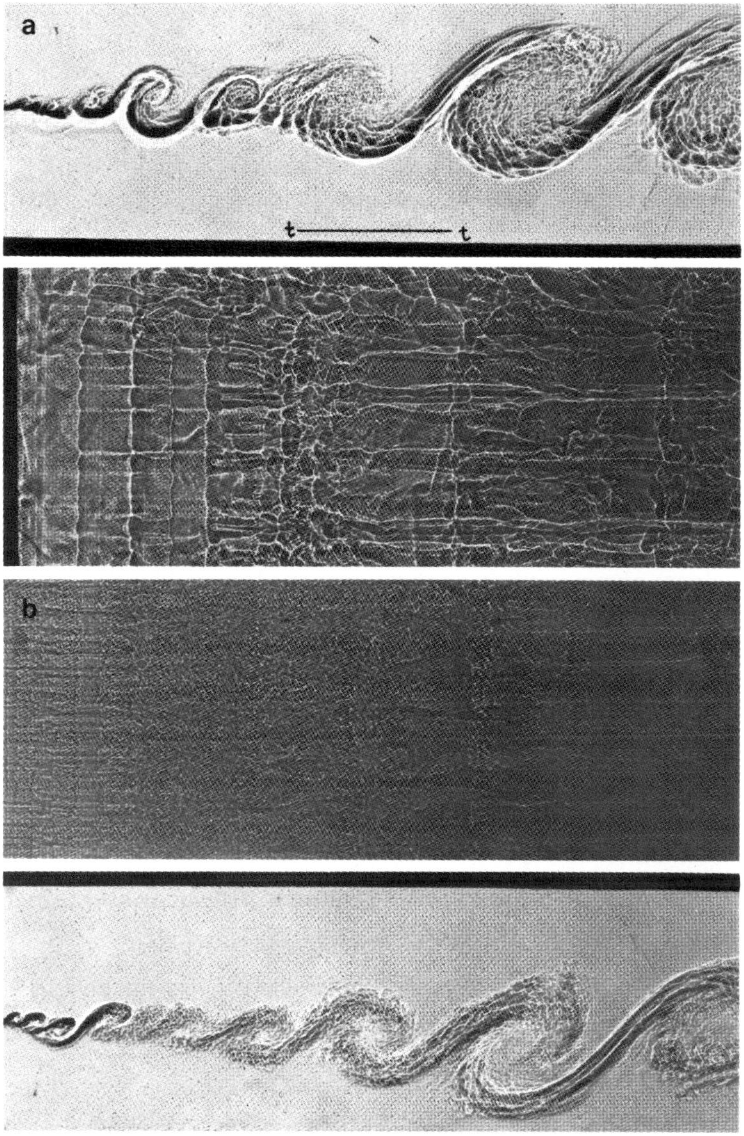

Fig. 2. (**a**) Spark shadowgraph showing simultaneous edge and plan views of mixing layer between streams of helium at velocity U_1 = 1,000 cm/s and nitrogen at velocity U_2 = 380 cm/s. Pressure = 4 atm. (**b**) Spark shadowgraph of mixing layer in flow at uniform density. U_1 = 1,000 cm/s, U_2 = 380 cm/s. Pressure = 4 atm

view by noting the correspondence between the edge of a vortex in the edge view and a spanwise line in the plan view. The latter is almost unbroken at the upstream side of the flow but is quite disturbed by small-scale structure on the downstream side. Nevertheless the spanwise coherence in the large is quite clear. The small-scale turbulence first appears well established at about 7 cm from the trailing edge of the splitter plate, which is at the left hand edge of the picture.

The most remarkable feature on the plan view is the streamwise streakiness which begins a short distance downstream of the splitter plate and persists through the regions of small-scale turbulence to the downstream side of the picture.

The pictures discussed so far are of mixing layers between streams of helium and nitrogen, for which of course there is a difference in densities and viscosities in the two streams. The ratio of densities ρ_2/ρ_1 in the low- and high-speed streams, respectively, is 7 and the ratio of viscosities is 0.90. The large difference in optical properties is very helpful for the flow visualization. To get pictures of the structure in a flow of uniform density, a mixture of helium and argon proportioned to give a density equal to that of nitrogen was used for the low speed stream and nitrogen was used for the high speed side. A spark shadowgraph obtained in this way is shown in Fig. 2b, for which the velocities and pressures are the same as in 2a. The features for Fig. 2a are also evident in 2b, but the development of small-scale structure and streakiness begins earlier (closer to the splitter plate) in the latter case. This is consistent with the fact that the Reynolds number is higher for the flow in Figure 2b, the value of the kinematic viscosity ν_1 being only 1/8 as large as the value for Fig. 2a.

Mixedness

In the mixing layer the fluids from the two sides are brought together and mixed, first grossly (enfolded) and ultimately intimately (at the molecular level). The last stage is accomplished by diffusion. In contemplating what might be the effects of Reynolds number on the latter process one sees two competing effects. Increasing Reynolds number reduces diffusivity but on the other hand it tends to produce smaller structure and thus reduce the distances across which diffusion must take place. Thus the overall effect of Reynolds number is not evident *a priori.*

In an experiment to explore this [5], a concentration probe was used to measure the time histories of helium concentration at various points in mixing layers like those in Fig. 1 and Fig. 2a. From these it was possible to infer the probabilities for He and N_2 to be in a molecularly mixed state at those points and to define this mixedness in terms of a simple reaction product that would be formed if the two streams were carrying reactive components. By integrating normal to the plane of the mixing layer the mean reaction product thickness P for each streamwise position x was defined. When normalized by the local vorticity thickness, $\delta(x)$, it gave a measure of the extent of mixedness, at the molecular level, as a function of x. The result for the flow depicted in Figs. 1 and 2a is shown in Fig. 3.[2] Here the data are plotted against local Reynolds number based on the vorticity thickness $\delta(x)$, the velocity difference U_1-U_2, and the kinematic viscosity ν_2 in the low speed (nitrogen) stream. If x instead of δ were used to define the Reynolds number, the latter would be 10 times greater than in Fig. 3. It is seen that, instead of a gradual change of mixedness with Re, there is a kind of transition after which there is no further effect of Re. The transition region defined by Fig. 3 is marked on

2 For equivalence ratio $\alpha \doteq 10^2$.

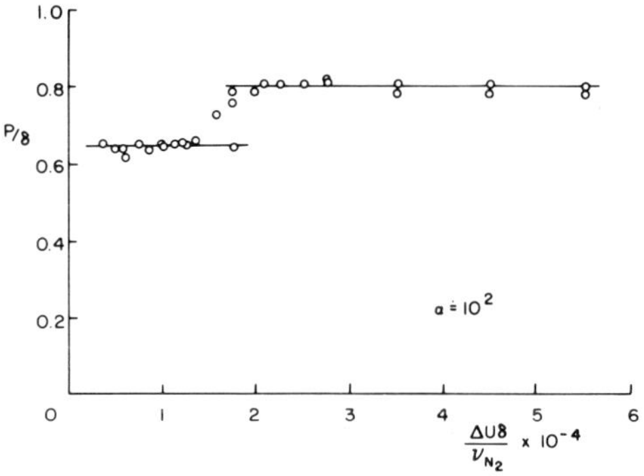

Fig. 3. Mixing transition in helium-nitrogen mixing layer

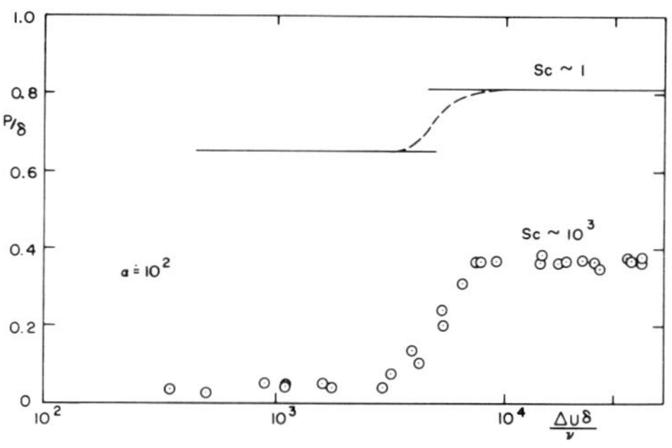

Fig. 4. Comparison of mixing in gaseous and aqueous mixing layers

Fig. 2a by the symbol $t-t$. It is seen to correlate well with the appearance of well developed small scale structure. For lower and higher pressures (Figs. 1a and 1b), all other parameters remaining the same, the transition region moves downstream and upstream, respectively.

A more direct measure of mixedness was obtained in another investigation [6] in which an actual chemical product was formed in the mixing layer between two streams of water carrying weak solutions of a base (sodium hydroxide) and an indicator (phenolphthalein). In this case the density is, of course, uniform. The greatest difference from the previous example is in the Schmidt number Sc, which is about three orders of magnitude higher than in the gas flow. A point of interest is whether this difference affects the extent of mixing at the molecular level. The result, presented in a form similar to that in Fig. 3, is shown in Fig. 4. The curve for the gaseous mixing layer has been transposed from Fig. 3 by calculating the Reynolds number with a value of kinematic viscosity weighted between the values for nitrogen and air (weights 0.2 and 0.8, respectively).

With this, the location of the mixing transition in water corresponds well with that in the He/N_2 mixing layer. However, the levels of mixedness are different. Before the transition the relative product thickness in the aqueous shear layer is very much less than in the gaseous layer, by a factor of approximately $Sc^{1/2}$, which would be the result for a simple laminar flow. Actually the mixing layer has already evolved through the Kelvin-Helmholtz instability and formed vortices which are interacting with each other as may be seen in Fig. 6, on which the beginning of the mixing transition region has been marked.

After the transition there is a large increase in the mixedness of the aqueous layer to a value which is within about a factor of 2 of the gaseous value. $Sc^{1/2}$ is evidently not the relevant factor here. In fact, the two values may be even closer, since the measurement in the gas with a probe is expected to be too high due to the smoothing effect of a probe of finite dimensions and the measurement in water is expected to be too low due to some reversibility in the chemical reaction. The much decreased, possibly zero effect of Schmidt number implies that the amount of product is limited not by the molecular diffusivity but by the large-scale motions, i.e., it is entrainment limited.

If, as implied by Figs. 3 and 4, there is no further increase of mixedness with Reynolds number with distance downstream, the asymptotic condition reached after the mixing transition may coincide with the attainment of self preservation in the velocity structure. Relevant to this are the observations of *Bradshaw* [7], who studied the development of Reynolds stress $(\overline{-u'v'})$ along a mixing layer on the edge of a free jet. He found that, in a mixing layer which was initially laminar, $(\overline{-u'v'})$ increased with distance downstream to a maximum value considerably higher than the final equilibrium value. He associated high Reynolds stresses with the early stages of vortex pairing, which he called the "confluence of vortex rings".

Bradshaw discussed the streamwise variations in terms of x/θ where θ is the initial momentum thickness. He found that the maximum in $(\overline{-u'v'})$ occurs at about $x/\theta \doteq 150$ and that the final equilibrium value is reached at $x/\theta \doteq 1{,}000$ which, in his case, $(U_2 = 0)$, corresponded to $U_2 x/\nu \doteq 7 \times 10^5$. This may be compared, in Fig. 4, with a value of $\Delta U x/\nu \doteq 10^5$ for completion of transition. For this case we estimate the corresponding $x/\theta \doteq 500$. Further remarks about comparison based on the parameters will be found in what follows.

The Streamwise Streaks

What is the nature and role of the streamwise streaky structure which appears in plan views (Fig. 2) of the mixing layer? How are they related to the enhanced internal mixing which occurs at the transition discussed above? Examination of repeated pictures indicated that their spacing does not vary much and that even their spanwise location is fairly well fixed for a given flow, but that it does vary with flow conditions. The streaks tend to persist for rather long distances downstream, surviving through one or more of the large vortex structures. There is a tendency for spanwise spacing to increase at some distance downstream (see Fig. 2a), i.e., the number of streaks across the span decreases.

By taking long time pictures, or by superimposing many spark exposures on one picture, mean streak patterns like the one shown in Fig. 5 were revealed. This one is for a flow close to the conditions of Fig. 2a; the estimated position of the transition region is marked on the picture. It is seen that the amplitude of this mean streak pattern increases from some position well ahead of the mixing transition, reaches a maximum value at approximately the beginning of the transition region, and is much reduced by the completion of the transition. The later must be an effect of nonsteadiness which has developed, since on instantaneous pictures (Fig. 2a) the streaks are still quite sharp well downstream of the transition region.

The effect on the mean streak pattern of varying flow parameters was studied. Briefly the following effects were observed. With increase of unit Reynolds number, e.g., due to increase of the pressure, the pattern shifts upstream and the spacing decreases, suggesting a correlation with the initial shear layer thickness. Indeed, the spanwise streak spacing is approximately equal to the streamwise spacing of the initial Kelvin-Helmholtz waves. With a change of geometry, namely an increased length of splitter plate, other conditions fixed, the streak spacing increases. This is consistent with the postulated dependence on initial shear layer thickness. However the streamwise location of the pattern does not necessarily shift downstream. This is because not only x/θ but also $U_1\theta/\nu$ are parameters of the streak phenomena and the mixing transition, as discussed in [6].

The output of a concentration probe placed at various locations in the streak pattern shows that the bright streaks in Fig. 2a correspond to locations where N_2 is strongly entrained from the low speed side. Also, correlations obtained from two probes deployed spanwise show variations which correlate with the streak spacing.

Our various observations lead us to suppose that the streaks mark the presence of counter rotating pairs of streamwise vortices which are superimposed on the main quasi two dimensional structure as a kind of secondary flow structure. We had speculated earlier [5] that they are due to a Rayleigh-Taylor type of instability connected with the curvature in the portions of shear layer (braids) between the large vortex structures. Indeed, the streaks are most sharply visible in the braids. The resulting streamwise vorticity in this secondary motion would be strengthened (vortex stretched) in the high strain field that exists in the braids. How these

Fig. 5. Plan view obtained from 200 spark schlieren illuminations. Helium stream at U_1 = 752 cm/s, nitrogen stream at U_2 = 227 cm/s. Pressure = 4 atm

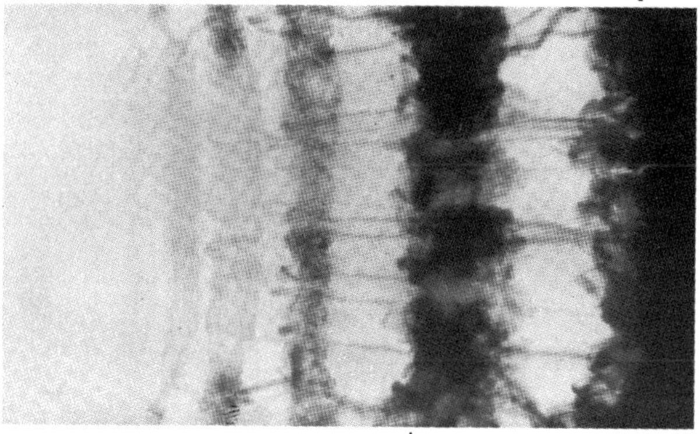

Fig. 6. Plan view showing mixing transition. Mixing layer in water with U_1 = 43 cm/s, U_2 = 16 cm/s

streamwise vortices interact with the large spanwise ones is not quite clear; precession effects may produce further instability and may account for the appearance of smaller scale structure and the associated mixing transition described earlier.

A new view of the origins of the streakiness has been obtained in the chemical-product experiments [6]. A plan view obtained in those experiments is shown in Fig. 6, in which the transition region is marked. From this and other pictures it can be seen that the streaks sometimes originate from a sinusoidal pattern that develops in the spanwise vortices which emerge from the Kelvin-Helmholtz instability. Amplication of the pattern's amplitude, possibly due to straining, as it convects downstream with the vortices tends to stretch out the streamwise oriented segments of the pattern; the streaks apparently result from this stretching. But this is not always nor at all conditions the only way in which three dimensionality is introduced into the flow [6]. Figure 6 also illustrates that after two pairings the spanwise coherence of the main vortices is still very good even though the samll scale three dimensional motion must be well developed, as indicated by the large amount of product that has been formed.

Discussion

This paper is concerned principally with the mixing transition of Figs. 3 and 4. One of the main findings is that during the transition the mixedness inside the mixing layer is considerably increased by the action of small-scale, three dimensional motions, which themselves result from an earlier instability that produces organized secondary motions with components of vorticity in the streamwise direction. It is probable that the asymptotic state of mixedness reached after the transition coincides with the state of equilibrium attained by the Reynolds stress in Bradshaw's experiments, but one cannot be conclusive since measurements of both quantities in a given flow have not been made so far. This is an important point in the discussion as to whether the final equilibrium state of a mixing layer is still dominated by the large, spanwise organized structures. Our pictures show that those structures still exist downstream of the mixing transition. Figure 2 gives two examples and others, for which values of unit Reynolds number and of x/θ were even higher, may be found in [8] and [9].

In making comparisons of the development of mixing layers it is important to note that, in addition to flow parameters such as velocity ratio U_2/U_1, density ratio ρ_2/ρ_1, and kinematic viscosity ratio ν_2/ν_1, as well as geometrical parameters, there are two development parameters [6]. One of these, x/θ, measures the distance downstream in terms of the initial thickness. It would be the only development parameter needed if the flow development were inviscid, as can indeed be assumed for much of it, especially the large, quasi two-dimensional processes. However, the initiation of secondary instabilities which lead to three dimensional motions, which in turn affect the large organized structure, are Reynolds-number dependent and an additional parameter such as $U_1 x/\nu$ or $U_1 \theta/\nu$ is needed. For example, when θ is increased by decreasing unit Reynolds number, the transition region moves downstream but when θ is increased by increasing the length of splitter plate the direction of change cannot be predicted without more detailed information about the particular flow.

Thus, comparisons between experiments with different flow parameters and in different geometries are difficult. Nevertheless much can be inferred by noting the occurence of similar phenomena, such as the first appearance of Kelvin-Helmholtz structure, development of streamwise structure, etc. Thus Bradshaw's observation from flow pictures that "breakdown to turbulence occurs via the longitudinal vortices predicted by Benney" suggests that he may have been observing the streaky structure of our experiments. Bradshaw used the term "breakdown to turbulence" synonymously with approach to final equilibrium. But the organized structures persist long after the streaks have developed and after the mixedness has reached an asymptotic value, suggesting that the final state is one which is still dominated by spanwise organized structures, which contain well developed small scale "turbulence".

In the work described here we have been concerned mainly with small-scale instabilities and development of small-scale structure. The term "breakdown to turbulence" no doubt implies those processes but also, perhaps, instabilities on a large scale. That the large quasi two dimensional vortices tend to survive strong external disturbances has been demonstrated by *Wygnanski* et al. [10], who introduced various disturbances into the free stream and into the splitter-plate boundary layer of a mixing layer of rather large scale and Reynolds number. Their observations included flow visualization and spanwise correlations of velocity fluctuations. *Browand* and *Troutt* [11] also studied spanwise correlations in a mixing layer at streamwise distances up to 92 cm and values of $\Delta U\delta/\nu$ up to nearly 2×10^5. They found that instantaneous visualization from the hot-wire outputs show the large structure to have spanwise orientation. A $z-t$ plot of such correlations shows parallel crests and valleys, but they contain phase discontinuities and dislocations which in long-time averages contribute to decrease of correlation and tend to give a misleading impression of lack of spanwise structure.

There may be some semantic differences as to what is meant by turbulence, final state of turbulence, etc., but there is also genuine disagreement in the interpretation of the various experiments and in conclusions about their implications. Considering the number of investigators that are working on various aspects of the problem a resolution of these differences is probably near at hand.

Acknowledgments. The various experiments on which this paper is based were made possible by the financial support of the Office of Naval Research, through Project SQUID and through its Fluid Dynamics Program, and of the Air Force Office of Scientific Research through the Air Force Weapons Laboratory.

References

1. C. Chandrsuda, R. D. Mehta, A. D. Weir, P. Bradshaw: Effect of free-stream turbulence on large structure in turbulent mixing layers. J. Fluid Mech. **85**, 693 (1978)
2. L. D. Landau, E. M. Lifshitz: *Fluid Mechanics* (pergamon Press 1959) p. 106
3. C. D. Winant, F. K. Browand: Vortex pairing: The mechanism of turbulent mixing-layer growth at moderate Reynolds number. J. Fluid Mech. **63**, Part 2, 237–255 (1974)
4. G. L. Brown, A. Roshko: On density effect and large structure in turbulent mixing Layers. J. Fluid Mech. **64**, Part 4, 775–816 (1974)
5. J. H. Konrad: "An Experimental Investigation of Mixing in Two-Dimensional Turbulent Shear Flows with Applications to Diffusion-Limited Chemical Reactions". Ph. D. Thesis, California Institute of Technology, Pasadena, California (1977) and Project SQUID Report, CIT-8-PU (1976)
6. R. E. Breidenthal, Jr.: "A Chemically Reacting, Turbulent Shear Layer". Ph. D. Thesis, California Institute of Technology, Pasadena, California (1978) and J. Am. Inst. Aeronaut. Astronaut. **17**, No. 3, 310–311 (1979)
7. P. Bradshaw: The Effect of initial conditions on the development of a free shear layer. J. Fluid Mech. **26**, Part 2, 225–236 (1966)
8. A. Roshko: Structure of Turbulent Shear Flows: A New Look. J. Am. Inst. Aeronaut. Astronaut. **14**, No. 10, 1349–1357 (1976)
9. P. E. Dimotakis, G. L. Brown: The mixing layer at high Reynolds number: Large-structure dynamics and entrainment. J. Fluid Mech. **78**, Part 3, 535–560 (1976)
10. I. Wygnanski, D. Oster, H. Fiedler, B. Dziomba: "On the Perseverance of a Quasi Two-Dimensional Eddy-Structure in a Turbulent Mixing Layer". Tel-Aviv University, 1978 (private communication)
11. F. K. Browand, T. Troutt: A note on spanwise structure in the two-dimensional mixing layer. J. Fluid Mech. **97**, Part 4, 771–781 (1980)

A Forced, Plane, Turbulent Mixing-Layer: A Challenge for the Predictor

I. Wygnanski and D. Oster

School of Engineering, Tel-Aviv University,
Ramat-Aviv, Israel

H. Fiedler

Hermann-Fottinger Institut für Thermo- und Fluiddynamik,
Technische Universität Berlin
D-1000 Berlin 12, Fed. Rep. of Germany

Abstract

Measurements were carried out in an isothermal plane mixing-layer in absence of pressure gradient and under controlled initial conditions. Disturbances of known frequency and amplitude were introduced into the flow at its origin while measurements were conducted further downstream. Constant temperature hot-wire anemometers were used to measure the mean velocity, the shear stress and the three components of turbulent intensity for various initial conditions. The flow is not self preserving in the range of Reynolds numbers considered. The initial spreading rate of the mixing layer depends on the amplitude and the frequency of the imposed oscillations. At a given frequency, the rate of spreading increases initially with the amplitude of the imposed oscillation but a saturation region occurs beyond which one can no longer control the flow. The spreading rate in the saturation region is less than for the undisturbed mixing layer. Some important characteristics of the mixing layer are intimately related to the behaviour of the large quasi-two-dimensional eddies which engulf fluid into their fold while being convected downstream.

Nomenclature

A	Amplitude of forcing
b	Local width of the mixing layer defined by $b = y_{0.1} - y_{0.95}$
f	Frequency of forcing
Re	Reynolds number $\mathrm{Re} = U_E\, x/\nu$
T_E	Exit temperature at the plane of the nozzle
$\overline{T}$	Mean temperature
T'	Temperature fluctuation
U_E	Exit velocity at the plane of the nozzle
$\overline{U}$	Mean velocity in the direction of streaming
$\overline{V}$	Transverse component of the mean velocity in the y direction
$\overline{u'^2}, \overline{v'^2}, \overline{w'^2}$	Three components of the turbulent intensity in the x, y, z direction respectively
x_0	The location of the hypothetical origin from the nozzle
x	Distance in the direction of streaming
y	Transverse length measured from the splitter plate

$y_{0.1}$ The y location at which the velocity is equal to 10% of the free stream velocity
$y_{0.5}$ The y location at which the velocity is equal to 50% of the free stream velocity
$y_{0.95}$ The y location at which the velocity is equal to 95% of the free stream velocity
$\theta_{0.1}$ Partial momentum thickness,

$$\theta_{0.1} = \int_{-\infty}^{y_{0.1}} \frac{U}{U_E} \left(1 - \frac{U}{U_E}\right) dy$$

η Similarity parameter $\eta = \dfrac{y - y_{0.5}}{x - x_0}$

σ Spreading parameter $\sigma = \sqrt{\pi} \dfrac{\partial}{\partial \eta} \left(\dfrac{U}{U_E}\right)_{max}$

Introduction

The two-dimensional mixing layer is sensitive to small amplitude periodic surging which is applied at the origin of the flow. The effect of surging on the mean velocity field was investigated [1] but only cursory observations were made of the longitudinal component of the turbulent intensity. In a continuing effort to understand and control the mixing process, it was decided to measure the mean velocity field, the three components of the turbulent intensity and the shear stress, for various forcing conditions. In this paper, conventional long time averaged measurements are presented to indicate that the two dimensional mixing layer is susceptible to initial conditions, even though the boundary layer on the splitter plate was fully turbulent without the forcing being applied.

Experimental Apparatus

The measurements were carried out in a low speed mixing layer facility in which the flow was allowed to mix with nominally quiescent surrounding fluid. The facility is enclosed by screens in order to avoid room drafts arising from open doors, air-condition vents, etc. The free stream velocity was 10 m/s. and the Reynolds number $\leqslant 4 * 10^5$. Small, two-dimensional, sinosoidal oscillations normal to the direction of streaming perturbed the mixing layer at its origin. The source of the orderly perturbation was located down stream of a trip wire and a coarse sand paper which ensured that the flow in the boundary layer was fully turbulent, however, the removal of the tripping devices had only a minor effect on the flow. Although the frequency of the oscillation was variable the results presented were taken at $f = 15$ Hz and three amplitudes corresponding to peak velocity fluctuations measured at $x = 10$ mm and characterized by:

	A	$(\overline{u'^2}/U_E^2)_{max}$	$(\overline{v'^2}/U_E^2)_{max}$
I)	0	0.016	0.009
II)	0.3	0.023	0.009
III)	2.5	0.132	0.012

The flow at the initiation of mixing ($x = 10$ mm) was fully turbulent in cases (I) and (II) but became intermittent in case (III) whereupon the turbulence was confined to a fraction of the forcing cycle only.

Three components of the turbulent intensity and shear stress were measured using constant temperature hot wire anemometers. The x-wire probes were calibrated for yaw angles not exceeding $\pm 25°$. The data was acquired using conventional analog circuitry comprising of squarers, multipliers, amplifiers and integrators.

The air flow in the wind tunnel could also be heated to 25 °C above the room temperature. The temperature was measured using 0.63 micron, platinum-rhodium wires as resistance thermometers. Some measurements of temperature are presented in this paper as they clarify the structural aspects of the flow. In particular, a rake of 9 temperature sensors was used to observe the large coherent eddies existing in the forced mixing layer. A more detailed description of the apparatus is given by *Fiedler* et al. [2].

Discussion of Results

Measurements of Velocity

The rate of growth of the mixing layer can be determined from the local width of the mean velocity profile. Assuming that the velocity profiles are similar to one another and the isotachs are straight lines intersecting in the virtual origin of the flow, one can define the local width of the mixing layer as the transverse distance between the location at which the velocity is equal to 95% of the free stream velocity and the location at which it is equal to 10% of the free stream (Fig. 1a).

The spreading rate of the unforced mixing layer is constant;

$$db/dx = (y_{0.1} - y_{0.95})/(x - x_0) = 0.218.$$

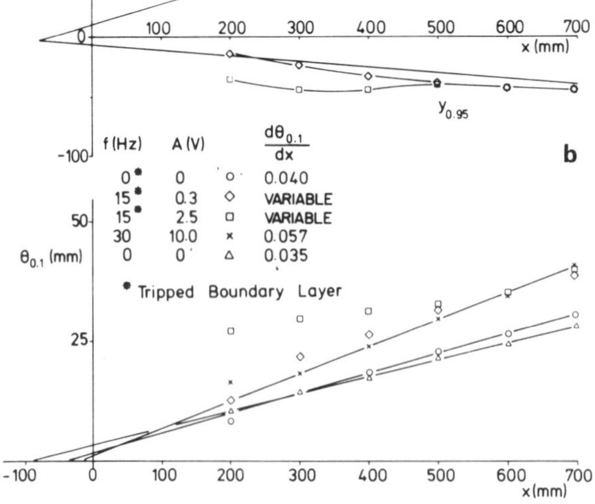

Fig. 1a, b

This result is in good agreement with previous measurements by *Batt* [3], *Oster* et al. [4], and by *Wygnanski* and *Fiedler* [5]. In the absence of similarity, the above mentioned definition of growth-rate loses some of its meaning. Nevertheless, the application of the definition to the forced case indicates that forcing at $f = 15$ Hz increases the width of the mixing layer. The spreading rate of the mixing layer does not vary linearly with the amplitude of the imposed oscillations. The difference between weak forcing (case II) and the strong forcing (case III) is small for downstream distances exceeding 500 mm. The effects of forcing are more readily visible on the low velocity side where the isotach $y_{0.1}$ has initially different slopes for the forced cases. However, at $x > 400$ mm the slope levels off, and approaches the slope of $y_{0.1}$ in the unforced cases. Strong forcing is required to affect the isotach $y_{0.95}$ on the high velocity side of the flow, and for $x > 500$ mm forcing has hardly any effect on this particular isotach.

A more reasonable length scale to be used in a case for which the velocity profiles are not similar, is an integral length-scale. The momentum thickness θ is an obvious choice. Since there are difficulties in measuring the low velocities at the outer part of this mixing layer, one often has to extrapolate the results to zero beyond the isotach $y_{0.1}$, or else define an incomplete momentum integral scale $\theta_{0.1}$ [6]. The above mentioned difficulty does not exist when the mixing occurs between parallel streams. The results shown in Fig. 1b, indicate that the momentum thickness varies linearly with downstream distance in absence of forcing with or without tripping. The slope of $\theta_{0.1}$ changes with the amplitude and frequency of forcing. The higher the frequency, the shorter the distance at which saturation occurs (i.e. $d\theta/dx$ decreases before the mixing layer resumes what appears to be its normal rate of growth).

Mean velocity profiles taken at 6 downstream stations for the three cases considered are plotted in Fig. 2.

The ordinate in this figure is a dimensionless velocity while the abscissa is the distance above or below the backward-facing step from which the mixing layer originates. For the un-

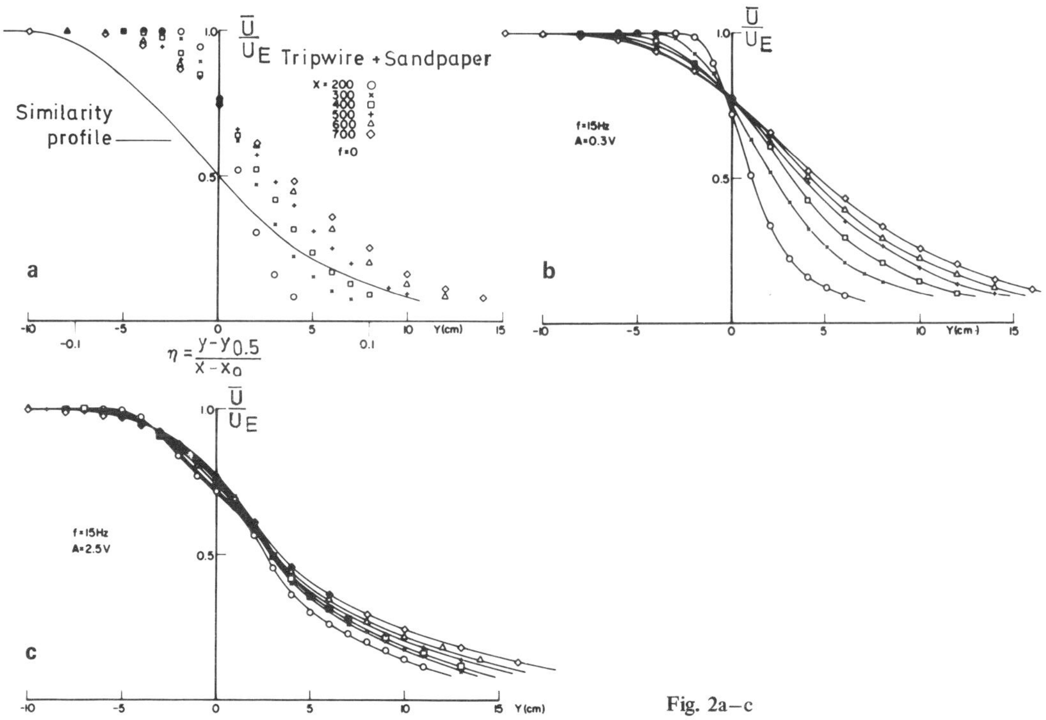

Fig. 2a–c

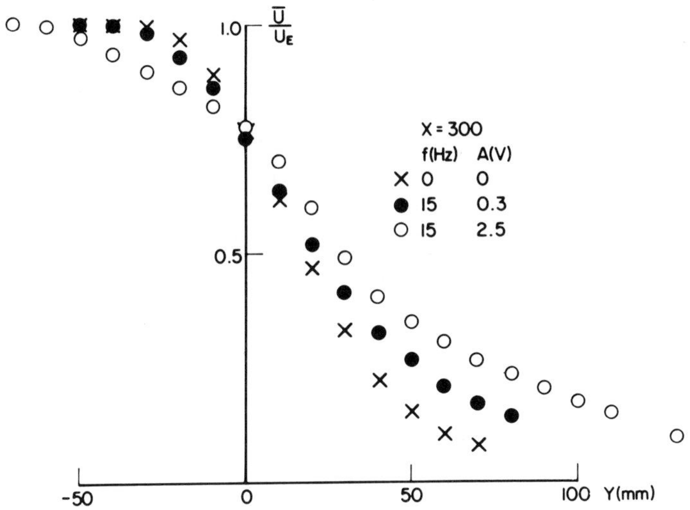

Fig. 3

forced case (Fig. 2a) as well as the weakly forced case (Fig. 2b) the flow accelerates at $y < 0$, indicating a transport of momentum from the free stream to the surrounding fluid, and suggesting that the slope of the transverse velocity ($d\bar{V}/dy$) must change sign around $y = 0$. The profiles shown in Fig. 2a are self-similar for $x \geqslant 300$ mm. The spreading parameter is $\sigma = 9.01$ and is in good agreement with the value measured by *Wygnanski* and *Fiedler* [5] and *Batt* [3]. Weakly forced mixing layer (Fig. 2b) exhibits similar characteristics to the unforced case in a sense that the flow decelerates on the high velocity side ($y < 0$) and accelerates on the low velocity side ($y > 0$). Although the flow is no longer self-preserving the velocity profile retains its basic form and does not deviate much from the accepted "error-function" profile. Large amplitudes of surging (Fig. 2c) result in a different behaviour of the mean flow field, in particular the initial growth of this mixing layer is quite spectacular. The value of θ attained at $x = 200$ mm is only attained in the unforced case at $x = 600$ mm. The initial rate of spread is not maintained at distances exceeding 300 mm with most of the growth occurring on the low velocity side. The shape of the velocity profile for $x < 500$ mm is also quite different. At large negative values of y the velocity decreases rapidly near $y = 0$ the slope $d\bar{U}/dy$ decreases, creating a local hump in the velocity profile, before increasing again at the outer part of the flow. The profile thus contains three inflection points, rather than the usual one observed for the unforced or weakly forced cases. The effect of forcing on the shape of the velocity profile can be seen in Fig. 3, where the results obtained at $x = 300$ mm, were cross plotted for various degrees of forcing. It should be remarked at this point, that the effect of tripping the boundary layer is secondary in comparison with the effects of orderly forcing.

Turbulent Intensities and Shear Stress

In the absence of forcing, the distribution of $\overline{u'^2}$, $\overline{v'^2}$ and $\overline{w'^2}$ attained self preservation at $x > 300$ mm and agreed fairly well with previously documented results. Forcing the mixing layer resulted in some unexpected behaviour of the various components of the turbulent intensity. A sample of the results measured at $x = 500$ mm are shown in Fig. 4. The maximum energy content in the streamwise fluctuations decreased roughly by 25% when forcing was applied, but the amplitude of forcing had little effect on the maximum level of $\overline{u'^2}$ (Fig. 4a). The major difference between the two forced cases is in the lateral distribution of the fluctuations. In the

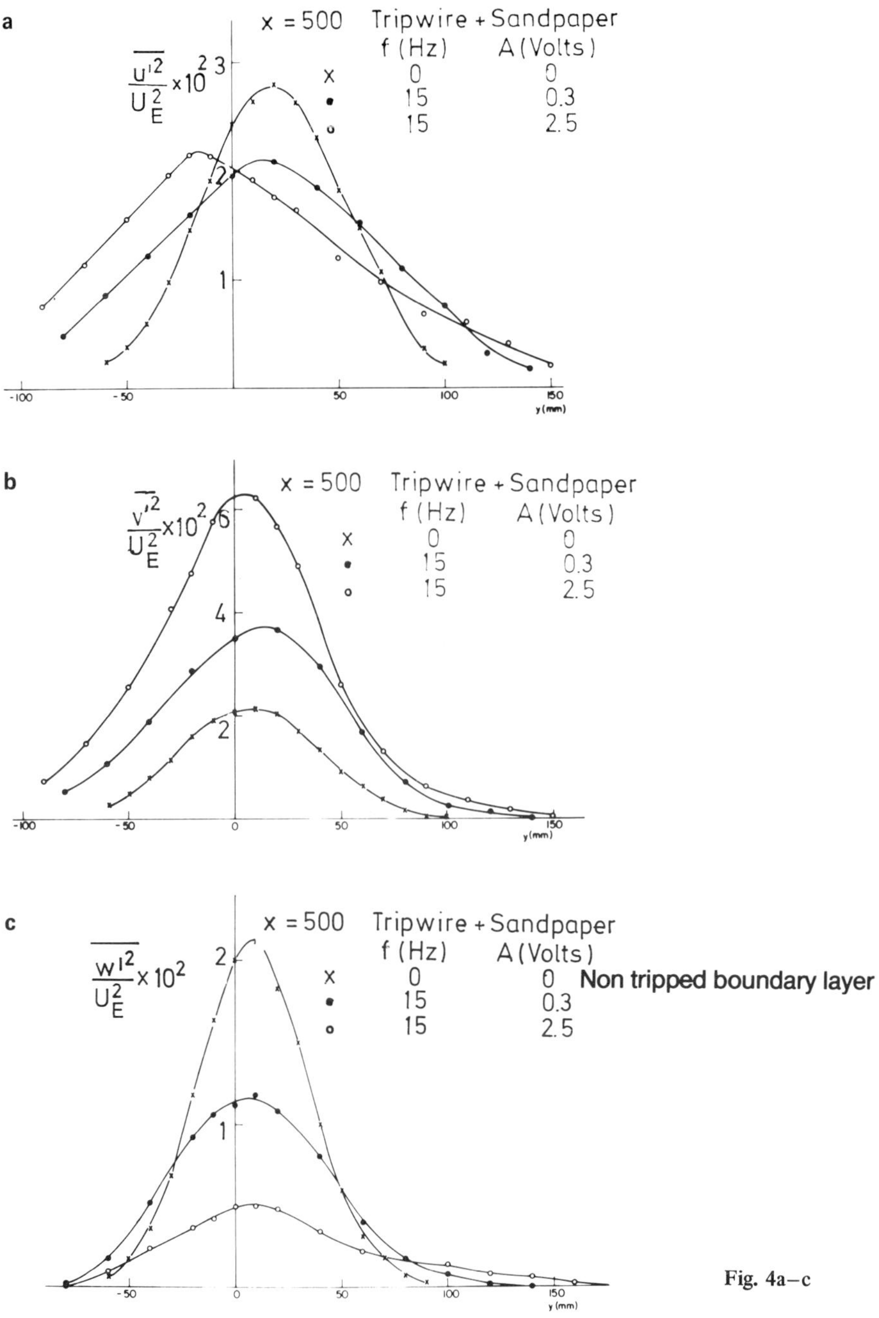

Fig. 4a–c

weakly forced case as well as in the unforced case, the peak in the streamwise component of the turbulent intensity occurs at positive values of y, while for strong forcing the location of the peak shifts to negative values of the lateral coordinate. There is also a difference in the shape of the intensity distribution which changes from an exponential distribution in the unforced case, to a triangular one in both forced cases. The intensity of the v' fluctuations increases with increasing the amplitude of forcing. This is to be expected since the forcing is applied in the lateral direction. The lateral distribution of $\overline{v'^2}$ is hardly influenced by the application of forcing, only the maximum intensity changes as forcing is applied. The maximum value of $\overline{v'^2}$ in the strongly forced case is roughly 3 times larger than the intensity in the unforced case (Fig. 4b). In the absence of external perturbations the distribution of a $\overline{w'^2}$ is roughly similar to the distribution of $\overline{v'^2}$ with peak intensity of $\overline{w'^2}/U_E^2 = 2.1 \times 10^{-2}$ (Fig. 4c). The peak intensity drops by a factor of 2 when weak forcing was applied and by a factor of 4 when strong forcing was applied, relative to the unforced case. The suppression of the spanwise component of the turbulent intensity in response to the two dimensional surging indicates that the mixing layer becomes more two-dimensional under the influence of forcing. The total turbulent intensity in the mixing layer changes rather little as a result of forcing, since most of the energy removed from the spanwise component is added to the lateral component of the fluctuations. The turbulent energy is redistributed in a manner which makes the flow more orderly and more two-dimensional than it naturally is. The distribution of energy among the various components of the turbulent intensity varies with the downstream distance.

The distribution of Reynolds stresses for various degrees of forcing is shown in Fig. 5. In the unforced case (Fig. 5a) the $u'v'$ fluctuations are self preserving for $x > 400$ mm (the abscissa in this figure is the similarity coordinate η). When weak forcing was applied, the maximum shear stress at $x = 200$ mm exceeded the shear stress for the unforced case by approximately 50%. At larger downstream distances the local maxima of Reynolds stress dropped until they reached a lowest value of

$$(\overline{u'v'}/U_E^2 * 10^2)_{max} = 0.69 \text{ at } x = 600 \text{ mm};$$

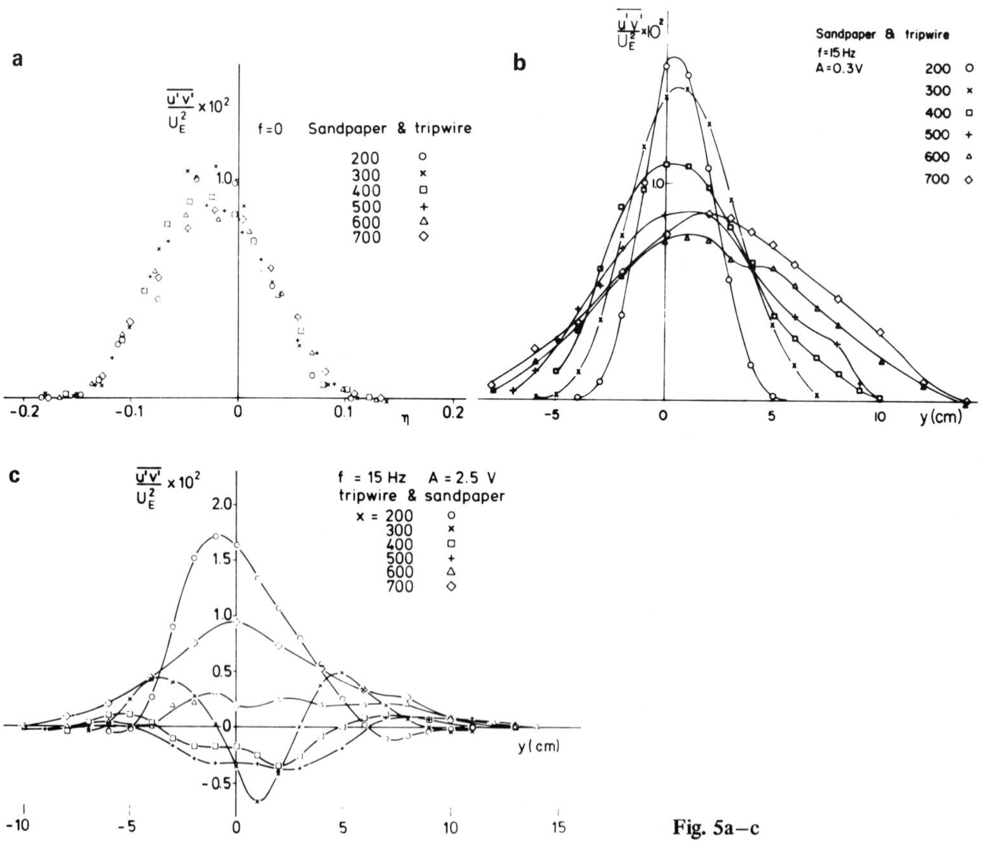

Fig. 5a–c

further downstream the value $\overline{u'v'}$ increased with increasing downstream distance and at $x = 700$ mm the value of

$$(\overline{u'v'}/U_E^2 \times 10^2)_{max} > 0.8.$$

At $x = 200$ mm, the distribution and maximum intensity of $\overline{u'v'}$ is not sensitive to the amplitude of forcing, however, 300 mm downstream of the nozzle the distribution of Reynolds stress for strong forcing is entirely different. The Reynolds stress is positive at the outer edges of the flow on both sides of the mixing layer. In the center, however, the Reynolds stress

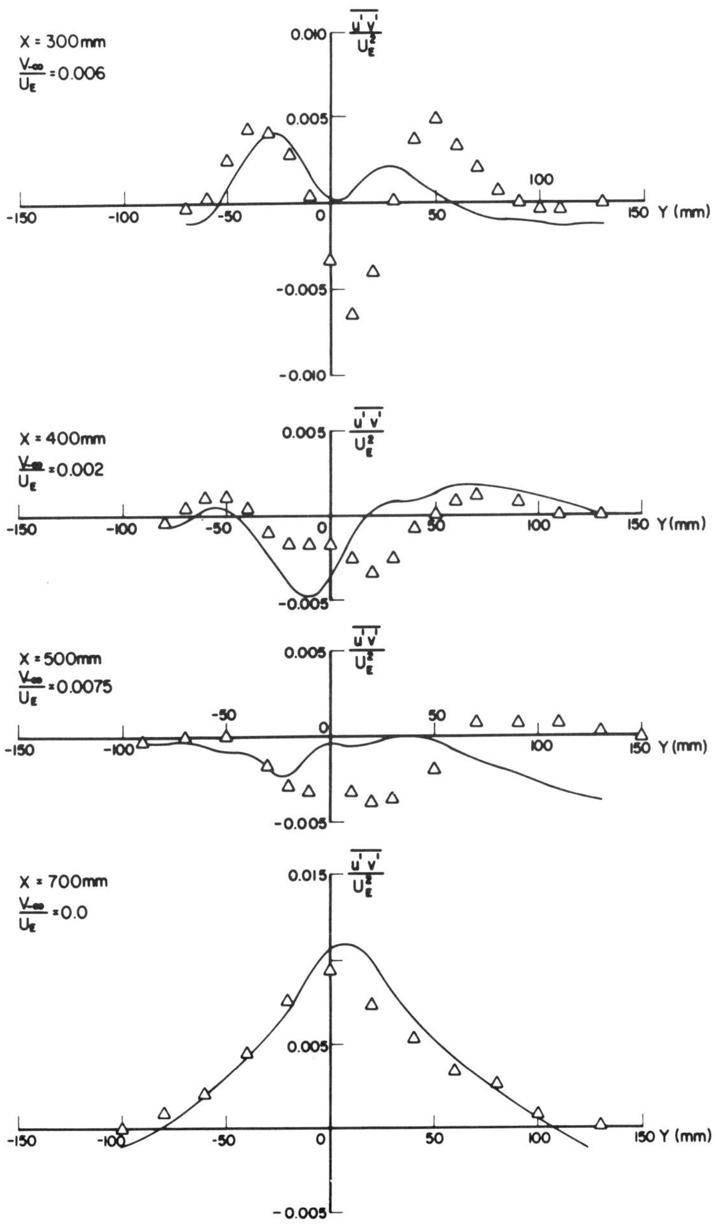

Fig. 6

becomes negative. The negative peak in the Reynolds stress for the strongly forced case is of the same order of magnitude as the maximum Reynolds stress attained in the unforced mixing layer. For $350 < x < 600$ the distribution of $\overline{u'v'}$ is almost uniform across the entire mixing layer, the level of the $\overline{u'v'}$ is very small and mostly negative. Only at $x > 600$ mm does the shear stress increase again, and at $x = 700$ mm it approaches the value of the unforced self preserving layer. Thus, the strongly forced mixing layer contains a large region for which turbulent production is negative. The negative production term may be responsible for the reduction of the longitudinal component of the turbulent intensity. The lateral distribution of $\overline{u'v'}$ shown in Fig. 5c, when viewed together with the mean velocity profiles (Fig. 2c) negates some of the early concepts of modeling turbulence. Thus, the lateral distribution of $\overline{u'v'}$ was calculated directly from the equation:

$$\frac{\overline{u'v'}}{U_E^2} = \frac{1}{U_E^2} \int_{-\infty}^{y} \left| \overline{U} \frac{\partial \overline{U}}{\partial x} + \overline{V} \frac{\partial \overline{U}}{\partial y} + \frac{\partial}{\partial x} (\overline{u'^2} - \overline{v'^2}) \right| dy$$

where

$$\overline{V} = - \int_{-\infty}^{y} \frac{\partial \overline{U}}{\partial x} dy + \overline{V}_{-\infty},$$

and all the derivatives were calculated directly from measurements. One has to assume a value of the lateral mean velocity outside the measured control volume ($\overline{V}_{-\infty}$) which is an arbitrary constant arising from the fact that the location of the dividing stream line cannot be determined from the boundary-layer equations. Using the value of $\overline{V}_{-\infty}/U_E < 0.01$, the results of the calculations for $x = 300$, 400, 500 and 700 mm are shown and compared with measured points (Fig. 6). In spite of the discrepancies between the measured and calculated values resulting from errors incurred by numerical differentiation and measurements of velocities at high turbulence levels, there is little doubt about the validity of the results.

Some Observations of the Strucutre of the Forced Mixing Layer

In this section we shall speculate about the possible structure of the mixing layer under various forcing conditions, and the possible relationship between the structure and the time averaged quantities. Because the degree of order in the strongly forced flow is highest we shall discuss this case first.

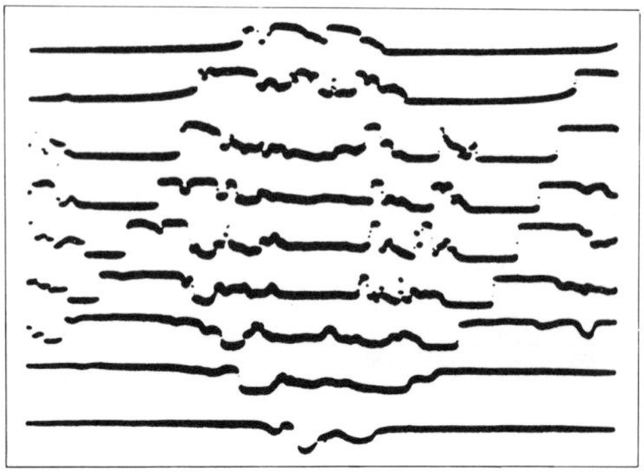

Fig. 7

Figure 7 was obtained by heating the flow above the room temperature and placing a rake of resistance thermometers across the mixing layer. The figure shows a time history of temperature during a passage of a large coherent eddy. The abscissa in this figure is time, increasing from left to right. The ordinate for each trace is proportional to the temperature with the lowest trace corresponding to a sensor placed primarily in the hot stream. Each trace is displaced vertically according to the position of the sensor in the flow. All nine traces combined, give a visual record of a large coherent eddy which resembles the smoke pictures shown in [1]. The strongly forced mixing layer consists of an array of vortices connected by a thin vortex sheet. The thin vortex sheet between the lumps of vorticity separates the fluid in its unmixed state as may be seen from the temperature jump across it. The engulfment of the hot fluid into the lump of vorticity can be detected following the vortex sheet on the bottom left hand side of the picture. Cold fluid, on the other hand, is engulfed into the eddy from the right, top-corner of the picture and is brought all the way to the high velocity side of the mixing layer. Turbulent fluctuations of temperature are mostly visible on the peripheries of the vortex lump. The central core of the vortex is relatively quiescent and fairly well mixed. The temperature in the central core, is almost an average temperature between the hot and the cold stream. This suggests that the lumpy structure resembles a hellical spring along whose surface the hot and the cold fluid are brought together increasing the contact area between them by many folds. The repeatability of the observation indicates that in the case of strong forcing, amalgamations of adjacent vortices are rare at $x \approx 500$ mm. The occurrence of amalgamation is more frequent when the degree of forcing is reduced. Three different events observed at one point in the flow are shown in Fig. 8. In the uppermost trace the lumps of vorticity correspond to the frequency of the forcing, in the middle trace an amalgamation was about to occur, the two vortices being still separated by a cold region of short duration. In the lowest trace the amalgamation occurred

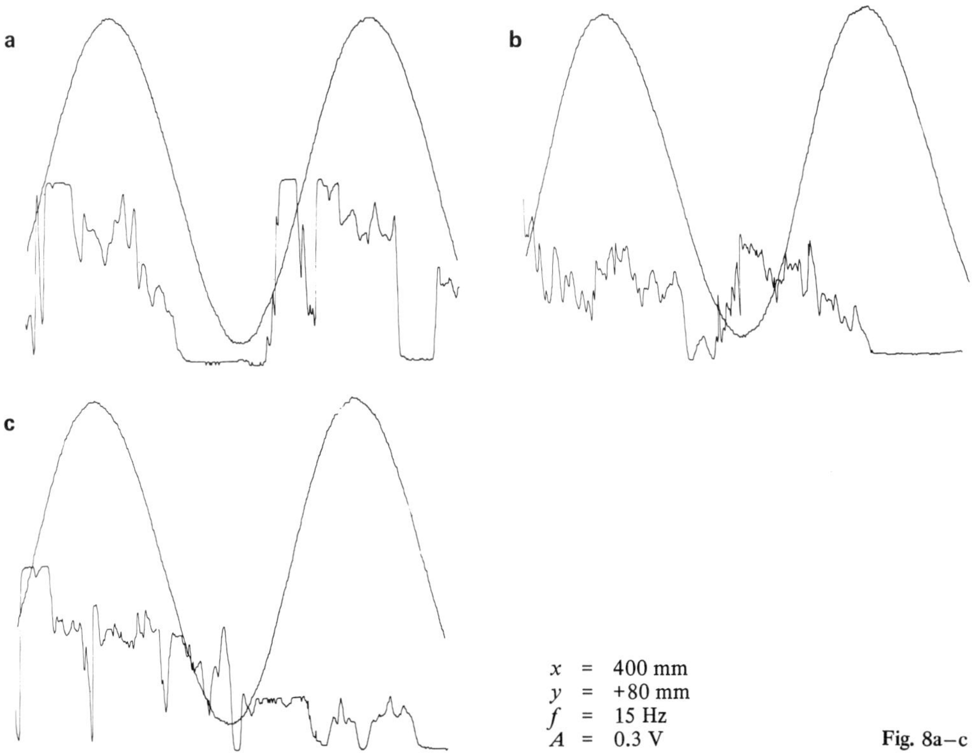

Fig. 8a–c

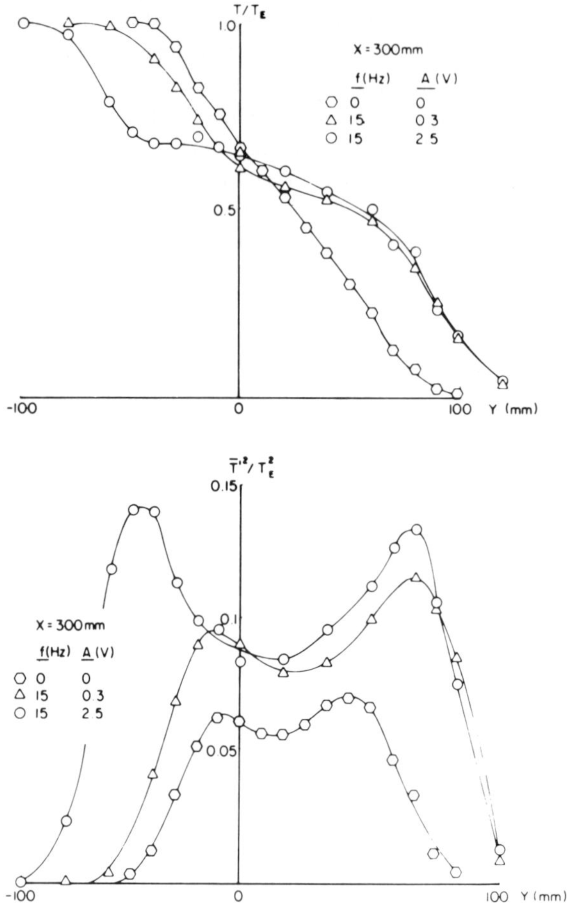

Fig. 9

earlier in time, the vortex lump was large and the temperature during its passage varied almost coninuously from the hot temperature of the free stream to the cold ambient temperature seen on the right-hand side of the trace. The weakly-forced case locks the large eddies into time scales which are mostly multiples of the forcing frequency; in the unforced case the time scales vary over a less discrete spectrum. Most amalgamations occur on the low velocity side of the mixing layer. The mode of amalgamation observed most often, is one in which adjacent eddies touch one another with the sheet of vorticity joining them totally disappearing. Only very seldom a trace of temperature was observed which could be interpreted as a "pairing process", in which one vortex lump rotates around the other during an amalgamation.

Conventionally averaged temperature distribution, measured at $x = 300$ mm for the three levels of forcing considered, and the intensity of the temperature fluctuations, are shown in Fig. 9. The distribution of temperature across the flow in a strongly forced mixing layer has three inflection points, with the temperature dropping from the hot stream to an intermediate temperature, which is fairly constant across the flow, and then dropping again to the ambient cold temperature. An orderly array of vortices whose core is fairly well mixed could give rise to such a temperature distribution. Temperature fluctuations have two local maxima at the peripheries of the mixing layer with a local minimum in the centre. The distribution of the temperature fluctuations is very consistent with the observed structure of the flow, and it is also consistent with the concentration distribution of species measured by *Konrad* [7] or *Batt* [3]. It should be noted, that even in the unforced case the profile of the temperature fluctuations has two local peaks at the peripheries of the mixing layer. Thus, forcing at a prescribed frequency introduces a length scale to which the mixing layer responds but which does not fundamentally change the structure of the flow. It should be mentioned, that the large eddies are essentially two-dimensional in the range of Reynolds numbers considered. The degree of

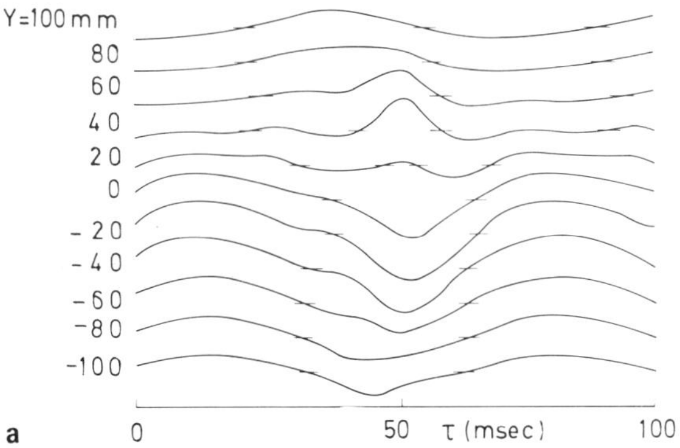

a

Ensemble average of u'v': X = 500, f = 15Hz, A = 2,5 V

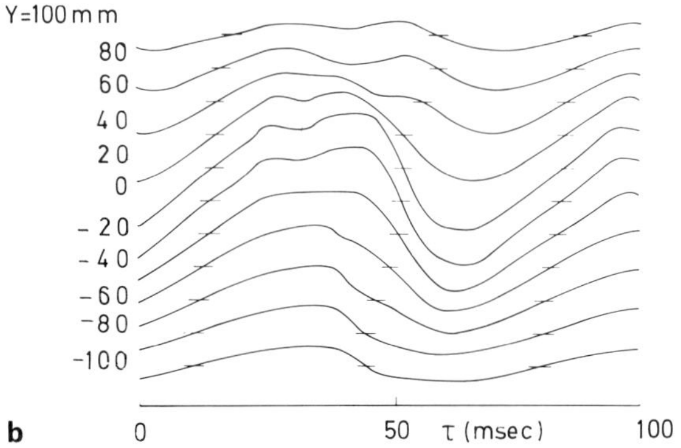

b

Ensemble average of u'v': X = 500, f = 15Hz, A = 2,5 V

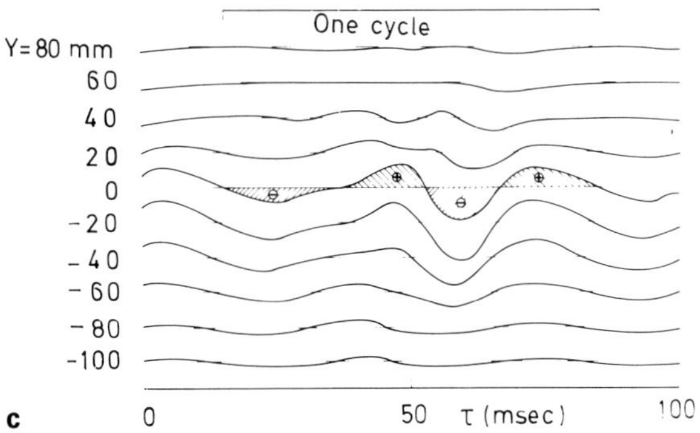

c

Ensemble average of u'v': X = 500, f = 15Hz, A = 2,5 V

Fig. 10a–c

two dimensionality is reduced whenever amalgamations are involved, since the structures may join together at one laction on their span, before doing so at another. Although the structure of the flow was discussed in terms of temperature fluctuations, heating was used as a passive contaminant mainly for the purpose of observations.

Phase-averaged data of the u' (not rms values) v', and their instantaneous product $u'v'$, shown in Fig. 10 was taken for the strongly forced case without any preselection of data (i.e. – all the data was taken). The degree of repeatability of the large structures can be deduced from the fact that the average u' and v' taken during many cycles did not vanish. The ensemble averaged longitudinal fluctuations taken at $y = 0$ mm when compared with data at $y = +40$ mm suggest that there is a change in phase of the u' fluctuations. This would naturally occur if the flow field would consist predominantly of an array of vortices, rotating in the same direction. The v' fluctuations on the other hand are all in phase across the entire flow. This is consistent with visual observations of the flow. One can see that the shear stress fluctuations are indeed the products of the two components. The area under each trace taken during one cycle is an indicator of the mean Reynolds stress measured in this flow. At most locations the area is small and negative, confirming the $u'v'$ distribution shown in Fig. 5.

Conclusions

The strongly forced turbulent mixing layer has a deterministic, two dimensional, vortex structure easily amenable to measurement. The mean velocity field, turbulent intensity and Reynolds stresses can be explained physically on the basis of the observed structures, but are inconsistent with some of the basic assumptions, used in modeling turbulent shear flows. Consequently, a strongly forced two-dimensional mixing layer offers an opportunity to study a turbulent flow in a deterministic fashion rather than in a context of a statistical approach.

It may thus serve as a good test case for modeling turbulent shear flows.

Acknowledgement. The research was sponsored in part by a grant from D.F.G., and done in collaboration between the Hermann Fottinger Institute in Berlin and the School of Engineering in Tel-Aviv.

References

1. D. Oster, I. Wygnanski, B. Dziomba, H. Fiedler: "On the Effect of Initial Conditions on the Two Dimensional Turbulent Mixing Layer" in *Structure and Mechanisms of Turbulence I,* ed. by H. Fiedler, Lecture Notes in Physics, Vol. 75 (Springer, Berlin, Heidelberg, New York 1978) pp. 48–64
2. H. Fiedler, D. Korschelt, P. Mensing: "On Transport Mechanism and Structure of Scalar Field in a Heated Plane Shear Layer", in *Structure and Mechanisms of Turbulence II,* ed. by H. Fiedler, Lecture Notes in Physics, Vol. 76 (Springer, Berlin, Heidelberg, New York, 1978) pp. 58–71
3. R. G. Batt: Turbulent mixing of passive and chemically reacting species in a low-speed shear layer. J. Fluid Mech. *82,* Part 1, 53–95 (1977)
4. D. Oster, I. Wygnanski, H. Fiedler: "Some Preliminary Observations on the Effect of Initial Conditions on the Structure of the Two-Dimensional Turbulent Mixing Layer", in *Turbulence In Internal Flows,* ed. by S. N. B. Murthy (Hemisphere, Washington 1977) pp. 67–84
5. I. Wygnanski, H. Fiedler: The two-dimensional mixing region. J. Fluid Mech. **41,** Part 2, 327–361 (1970)
6. A. K. M. F. Hussain, M. F. Zedan: Effect of initial condition on the axisymmetric free shear layer: Effect of the initial momentum thickness. Phys. Fluids **21,** No. 7, 1100–1112 (1978)
7. J. H. Konrad: " An Experimental Investigation of Mixing in Two-Dimensional Turbulent Shear Flows with Applications to Diffusion-Limited Chemical Reactions". Project SQUID Technical Report CIT-8-PU, California Institute of Technology, Pasadena, Calif. (1976)

The Mechanics of Vortex Pairing in an Axisymmetric Mixing Layer

K. B. M. Q. Zaman and A. K. M. F. Hussain

Department of Mechanical Engineering, University of Houston
Houston, TX 77004, USA

Abstract

The conditions most favorable for vortex pairing in the near field of a circular jet under controlled axisymmetric excitation have been investigated through flow visualization and hot-wire measurements. Two distinct modes of vortex pairing, inferred from the subharmonic spectral component of the *u*-signal, have been identified. The first mode is attributed to the exit shear layer instability which scales on the exit momentum thickness θ_e, and the second (jet column) mode scales on the jet diameter. Strongest pairing occurs at $St_\theta \simeq 0.012$ in the "shear layer mode" and at $St_D \simeq 0.85$ in the "jet column mode".

A typical flow under the jet column mode of vortex pairing, at $Re_D = 32{,}000$ in a 7.62 cm diameter jet, is then investigated in some detail by the method of phase-averaging. The trajectories of the two vortex centers and their longitudinal and transverse convection velocities during the pairing process are measured. Spatial distributions of the phase-average velocity, vorticity, coherent Reynolds stress, streamlines and pseudo-streamfunction are educed at a particular phase of the pairing process. These data represent field measurements at one particular phase and do not involve use of the Taylor hypothesis. The spatial distributions of the velocities and the coherent Reynolds stress show that significant contribution to the total time average Reynolds stress occurs in the region between the two pairing vortices.

Introduction

The developing regions of free turbulent shear flows are dominated by large-scale coherent structures which appear to play the key role in nearfield gross features like entrainment, mixing and aerodynamic noise production [1–3]. The evolution of the large-scale structures appears to occur through interactions (like pairing) of large-scale vortical motions [1–7], an anticascade phenomenon occurring simultaneously with the evolution of the smallscale motions through vortex stretching. When the initial free shear layer is laminar, the initial vortical structure results from the (inviscid) instability and roll-up of the free shear layer, the initial size and spacing being determined by the most unstable eigenmode of the profile. It is possible that a free shear layer resulting from an initially turbulent boundary layer can also roll up into organized vortical structures which then evolve not unlike the initially laminar case [7].

The role of vortex pairing in mixing layer growth has been the subject of extensive recent studies. *Winant* and *Browand* [6] showed that any non-uniformity in strengths or spacing of two (or three) adjacent rolled up line vortices caused them to undergo successive stages of pairing to form progressively larger coherent structures downstream. They inferred that ingestion of nonturbulent fluid during the pairing process was the primary mechanism for entrainment and thus, growth of the shear layer. *Browand* and *Wiedman* [8] have shown that significant Reynolds stress production is associated with the vortex pairing process.

Browand and *Laufer* [1], from a flow visualization experiment in a circular water jet, hypothesized that the large-scale structures just downstream of the potential core can be traced back to the upstream vortices whose generation depended on the exit shear layer instability. They observed difference in the statistical behavior (e.g. passage frequency, spatial coherence) between these initial rolled up vortices and the downstream large-scale structures, and conjectured that independence of the large-scale structures further downstream from the initial shear layer structures was achieved by successive stages of vortex pairing.

Crow and *Champagne* [9] observed the existence of orderly "vortical puffs" in axisymmetric jets, in addition to the slender vortex rings formed near the exit due to exit shear layer instability and roll up. They destroyed the "surface ripples" by tripping the exit boundary layer and studied the effect of controlled excitation on the "puff mode". They found that the "preferred mode", i.e., the one producing maximum disturbance amplification downstream, occurred at $St_D = 0.30$. They also noticed pairing of the vortex puffs giving rise to a strong subharmonic structure when driven at twice the preferred mode, namely, at $St_D = 0.60$.

We had observed before [3] that excitation at $St_D \cong 0.85$ results in even stronger subharmonic structure due to the occurrence of pairing of the vortex rings regularly in space and time. We shall show here that for a fixed Reynolds number, there are actually two Strouhal numbers of excitation that result in such strong, "stable" pairing process in the near field of a circular jet. A typical pairing process under one of these excitation conditions will then be explored in detail. Note that even under excitation, the pairing phenomenon may occur at random in space and time and only specific excitation conditions can induce "stable" vortex pairing, when a specific phase of the event may be captured pariodically at a given point in the flow, except for some unavoidable jitter.

Such a "stable" pairing phenomenon permits one to apply the method of phase-averaging [10] to study the dynamics of the structures during the pairing event. The significant advantage of this method is that only the necessary data at the phase of interest need to be sampled. Since the phase-average at a point in the flow can be obtained on-line, the procedures can be repeated at a number of spatial points at the same phase and thus a true spatial distribution of the phase-average data can be obtained in a reasonable amount of time. These spatial distributions would not involve use of the Taylor hypothesis, whose application to shear flows such as the present one is questionable.

In the presence of the coherent structures induced by controlled excitation, the instantaneous velocities $(\boldsymbol{u}, \boldsymbol{v})$ can be decomposed into time means (U, V), the coherent structure (i.e., periodic) oscillations (u_p, v_p) and the superimposed random fluctuations (u_r, v_r). Thus,

$$u(\boldsymbol{x}, t) = U(\boldsymbol{x}) + u_p(\boldsymbol{x}, t) + u_r(\boldsymbol{x}, t)$$

$$v(\boldsymbol{x}, t) = V(\boldsymbol{x}) + v_p(\boldsymbol{x}, t) + v_r(\boldsymbol{x}, t).$$

Phase-averages of the instantaneous signals give, $\langle u \rangle = U + \langle u_p \rangle$, $\langle v \rangle = V + \langle v_p \rangle$; so that, $\langle (u - U)(v - V) \rangle = \langle uv \rangle = \langle u_p v_p \rangle + \langle u_r v_r \rangle$. Thus, $\langle u_p v_p \rangle$ represents the phase-average of the Reynolds stress due to the periodically induced coherent motions, and $\langle u_r v_r \rangle$ that due to the background turbulence. The phase average of the azimuthal component of vorticity is then, $\Omega = \partial \langle v \rangle / \partial x - \partial \langle u \rangle / \partial y$.

The phase-average streamlines ψ_p at a particular phase can be drawn from the spatial distribution of the phase-average velocity vectors $(\langle u \rangle, \langle v \rangle)$ by the method of isoclines. Assuming axisymmetry of the phase-average velocity field, one can also compute the spatial distribution of streamfunctions $\langle \psi \rangle$ from the $\langle u \rangle$ data alone. The nondimensional streamfunction is defined as,

$$\langle \psi \rangle(x, y) = \frac{1}{DU_e} \int_0^y r[\langle u \rangle(x, r) - U_{ref}]dr + \langle \psi_0 \rangle,$$

where D is the diameter of the jet and U_e the exit velocity. The constant of integration $\langle \psi_0 \rangle$ which is the streamfunction on the jet centerline will be assumed zero. Because of this assumption and the assumption of axisymmetry of the phase-average velocity field, $\langle \psi \rangle$ will be referred to as *pseudo-streamfunction.* Note that both ψ_p and $\langle \psi \rangle$ depend on the velocity U_{ref} of the reference frame, while Ω is independent of the reference frame.

For a typical "jet column mode of pairing", the phase-average spatial distribution of the velocity, vorticity, coherent and background Reynolds stresses, streamlines and pseudo-streamfunctions are reported here for one particular phase of the pairing process. The measurements were carried out in a region where the two vortices undergo pairing. The chosen phase is such that the two vortices are approximately in the "radial configuration", i.e., when one vortex is inside the other during the one-stage leapfrog motion. These data reveal heretofore unreported aspects of the coherent vortical motions during pairing in a circular jet.

Experimental Procedures

The experiments have been carried out in a circular air jet facility consisting of two settling chambers in sequence. Sinusoidal perturbations in the exit profile are introduced at controlled frequencies and amplitudes with the help of a loudspeaker attached to the wall of the first chamber. The flow from the first settling chamber goes through a contraction and diffuser into the second settling chamber before exiting through the nozzle into a large room with controlled temperature and humidity. The two settling chamber arrangement was introduced for eliminating any possible asymmetry introduced by the speaker. The exit mean and turbulence profiles were checked to be axisymmetric. The absence of any harmonic of the excitation frequency in the spectra of the exit velocity signal confirmed that the excitation was indeed pure tone (sinusoidal). The study was carried out with two nozzles (of Batchelor-shaw contour) of diameters 2.54 and 7.62 cms. A 4 μm dia., 2.5 mm long tungsten hot-wire along with DISA equipment was used to obtain the instantaneous velocity signal. A Spectrascope SD335 spectrum analyzer (500 frequency lines) was used to obtain the velocity spectrum. The amplitudes (in dB) and frequencies (Hz) of the spectral peaks were read directly from the spectrum analyzer.

For the flow-visualization experiments, two streaks of titanium oxide smoke were introduced at the nozzle inlet at two diametrically opposite points in the vertical plane. Viewed horizontally under stroboscopic illumination, the smoke streaks provided a "cross-sectional" view of the ring vortices until the smoke tracers became too diffuse at $x/D \gtrsim 3.5$. The pairing activity roughly covered the range $0.5 \lesssim x/D \lesssim 2.5$.

For the specific phase of the vortex pairing event chosen for detailed field measurements, the spatial field, on a radial (x, y) plane, extended over $1.1 \lesssim x/D \lesssim 2.1$ and $0 \leqslant y/D \lesssim 1.0$. Measurements were carried out at a number of spatial grid points (15 in x and 30 in y) within this selected region. The region of measurement and the relative vortex locations for the particular phase are schematically shown in Fig. 1.

In Fig. 1, the general probe arrangements and instrumentation are also schematically shown. A reference probe (single wire) placed at $x/D \cong 1.5$ and $y/D \cong 0$ provided a periodic signal at the passage frequency of the paired vortices (i.e., at $f_p/2$, where f_p is the excitation frequency).

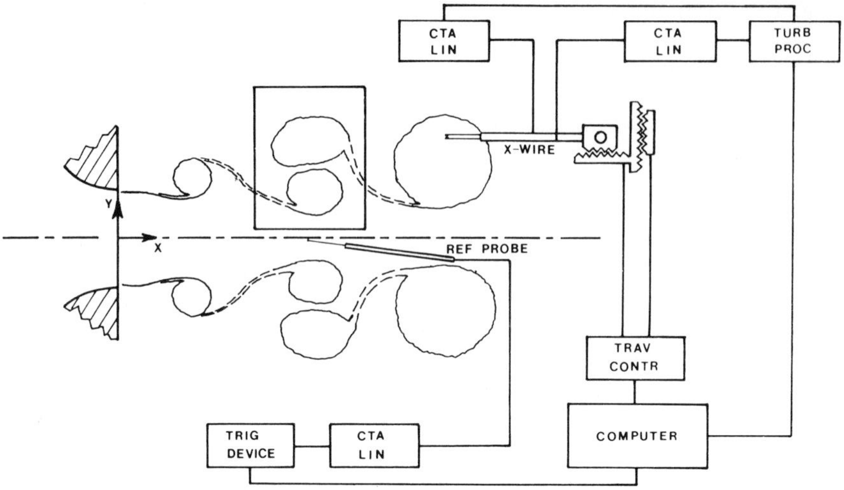

Fig. 1. Schematic of the flow and instrumentation for phase-average measurements

The reference probe signal activated a triggering device; its output (the triggering signal) was a square-wave (of frequency $f_p/2$) which in turn activated the A/D converter of the laboratory minicomputer (HP 2100S). Every time the triggering signal dropped from a set high level to the zero voltage level, three signals, namely, $u(t)$, $v(t)$ and $uv(t)$ were sampled from an x-wire probe. The "dwell" of the triggering signal at the high voltage level could be adjusted by a potentiometer, thus permitting selection of a desired phase for data sampling. The movement of the x-wire (i.e., measurement probe) in the x–y plane was facilitated by an automated traversing mechanism controlled by the computer. At a given grid location, typically an ensemble size of 1920 was used to compute the desired phase-averages.

For the contour plots and vorticity calculation, a data smoothing scheme was employed using a dynamic least-squares fit technique. For a given array of the data, say $\langle u\rangle(y)$ at a given x, a third order polynomial was fitted through five contiguous data points (after locating the coordinate origin at the middle one of the five points) and then the value at the mid-point calculated from this polynomial was substituted for the old value. Slopes of these fitted polynomials at the middle point for arrays $\langle u\rangle(y)$ yielded $\partial\langle u\rangle/\partial y$ and that for arrays $\langle v\rangle(x)$ yielded $\partial\langle v\rangle/\partial x$. These components at a point gave the phase-average vorticity Ω. Progression of the curve fitting through the entire 2-*D* array yielded smoothed values of $(\langle u\rangle, \langle v\rangle)$ and the vorticity.

Results and Discussion

Figure 2 shows the one-dimensional frequency spectra of the longitudinal velocity u at the indicated off-centerline probe locations for the cases when vortex pairing can be inferred from the subharmonic amplitude. Note that for each location, the second peak (from left) in the spectra corresponds to the frequency of excitation while the subharmonic component can be attributed to the occurrence of pairing. Note that for the five cases in Fig. 2 representing occurrence of vortex pairing (in the shear layer mode), the jet Strouhal number St_D ranges from 0.85 at $U_e = 5.21$ m/s to 2.65 at $U_e = 33.2$ m/s. However, the Strouhal number St_θ based on the shear layer exit momentum thickness θ_e is about 0.012 for all the cases. Similar data with

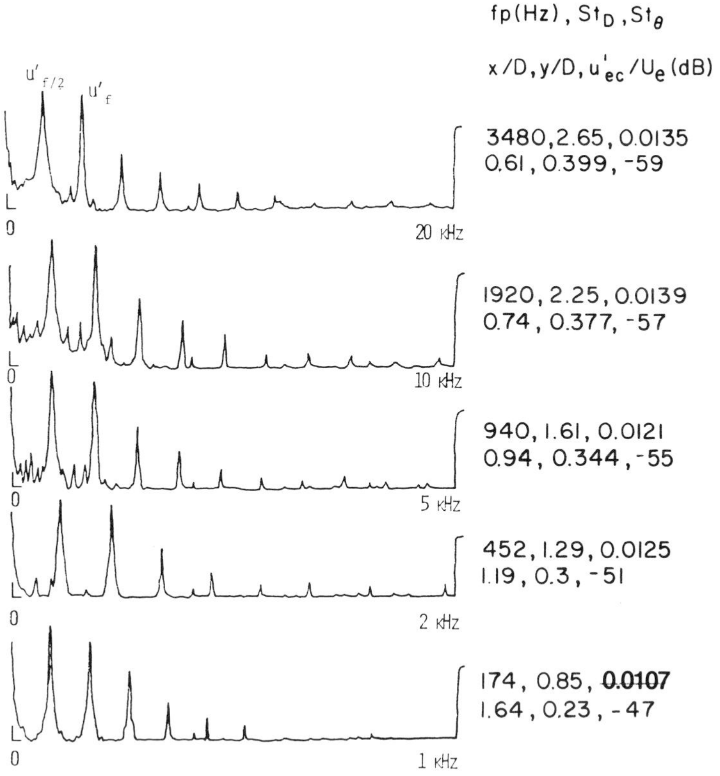

Fig. 2. Frequency spectra of the longitudinal velocity signal for shear layer mode excitations; jet dia: 2.54 cm. Vertical scales (log) are arbitrary and horizontal scales (linear) are indicated. The second peak in each plot represents the fundamental (f_p). Related information is indicated for each case

the 7.62 cm dia jet revealed occurrence of vortex pairing in the shear layer at a jet Strouhal number as high as $St_D = 8.0$ while the shear layer Strouhal number St_θ remained about 0.012. This is the "shear layer mode" and is the same phenomenon studied by *Freymuth* [5], *Winant* and *Browand* [6], *Brown* and *Roshko* [2] and many others.

On the other hand, as first shown in [3], stable vortex pairing occurs in the jet at the jet Strouhal number $St_D = 0.85$ over a large range of jet Reynolds number (i.e. independent of St_θ). This is the "jet column mode" and we have confirmed its occurrence at $St_D = 0.85$ over a St_θ range of 0.001–0.015.

These are the two "modes" for a circular jet in which controlled excitation induces strong vortex pairing in the near field. We have not observed any indication of vortex pairing with excitations at Strouhal numbers significantly different from either $St_D = 0.85$ or $St_\theta = 0.012$. In the following sections we present data to establish the occurrence as well as the characteristics of these two modes.

A series of experiments using hot-wire techniques were undertaken to find the nondimensional characteristic parameters controlling velocity fluctuation amplification and vortex pairing in circular jets. The experiments were carried out for each jet according to the following scheme. For a fixed excitation frequency f_p and probe location in the potential core, the amplitudes of the different harmonic contents in the longitudinal one-dimensional velocity spectrum were found out as a function of the exit speed. The relative exit excitation level

u'_e/U_e was set at a preselected value, at a speed that resulted in the shear layer mode pairing; this level varied slightly with the exit speed and was higher at lower speeds and lower at higher speeds. The detailed considerations employed in choosing the excitation level and the probe locations will not be discussed here because they are not central to the results presented here. For a limited number of cases, similar amplitude variation data were obtained keeping the exit speed and relative excitation level the same but varying the excitation frequency. This latter scheme could not be used at all exit speeds because sinusoidal surging of adequate level could not be obtained at appropriate frequencies corresponding to each speed U_e.

Figure 3a shows a typical plot of the fundamental and subharmonic amplitudes as a function of the exit speed U_e; the total turbulence intensity variation is shown for comparison. Note that the subharmonic amplitude peaks at two exit speeds. The total rms intensity, being a result of all the harmonic contents, also shows peaks at these conditions indicating that at these two states the subharmonic component is extremely strong and the rms value of the entire signal is essentially due to the subharmonic. The fundamental component varies much more monotonically. As we will see, whenever pairing occurs, the measured peak turbulence intensity is mostly due to the motions associated with vortex pairing. Thus, very little can be inferred about the jet sensitivity from the fundamental amplitude variation.

The striking feature of these two peaks in all similar plots is that the one on the left (Mode I) corresponds to a value of St_θ in the range 0.010–0.014 while the Mode II corresponds to a St_D value in the range 0.75 to 1.0. It is to be emphasized that, these two parameters could be varied over a wide range; viz., St_θ from 0.0001 to 0.20 and St_D from 0.02 to about 50.0. However, spectral data indicated that vortex pairing could be induced in the jet only when either $\mathrm{St}_\theta = 0.012$ or $\mathrm{St}_D = 0.85$ was satisfied.

$\mathrm{St}_\theta \cong 0.012$ has been found by us to be also the natural roll-up frequency of the laminar exit shear layer [11]. Note that this value is substantially lower than the corresponding maximally unstable frequency predicted by the spatial stability theory [12]. Similar lower natural roll-up frequency (at $\mathrm{St}_\theta = 0.0128$) was observed by *Pfizenmaier* [13] although his sound excitation experiments supported the prediction of the spatial stability theory, i.e., that the maximum growth rate of a disturbance occurs at a higher frequency (at $\mathrm{St}_\theta \simeq 0.017$). This appears to be an anomaly and is being further investigated in our laboratory. However, the present data show that excitation at the natural roll up frequency (i.e. at $\mathrm{St}_\theta \simeq 0.012$) results in strong pairing activity. This, together with the fact that a constant St_θ is associated with the first peak (Fig. 3a), indicates that the first mode is associated with the jet exit shear layer instability and hence the name "shear layer mode". The second mode of vortex pairing, on the otherhand, always occurs at $\mathrm{St}_D \simeq 0.85$, scaling on the jet diameter D and hence the name "jet column mode".

The shear layer mode, when St_D is high, is found to involve more than one stage of pairing; the peak of the subharmonic amplitude curve in such a case no longer corresponds to $\mathrm{St}_\theta \simeq 0.012$. However, the peak of the lowest subharmonic amplitude variation in such a case corresponds to $\mathrm{St}_\theta \simeq 0.012$. This is shown in Fig. 3b for the 7.62 cm jet at $f_p = 70$ Hz; the probe

---

Fig. 3. (a) Variations of the fundamental (u'_f) and subharmonic ($u'_{f/2}$) amplitudes and the total rms turbulence intensity (u'_t) as a function of exit speed U_e for the 2.54 cm jet; $f_p = 1004$ Hz; $u'_{ec}/U_e = 1\%$; $U_e = 19.0$ m/s; probe at $x/D = 2.0$ on the centerline. **(b)** Amplitude variations of u'_f, $u'_{f/2}$ and $u'_{f/4}$ with exit speed U_e for the 7.62 cm jet; $f_p = 70$ Hz; $u'_{ec}/U_e = 3\%$, at $U_e = 2.4$ m/s; probe at $x/D = 0.67$ on the centerline. **(c)** Amplitude variation of u'_f, $u'_{f/2}$, $u'_{f/4}$ and $u'_{f/8}$ with exit speed U_e for the 7.62 cm jet; $f_p = 70$ Hz; $u'_{ec}/U_e = 3\%$ at $U_e = 2.4$ m/s; probe at $x/D = 2.0$ on the centerline. **(d)** Amplitude variations of u'_f and $u'_{f/2}$ with excitation frequency f_p for the 7.62 cm jet at constant $U_e = 4.1$ m/s for two probe locations: *(I)* $x/D = 0.20$ and centerline (solid data points) *(II)* $x/D = 0.20$ and $U/U_c = 0.95$ point near the shear layer (open data points) ▶

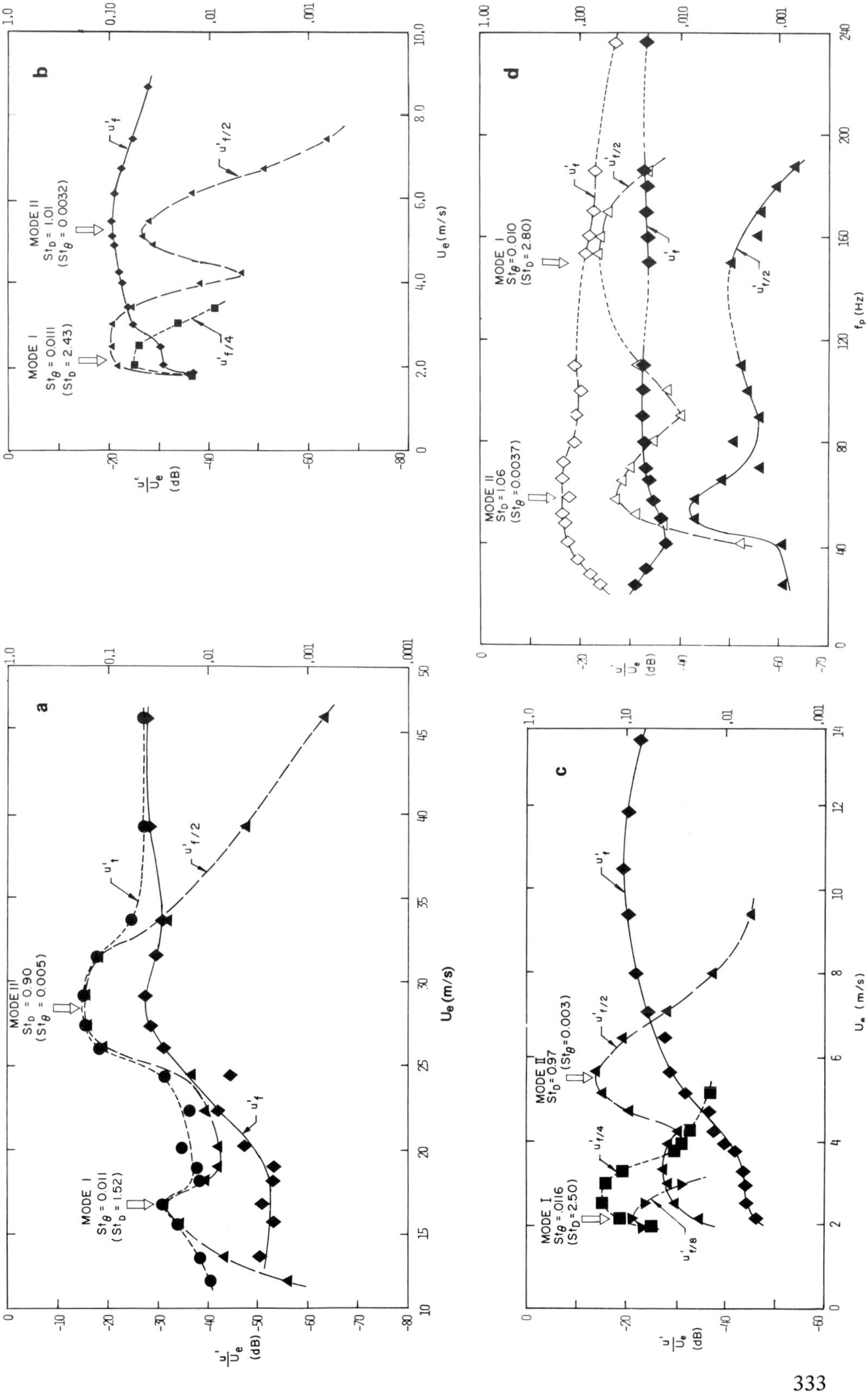

a
MODE I
St_θ = 0.011
(St_D = 1.52)
MODE II
St_D = 0.90
(St_θ = 0.005)
u'_f
u'_f/2
U_e (m/s)
u'/U_e (dB)
b
MODE I
St_θ = 0.0111
(St_D = 2.43)
MODE II
St_D = 1.01
(St_θ = 0.0032)
u'_f
u'_f/2
u'_f/4
U_e (m/s)
u'/U_e (dB)
c
MODE I
St_θ = .0116
(St_D = 2.50)
MODE II
St_D = 0.97
(St_θ = 0.003)
u'_f
u'_f/2
u'_f/4
u'_f/8
U_e (m/s)
u'/U_e (dB)
d
MODE II
St_D = 1.06
(St_θ = 0.0037)
MODE I
St_θ = 0.010
(St_D = 2.80)
u'_f
u'_f/2
f_p (Hz)
u'/U_e (dB)

position is such as to bear identical geometric location as that used for the data in Fig. 3a. Note that the $u'_{f/4}$ spectral component has its peak at a velocity corresponding to $St_\theta \simeq 0.012$. As we shall see later, the $u'_{f/4}$ component extracts energy from the $u'_{f/4}$ component as a result of which the peak of the subharmonic $u'_{f/2}$ in such a case is flattened out and its mean no longer corresponds to $St_\theta = 0.012$. Note that the second peak in Fig. 3b corresponds to $St_D = 1.0$.

The optimum speeds for the formation of the two modes as found from the two peaks in Figs 3a, b are not artifacts of the probe locations used. These peaks occur, of course with varying relative amplitudes, independent of the location of the probe. Fig. 3c shows similar data obtained with the probe located on the centerline for the same flow condition as for Fig. 3b, but with the probe located at $x/D = 2.0$ instead of at $x/D = 0.67$. The amplitudes of the different spectral components are plotted as a function of the exit speed. In this case however, a third stage of pairing associated with the shear layer mode occurs for a small range of velocities and the peak of the $u'_{f/8}$ component now corresponds to $St_\theta = 0.0116$, while the $u'_{f/4}$ and $u'_{f/2}$ components have broader peaks.

The jet mode does not show any successive subharmonic beyond the first indicating that either only one pairing occurs in the jet mode or that a second pairing occurs so far downstream that no significant motion is induced by the pairing process at $x/D = 2.0$. It is not going to be easy to resolve this question as the random turbulence becomes so dominant after the first jet mode pairing that any other stage of pairing, even if present, will be obscured.

Data (similar to those in Figs. 3a–c) with different relative excitation amplitudes but for a fixed probe location show (not presented here) that the Strouhal numbers for the two modes are the same as indicated above i.e., $St_\theta \simeq 0.012$ and $St_D \simeq 0.85$. Note that the spectral components, including the fundamental, cannot he distinguished clearly from the background turbulence after 5 diameters downstream. Very near the exit, on the other hand, the subharmonic peaks become too weak to be distinguished from the background turbulence and electronic noise.

In order to further confirm the Strouhal numbers associated with the two modes of vortex pairing it is desirable to show the jet response for fixed velocities but varying frequencies of excitation. However, the frequencies of excitation, being determined by the cavity resonance frequencies of the settling chamber combination, cannot be varied continuously; also, the amplitude of excitation achievable depends on settling chamber resonance modes. However, for a limited number of cases at relatively lower frequencies (f_p), it was possible to obtain such data which also showed the existence of the two modes. Figure 3d shows the (7.62 cm) jet response data as a function of excitation frequency at a fixed speed $U_e = 4.1$ m/s; the dashed portion indicates the frequency range where the necessary exit excitation level was unattainable. A second probe was placed at the jet exit and the relative excitation level was adjusted to the same value at each frequency. This figure covers two experiments with the probe at two transverse locations but at the same downstream distance from the exit – one at the centerline (open data points) and the other near the shear layer at $U/U_c = 0.95$ (solid data points). Amplitude variation for only the fundamental and subharmonic frequencies are recorded and no noticeable $u'_{f/4}$ component could be discerned in the spectra. As is evident, the subharmonic curves are characterized by two distinct peaks – the one on the right now represents the shear layer mode and the other the jet mode.

From several plots similar to the ones in Figs. 3a–c, the locii of the shear layer and the jet mode subharmonic peaks are shown on a St_D vs $Re_D^{1/2}$ plot in Fig. 4. For the jet mode, the region is terminated on the right because the second peak (Figs 3a–c) becomes progressively broader with increasing Re_D and location of the second peak becomes ambiguous in such cases. Both the regions terminate on the left (Fig. 4) due to difficulty in measurements at low velocities and/or non-availability of very low excitation frequencies. The shear layer mode plot is

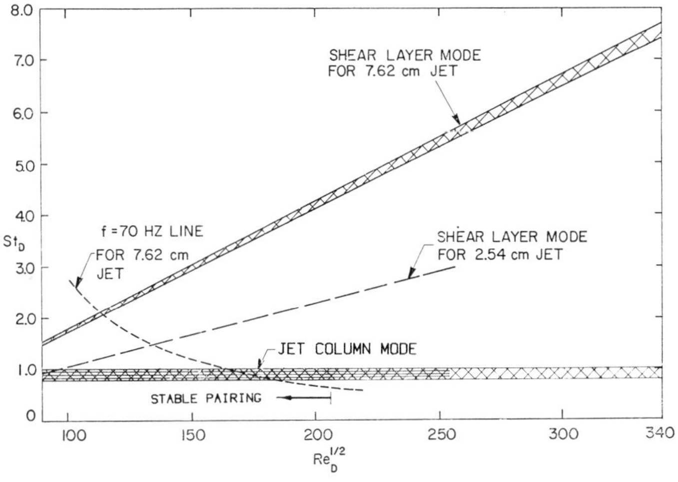

Fig. 4. Conditions for strong vortex pairing in the shear layer mode and the jet column mode

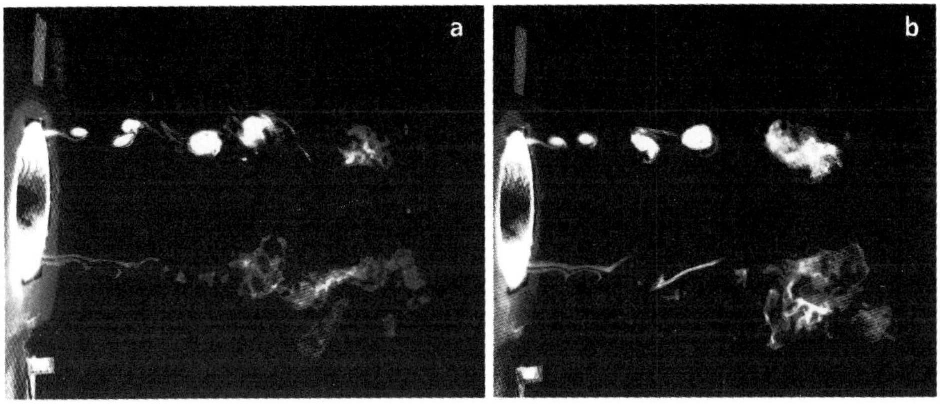

Fig. 5. Visualization pictures of the shear layer mode pairing. 7.62 cm diameter jet; $Re_D = 11{,}300$; $St_\theta = 0.011$

limited on the right by the maximum speed available and/or maximum available excitation frequency of sufficient amplitude. Note that on a St_θ vs Re_θ plot, while the shear layer mode curves for the two jets would coincide, the jet mode curves for the two jets would be apart.

Figures 5a, b show the flow-visualization pictures for a typical shear layer mode of pairing. These Figures represent two different phases corresponding to the subharmonic amplitude peak situation in figure 3(b). Note that in Figs 5a, b two stages of pairing are evident. The flow-visualization pictures of a typical jet column mode pairing are shown in Figs 6a, b. The jet column mode of pairing is investigated in detail by phase-averaging techniques and discussed in the following.

All the data presented subsequently pertain to a 7.62 cm diameter air jet at $Re_D = 32{,}000$, excited at $St_D = 0.85$ $(= f_p D/U_e)$ with exit excitation level $u_e'/U_e = 3\%$. For the unexcited jet at this Reynolds number, the exit boundary layer was laminar, the exit turbulence intensity varied from 0.35% at the centerline to about 0.5% in the boundary layer, the peak occurring about one displacement thickness away from the nozzle wall. Under excitation, the exit plane fluctuation intensity increased from 3% at the centerline to about 8% in the boundary layer, the peak in this case also occurring about one displacement thickness from the wall. The mo-

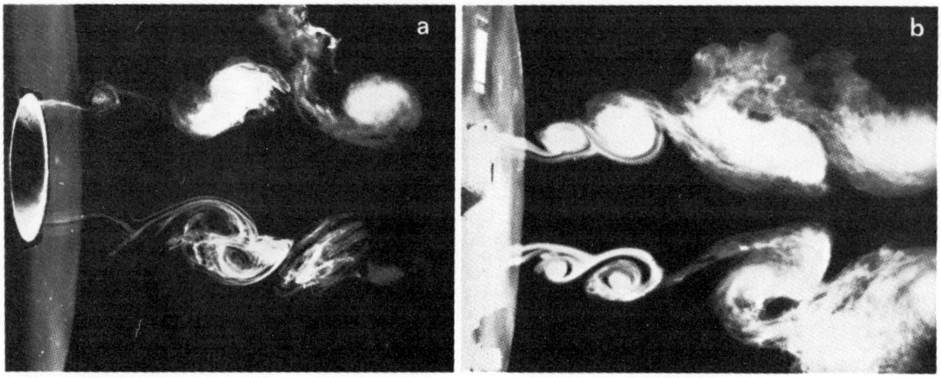

Fig. 6. Visualization pictures of the jet column mode pairing. 7.62 cm diameter jet; $Re_D = 32{,}000$; $St_D = 0.85$

mentum and the displacement thicknesses of the exit boundary layer under excitation were the same as those for the unexcited case.

Figures 6a, b show the flow-visualization pictures at two successive phases of the pairing process under study. The two vortices shortly before the "radial configuration" are captured in (a), while (b) shows them at a later stage of the pairing. The following pertinent observations were made from the flow visualization experiments. The smoke streaks roll up into distinct vortex structures about 0.2 diameters from the nozzle exit. Alternate vortices soon start attaining different characteristics as the train moves downstream. One vortex starts accelerating and shrinking in its toroid diameter while no noticeable change occurs in its core cross-sectional diameter. This will be identified as "vortex 1" (type 1). The other of the pair starts decelerating and undergoes increases in both the toroid and the core cross-sectional diameter; this will be identified as "vortex 2" (type 2). Around $x/D = 1.50$, the two vortices are one inside the other i.e., in the "radial configuration". Shortly after the "radial configuration", the inner vortex undergoes a sudden transition and diffuse smoke traces appear to unwind and wrap around vortex 2, which now is behind vortex 1. In the process, vortex 1 center is seen to recede i.e., move away from vortex 2 center until transition sets in the entire cores of the two vortices and the two can no longer be distinguished as separate entities. Around $x/D = 2.5$, traces of diffuse smoke are seen to sprout out in jet-like large-scale violent motions. This is apparently associated with the evolution of azimuthal secondary structures, three-dimensionality, and turbulent breakdown [16, 17]. The smoke traces continue to mark the cross-section of a vortical structure resulting from pairing, till about $x/D = 3\ 1/2$; the visualization technique failed to reveal any coherent structure farther downstream.

By proper stroboscopic illumination of the smoke traces, the trajectories of the centers of the two pairing vortices could be traced in the flow. A needle mounted on an $x - y$ traverse mechanism was used to accurately locate the centers of the vortices at a stroboscopically "frozen" location. The trajectories were determined by these measurements at successive small phase differences (Fig. 7a); the centers of the two vortices at each phase are shown joined by a line. Thus if one considers a diametral plane cut through one side of the two vortices and defines a "dumbbell" with the two vortex cross-section at its ends, then Fig. 7a shows the motions of this dumbbell in the $x - y$ plane as the vortex pair undergoes the leapfrog motion and merger while moving downstream. A line connecting the triangular data points will thus be the phase-average trajectory of the center of vortex 1 and the line through the circular data points will be the phase-average trajectory of the center of vortex 2.

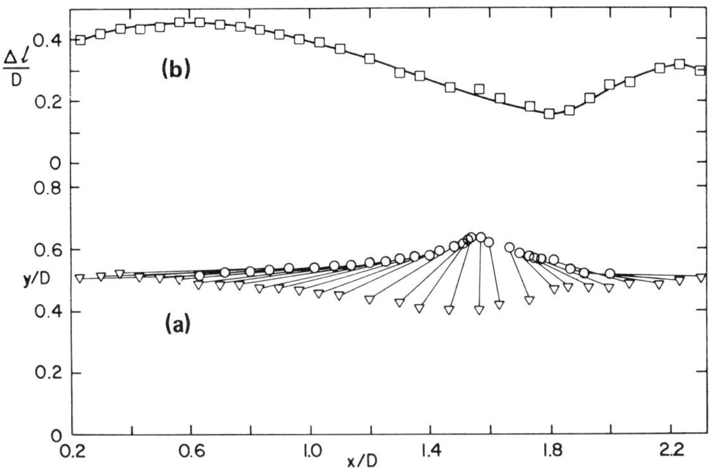

Fig. 7. (a) Trajectories of vortex centers. ▿, vortex 1; ○, vortex 2; **(b)** Δl variation

The curve in Fig. 7b shows the distance Δl of the centers of the two vortices (or the dumb-bell axis length) as a function of the axial location of vortex 1. Note that minimum Δl occurs shortly after the radial configuration where vortex 1 is seen to undergo an explosive transition. From this latter axial location onward, while smoke traces from vortex 1 wrap around vortex 2, the still identifiable center of the former is seen to recede from the latter; this is indicated by the increase in $\Delta l(x)$ after $x/D \cong 1.75$.

The vortex trajectories of Fig. 7a combined with the phase lag data for the triggering signal activating the stroboscope, permitted the computation of both the axial and the transverse components of the convection velocities of the two vortices. The signal triggering the stroboscope was obtained from the reference probe as shown in Fig. 1 and the relative phase readings were obtained by a lock-in-amplifier (PAR 129A). Further details of these measurements can be found in [15].

The radial convection velocity $v_c(x)/U_e$ for the two vortices are shown in Fig. 8a. Since the radial displacements of the center of a particular vortex for a small phase difference is very small, there are large uncertainties in $v_c(x)$; Fig. 8a shows an average smooth line through the data. $v_c(x)$ for the vortices change in sign roughly at the location of the "radial configuration". The faster moving vortex 1 reaches large radial velocities v_c/U_e of -0.20 and 0.20 shortly before and after $x/D = 1.5$ (the location of the "radial configuration"), thus undergoing large radial acceleration at $x/D = 1.5$. The radial velocity of vortex 2 is everywhere opposite in sign to that of vortex 1 and the magnitudes are relatively lower than that of the latter.

The streamwise convection velocity $u_c(x)/U_e$ for the two vortices are plotted in Fig. 8b as a function of x/D for $x/D \lesssim 4$. Note that for this figure the above mentioned phase-locked visualization technique has been used to obtain the data up to $x/D = 2.5$; the individual identities of the two vortices are lost farther downstream, and u_c/U_e in this region was computed by phaselocked measurements based on analog signals by treating each paired vortex as a single entity.

The present data show that both the vortices accelerate after rolling up, then as the interactions begin, vortex 1 continues to accelerate but vortex 2 starts decelerating. At $x/D = 1.5$, the former reaches its peak velocity while the latter reaches its lowest. At this point, vortex 1 has a velocity even higher than the jet core mean velocity by about 25%, while vortex 2 velocity is about 1/4th that of vortex 1. The little knot is Fig. 8b after $x/D = 2$ is not unambiguous

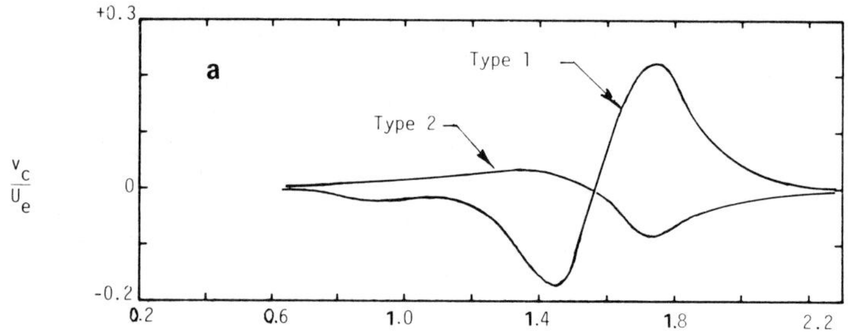

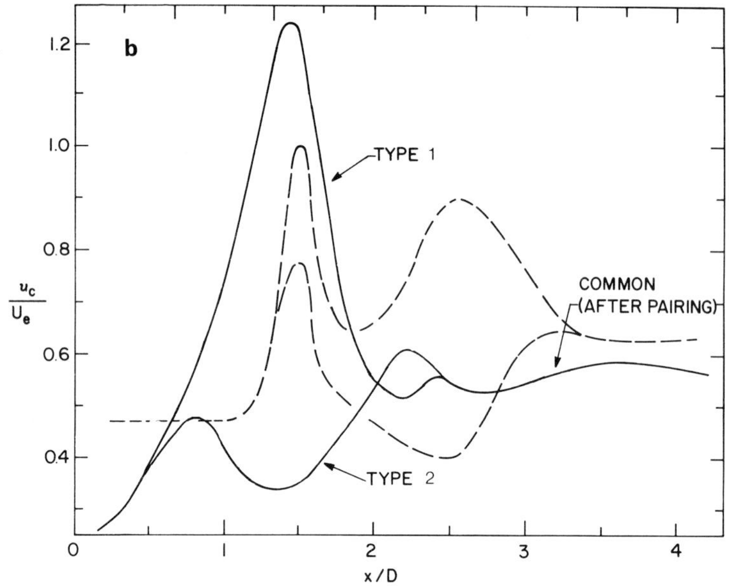

Fig. 8 a, b. Convection velocities of the vortex centers **(a)** v_c; **(b)** u_c; --- data of [1]

but can be ascribed to the continued "leapfrogging" tendency between the two vortices. The accuracy of the measurements for u_c is about ±5% of U_e, and the uncertainty in the data resulted mainly from errors in determining the locations of the centers of the vortices.

The dashed lines in Fig. 8b represent *Browand* and *Laufer*'s [1] data in a circular water jet at $Re_D = 5{,}000$. There are large differences between these and the present data which could be due to a number of reasons. First, the Reynolds number dependence. *Davies* and *Baxter* [17] have shown that with increasing Re_D, the roll up of the vortex rings occur progressively closer to the lip. It appears therefore that the location of pairing, say as denoted by the location of maximum velocity difference between the two vortices, should also move upstream with increasing Re_D. Second, the boundary condition effect. The experiment in [1] involved a confined jet in which the confinement effect on the behavior of the vortices may not be negligible [18]. Third, the pairing phenomenon in the present study is under controlled excitation. Although it is not clear whether the pairing mechanism is somewhat affected by the excitation and whether it is different from one occurring naturally, the present data, being phase-averages, are representative of all the pairing vortices under this flow condition. It appears that the convection velocities shown in [1] pertain to one realization of the interaction and thus may not be representative of all pairing vortices.

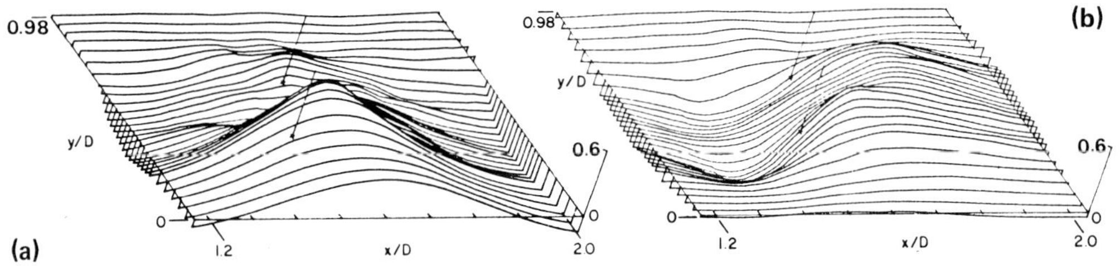

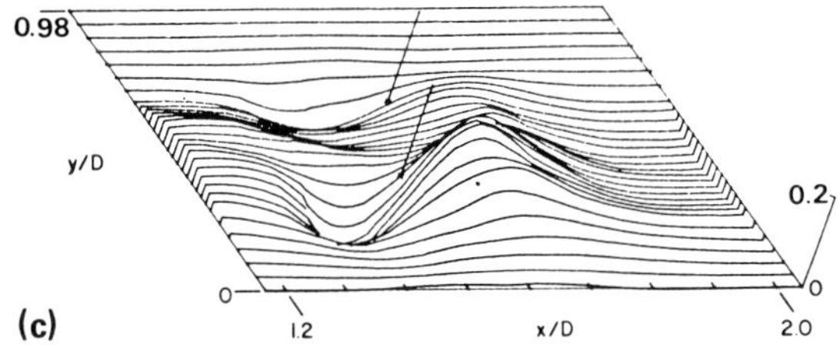

Fig. 9 a–c. Variations of: **(a)** $\langle u_p \rangle / U_e$; **(b)** $\langle v_p \rangle / U_e$; and **(c)** $\langle uv \rangle / U_e^2$

The phase-average $\langle u_p \rangle$ distributions as functions of x are shown in Fig. 9a. The corresponding $\langle v_p \rangle$ profiles are presented in Fig. 9b. The profiles are staggered in y and presented at an oblique view in order to provide a clear perception of the velocity distributions. The x and y scales are indicated while the vertical bar on the right hand side corner gives the velocity scale, i.e., at any (x, y), the $\langle u_p \rangle$ or the $\langle v_p \rangle$ value is the vertical distance from the base plane to the respective curve according to this scale. The centers of the two pairing vortices as determined from the vorticity contours are shown by the vertical lines with a cross-sign in the base plane. Curves at larger y stations crossing those at smaller y stations are not shown for clarity.

Note that the phase-averaging technique produces data with smooth spatial variations even though the instantaneous signals contain large-amplitude random fluctuations and thus sift out the coherent structure signatures. Although due to the presence of the two interacting vortices, the $\langle u_p \rangle(x)$ and $\langle v_p \rangle(x)$ variations are complicated, except in the region between the two vortices, the $\langle u_p \rangle(x)$ profiles appear symmetric about a transverse plane containing the vortex centers, with respect to which $\langle v_p \rangle(x)$ profiles appear antisymmetric. Thus in the regions except in between the two vortices, the $\langle u_p v_p \rangle(x)$ profiles would be expected to be antisymmetric about the transverse plane containing the vortex centers, and with a spatial frequency twice that of either $\langle u_p \rangle(x)$ or $\langle v_p \rangle(x)$. This is indeed observed in the $\langle uv \rangle(x)$ profiles in Fig. 9c. Note that $\langle u_r v_r \rangle$ in these regions being negligible compared to $\langle u_p v_p \rangle$, the $\langle uv \rangle$ values in Fig. 9c very closely approximate $\langle u_p v_p \rangle$ values. In the intermediate region between the two vortices, where the induced velocities of the two act in opposite directions and hence there being high shear, $\langle uv \rangle$ values are mostly positive over the extent of the vortices. Thus a "zone averaging" (i.e., averaging $\langle u_p v_p \rangle$ data in x over the extent of the vortices) shows large average correlations in those regions; see [19] for the data. This would indicate that the pairing activity contributes significantly to the total (time average) Reynolds stress.

The contours of constant phase-average vorticity Ω are shown in Fig. 10. The vorticity data have been non-dimensionalized by the excitation frequency f_p. The outermost contours in Fig. 10 (i.e., $\Omega / f_p = 0.5$) has considerable scatter and thus has been shown by a dotted line. On the low-speed side, the vorticity contours are marked by a "dip" which is a result of the combined effects of triggering signal jitter and flow-reversal occurring in those locations [19].

The smoke trace pictures in Fig. 6 show the "streaklines" at particular instants of time. The smoke traces introduced within the boundary layer upstream from the lip marks vortical

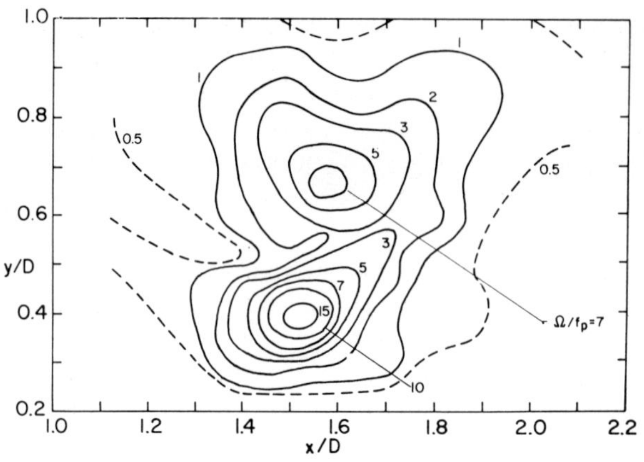

Fig. 10. Contours of phase-average vorticity Ω

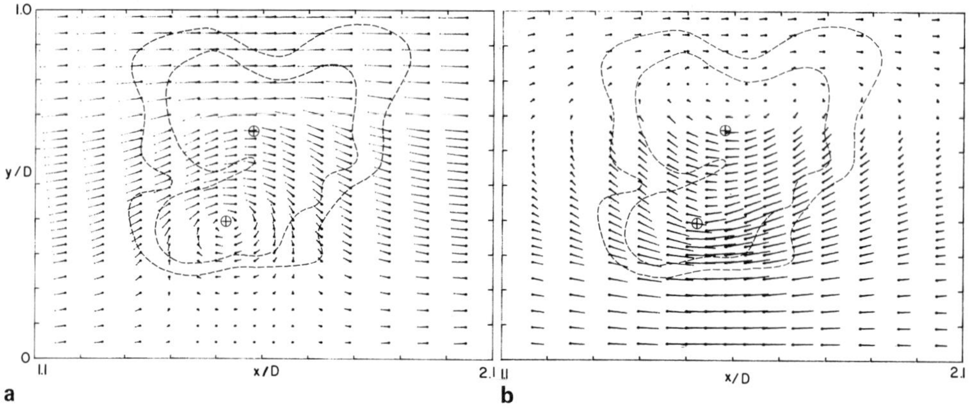

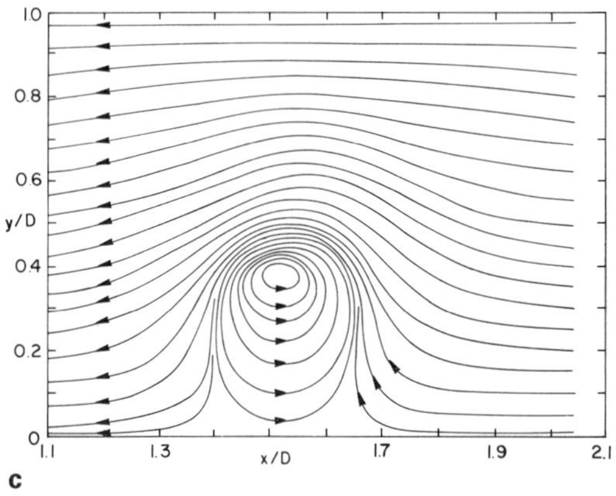

Fig. 11 a–c. Phase-average velocity vectors: **(a)**, $U_{ref} = 1.25\ U_e$; **(b)** $U_{ref} = 0.35\ U_e$; **(c)**, the streamlines ψ_p corresponding to (a)

fluid emanating from the jet. Since the vortical fluid elements accumulate in the vortex core the smoke tracers remain in the core and thus continue to mark the vortex core, even though its concentration can decay gradually due to diffusion or undergo obliteration when transition sets in. The visualization pictures thus reasonably show the shapes and sizes of the vortex cores. Indeed, the phase-average vorticity contours, except for the low-speed side distortions, bear excellent resemblance to the smoke trace pictures. For example, compare Fig. 10 with Fig. 6a.

The spatial distribution of the phase-average velocity vectors are shown in Fig. 11a, which provides a qualitative picture of the phase-average streamline patterns (ψ_p). The iso-vorticity contours from Fig. 10 (for $\Omega/f_p = 1$ and 2) are shown in this figure as dotted lines for reference; the vortex centers are marked by the ⊕ sign. These streamline patterns, unlike the vorticity contours, are not invariant under Galilean transformation. That is, the pattern would be entirely different if they are drawn with respect to a different reference frame. The coordinate translation velocity used in Fig. 11a is $U_{ref} = 1.25\ U_e$ which is roughly the longitudinal convection velocity of vortex 1 at this location and phase. Thus, closed streamlines appear around vortex 1 only, while vortex 2 remains hidden. On the other hand, when $U_{ref} = 0.35\ U_e$ (the convection velocity of vortex 2) is used, closed streamlines appear approximately around the vortex 2 center. This can be seen in Fig. 11b where vortex 1 is completely obliterated. The much higher convection velocity of vortex 1 influences the streamlines in the latter figure resulting in an elongated and somewhat elliptical closed contours around vortex 2. In order to provide an example of the nature of the streamlines drawn from the spatial distribution of the velocity vectors, the streamline pattern corresponding to Fig. 11a are shown in Fig. 11c.

If the streamline patterns were drawn with respect to the fixed laboratory frame, they would essentially be parallel straightlines directed downstream. The streaklines shown in Fig. 6, however, resemble closely the vorticity contours shown in Fig. 10. We observe from this that the streaklines in the smoke pictures, even in this complicated flow situation, closely approximate the vorticity contours and are therefore independent of Galilean transformation.

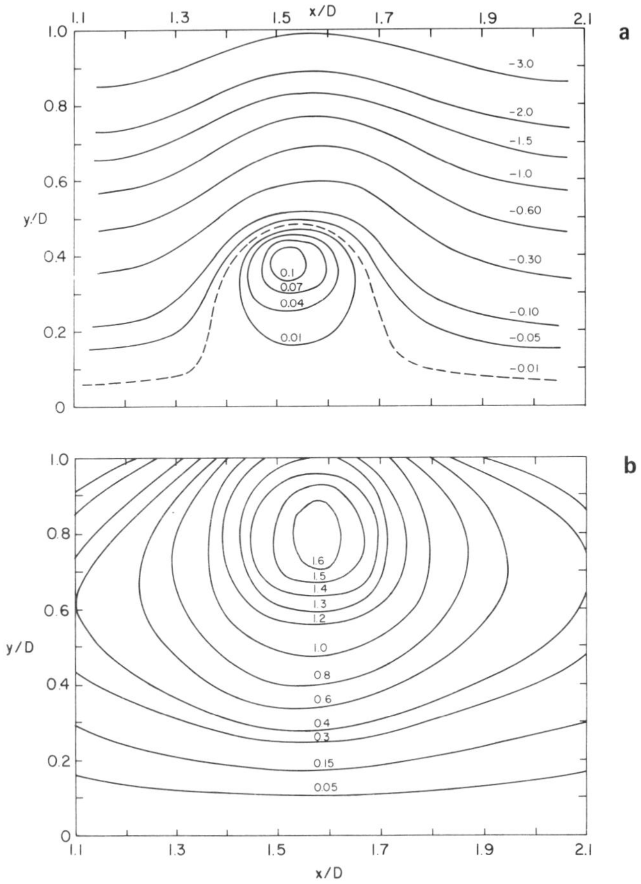

Fig. 12 a, b. Contours of the pseudo-streamfunction $\langle\psi\rangle$: **(a)**, $U_{ref} = 1.25\ U_e$; **(b)** $U_{ref} = 0.35\ U_e$

Thus the shapes and sizes of the vortices indicated by the smoke pictures should remain unaltered even if the camera is moved with respect to the vortices.

The contours of the phase-average pseudo-streamfunction $\langle\psi\rangle$ corresponding to the situations in Fig. 11a, b are shown in Figs. 12a, b, respectively. The agreement between the streamline patterns and the $\langle\psi\rangle$ contours is satisfactory on the high-speed side but large differences exist on the low-speed side. Thus, even though the closed contours in Fig. 12a and in Fig. 11a [or (c)] agree reasonably well (since the vortex is situated on the high-speed side), the agreement between the pattern in Fig. 11b and the contours in Fig. 12b is not as satisfactory. While the velocity vectors on the low-speed side of Figs. 11a, b essentially indicate parallel, straight streamlines, an outward bulge is indicated by the $\langle\psi\rangle$ contours, especially on the low-speed side and around a transverse line passing through the vortex centers. This discrepancy has been shown to be at least partially due to flow reversal occurring on the low-speed side where the hot-wire rectified the velocity signal corresponding to instants of upstream motion [19]. The general agreement between the ψ_p and $\langle\psi\rangle$ contours, however, shows that streamline patterns obtained by single wire measurements can be reliable.

Concluding Remarks

Vortex pairing in circular jets can occur in two distinct modes: the shear layer mode and the jet column mode. The shear layer mode is associated with the exit shear layer instability mechanism and the strongest pairing occurs when the excitation frequency matches the natural roll-up frequency. Strouhal number based on exit boundary layer momentum thickness θ_e for this mode of vortex pairing is found to be $St_\theta \simeq 0.012$. In the jet column mode of vortex pairing, the phenomenon scales on the jet diameter and most pronounced pairing occurs at $St_D \simeq 0.85$. The velocity fluctuation intensity in the near field of a round jet reaches relative maxima when it undergoes vortex pairing in either of the two modes.

Some details of the vortex pairing mechanism in a circular jet are unveiled by phase-average spatial measurements. The convection velocity measurements during pairing for the two vortices indicate that they undergo large relative motions and as a result, the fluid in between the two vortices undergo large shear. Distribution of the phase-average longitudinal and transverse velocities as well as the phase-average Reynolds stress indicate that a significant contribution to the total (time-average) Reynolds stress occurs from the high shear of fluid in between two pairing vortices. The phase-average vorticity contours bear excellent resemblance to the smoke trace pictures at the corresponding time and thus it follows that the smoke trace pictures should be independent of the reference frame velocity. Comparison of the pseudo-streamfunction contours, computed from the spatial distribution of $\langle u\rangle$, with the streamline patterns drawn from the distribution of the actual velocity vectors show reasonable agreement with each other. This report highlights a few aspects of the results of our continuing investigation in the field.

Acknowledgements. This research is being supported by the NASA Langley Research Center under grant NSG-1475 and the Office of Naval Research under grant 76-C-9128.

References

1. F. K. Browand, J. Laufer: *Symposium on Turbulence,* Vol. 5 (University of Missouri-Rolla 1975) p. 333
2. G. Brown, A. Roshko: J. Fluid Mech. **64,** 775 (1974)
3. A. K. M. F. Hussain, K. B. M. Q. Zaman: *Proc. 3rd Interagency Symposium on Transportation Noise* (University of Utah 1975) p. 314
4. A. Roshko: AIAA J. **10,** 1349 (1976)
5. P. Freymuth: J. Fluid Mech. **25,** 683 (1966)
6. C. D. Winant, F. K. Browand: J. Fluid Mech. **63,** 237 (1974)
7. A. R. Clark: Ph. D. Thesis, University of Houston (1979)
8. F. K. Browand, P. D. Wiedman: J. Fluid Mech. **76,** 127 (1976)
9. S. C. Crow, F. H. Champagne: J. Fluid Mech. **48,** 547 (1971)
10. A. K. M. F. Hussain, W. C. Reynolds: J. Fluid Mech. **41,** 241 (1970)
11. A. K. M. F. Hussain, K. B. M. Q. Zaman: 1978 J. Fluid Mech. **87,** 349 (1978)
12. A. Michalke: J. Fluid Mech. **23,** 521 (1965)
13. E. Pfeizenmaier: Doktor-Ingenieur Thesis, Technische Universität Berlin (1973)
14. K. B. M. Q. Zaman, A. K. M. F. Hussain: (Submitted to the 1st International Symposium on Turbulent Shear Flows, Pennsylvania, USA 1977)
15. K. B. M. Q. Zaman, A. K. M. F. Hussain: (Submitted to J. Fluid Mech.)
16. A. J. Yule: J. Fluid Mech. **89,** 413 (1978)
17. P. O. A. L. Davies, D. R. J. Baxter: "Transition in Free Shear Layers", in *Structure and Mechanisms of Turbulence I,* ed. by H. Fiedler, Lecture Notes in Physics, Vol. 75 (Springer, Berlin, Heidelberg, New York 1978) pp. 125–135
18. H. A. Becker, T. A. Massaro: J. Fluid Mech. **31,** 435 (1978)
19. A. K. M. F. Hussain, K. B. M. Q. Zaman: (Submitted to J. Fluid Mech.)

Part V

Environmental Flows

Geophysical Flows – A Brief Introduction and Some General Remarks

J. C. André

Direction de la Météorologie, EERM/GMD, 73 rue de Sèvres,
F-92100 Boulogne, France

In the 2nd Symposium on Turbulent Shear Flows, two separate sessions on environmental flows were included for the first time and this reflects the increasing impetus being given to studies of turbulence phenomena in the atmosphere and oceans. The field encompasses a very wide range of topics and those workers concerned with environmental flows have contributed over a long period to general progress in turbulence research. As examples, we can refer to the early work of *Lilly* [1] and *Deardorff* [2] on sub-grid modelling and to the work of *Lumley* [3] and *André* [4] and their co-workers on convectively driven turbulence. It is perhaps worth noting that this last topic was also the subject of three papers in the preceding volume of this series through the contributions of *Zeman* and *Lumley* [5], *André* et al. [6] and *Wyngaard* and *Sundararajan* [7]. Thermally driven turbulence is also of interest in many geophysical problems and the recent paper of *Hopfinger* et al. [8] gives a review of work in this area.

At the present time, it seems that the methods of predicting the development of turbulent flows, such as the so-called higher order closure models, have been developed to a point where they may now be applied to a variety of interesting environmental flows. It is to be hoped that this will lead not only to a better understanding of the dynamical and diffusive properties of these flows but it may also provide new opportunities to test these prediction methods in situations outside the scope of more traditional turbulence research. For example, these could range from flows at very high Reynolds numbers and Rayleigh numbers on the one hand to those involving significant Coriolis forces on the other.

The three papers which have been selected for this chapter deal with a diversity of topics. In the first paper by *Wyngaard,* the various features of the convective atmospheric boundary layers under stable und unstable conditions are discussed on the basis of both theoretical arguments and experimental evidence. Some suggestions are offered about possible developments in turbulence modelling to account for the observed features of the atmospheric boundary layer. In the same field *Lewellen, Teske* and *Sheng* in the second paper describe the application of a particular second-order closure model to a variety of geophysical flows ranging from a gust front arising from a thunderstorm to a fairly detailed discussion of the predicted longitudinal roll vortices that occur in the unstable planetary boundary layer. Finally, and in complete contrast to the first two papers, *Chatwin* and *Sullivan* discuss the structure of diffusing clouds.

Present Studies and Extensions

The two most important processes which affect and modify environmental flows are rotation and buoyancy. With respect to the latter phenomenon, the combined effects of gravity and density variations can give rise to two very different situations, namely unstable and stable stratification. A whole book has been devoted to this subject [9] and we shall only here comment briefly on some specific points.

In the unstable case, it is now recognized that in addition to the general three-dimensional turbulent motions, larger scale quasi-two-dimensional structures (or eddies, or rolls) are also present. These structures have been observed on numerous occasions by meteorologists and appear very clearly in satellite pictures as the so-called "cloud streets". This two-fold structure has been documented, for various geographical locations, by *LeMone* [10]. Based on micrometeorological measurements, *Wyngaard* proposes here quantitative estimates of the influences of the large scale eddies on the turbulence characteristics near the ground. It is essentially confirmed that the height of the planetary boundary layer has to be included for a proper parameterization of the statistics involving the fluctuations of horizontal wind components. *Wyngaard* recommends also the use of relatively sophisticated turbulence models for the simulation of the unstable planetary boundary layer (see also [11]) and makes some specific suggestions for the treatment of pressure fluctuations and third-order terms. These suggestions confirm rather well the theoretical arguments developed in *Turbulent Shear Flows I* ([7, 8]) for the description of the counter-gradient flux of eddy kinetic energy. They are also consistent with results from laboratory experiments in water tanks by *Willis* and *Deardorff* [12] and *Adrian* and *Ferreira* [13]. *Lewellen* et al. present a numerical simulation of the unstable planetary boundary layer with the aid of a second-order turbulence model discretized on a two-dimensional grid. The large scale eddies, i.e. the horizontal rolls, are indeed recovered and it is confirmed that they contribute largely to second order correlations and particularly to the horizontal wind variance. In the studies by *Wyngaard* and *Lewellen* et al., the appearance of the horizontal rolls is attributed to convective instability. It should however be kept in mind that other mechanisms (i.e. the so-called "inflexion point" instability) can also account for the generation of horizontal rolls (see e.g. [14]) and that numerical two-dimensional simulations in the neutral case (see e.g. [15]) also exhibit large scale structures of the same type. These two kinds of instability studies need to be extended in order to include more comprehensive turbulence models and this might possibly be achieved along the lines outlined by *Etling* and *Wippermann* [16, 17].

The alternative situation of stable stratification is also of interest in the study of the planetary boundary layer. One has to account for the mechanisms by which the stable stratification progressively destroys the turbulence propagating from below and whose destruction rate is directly related to the underlying mixed layer growth rate. Since the early experimental work by *Rouse* and *Dodu* [18], this subject has received much attention from oceanographers and meteorologists. Many experiments have been conducted and many theories have been proposed, some of which have been recently reviewed and compared ([1, 19, 20]). In authors's opinion, and despite the important progress which has been made during the last ten years, this question is still open and much work remains to be done. Another closely related problem concerns the structure and evolution of the nocturnal boundary layer, where turbulence generated by wind shear near the ground has to survive against the stable stratification created by radiative emission of the ground and of the atmospheric water vapour. The basic difficulties concern not only the experimental approach – turbulence intensity is very low in stable situations- but they are also of a theoretical nature and involve complex phenomena such as the interaction of breaking waves with turbulence. These difficulties and uncertainties account for the abundance and variety of papers which have been written on this subject. Among these one should recall the paper by *Zeman* and *Lumley* [6] which appeared in the preceding proceedings and the four papers by *Brost* and *Wyngaard* [21], *André* et al. [22], *Zeman* [23] and *Caughey* et al. [24]. These last four papers appeared in the same journal in less than one year!

An important topic in the study of environmental flows is turbulent diffusion of contaminents, either passive or reactive, in the atmospheric boundary layer. There exist two main approaches to this problem, namely the "Gaussian plume model" based on hypotheses about the statistical distribution of the transported property and the "closure theory" in which rate equations for the various moments are solved with the aid of more or less sophisticated closure assumptions. These two approaches have been related to each other by a few authors (see e.g. [25] and [26]), but only in a limited number of cases. With respect to this matter, the study by *Chatwin* and *Sullivan* does represent a new approach. They indeed show, by an order of magnitude analysis of the rate equations for mean concentration and concentration variance, that a cloud of passive contaminant does not diffuse uniformly but, on the contrary, evolves with a spatial "core-bulk" structure.

New Directions and Some Unsolved Issues

The other phenomenon of interest for environmental flows is rotation. It is now well known that Coriolis forcing restricts, in some sense, the atmospheric flow to be quasi-two-dimensional. More precisely the large scale geophysical flows do posess the same kind of conservation properties [27] that characterize purely two-dimensional flows, namely energy – square of velocity- and enstrophy – square of vorticity. The consequences of two-dimensionality have been studied in some detail, almost always in the framework of two-points closures [28], but one can nevertheless find general reviews of the essential differences between small and large scale geophysical turbulence [29, 30]. On the other hand, the computation of real large scale, i.e. quasi-two-dimensional, flows requires the same kind of studies but, of course and for practical reasons, in the framework of one-point closures. Apart from very simple and early proposals, like for example the one by *Leith* [31] based on the eddy viscosity concept, there is unfortunately no model presently available for the solution of these problems.

As a last point concerning the necessary development, or improvment, of our understanding of turbulent processes in environmental flows, we shall mention here that turbulence is also of importance when it interacts with other physical phenomena. Among these, radiation is one of the most important. Despite this "old" astrophysical problem, a lot of work remains, and needs, to be done in the field of turbulence and radiative transfer interaction (see e.g. *Coantic* [32] for review). Another problem of interest in meteorology concerns water vapor condensation in turbulent atmospheric flows [33], the solution of which could possibly require the knowledge, and therefore modelisation, of high order statistics of temperature and humidity [34].

Hopefully, this short presentation has shown that turbulence in environmental flows is an attracting and interesting subject of fundamental importance. It is also hoped, and believed, that the fruitful cooperation between fluid dynamicists, meteorologists, oceanographers and astrophysicists will continue to lead to an original and creative contribution in the field of turbulence.

References

1. D. K. Lilly: "The Representation of Small-Scale Turbulence in Numerical Simulation Experiments" in Proc. IBM Scientific Computing Symposium on Environmental Science, White Plains, N. Y. (1967) p. 195
2. J. W. Deardorff: A numerical study of three-dimensional turbulent channel flow at large Reynolds number. J. Fluid Mech. **41**, 453 (1970)
3. J. L. Lumley, O. Zeman, J. Siess: The influence of buoyancy on turbulent transport. J. Fluid Mech. **84**, 581 (1978)
4. J. C. André, G. De Moor, P. Lacarrère, R. du Vachat: Turbulence approximation for inhomogeneous flows – Parts I and II. J. Atmos. Sci. **33**, 476 (1976)
5. O. Zeman, J. L. Lumley: "Buoyancy Effects in Entraining Turbulent Boundary Layer: A Second-Order Closure Study", in *Turbulent Shear Flows I,* ed. by F. Durst, B. E. Launder, F. W. Schmidt, J. H. Whitelaw (Springer, Berlin, Heidelberg, New York 1979) p. 295
6. J. C. André, G. De Moor, P. Lacarrère, G. Therry, R. du Vachat: "The Clipping Approximation and Inhomogeneous Turbulence Simulations", in *Turbulent Shear Flows I,* ed. by F. Durst, B. E. Launder, F. W. Schmidt, J. H. Whitelaw (Springer, Berlin, Heidelberg, New York 1979) p. 307
7. J. C. Wyngaard, A. Sundararajan: "The Temperature Skewness Budget in the Lower Atmosphere and Its Implications for Turbulence Modeling", in *Turbulent Shear Flows I,* ed. by F. Durst, B. E. Launder, F. W. Schmidt, J. H. Whitelaw (Springer, Berlin, Heidelberg, New York 1979) p. 319
8. E. J. Hopfinger, P. Atten, F. H. Busse: Instability and convection in fluid layers: A report on Euromech 106. J. Fluid Mech. **92**, 217 (1979)
9. J. S. Turner: *Buoyancy Effects in Fluids* (Cambridge University Press, 1973) p. 367
10. M. A. LeMone: Modulation of turbulence energy by longitudinal rolls in an unstable planetary boundary layer. J. Atmos. Sci. **33**, 1308 (1976)
11. S. R. Hanna: Summary of the 3rd symposium on atmospheric turbulence, diffusion, and air quality. Raleigh, N. C. Bull. Amer. Meteorol. Soc. **58**, 242 (1977)
12. G. E. Willis, J. W. Deardorff: A laboratory model of the unstable planetary boundary layer. J. Atmos. Sci. **31**, 1297 (1974)
13. R. J. Adrian, R. T. S. Ferreira: "Higher Order Moments in Turbulent Thermal Convection" (Submitted to the 2nd Symposium on Turbulent Shear Flows, London 1979)
14. M. A. LeMone: The structure and dynamics of horizontal roll vortices in the planetary boundary layer. J. Atmos. Sci. *30,* 1077 (1973)
15. P. J. Mason, R. I. Sykes: "A Two-dimensional Numerical Study of Horizontal Roll Vortices in the Neutral Atmospheric Boundary Layer" (Submitted to the 2nd Symposium on Turbulent Shear Flows, London 1979)
16. D. Etling, F. K. Wippermann: On the instability of a planetary boundary layer with Rossby number similarity. Boundary-Layer Meteorol. **9**, 341 (1975)
17. F. K. Wippermann, D. Etling, H. J. Kirstein: On the instability of a planetary boundary layer with Rossby number similarity. Part II: The combined effect of inflection point instability and convective instability. Boundary-Layer Meteorol. **15**, 301 (1978)
18. H. Rouse, J. Dodu: Turbulent diffusion across a density discontinuity. Houille Blanche **10**, 289 (1955)
19. M. A. Artaz, J. C. André: Similarity studies of entrainment in convective mixed layers. Boundary-Layer Meteorol. (to appear)
20. J. W. Deardorff: "Progress in Understanding Entrainment at the Top of Mixed Layers" in *Proc. Planetary Boundary Layer Workshop,* Boulder, August 1978 (To be published by Amer. Meteorol. Soc.)
21. R. A. Brost, J. C. Wyngaard: A model study of the stably-stratified planetary boundary layer. J. Atmos. Sci. **35**, 1427 (1978)
22. J. C. André, G. De Moor, P. Lacarrère, G. Therry, R. du Vachat: Modeling the 24-hour evolution of the mean and turbulent structures of the planetary boundary layer. J. Atmos. Sci. **35**, 1861 (1978)
23. O. Zeman: Parameterization of the dynamics of stable boundary layers and nocturnal jets. J. Atmos. Sci. **36**, 792 (1979)
24. S. J. Caughey, J. C. Wyngaard, J. C. Kaimal: Turbulence in the evolving stable boundary layer. J. Atmos. Sci. **36**, 1041 (1979)
25. O. G. Sutton: *Micrometeorology* (McGraw Hill, New York 1953) p. 273
26. J. W. Deardorff: Closure of second- and third-moment rate equations for diffusion in homogeneous turbulence. Phys. Fluids **21**, 525 (1978)
27. J. G. Charney: Geostrophic turbulence. J. Atmos. Sci. **28**, 1087 (1971)

28. J. R. Herring: "Statistical Theory of Two- and Quasi-Two-Dimensional Turbulence". Euromech 105 on Two-dimensional and Quasi-two-dimensional turbulence, Grenoble, September 1978
29. J. C. André: La turbulence atmosphérique. J. Phys. Paris **37**, C1–163 (1976)
30. H. Tennekes: Turbulent flow in two and three dimensions. Bull. Amer. Meteor. Soc. **59**, 22 (1978)
31. C. E. Leith: "Two-Dimensional Eddy Viscosity Coefficients", in *WMO/IUGG Symposium on Numerical Weather Prediction,* Tokyo (1968) p. 41
32. M. F. Coantic: "An Introduction to Turbulence in Geophysics and Air-Sea Interactions". AGARDograph Ser. No. 232, NATO (1978)
33. G. Sommeria, J. W. Deardorff: Subgrid-scale condensation in models of nonprecipitating clouds. J. Atmos. Sci. **34**, 344 (1977)
34. P. Bougeault, G. Sommeria, J. C. André: Méthode de modélisation unidimensionnelle de la couche limite planétaire nuageuse. Comparaison avec un modèle tridimensionnel. J. Rech. Atmos. **13**, 249 (1979)

The Atmospheric Boundary Layer-Modeling and Measurements

J. C. Wyngaard

National Center for Atmospheric Research[1], Boulder, CO 80307, USA

Abstract

The turbulence in the atmospheric boundary layer is strikingly different under convective and stably-stratified conditions. The dominant eddies under convective conditions fill the boundary layer, but these are the eddies most effectively damped by stable stratification.

Pronounced structural differences between unstable and stable conditions exist even in the surface layer. Thus, a unified treatment surface-layer turbulence, as described by the Monin-Obukhov similarity hypothesis, is not completely successful.

In this paper, we show that the atmospheric boundary layer presents some second-order turbulence modeling problems which do not seem amenable to the usual approaches. One current trend is to make some closures at third order, and we examine the budget of the third moment of vertical velocity in this context.

Nomenclature

C_1, C_2	Constants
D	Viscous destruction rate of $\overline{u_3^3}$
g_i	Gravity vector, $(0, 0, -g)$
k	Von Karman constant
l	Turbulent integral scale
L	Monin-Obukhov length
p	Fluctuating kinematic pressure
Q_0	Surface temperature flux
q^2	Squared turbulent velocity, $u_i u_i$
T_0	Mean temperature
T_*	Scaling temperature Q_0/u_*
T_f	Local free convection temperature scale
u_i	Fluctuating velocity
U_i	Mean velocity
u_*	Friction velocity
u_f	Local free convection velocity scale
w_*	Free convection velocity scale
z, x_3	Vertical coordinate
z_i	Convective ABL depth

Greek Symbols

δ_{ik}	Kronecker delta
ϵ	Viscous destruction rate of $\overline{q^2}$
θ	Fluctuating temperature
θ_*	Free convection temperature scale
κ_i	Wavenumber vector
π_{ik}	Intercomponent pressure transfer
σ	Standard deviation
ν	Kinematic viscosity

---

[1] The National Center for Atmospheric Research is sponsored by the National Science Foundation.

1. Introduction

The atmospheric boundary layer (ABL) was one of the first geophysical flows to be modeled with the second order closure technique. Early ABL models borrowed closures first developed for shear flows, and while many models still use shear-flow closures there is a growing recognition that the buoyancy effects which dominate the ABL require special consideration. As a result, sophisticated transport (third-moment divergence) closures tailored for convectively-driven turbulence have recently been developed [1–3].

These new transport closures for convective turbulence were stimulated by the failure of shear-flow closure models to reproduce many observed features of the convective ABL. Models with these new transport closures do seem to represent some aspects of buoyant transport much more accurately than do their predecessors.

In spite of these advances, even the best of the contemporary models still have inadequacies, probably stemming mainly from their pressure covariance and dissipation rate closures. With the hope of stimulating yet further improvements in models, we present here some important structural and dynamical features of the ABL, emphasizing particularly the implications for the second-order closure problem.

2. The Convective Boundary Layer

The convective ABL is usually capped by an inversion whose base at z_i establishes the ABL depth. We consider conditions approaching free convection, where the turbulence is produced primarily by the buoyancy arising from the heat (and/or moisture) transfer from the surface. This is common over land during sunny days under a high pressure center, and is the prevailing case over the world's oceans. Over land, the buoyancy usually is dominated by heat transfer; over water, both heat and moisture are important.

Laboratory free convection experiments [4, 5] as well as those in the atmosphere [6], together with three-dimensional numerical calculations [7] confirm that the turbulence characteristic of the entire convective boundary layer has length, velocity, and temperature scales l, w_*, and θ_* defined by

$$\begin{aligned} l &= z_i \\ w_* &= \left(\frac{g}{T_0} Q_0 z_i \right)^{1/3} \\ \theta_* &= Q_0/w_*, \end{aligned} \tag{1}$$

where Q_0 is the surface temperature flux (in the moist case, the virtual temperature flux) ($K\,\mathrm{ms}^{-1}$), and the gravity vector $g_i = (0, 0, -g)$. Although Q_0 and z_i often vary with time, this scaling still holds unless the variations are unusually rapid.

2.1. The Surface Layer

The lowest 10 or 20% of the ABL (i.e., the surface layer) is the region of Monin-Obukhov (M–O) similarity [8]; governing parameters are the friction velocity u_*, distance from the surface $x_3 = z$, the buoyancy parameter g/T_0, and Q_0. In the free convection limit the mean wind and hence u_* vanishes, so that in the M–O hypothesis the scales become [9]

$$l = z$$

$$u_f = \left(\frac{g}{T_0} Q_0 z \right)^{1/3} \tag{2}$$

$$T_f = Q_0/u_f.$$

While there is invariably some mean wind in the ABL, its effects are small if Q_0 is large enough. More precisely, we say that free convection is approached for large $-z_i/L$ values, where $L = -u_*^3 T_0/kgQ_0$ is the Monin-Obukhov length and k is the von Karman constant. The ABL depth z_i is not always measured in experiments, so the criterion used for very convective conditions in the surface layer is $-z/L \gg 1$; this is called "local free convection."

Many surface-layer statistics follow M—O similarity, and some (e.g. vertical velocity variance $\overline{u_3^2}$, temperature variance $\overline{\theta^2}$) also show clear evidence of following local free convection predictions [9]. Others, most conspicuously the horizontal wind variances $\overline{u_1^2}$ and $\overline{u_2^2}$, do not follow M—O similarity. The evidence indicates they are sensitive to w_*, even in the surface layer; thus they depend on z_i, which is not one of the M—O parameters.

Figures 1 and 2 (adapted from [10]) contrast the behavior of the vertical and horizontal wind variances in the unstable surface layer. Under local free convection we expect $\sigma_w = (\overline{u_3^2})^{1/2} \sim u_f$, while near neutral ($|z/L| \ll 1$) we can expect $\sigma_w \sim u_*$; this implies σ_w/u_* is con-

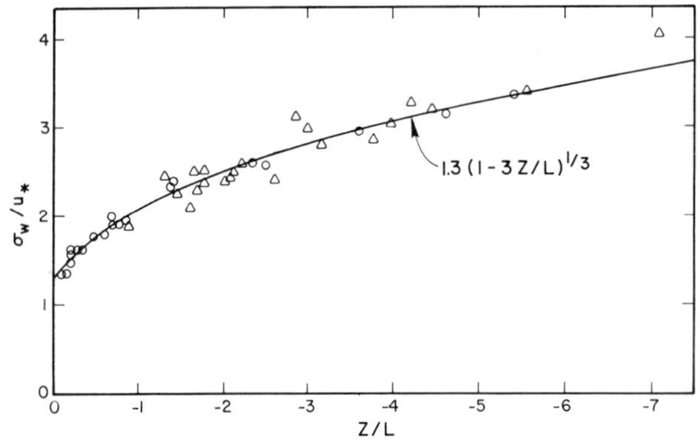

Fig. 1. Surface-layer measurements, from airplane (△) and tower (○), of the standard deviation of vertical velocity, normalized by u_*, as a function of z/L. From [10]

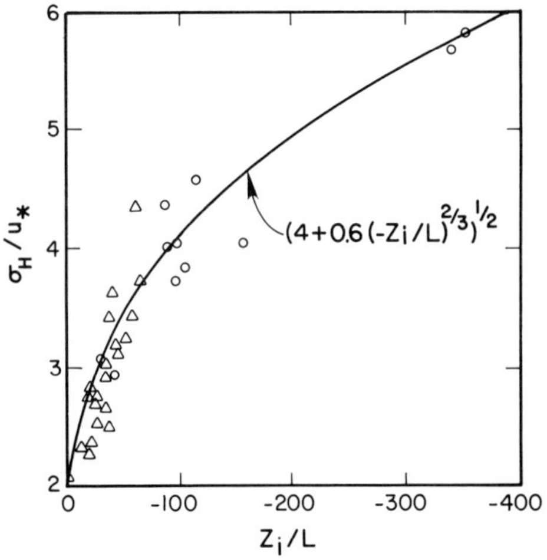

Fig. 2. Surface-layer measurements, from airplane (△) and tower (○), of the standard deviations of horizontal velocity components, normalized by u_*, as a function of z_i/L. From [10]

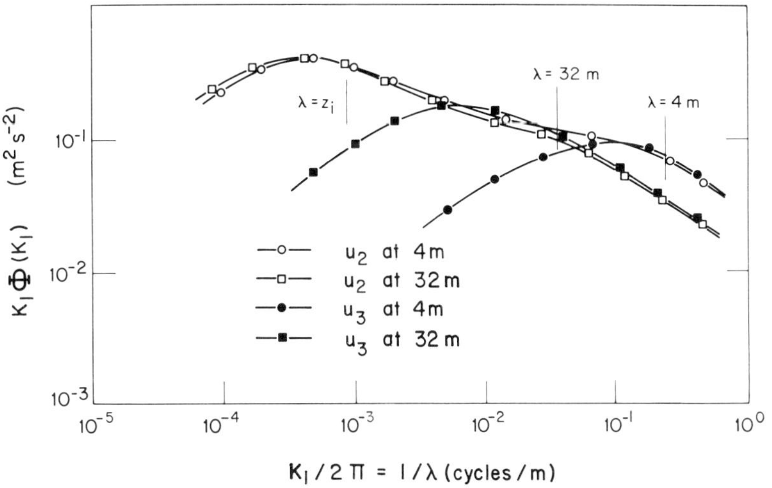

Fig. 3. Surface-layer measurements of streamwise spectra of lateral (u_2) and vertical (u_3) velocity spectra under convective conditions. From [11]

stant at neutral, and varies as $(-z/L)^{1/3}$ under very unstable conditions. Figure 1 confirms this. For $\sigma_H = (\overline{u_1^2})^{1/2}$ or $(\overline{u_2^2})^{1/2}$, we expect proportionality to u_* at neutral and to w_*, not u_f, under very unstable conditions; this implies σ_H/u_* constant at neutral, and σ_H/u_* approaching $(-z_i/L)^{1/3}$ with increasing instability. Figure 2 confirms this.

Figure 3 (from [11]) shows measured one dimensional (streamwise) spectral levels of u_2 and u_3 in the unstable surface layer and the striking contrast in the dominant length scales. The u_2 spectrum at $x_3 = 4$ m and 32 m peaks at streamwise wavelengths of the order of z_i, while the u_3 spectral peak occurs at wavelengths of the order of z. Thus the dominant wavelengths of horizontal velocity fluctuations are typically 1–3 orders of magnitude larger than those of the vertical velocity fluctuations. Since in the convective limit we saw that $\overline{u_1^2} = \overline{u_2^2} \sim w_*^2$, while $\overline{u_3^2} \sim u_f^2$, we also have $\overline{u_1^2}/\overline{u_3^2} \sim (z_i/z)^{2/3} \gg 1$. Note, however, from Fig. 3 that the inertial ranges of the two spectra are consistent with equal viscous dissipation rates of $\overline{u_2^2}$ and $\overline{u_3^2}$.

In the convective limit of a horizontally homogeneous surface layer, the component velocity variance equations reduce to, in the Boussinesq approximation,

$$\frac{1}{2}(\overline{u_1^2})_{,t} = -\frac{1}{2}(\overline{u_3 u_1^2})_{,3} - \overline{p_{,1} u_1} - \epsilon/3$$

$$\frac{1}{2}(\overline{u_2^2})_{,t} = -\frac{1}{2}(\overline{u_3 u_2^2})_{,3} - \overline{p_{,2} u_2} - \epsilon/3 \tag{3}$$

$$\frac{1}{2}(\overline{u_3^2})_{,t} = -\frac{1}{2}(\overline{u_3^3})_{,3} - \overline{p_{,3} u_3} + \frac{g}{T_0}\overline{\theta u_3} - \epsilon/3.$$

In (3), and hereafter, we use kinematic pressure. Measurements [4, 12–14] indicate unambiguously that the sum of the three transport terms in (3) [i.e., $-\frac{1}{2}(\overline{u_3 q^2})_{,3}$] is a loss, representing an export of turbulent kinetic energy to upper levels. These data also indicate that while all three energy components share in this loss, $\overline{u_3^2}$ experiences most of it. Thus, we conclude that the pressure covariances are the only source of $\overline{u_1^2}$ and $\overline{u_2^2}$ in the convective limit.

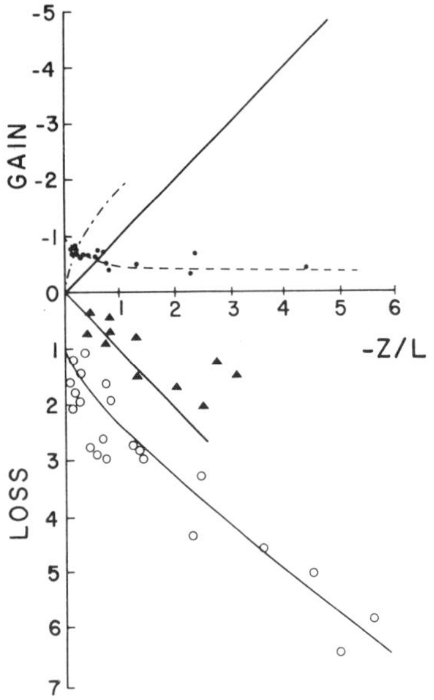

Fig. 4. Surface-layer measurements of the turbulent kinetic energy budget normalized by u_*^3/kz, under convective conditions. (—) buoyant production; (•) shear production; (▲) turbulent transport; (○) viscous dissipation; (- — - —) imbalance, most likely pressure transport. From [13]

Summing the component equations in (4) and using incompressibility and horizontal homogeneity gives

$$\tfrac{1}{2}\,(\overline{q^2})_{,t} = -\tfrac{1}{2}\,(\overline{u_3 q^2})_{,3} - (\overline{pu_3})_{,3} + \frac{g}{T_0}\,\overline{\theta u_3} - \epsilon. \tag{4}$$

Measurements in local free convection [12, 13] of all but the pressure term in (4) indicate a strong imbalance, with the only plausible explanation being a large pressure term (which we call pressure transport) acting as a source of $\overline{q^2}$. Recent results are shown in Fig. 4.

Thus, the evidence suggests that pressure correlations are the sole source of $\overline{u_1^2}$ and $\overline{u_2^2}$ in the convective surface layer limit. Because the pressure correlation terms in the three component energy budgets clearly do not sum to zero, however, this energy to $\overline{u_1^2}$ and $\overline{u_2^2}$ is not simply transferred from $\overline{u_3^2}$; instead, there is a net energy source due to pressure effects.

We turn now to dissipation. In second-order modeling it is common to express dissipation through a dissipation length scale l_ϵ:

$$\epsilon = (\overline{q^2})^{3/2}/l_\epsilon. \tag{5}$$

In the neutral (or slightly stable) surface layer, it is believed that $\overline{q^2} = Au_*^2$, where A is a constant between about 7.5 and 15, somewhat larger than its value in the laboratory case [15]. The Kansas data [16] suggest $A \sim 11$, which is consistent with recent measurements in Minnesota [17]. Since $\epsilon = u_*^3/(kz)$, we have the expected result that l_ϵ is proportional to z:

$$l_\epsilon = A^{3/2} kz \quad \text{(stable, neutral)}. \tag{6}$$

In the convective surface layer, however, $\overline{q^2} \sim w_*^2$, which is typically much larger than its value at neutral. It seems that $\epsilon \sim u_f^3/z$, so that $\Phi_\epsilon = kz\epsilon/u_*^3 \sim (-z/L)$; this is consistent with the measurements in Fig. 4. Thus, in the convective surface layer we have the unexpected result

$$l_\epsilon \sim w_*^3/(u_f^3/z) \sim z_i \quad \text{(unstable)}. \tag{7}$$

The difference between (6) and (7) is striking. One interpretation is that the large, convectively-driven eddies which dominate q^2 in the unstable surface layer interact only weakly with the smaller-scale turbulence, since their length-scale ratio is of order z_i/z; hence the large eddies do not significantly influence inertial transfer or the dissipation rate. This would suggest that a more useful parameterization of ϵ might be

$$\epsilon = \sigma_w^3/l_\epsilon, \tag{8}$$

since, as we saw, σ_w does not scale with w_*. From Figs. 1 and 4 we can quickly confirm that (8) does imply $l_\epsilon \sim z$ under both neutral and unstable conditions, which is an improvement over (5). We find, however, that the proportionality constant $B = l_\epsilon/(kz)$ varies with stability; $B = 2.2$ at neutral, 4.7 at $z/L = -2$ and 6.9 at $z/L = -5$.

Summarizing, the evidence suggests that convection radically changes the structure of the surface layer from that found in the neutral or stably-stratified case. The horizontal length scale and the velocity scale of the horizontal wind fluctuations become much larger than at neutral, but the scales of the vertical fluctuations do not. The horizontal turbulent energy components receive a large gain due to pressure transport, while energy is exported from the surface layer by turbulent transport; most of this export loss is experienced by $\overline{u_3^2}$. This export is up-gradient, since both $\overline{u_3^3}$ and $(\overline{u_3^2})_{,3}$ are positive. Finally, the multiple scales make a simple dissipation parameterization of the form $\epsilon \sim q^3/l$ invalid for arbitrary stability.

2.2. Upper Levels

Measurements above the surface layer are much more difficult to make and to interpret. To cite only one complicating factor, it is now evident that entrainment processes at z_i can influence structure well down into the boundary layer. Nonetheless, some broad features of mid-layer structure can be discussed.

The strong surface-layer anisotropy in energy components disappears by mid layer; at $z = 0.5\, z_i$, $\overline{u_1^2} \simeq \overline{u_2^2} \simeq 0.3\, w_*^2$, $\overline{u_3^2} \simeq 0.4\, w_*^2$, so that $\overline{q^2} \simeq 1.0\, w_*^2$. The dissipation rate $\epsilon \simeq 0.5 \frac{g}{T} Q_0 = 0.5\, w_*^3/z_i$, although it can be larger if there is significant shear production, as can be the case in a baroclinic flow. If $\epsilon \sim w_*^3/z_i$, $\overline{q^2} \sim w_*^2$, then $l_\epsilon \sim z_i$.

The buoyant production term, $\frac{g}{T}\,\overline{\theta u_3}$, changes sign and becomes an energy sink typically near $z \sim 0.8\, z_i$; the location of the crossover and the maximum negative flux depend on entrainment parameters. The $\overline{pu_3}$ covariance, which starts from zero at the (rigid) surface and becomes negative, crosses zero roughly where $\overline{\theta u_3}$ does, according to Deardorff's 3-D numerical calculations [18], and reaches a positive maximum near z_i. This indicates a pressure transport loss in upper levels, but according to these 3-D calculations this loss is small, amounting at most to about 20% of the gain by turbulent transport. There seems to be no experimental confirmation, either direct or indirect, of these results for pressure transport.

3. The Stably-Stratified Boundary Layer

While the convective ABL can exist in the absence of wind shear, the stable ABL cannot; wind shear provides its sole source of turbulent kinetic energy. The outer edge of the stable ABL, which is usually quite apparent in acoustic sounder records, is the level at which turbulence can no longer be supported against the losses to buoyancy and dissipation.

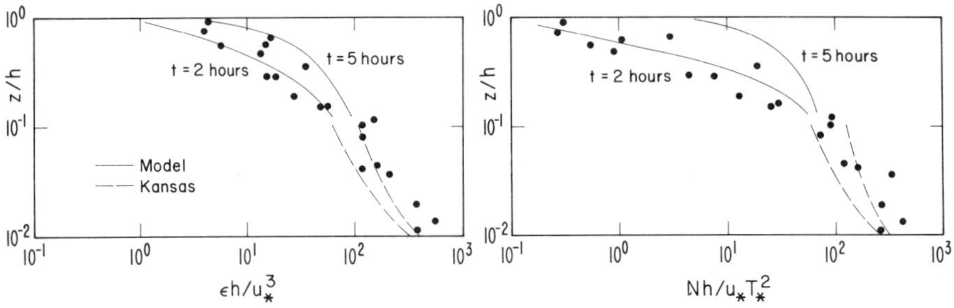

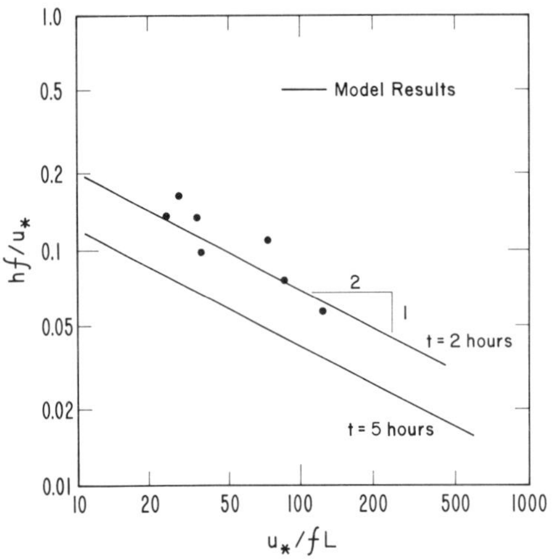

▲ **Fig. 5.** Measurements and model predictions [19] of ϵ and N (half the molecular destruction rate of temperature variance) in a stable boundary layer of depth h, for two times after transition. (•) Minnesota data; (– –) envelope of Kansas data. From [17]

◀ **Fig. 6.** Measurements and model predictions [19] of stable boundary layer depth h for two times after transition; f is the Coriolis parameter. •, Minnesota data. From [17]

Our experimental understanding of the structure of the stable ABL lags well behind that of the convective ABL. In a recent experimental program [17] designed to study ABL structure, the stably-stratified cases which evolved after sunset were found to be very time-dependent and short-lived.

There is evidence, however, that relatively simple second-moment closures can account for many observed features of the stable ABL. The Brost-Wyngaard [19] model is transport free, and uses a dissipation length scale prescribed to vary as z near the surface and as σ_w/ω_B, where $\omega_B^2 = g\Theta_{,3}/T$ is the Brunt-Vaisala frequency, under very stable stratification. This length scale was also used in the pressure covariance closure. Such a model reproduces well the observed Monin-Obukhov similarity in the stable surface layer and gives the observed monotonically decreasing turbulence profiles and monotonically increasing wind and temperature profiles above the surface layer. In the original paper [19] some good comparisons with experimentally-determined eddy diffusivity profiles are also shown.

Figure 5 and 6 show some comparisons of the model of [19] with some measurements in the Minnesota experiments [17]. The measurements occurred over a time interval centered about one hour after transition. Figure 5 shows measurements of ϵ and N (half the molecular destruction rate of temperature variance) compared with model predictions for 2 and 5 hours after transition (earlier runs were judged too sensitive to the initial conditions, some of which had to be assumed). Figure 6 compares measurements of the ABL depth h with model predictions, again at 2 and 5 hours. The overall agreement is quite good.

4. Pressure Covariances in the Convective ABL

The fluctuating kinematic pressure satisfies the Poisson equation

$$-p_{,ii} = u_{i,j}u_{j,i} - \overline{u_{i,j}u_{j,i}} + 2\,U_{i,j}u_{j,i} + \frac{g_i}{T_0}\,\theta_{,i}. \tag{9}$$

We can neglect the mean shear term (third on the right) in the convective surface layer limit. This does not require $U_i = 0$; since the large convective eddies induce horizontal speeds of order w_* in the surface layer, we have the weaker requirement that $|U_i| \ll w_*$. For strong convection ($z_i \sim 3$ km, $Q_0 \sim 0.3$ ms^{-1} K), $w_* \sim 3$ ms^{-1}.

Consider first the contribution of the first pair of terms in (9) – called the return-to-isotropy terms – to the pressure covariances in the $\overline{u_i u_k}$ equations. There are two approaches to writing these covariances. In one approach (e.g., [20]) one writes

$$(\overline{p_{,i}u_k} + \overline{p_{,k}u_i}) = -\overline{p(u_{i,k} + u_{k,i})} + (\overline{pu_i})_{,k} + (\overline{pu_k})_{,i} = -P_{ik} + T_{ik} \tag{10}$$

identifying P_{ik} as intercomponent transfer (since its trace vanishes) and T_{ik} as transport (although technically it is not). The return-to-isotropy portion of P_{ik} is invariably parameterized as

$$-P_{ik}^{\mathrm{RI}} = \overline{-p(u_{i,k} + u_{k,i})} = C_1\epsilon(\overline{u_i u_k}/\overline{q^2} - 1/3)/2 \tag{11}$$

with C_1 a positive constant. In a horizontally homogeneous flow (10) and (11) imply

$$\overline{p_{,1}u_1}^{\,\mathrm{RI}} = C_1\epsilon(\overline{u_1^2}/\overline{q^2} - 1/3)/2. \tag{12}$$

We showed earlier that in the convective surface layer limit $\overline{u_1^2} = \overline{u_2^2} \gg \overline{u_3^2}$; hence $\overline{u_1^2}/\overline{q^2} > 1/3$, and the parameterization (12) for $\overline{u_1^2}$ (and its counterpart for $\overline{u_2^2}$) predict a return-to-isotropy loss of $\overline{u_1^2}$ and $\overline{u_2^2}$ by transfer to $\overline{u_3^2}$, in clear conflict with the data. This wrong-direction transfer can be reduced by allowing C_1 to depend on the anisotropy (see [21]).

An alternative decomposition is suggested in [22]:

$$(\overline{p_{,i}u_k} + \overline{p_{,k}u_i}) = [\overline{p_{,i}u_k} + \overline{p_{,k}u_i} - \tfrac{2}{3}\,\delta_{ik}(\overline{pu_j})_{,j}] + \tfrac{2}{3}\,\delta_{ik}(\overline{pu_j})_{,j}\,. \tag{13}$$

The first portion of (13), which we call π_{ik}, is again trace free and represents intercomponent transfer; the second is a true transport term. The return-to-isotropy contribution to π_{ik} is parameterized in the usual fashion

$$\pi_{ik}^{\mathrm{RI}} = C_1\epsilon(\overline{u_i u_k}/\overline{q^2} - \delta_{ik}/3). \tag{14}$$

Because the trace is added and subtracted in (13), the effectiveness of (14) depends on how $(\overline{pu_3})_{,3}$ is parameterized.

Consider next the buoyant contribution to the pressure covariances. *Launder* [23] proposed the parameterization

$$-P_{ik}^{\mathrm{B}} = -C_2\left(\frac{g_i}{T_0}\,\overline{\theta u_k} + \frac{g_k}{T_0}\,\overline{\theta u_i} - \tfrac{2}{3}\,\delta_{ik}\,\frac{g_j}{T_0}\,\overline{\theta u_j}\right), \tag{15}$$

where $C_2 = 0.6$. This implies that

$$\overline{p_{,1}u_1^B} = -\frac{C_2}{3}\frac{g}{T_0}\overline{\theta u_3}. \tag{16}$$

In the alternative decomposition, *Zeman* and *Lumley* [1] use the parameterization

$$\pi_{ik}^B = -\frac{3}{10}\left(\frac{g_i}{T_0}\overline{\theta u_k} + \frac{g_k}{T}\overline{\theta u_i} - \tfrac{2}{3}\delta_{ik}\frac{g_j}{T_0}\overline{\theta u_j}\right). \tag{17}$$

The parameterization (16) amounts simply to transferring some of the direct buoyant production of $\overline{u_3^2}$ directly to $\overline{u_1^2}$ and $\overline{u_2^2}$. It is not clear that in strong convection this can provide enough transfer to $\overline{u_1^2}$ and $\overline{u_2^2}$ to balance the observed loss to transport and dissipation. More seriously, since this transfer to $\overline{u_1^2}$ or $\overline{u_2^2}$ is presumably less than goes directly into $\overline{u_3^2}$, it is not clear whether a model using (16) would create the large values of $\overline{u_1^2}/\overline{u_3^2}$ found in a very convective surface layer. Again, we cannot assess the alternative parameterization (17) without knowing how $(\overline{pu_3})_{,3}$ is handled.

The *Zeman-Lumley* [1] approach to parameterizing $(\overline{pu_3})_{,3}$ proceeds as follows. One assumes an unbounded turbulence field, and writes the Poisson equation (9) for p in its wave-number space form by using Fourier representations. One then has three contributions to $\overline{pu_k}$: one from return-to-isotropy, and two from a pair of "rapid" terms, one due to mean shear and one due to buoyancy. The return to isotropy term is

$$\overline{pu_k}^{RI} = -\int \frac{\kappa_i\kappa_j}{\kappa^2}\phi_{ijk}\,d\kappa = I_{ijijk}, \tag{18}$$

where ϕ_{ijk} is the spectrum of $\overline{u_iu_ju_k}$. One then parameterizes I_{ijklm} as an isotropic tensor function of $\overline{u_iu_ju_k}$, and determines the form of this function and the free constants through the symmetry properties of (18). The result [1] is $\overline{pu_k}^{RI} = -\overline{q^2u_k}/5$. By the same process it is found that the rapid contributions to $\overline{pu_k}$ vanish. Thus, the Zeman-Lumley expression for the pressure flux divergence is simply

$$(\overline{pu_3})_{,3} = -\tfrac{1}{5}(\overline{q^2u_3})_{,3}. \tag{19}$$

Implementing (19) requires third-order closure, as we will discuss in the next section. In the meantime, we point out that (19) is broadly consistent with convective surface layer data. From the free convection energy budget (4),

$$(\overline{pu_3})_{,3} = -\tfrac{1}{2}(\overline{q^2u_3})_{,3} + \frac{g}{T_0}\overline{\theta u_3} - \epsilon. \tag{20}$$

Data from very unstable conditions (but with some wind shear), where all three terms on the right side of (20) were measured [12, 13], suggest that the proportionality between $(\overline{pu_3})_{,3}$ and $(\overline{q^2u_3})_{,3}$ is, approximately (see Fig. 4)

$$(\overline{pu_3})_{,3} \sim -(\overline{q^2u_3})_{,3}. \tag{21}$$

The proportionality factor in (21) is uncertain to within a factor of perhaps two or three, because of the accumulated experimental uncertainties in the turbulent energy budget. We note that (19) and (21) do agree in sign, however.

Deardorff's 3-D simulations of the convective ABL [18, 24] also indicate that $(\overline{pu_3})_{,3}$ and $(\overline{q^2u_3})_{,3}$ have opposite signs near the lower surface. However, near the surface (and near z_i) the contributions to $\overline{pu_3}$ and $\overline{q^2u_3}$ made by the unresolvable, sub-grid scale motion are important, and for just the reasons we are discussing the subgrid parameterizations may not be particularly trustworthy under convective conditions.

Deardorff's recent 3-D simulations [24] of the cloud-topped convective ABL show a significant $(\overline{pu_3})_{,3}$ term near cloud base. In this case, however, the results show that the relation (19) is not followed; both pressure transport and turbulent transport are of the same sign (gains), with pressure transport being the larger. One possibility is that buoyancy, rather than return-to-isotropy, dominates the pressure field which contributes to $\overline{pu_3}$ in this case. If so, we have a parameterization problem, because the arguments which gave the closure (19) predict that this buoyancy contribution vanishes identically.

Finally, the 3-D simulations indicate that near the convective ABL top the relation (19) is again qualitatively correct for both the clear [18] and cloud-topped [24] cases, although there are substantial deviations from the factor 1/5 in (19).

Gibson and *Launder* [25] have shown, through the exact solution to the pressure equation (9), that the lower boundary affects the pressure field in the surface layer; they argue that pressure term closures should include "wall effects." In view of our previous discussion of the behavior of $\overline{pu_3}$ near the inversion lid at z_i, one could also argue that one should include "lid effects" as well. As with most closure problems, however, it is difficult to know how to proceed. The wall effect corrections proposed by *Gibson* and *Launder* [25], which build on the base of (11) and (15), seem not to remedy the deficiencies these expressions exhibit in the convective surface layer.

The fact remains, however that the closures we have discussed neglect any influence of the flow geometry in the convective ABL. Closure (17), for example, results from modeling π_{ik}^{B} as an isotropic tensor function of g_i/T_0 and $\overline{\theta u_k}$. This does not imply that P_{ik}^{B} is isotropic, but only that the functional relation is isotropic; thus, any anisotropy in P_{ik}^{B} is induced only by the anisotropy in g_i and in $\overline{\theta u_k}$. Perhaps this is too restrictive near the lower surface; possibly axisymmetry about the vertical coordinate is a more reasonable assumption. In any event, it is clear that convective surface layer turbulence structure is strongly influenced by the flow geometry; we showed that z_i directly influences $\overline{u_1^2}$ and $\overline{u_2^2}$, but not $\overline{u_3^2}$, and how as a result they differ strongly in intensity and scale.

With this in mind, it is instructive to examine the exact expression for the buoyant contribution to $\overline{p_{,1}u_1}$, for example:

$$\overline{p_{,1}u_1}^{\mathrm{B}}(x) = \frac{1}{4\pi}\int \frac{g}{T_0}(\overline{\theta' u_1})_{,1'3'}\left(\frac{1}{r}+\frac{1}{r^*}\right)dy, \tag{22}$$

where $r = |\boldsymbol{x}-\boldsymbol{y}|$, $r^* = |\boldsymbol{x}-\boldsymbol{y}^*|$, primed quantities refer to $\boldsymbol{y}$, and $\boldsymbol{y}^*$ is the image (through the surface) of $\boldsymbol{y}$; we have written (22) in the form suggested by [20]. If we assume that in a large turbulent Reynolds number surface layer

$$\overline{\theta' u_1} = (\overline{\theta^2}\,\overline{u_1^2})^{1/2} f_1(z, r, z_i), \tag{23}$$

then (22) should integrate to

$$\overline{p_{,1}u_1}^{B} = \frac{g}{T_0} (\overline{\theta^2}\,\overline{u_1^2})^{1/2} f_2(z/z_i). \tag{24}$$

It is interesting that (24) involves $(\overline{u_1^2})^{1/2}$. The usual parameterization involves $\overline{\theta u_3} \sim 0.5$ $(\overline{\theta^2}\,\overline{u_3^2})^{1/2}$; since in the surface layer $\overline{u_1^2}$ can be much larger than $\overline{u_3^2}$, (24) has the potential of providing part of the necessary large energy source for $\overline{u_1^2}$ and $\overline{u_2^2}$. That the rms value of u_1 appears, rather than the mean square, might also be significant. Since it appears in an equation for $(\overline{u_1^2})_{,t}$, it provides a relatively larger source at small values of $\overline{u_1^2}$ than at large values. This could give stable feedback which maintains $\overline{u_1^2}$ (and $\overline{u_2^2}$) in the surface layer at larger levels than $\overline{u_3^2}$. A similar, stable feedback results from the parameterization (5), $\epsilon \sim (\overline{q^2})^{3/2}/l_\epsilon$. Since this occurs in the $(\overline{q^2})_{,t}$ equation, it provides an excessive energy loss when $\overline{q^2}$ is larger than its equilibrium value; conversely, the loss is too small when $\overline{q^2}$ is too small. This allows $\overline{q^2}$ to set its own level in response to the mean shear and other external influences. It is not clear how to derive a result such as (24) with the techniques of invariant modeling.

5. Turbulent Transport in the Convective ABL

We have seen that turbulent transport [i.e. $(\overline{u_3 q^2})_{,3}$] is important in the convective surface layer turbulent kinetic energy budget (4), and also figures prominently in the parameterizations (11) and (15) for pressure covariances.

The typical shear-flow transport parameterization

$$\overline{u_i u_j u_k} = -K[(\overline{u_i u_j})_{,k} + (\overline{u_i u_k})_{,j} + (\overline{u_j u_k})_{,i}] \tag{25}$$

with K a (positive) transport parameter is not adequate in the presence of strong buoyancy effects. For example, in the unstable surface layer both $\overline{u_3^3}$ and $(\overline{u_3^3})_{,3}$ are positive, contrary to (25).

As a result of such failures some models [1–3] now use third moment closure for transport. Since third moment budgets are not yet well-understood, we will present the budget of $\overline{u_3^3}$. In a large turbulent Reynolds number, convective surface layer it reads

$$(\overline{u_3^3})_{,t} + (\overline{u_3^4})_{,3} - 3\,\overline{u_3^2}\,(\overline{u_3^2})_{,3} = -3\,\overline{p_{,3}u_3^2} + 3\,\frac{g}{T_0}\,\overline{\theta u_3^2} - 3\,\nu\overline{u_{3,i}(u_3 u_3)_{,i}}\,. \tag{26}$$

Data from the atmospheric surface layer and from laboratory free convection [5, 14] indicate that the Gaussian approximation $\overline{u_3^4} = 3\,\overline{u_3^2}\,\overline{u_3^2}$ is a good one; it is discussed further by [3]. Adopting this gives in quasi-steady state

$$(\overline{u_3^3})_{,t} = 0 = -3\,\overline{u_3^2}\,(\overline{u_3^2})_{,3} - P + 3\,\frac{g}{T_0}\,\overline{\theta u_3^2} - D, \tag{27}$$

where P and D denote the pressure and molecular terms in (25).

We can quickly compare the first and third terms on the right side of (27). We saw (Fig. 1) that $(\overline{u_3^2})_{,3}$ is positive; since $\overline{u_3^3}$ is also positive, we see that the first term represents a loss. The buoyant term is positive and hence is a gain. We take $(\overline{u_3^2})^2 = 3.6\,u_*^2(-z/L)^{2/3}$ (which fits the very unstable data in Fig. 1 well), and take [26] $\overline{\theta u_3^2} \sim 1.3\, u_*^2 T_* (-z/L)^{1/3}$ under very unstable conditions. Therefore, we have

$$\overline{u_3^2}\,(\overline{u_3^2})_{,3} \simeq 2.3\, \frac{g}{T_0}\, \overline{\theta u_3^2}, \tag{28}$$

which indicates that the loss due to the first term in the $\overline{u_3^3}$ budget (27) considerably exceeds the gain due to buoyant production. We also calculated (from the Kansas data, [16]) the term $[(\overline{u_3^4})_{,3} - 3\,\overline{u_3^2}\,(\overline{u_3^2})_{,3}]$ directly, without making the quasi-Gaussian assumption; this is more difficult because of scatter, but gives a multiplier on the right of (28) of about 1.5, rather than 2.3. It does appear, therefore, that $-(P+D)$ in (27) acts as a net gain.

The molecular term D can be written, from (26)

$$D = 3\,\nu \int \kappa^2\, C_{333}\, d\kappa, \tag{29}$$

where C_{ijk} is the spectrum of $\overline{u_i u_j u_k}$. Since (29) indicates that the large-wavenumber portion of C_{333} dominates D, it is tempting to use a locally isotropic form for C_{ijk}; this would make C_{ijk} zero, since it is odd, and hence would make D vanish. It is not clear, however, that C_{ijk} should tend to isotropy at large wavenumbers, since the anisotropic, low wavenumber structure can interact through the multiplication of u_j and u_k and produce anisotropic high wavenumber contributions to C_{ijk}. An alternative prediction is that of *Van Atta* and *Wyngaard* [27], who find a $\kappa^{-5/3}$ inertial range for C_{333} and its sign that of $\overline{u_3^3}$; thus D is a loss term. Thus, each argument is consistent with $D \geqslant 0$. For the case of the θ^3 budget, it has been shown directly [28] that the molecular term is a significant loss.

We conclude that the pressure term in the $\overline{u_3^3}$ budget is a net source in the convective surface layer. This pressure term has three parts: return-to-isotropy, rapid buoyant, and rapid mean shear. *Lumley* et al. [3] model the first as proportional to $-\epsilon\, \overline{u_3^3}/q^2$, which is a loss; their modeling of the rapid buoyant part also gives a loss. Since they considered free convection, their shear term vanished.

If one derives an invariant model of this rapid shear term, as was done for π_{ik} in (17), for $\overline{pu_3}$ in (19), and for other terms in [3], one finds

$$-3\,\overline{p_{,3} u_3^2}^{\,S} = -3\, U_{1,3}\, \overline{u_3^2 u_1}/5\,. \tag{30}$$

The moment $\overline{u_3^2 u_1}$, the vertical turbulent flux of $u_3 u_1$, vanishes in free convection, but under typical unstable conditions with wind shear it is of order $-u_*^3$ [26]. Thus by our parameterization (30), the mean shear contribution to the pressure term in the $\overline{u_3^3}$ budget is a source.

Laboratory free convection measurements [5, 14] also provide data on the behavior of the $\overline{u_3^3}$ budget. In the region $0.1\, z_i < z < 0.2\, z_i$, these data confirm that the quasi-Gaussian simplification of the $\overline{u_3^3}$ budget, which reduces (26) to (27), is a good approximation. In contrast with the atmospheric data, however, these results indicate that buoyant production much more closely balances the $\overline{u_3^2}\,(\overline{u_3^2})_{,3}$ loss term; their imbalance appears to be a slight gain. Possibly these differences in the $\overline{u_3^3}$ budgets are due to the presence of mean shear in the atmosphere, which according to (30) contributes a rapid term to P. The atmospheric data indicate that the rapid shear term could, with a larger constant on the right side of (30), provide the necessary additional source; recall that we also found, in (21), that a larger constant was necessary in the analogous expression (19).

Summarizing, the behavior of the $\overline{u_3^3}$ budget seems broadly similar to that of a second moment, with buoyancy, mean shear, turbulent transport, molecular destruction, and fluctuating pressure effects all probably important. It seems premature to conclude that all these terms are significant in all third moment budgets; quite possibly some simplifications will develop regarding the molecular and pressure terms, as do for second moments. Clearly more measurements would increase our understanding, but it remains to be seen whether third moment budgets will attract the interest of many experimentalists.

The attractions of third-order transport closure are, first, that at least some of the buoyancy effects which so complicate closure at second order appear explicitly in the third moment equations; and second, that the quasi-Gaussian approximation for fourth moments might well be a safer approximation than is gradient diffusion for third moments. On the first point, we add that since the third moment budgets are coupled through their buoyancy terms, one needs budgets for $\overline{u_3^2\theta}$ and $\overline{u_3\theta^2}$, for example, in order to use that for $\overline{u_3^3}$. This proliferation of equations can greatly complicate a model. However, gradient diffusion approximations for third moments do not always fail, so presumably one need not make all transport closures at fourth order. For example, while the conventional gradient-diffusion approximation for $\overline{u_3^2\theta}$ fails, that for $\overline{u_3\theta^2}$ succeeds [9]; an analysis of the production terms in their budgets strongly suggests why this happens. On the second point, we add that quasi-Gaussian transport closure is not without problems; *Deardorff* [29] has pointed out some of its disturbing properties in diffusion applications.

6. Conclusions

The convective ABL still presents strong challenges for second-order models, judging from our present inability to predict many observed features of its turbulence structure.

One recent advance, third-order transport closure, has the potential of alleviating many of the difficulties of the usual (gradient-diffusion) transport closure, which often fails in convective turbulence. There is yet no comparable approach to the closure problem for dissipation rate and pressure covariances in convective turbulence, for which the usual closures can also perform poorly.

There is evidence, both experimental and from 3-D models, that the pressure covariance (pressure transport) term in the turbulent energy equation can be large in the convective ABL. Buoyancy could be a major contributor to these covariances. The buoyancy effects, which come from the hydrostatic contribution to the fluctuating pressure, might be particularly difficult to parameterize in terms of local variables.

The traditional dissipation parameterization $\epsilon \sim q^3/l$ is less useful in the ABL because convection introduces multiple velocity and length scales. Even with q and l chosen such that the obvious effects of the large eddies are removed, the proportionality factor is found not to be constant.

Acknowledgements. I am grateful to Drs. Richard Brost and Otto Zeman for helpful discussions, and to Mrs. Pat Waukau for expertly typing the manuscript.

References

1. O. Zeman, J. L. Lumley: Modeling buoyancy driven mixed layers. J. Atmos. Sci. **33**, 1974–1988 (1976)
2. J. C. André, G. De Moor, P. Lacarrere, G. Therry, R. Du Vachat: Modeling the 24-hour evolution of the mean and turbulent structure of the planetary boundary layer. J. Atmos. Sci. **35**, 1861–1883 (1978)
3. J. L. Lumley, O. Zeman, J. Siess: The influence of buoyancy on turbulent transport. J. Fluid. Mech. **84**, 581–597 (1978)
4. G. E. Willis, J. W. Deardorff: A laboratory model of the unstable planetary boundary layer. J. Atmos. Sci. **31**, 1297–1307 (1974)
5. R. T. Ferreira: "Unsteady Turbulent Thermal Convection". Ph. D. Thesis, Mech. Eng. Dept., Univ. of Ill., Urbana-Champaign (1978)
6. J. C. Kaimal, J. C. Wyngaard, D. A. Haugen, O. R. Coté, Y. Izumi, S. J. Caughey, C. J. Readings: Turbulence structure in the convective boundary layer: J. Atmos. Sci. **33**, 2152–2169 (1976)
7. J. Deardorff: Convective velocity and temperature scales for the unstable planetary boundary layer and for Rayleigh convection. J. Atmos. Sci. **27**, 1121–1123 (1970)
8. N. E. Busch: "On the Mechanics of Atmospheric Turbulence", in *Workshop on Micrometeorology*, ed.by D. A. Haugen (AMS Boston, 1973) pp. 1–65
9. J. C. Wyngaard: "On Surface Layer Turbulence", in *Workshop on Micrometeorology*, ed. by D. A. Haugen (AMS, Boston, 1973) pp. 101–149
10. H. A. Panofsky, H. Tennekes, D. H. Lenschow, J. C. Wyngaard: The characteristics of turbulent velocity components in the surface layer under convective conditions. Boundary-Layer Meteorol. **11**, 355–361 (1977)
11. J. C. Kaimal: Horizontal velocity spectra in an unstable surface layer. J. Atmos. Sci. **35**, 18–24 (1978)
12. J. C. Wyngaard, O. R. Coté: The budgets of turbulent kinetic energy and temperature variance in the atmospheric surface layer. J. Atmos. Sci. **28**, 190–201 (1971)
13. S. J. Caughey, J. C. Wyngaard: The turbulent kinetic energy budget in convective conditions. Q. J. R. Meteorol. Soc. **105**, 18–24 (1979)
14. R. J. Adrian, R. T. S. Ferreira: "Higher Order Moments in Turbulent Thermal Convection" (Submitted to the 2nd Symposium on Turbulent Shear Flows, London 1979)
15. O. Zeman, H. Tennekes: A self-contained model for the pressure terms in the turbulent stress equations of the neutral atmospheric boundary layer. J. Atmos. Sci. **32**, 1808–1813 (1975)
16. Y. Ed. Izumi: "Kansas 1968 Field Program Data Report". Env. Res. Paper, No. 379, AFCRL, Hanscom AFB, Mass.
17. S. J. Caughey, J. C. Wyngaard, J. C. Kaimal: Turbulence in the evolving stable boundary layer. J. Atmos. Sci. **36**, 1041–1052 (1979)
18. J. W. Deardorff: Three-dimensional numerical study of turbulence in an entraining mixed layer. Boundary-Layer Meteorol. 7, 199–226 (1974)
19. R. W. Brost, J. C. Wyngaard: A model study of the stably-stratified planetary boundary layer. J. Atmos. Sci. **35**, 1427–1440 (1978)
20. B. E. Launder, G. J. Reece, W. Rodi: Progress in the development of a Reynolds-stress turbulence closure. J. Fluid Mech. **68**, 537–566 (1975)
21. J. L. Lumley: Computational modeling of turbulent flows. Adv. Appl. Mech. **18**, 123–176 (1978)
22. J. L. Lumley: Pressure-strain correlation. Phys. Fluids **18**, 750 (1975)
23. B. E. Launder: On the effects of a gravitational field on the turbulent transport of heat and momentum. J. Fluid Mech. **67**, 569–581 (1975)
24. J. W. Deardorff: Stratocumulus – Capped mixed layers derived from a three-dimensional model (to be published in Boundary-Layer Meteorol., 1980)
25. M. M. Gibson, B. E. Launder: Ground effects on pressure fluctuations in the atmospheric boundary layer. J. Fluid Mech. **86**, 491–511 (1978)
26. J. C. Wyngaard, O. R. Coté, Y. Izumi: Local free convection, similarity, and the budgets of shear stress and heat flux. J. Atmos. Sci. **28**, 1171–1182 (1971)
27. C. W. Van Atta, J. C. Wyngaard: On higher-order spectra of turbulence. J. Fluid Mech. **72**, 673–694 (1975)
28. J. C. Wyngaard, A. Sundararajan: "The Temperature Skewness Budget in the Lower Atmosphere and Its Implications for Turbulence Modeling", in *Turbulent Shear Flows I*, ed. by F. Durst, B. E. Launder, F. W. Schmidt, J. H. Whitelaw (Springer, Berlin, Heidelberg, New York 1979) p. 319
29. J. W. Deardorff: Closure of second- and third-moment rate equations for diffusion in homogeneous turbulence. Phys. Fluids **21**, 525–530 (1978)

Micrometeorological Applications of a Second-Order Closure Model of Turbulent Transport

W. S. Lewellen, M. E. Teske, and Y. P. Sheng

Aeronautical Research Associates of Princeton, Inc.
50 Washington Road, P. O. Box 2229, Princeton, N. J., 08540, USA

Abstract

Recent applications of the turbulent transport model originally developed by Donaldson to some problems of practical importance in micrometeorology are discussed. Four particular examples considered are the local boundary layer gust front emanating from a thunderstorm; the low-level wind and turbulent distributions of a tornado; the transport of momentum, heat and species within a surface layer canopy, and longitudinal roll vortices in the unstable planetary boundary layer. Results for the last example are discussed in some detail. Comparisons are made between a one-dimensional and a two-dimensional computation of this phenomenon.

List of Symbols

C	Mean species concentration
c	Species fluctuation
g	Gravitational acceleration
H	Height of vegetation canopy
h	Height of numerical grid
L	Monin-Obukhov length
P	Mean pressure
q	Root-mean velocity fluctuation
r	Radius
$\hat{r}$	Radius at which maximum swirl occurs
t	Time
U_i, U_j, U_k	Mean velocity components
u_i, u_j, u_k	Fluctuating velocity components
U_G	Geostrophic wind
U_*	Friction velocity
U	Mean longitudinal velocity
u	Fluctuating longitudinal velocity
V	Mean transverse velocity
v	Fluctuating transverse velocity
W	Mean Vertical velocity
w	Fluctuating vertical velocity
w_*	Characteristic velocity for free convection
x, y	Horizontal coordinates
z	Vertical coordinate
z_i	Height of mixed layer
z_0	Surface roughness height
α	Angle between the roll axis and the geostrophic wind
δ_{ij}	Kronecker delta
ϵ_{ijk}	Alternating tensor
Λ	Turbulent macroscale
λ	Wavelength of the roll
Θ	Mean temperature
θ	Temperature fluctuation
θ_*	$-\overline{w\theta} / U_*$
σ_u	Standard deviation of longitudinal wind velocity
σ_w	Standard deviation of vertical wind velocity
ν	Kinematic viscosity
ρ	Density
ψ	Stream function
Ω	Earth's rotation

Introduction

The basic turbulent transport model used in these applications is that outlined by *Donaldson* [1] and described in detail by *Lewellen* [2]. Simple closure approximations were made to the equations for the ensemble-averaged, single-point, second-order moments of the fluctuating variables. The theme of the present paper is to discuss the various applications rather than to rederive the model. Since invariance is an essential feature of the model, the same modeling coefficients are used for all the applications. The basic modeled equations are given in the Appendix. Although this simple closure cannot be expected to faithfully represent all of the information present in complex micrometeorological turbulent flows, the added physics incorporated in the second-order correlation equations permits an influence of such phenomena as buoyancy and rotation to be calculated directly without resorting to ad hoc eddy viscosity fixes.

The first three examples will only be discussed very briefly, but the fourth will be presented in greater detail because it not only helps to illuminate the influence that roll vortices exert on the turbulent transport of momentum, heat and species, but also the strong and weak points of the basic turbulence closure model.

Thunderstorm Gust Front

The gust front caused by the cold outflow emanating from a thunderstorm has been modeled with an axisymmetric implementation of our turbulence model [3]. The simulated flowfield is illustrated in Fig. 1. We idealize the outflow from the thunderstorm as a cold jet of temperature $\Theta(r)$ impinging normal to the ground, released at a height $z_{\max}$ with vertical velocity $W(r)$. The temperature defect below the ambient temperature is caused by evaporation of falling rain by relatively dry air at some altitude $z > z_{\max}$.

The primary variables in this simulation are the temperature defect (below ambient) of the jet, its diameter, and the height at which it is released. The larger scale updraft within which the downdraft is embedded will play a role in retarding the latetime gust front within the computational domain, but should not be a critical factor as long as the inflow velocity is smaller than the resulting internal simulated velocities. The other critical parameters are the surface temperature, the surface roughness, and the stability of the ambient atmosphere. In an effort to evaluate hazardous conditions for aircraft operations, a sensitivity analysis has been performed on the gust front as a function of five different dimensionless physical parameters [4, 5].

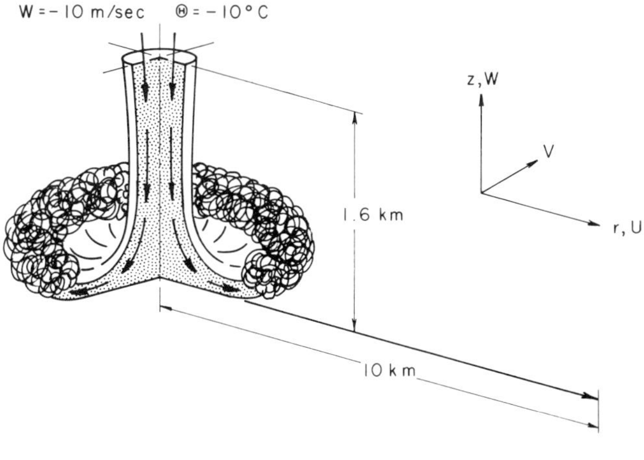

Fig. 1. Coordinate system for the axisymmetric radial thunderstorm gust front simulation

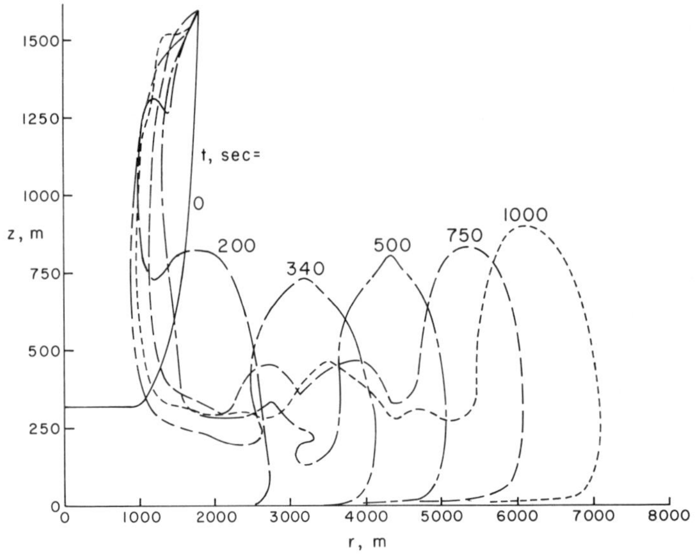

Fig. 2. Isopleths of constant temperature defect ($\Theta = -2°$ C) at several times after initialization

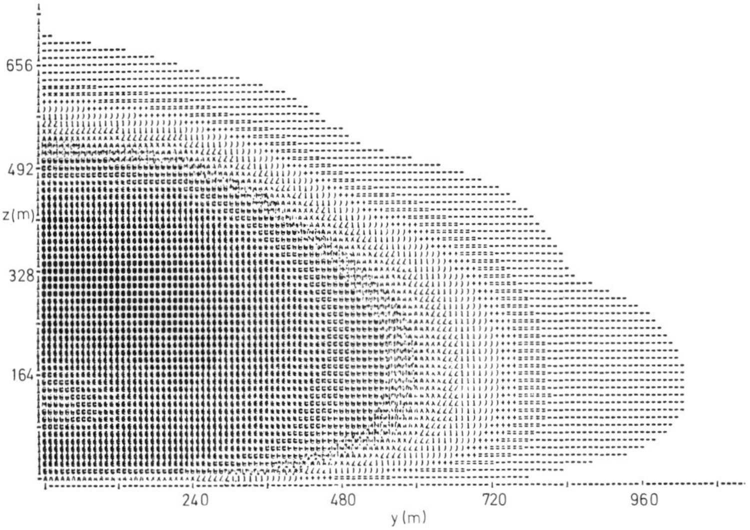

Fig. 3. Model simulation of the leading edge of a gust front when visualized in terms of temperature defect intensity

One way to follow the structure of the developing gust is to observe the movement of a fixed temperature value line, in this case $\Theta = -2$ °C, as shown in Fig. 2 for the simulation sketched in Fig. 1. At $t = 0$ the initial linear profile is shown, but by $t = 200$ s the structure has developed into a moving front. At $t = 340$ s the front has torn away from the downdraft region. At later times the strength of the gust decreases as the effect of the area change becomes more important. However, the height of the gust appears to grow slowly, so that by $t = 1{,}000$ s the $\Theta = -2$ °C line reaches nearly 1 km in altitude at a distance of nearly 8 km from the source centerline. The front is still quite strong at this point, and is moving outward at approximately 2.6 m/s.

Fig. 4. The leading edge of a thunderstorm gust front as marked by dust. (Photo by Andrew Watson, Miami, Florida, 1975)

When our numerically simulated front is visualized in terms of the temperature defect intensity, the leading edge appears as shown in Fig. 3. There is a very strong qualitative similarity between this picture and the gust front as visualized by dust in Fig. 4 (also see picture in [6]). The quantitative predictions of maximum mean velocity, velocity shear and variance also appear consistent with available observations.

Tornado Boundary Layer

The strongest winds occurring in nature are those in a tornado. In order to estimate wind loads on structures, we have modeled the tornado boundary layer [7] again using the axisymmetric version of our basic turbulent transport model. Here we have chosen a computer simulation which shows a resemblance to the 1974 Xenia, Ohio tornado. The domain's outer radius and top are both placed at 400 m. The surface roughness is taken to be large (0.4 m) to correspond to a relatively urban area. All components of the velocity are specified at the outer radius inflow boundary and relatively free zero slope conditions applied at the top of the domain.

The mean tangential and radial velocity field is given in Figs. 5 and 6. The velocities are normalized by the maximum tangential velocity which occurs at $r \approx 180$ m, $z \approx 100$ m. Cur-

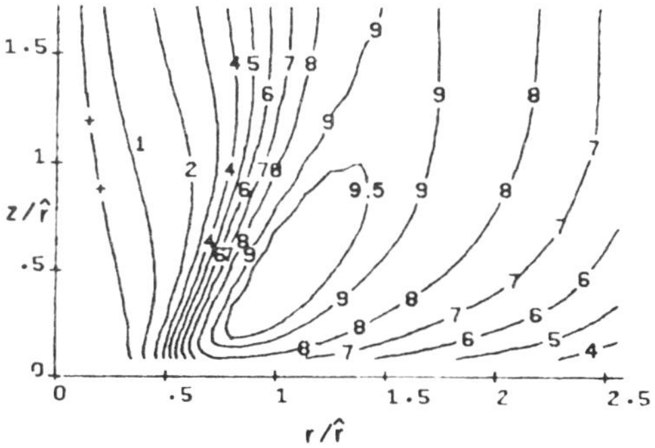

Fig. 5. Mean tangential velocity contours as predicted by an axisymmetric tornado model. Contours labeled in tenths of maximum V, and radial and vertical coordinate normalized by the radius at which V_{max} occurs

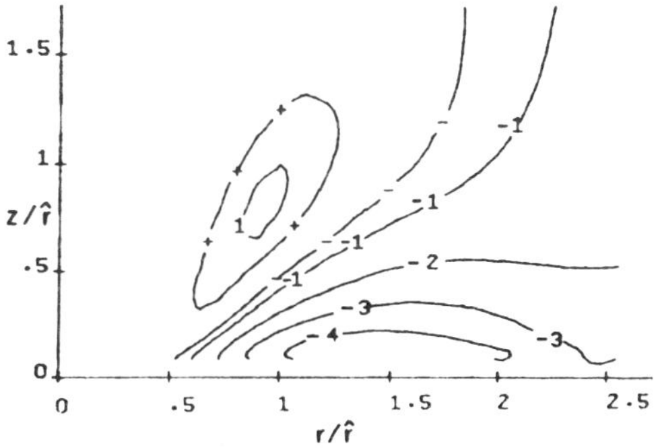

Fig. 6. Mean radial velocity contours for the same conditions as Fig. 5

rently, boundary conditions are not known precisely enough for the absolute value of the maximum velocity to be predicted by the model. However, from photogrammetric analyses, it appears the tangential velocity should be scaled by a maximum value of approximately 100 m/s.

A dominant feature of the flowfields computed to date [8, 9] is that the maximum winds occur at quite low altitudes. The surface layer plays the rather paradoxical role of increasing the velocity in the neighborhood of the ground at small radii, before actually reducing the velocity to zero at the surface. The strong radial inflow in the surface layer permits the streamlines to penetrate to smaller radii here than at higher altitudes. Even without total conservation of angular momentum, this permits higher swirling velocities to develop. The simulation also predicts that the r.m.s. average of the velocity fluctuations reaches values as high as 0.3 V_{max}. This large magnitude of the fluctuating velocities indicates that they should be considered in setting design criteria. The maximum damage will most likely occur where the fluctuating velocities add to the mean velocity.

Flow Within a Surface Canopy

A canopy of vegetation presents a complex lower boundary for atmospheric flows. For flow well above this canopy, it is usually adequate to characterize the boundary in terms of only an aerodynamic roughness, z_0. But for flow within the canopy itself, a more detailed representation is required. Our motivation for detailing this flow is to aid in the prediction of the dry deposition of gaseous SO_2 and particulate sulfate.

A second-order closure model for canopy flow has recently been presented by *Wilson* and *Shaw* [10]. The principal difference between their model and ours is that we consider heat and species transport as well as momentum transport. We also use a somewhat more general representation of the drag per unit volume of the vegetation. The comparison of the mean

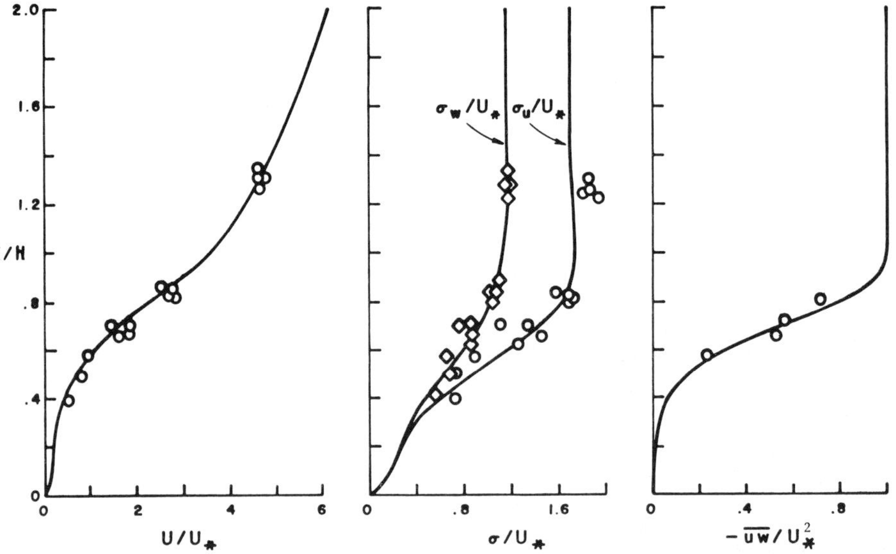

Fig. 7. Comparison of A.R.A.P. model prediction with data from *Shaw* et al. [11, 12] in and above a corn canopy; where H is canopy height, U_* the friction velocity above the canopy, U/U_* the normalized mean longitudinal wind velocity, σ_u/U_* and σ_w/U_* the dimensionless standard deviations of the longitudinal and vertical velocities, and $(-\overline{uw}/U_*^2)$ the dimensionless Reynolds stress

distributions of velocity, Reynolds stress and velocity variance predicted by our model are compared with some limited data for a corn canopy [11, 12] in Fig. 7. Although detailed plant area densities must be given for any type of vegetation, the model can predict the variation in surface layer heat and species transport as a function of surface Reynolds number $(U_* z_0 / \nu)$, Prandtl number and Schmidt number for any given plant area density distribution [13].

Roll Vortices in the Planetary Boundary Layer

The final example, which we will discuss in more detail, is that of computing the longitudinal roll vortices which often occur in the unstable planetary boundary layer [14]. This boundary layer feature can often be visualized as in Fig. 8 by the streets of clouds which occur at the top of the upflow region between the roll vortices. The simulation is performed on a two-dimensional grid, stretching over a wavelength λ in the y direction, and from the surface (represented by an effective hydrodynamic roughness z_0) to the total height h of the numerical grid. At a position $z_i < h$, the temperature profile is damped back to its background gradient, capping the domain and forcing z_i to become the inversion height. All mean variables and turbulence quantities are initialized with profiles from a compatible boundary layer solution of our one-dimensional code. The spatially homogeneous one-dimensional code necessarily treats the motion arising from the roll vortices as a part of the turbulence. Periodic boundary conditions are applied to the two-dimensional Cartesian version of our model. This permits horizontal roll vortices to appear as part of the ensemble mean motion when calculating the unsteady flow in the unstable boundary layer.

The numerical code is free to partition the energy between the mean background motion which is a function of the vertical coordinate only; the mean quasi-periodic, two-dimensional, roll vortex motion which is a function of z and y; and the more random turbulent motion which,

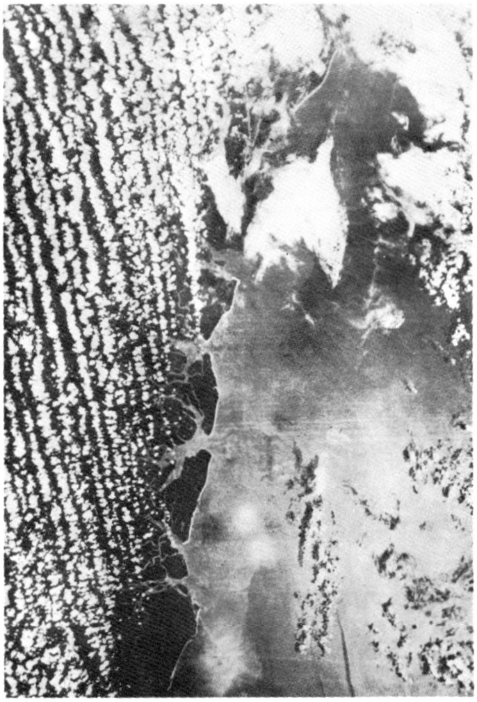

Fig. 8. View of cloud streets over the coast of Georgia as seen by Apollo [14]

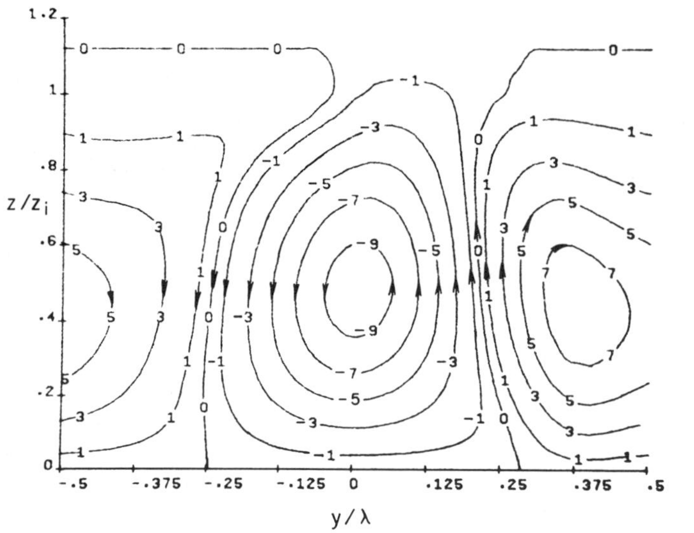

Fig. 9. Roll stream function contours for the example case of $\alpha = -10°$, $\lambda/z_i = 3$, $\Lambda_{max}/z_i = 0.1$. The profiles are normalized by a maximum value $\Delta\psi_{max}/w_* z_i = 0.287$ ($\Delta\psi = \psi - \langle\psi\rangle$), with ±9 denoting ±90% of maximum value; ±7, ±70%, etc. The arrows indicate flow direction

although three-dimensional in character, is only a function of y and z in the mean. The energy in the organized roll motion varies with the ratio of the wavelength, λ, of the roll to the inversion height z_i, the instability of the layer as measured by the ratio of the Monin-Obukhov length, L, to z_i, and the angle between the roll axis and the geostrophic wind. Figure 9 illustrates the typical cross-sectional structure of the stream function of the roll motion for an angle of $-10°$, $\lambda/z_i = 3$ and $L/z_i \approx -0.1$. Surface conditions are chosen appropriate for a sea surface at constant temperature. Results are scaled in terms of the characteristic velocity

$$w_* = \left(\frac{g}{\Theta_0} \langle \overline{w\theta}_0 \rangle z_i \right)^{1/3}$$

to minimize the influence of changes in $\overline{w\theta}_0$. (The angled bracket represents an average over y.) The geostrophic velocity is held at $U_G = 10$ m/s throughout the calculations. To maintain the roll structure, we impose periodic boundary conditions on all variables at $y = \pm\lambda/2$, so that whatever goes out one side will enter the other.

The corresponding cross-sectional structure of temperature, velocity, turbulent kinetic energy, and passive species released from the surface perturbations are shown in Figs. 10 through 13. The numerical results for the relative contributions of the mean roll motion and the roll modulated turbulence to the transport of momentum and heat compare reasonably (15) with the observations of *LeMone* [16].

Figures 10 through 13 show strongly similar results for differences above background average for U, Θ, q^2, and C. Here the background levels have been removed; hence Δq^2 shows some regions of negative values. What is obvious from all four plots is that the updraft region shown in Fig. 9 produces a band of less than background U and greater than background Θ, q^2 and C. The typical spread of this region is about 0.25 λ. The rest of the q^2 region is dominated by a much broader area of less than average values of Δq^2. Figure 7 shows a noticeable, almost jet-like character to U, although it must be realized that $\Delta U_{max} = 0.096 \langle U \rangle_{max}$ and may not be that easily detected. Consistent with this undershoot in ΔU is an overshoot in $\Delta\Theta$ (Fig 10), although there is also an overshoot coming down from near the inversion height on the downward side of the roll. A very broad region exists where $\Delta\Theta$ is nearly zero, implying little variation of Θ in y. These results are all consistent with LeMone's observation that the turbulence was largest where the vertical flux was upward. The heat flux is concentrated upward in the same region also, producing the temperature overshoot.

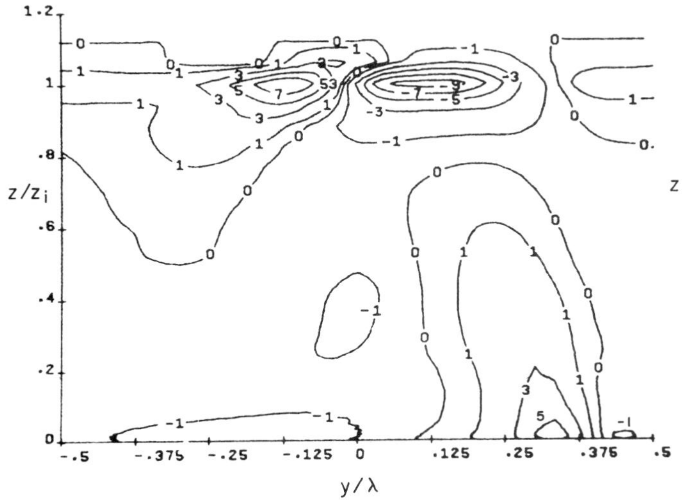

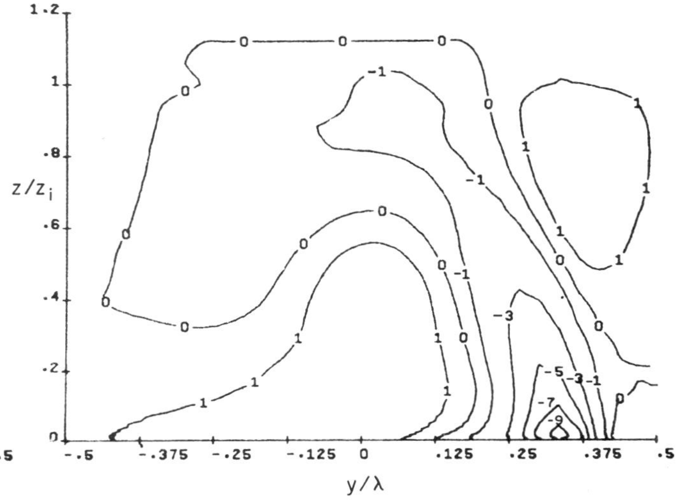

◀ **Fig. 10.** Perturbation temperature contours for the conditions of Fig. 9. The normalizing maximum value is $\Delta\Theta_{\max} w_* / \overline{w\theta_0} = 10.4$

Fig. 11. Perturbation axial velocity contours for the conditions of Fig. 9. The normalizing maximum value is $\Delta U_{\max}/w_* = 1.09$

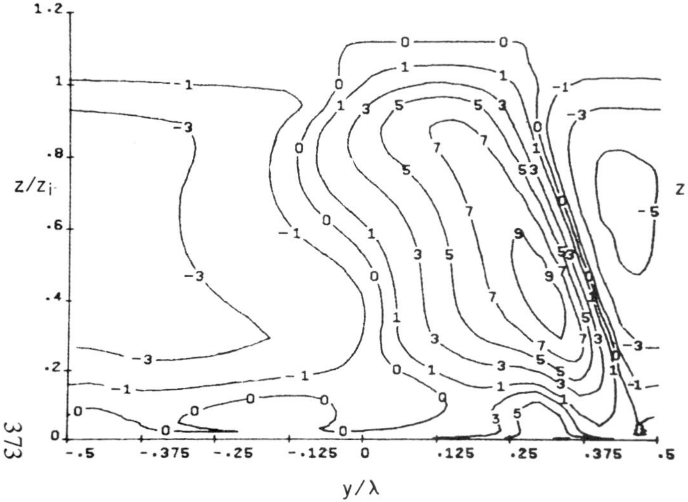

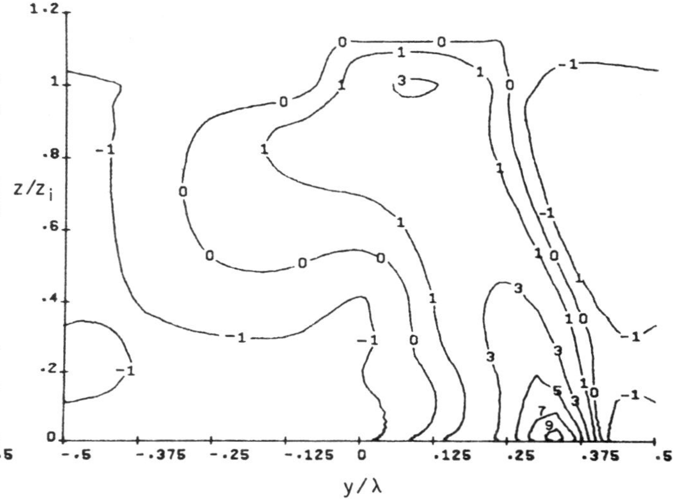

◀ **Fig. 12.** Perturbation turbulence contours for the conditions given in Fig. 9. Here $\Delta q^2_{\max}/w_*^2 = 0.4$ is the normalizing maximum value

Fig. 13. Roll species contours for the conditions of Fig. 2. The normalizing maximum value is $\Delta C_{\max} w_* / \overline{cw_0} = 6.62$

Fig. 13 shows the contour pattern of the roll-averaged ΔC, showing it to be qualitatively similar to the transport of $\Delta\Theta$ off the surface. No return transport exists, since C is assumed to go to a small value at the top. LeMone observed clouds near the top of the updraft regions of flow; but this observation should not be confused with our species results. The passive species is fairly uniform across the domain ($\Delta C_{max} = 0.050 \langle C \rangle_{max}$).

If the closure modeling were exact, the one-dimensional (horizontally homogeneous) analysis of this problem, which averages over the periodic large eddy roll structure and treats the roll energy as part of the turbulent kinetic energy, would yield the same answer as obtained by horizontally averaging over the two-dimensional result. Since the two-dimensional computation allows the dominant large eddy structure to be determined, the closure modeling should be less critical in this computation than it is in the analogous one-dimensional computation. For this phenomenon, which has a strong two-dimensional character, our model represents an intermediate step between depending completely on closure modeling and depending on sub-grid

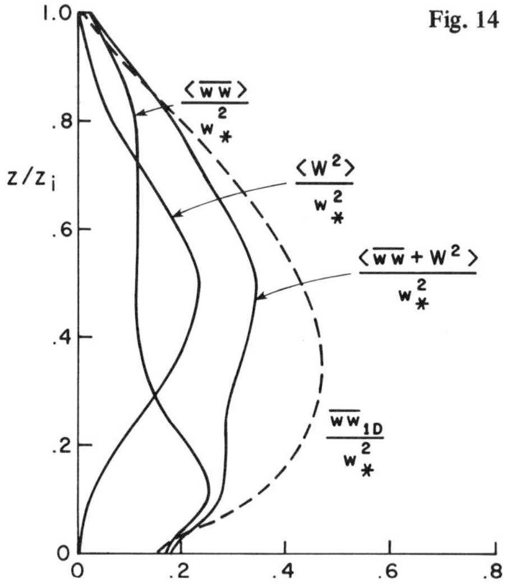

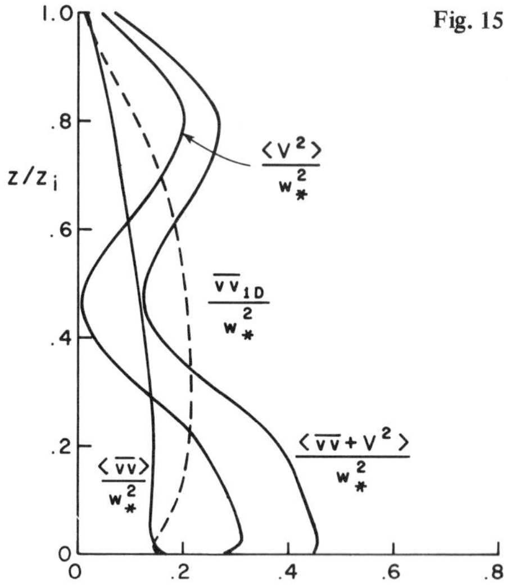

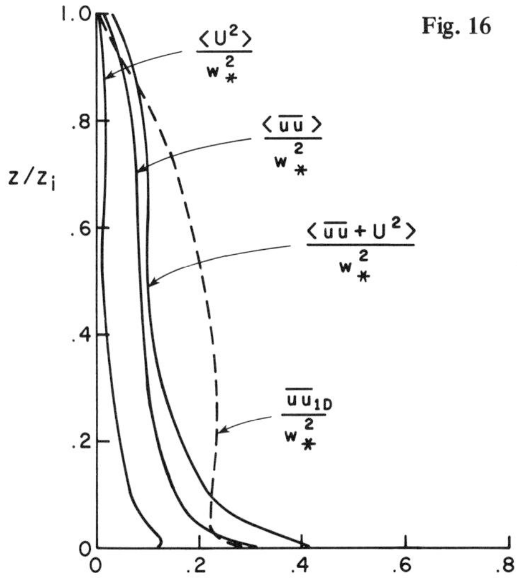

Fig. 14. Comparison between the one-dimensional and two-dimensional predictions for vertical velocity variance. For the two-dimensional calculation, the y-averaged, large scale roll contribution is denoted by $\langle W^2 \rangle$ and the smaller scale turbulence by $\langle \overline{ww} \rangle$

Fig. 15. Comparison between the one-dimensional and two-dimensional predictions for transverse velocity variance. For the two-dimensional calculation, the y-averaged, large-scale roll contribution is denoted by $\langle V^2 \rangle$ and the smaller scale turbulence by $\langle \overline{vv} \rangle$

Fig. 16. Comparison between the one-dimensional and two-dimensional predictions for longitudinal velocity variance. For the two-dimensional calculation, the y-averaged, large scale roll contribution is denoted by $\langle U^2 \rangle$ and the smaller scale turbulence by $\langle \overline{uu} \rangle$

closure modeling [17]. Thus, the two-dimensional result can be used to check strengths and weaknesses of the one-dimensional model.

The separate components of the velocity variance are shown in Figs. 14 through 16. There is a reasonable comparison between the vertical variance predicted by the one-dimensional model and the spatial average of the two-dimensional results. But the other two components show significant differences.

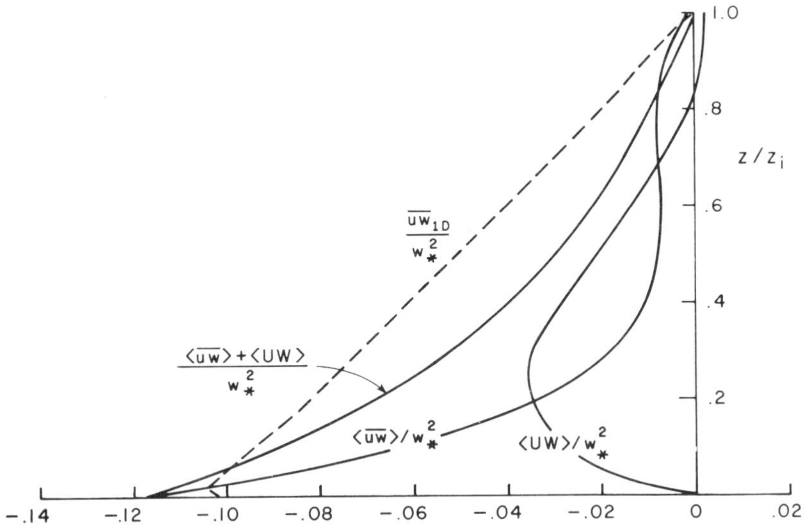

Fig. 17. Comparison between the one-dimensional and two-dimensional predictions for the vertical longitudinal momentum flux. The y-averaged roll contribution is denoted by $\langle UW \rangle$ and the turbulent transport by $\langle \overline{uw} \rangle$

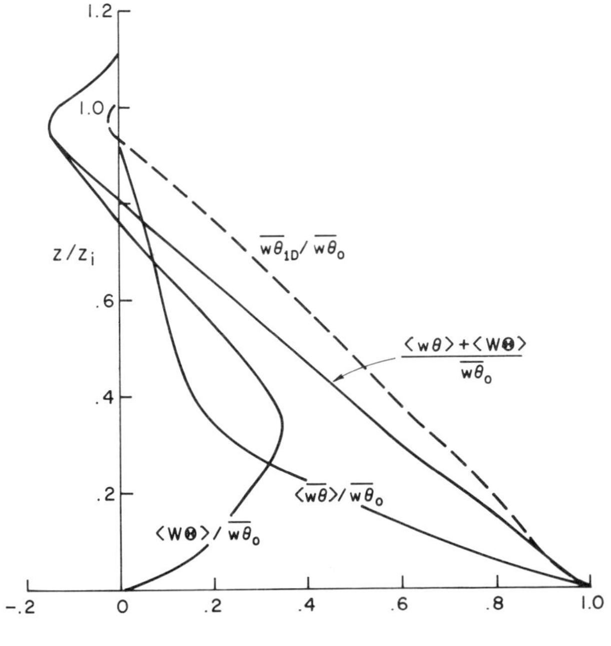

Fig. 18. Comparison between the one-dimensional and two-dimensional predictions for the vertical heat flux. The y-averaged roll contribution is denoted by $\langle W\Theta \rangle$ and the turbulent ransport by $\langle \overline{w\theta} \rangle$

The analogous comparisons of the vertical momentum flux and the vertical heat flux are shown in Figs. 17 and 18. In the middle of the mixed layer, a major portion of the turbulent transport is carried by the large roll structure. In spite of this, the one-dimensional model is able to model the turbulent transport reasonably well except in the vicinity of the upper inversion. Both the one-dimensional and two-dimensional models predict a heat flux which is counter gradient over approximately one-half of the mixed layer depth. The two-dimensional model predicts a larger undershoot in the heat flux near the inversion. This seems to be associated with the fact that the two-dimensional model can account for more influence of the wave motion in the stable region than is represented in the one-dimensional model. This is perhaps shown better in Fig. 19 which shows the temperature variance. By far, the largest contribution to temperature variance at the top of the mixed layer is the wave motions induced by the large rolls. However, even the small scale turbulent variance is approximately a factor of 2 larger than that predicted by the one-dimensional model at $z/z_i = 1$.

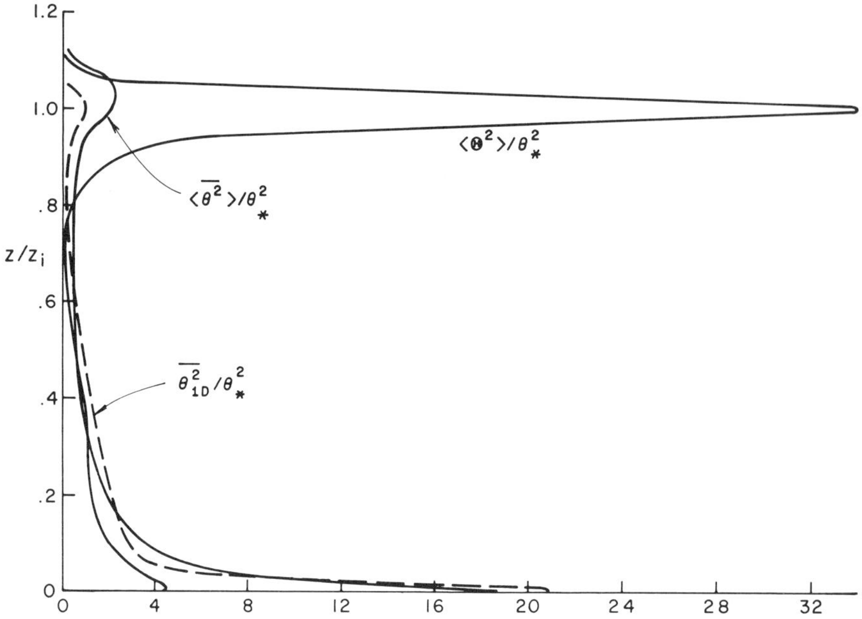

Fig. 19. Comparison between the one-dimensional and two-dimensional predictions for the temperature variance. The y-averaged roll contribution is denoted by $\langle \Theta^2 \rangle$ and the turbulent contribution by $\langle \theta^2 \rangle$

Figures 14 through 19 show that although the turbulent transport compares quite favorably for these two models, there is considerably more kinetic energy near the surface and near the inversion in the more correct two-dimensional model. The deficiencies of the one-dimensional model appear to reflect the difficulty of modeling with an isotropic turbulent macroscale. It is clear from the two-dimensional simulation as it is from field observations [18] that at both the top and bottom of the domain, the characteristic horizontal scale of the turbulence is much larger than the characteristic vertical scale. It appears that the next step in improving the model calls for incorporating some structural shape in the scale representation [19].

Appendix

Model Equations

The model equations of motion for an incompressible fluid in the presence of both a gravitational and a Coriolis body force, with the mean variables denoted by capitals and the turbulent fluctuations by lower-case, may be written in general tensor notation as follows.

$$\frac{\partial U_i}{\partial t} + U_j \frac{\partial U_i}{\partial x_j} = -\frac{\partial \overline{u_i u_j}}{\partial x_j} - \frac{1}{\rho}\frac{\partial P}{\partial x_i} + g_i \frac{(\Theta - \Theta_0)}{\Theta_0} - 2\epsilon_{ijk}\Omega_j U_k$$

$$\frac{\partial U_i}{\partial x_i} = 0$$

$$\frac{\partial \Theta}{\partial t} + U_j \frac{\partial \Theta}{\partial x_j} = -\frac{\partial \overline{u_j \theta}}{\partial x_j}$$

$$\begin{aligned}\frac{\partial \overline{u_i u_j}}{\partial t} + U_k \frac{\partial \overline{u_i u_j}}{\partial x_k} = & -\overline{u_i u_k}\frac{\partial U_j}{\partial x_k} - \overline{u_j u_k}\frac{\partial U_i}{\partial x_k} + g_i \frac{\overline{u_j \theta}}{\Theta_0} + g_j \frac{\overline{u_i \theta}}{\Theta_0} \\ & - 2\epsilon_{ikl}\Omega_k \overline{u_l u_j} - 2\epsilon_{jlk}\Omega_l \overline{u_k u_i} \\ & + 0.3\frac{\partial}{\partial x_k}\left(q\Lambda \frac{\partial \overline{u_i u_j}}{\partial x_k}\right) - \frac{q}{\Lambda}\left(\overline{u_i u_j} - \delta_{ij}\frac{q^2}{3}\right) - \delta_{ij}\frac{q^3}{12\Lambda}\end{aligned}$$

$$\begin{aligned}\frac{\partial \overline{u_i \theta}}{\partial t} + U_j \frac{\partial \overline{u_i \theta}}{\partial x_j} = & -\overline{u_i u_j}\frac{\partial \Theta}{\partial x_j} - \overline{u_j \theta}\frac{\partial U_i}{\partial x_j} + g_i \frac{\overline{\theta^2}}{\Theta_0} - 2\epsilon_{ijk}\Omega_j \overline{u_k \theta} \\ & + 0.3\frac{\partial}{\partial x_j}\left(q\Lambda \frac{\partial \overline{u_i \theta}}{\partial x_j}\right) - \frac{0.75q}{\Lambda}\overline{u_i \theta}\end{aligned}$$

$$\frac{\partial \overline{\theta^2}}{\partial t} + U_j \frac{\partial \overline{\theta^2}}{\partial x_j} = -2\overline{u_j \theta}\frac{\partial \Theta}{\partial x_j} + 0.3\frac{\partial}{\partial x_j}\left(q\Lambda \frac{\partial \overline{\theta^2}}{\partial x_j}\right) - \frac{0.45 q\overline{\theta^2}}{\Lambda}$$

$$\begin{aligned}\frac{\partial \Lambda}{\partial t} + U_j \frac{\partial \Lambda}{\partial x_j} = & \; 0.35\frac{\Lambda}{q^2}\overline{u_i u_j}\frac{\partial U_i}{\partial x_j} + 0.75q + 0.3\frac{\partial}{\partial x_i}\left(q\Lambda \frac{\partial \Lambda}{\partial x_i}\right) \\ & - \frac{0.375}{q}\left(\frac{\partial q\Lambda}{\partial x_i}\right)^2 + \frac{0.8}{q^2} q_i \frac{\overline{u_i \theta}}{\Theta_0}\end{aligned}$$

The overbar represents an ensemble average, and $q^2 = \overline{u_i u_i}$.

Along with the differing boundary conditions for the individual problems, an upper bound is placed on Λ depending upon the spread of the region of turbulence.

For the canopy flow example, variation with respect to one coordinate only is permitted. Also extra sink terms must be added to the equations for the mean variables, and both source and sink terms added to the Reynolds stress and heat flux equations (13). The other three examples permit dependence of variables on two coordinates. The flow may be either planar or axisymmetric.

Acknowledgment. The example investigations discussed in this paper have been supported by the National Aeronautics and Space Administration, the Nuclear Regulatory Commission, the Electric Power Research Institute and the Naval Air Systems Command.

References

1. C. duP. Donaldson: "Atmospheric Turbulence and the Dispersal of Atmospheric Pollutants", in *Proceedings of Workshop on Micrometeorology,* American Meteorological Society, ed. by D. A. Haugen (Science Press 1973) pp. 313–390
2. W. S. Lewellen: "Use of Invariant Modeling", in *Handbook of Turbulence,* Vol. I, ed. by W. Frost, T. H. Moulden (Plenum Publishing Corp. 1977) pp. 237–280
3. M. E. Teske, W. S. Lewellen: "Turbulent Transport Model of a Thunderstorm Gust Front", in *Proceedings of 10th Conference on Severe Local Storms,* American Meteorological Society, Omaha, NE, Oct. 18–21 (1977)
4. W. S. Lewellen, M. E. Teske, H. Segur: "Turbulent Transport Model of Wind Shear in Thunderstorm Gust Fronts and Warm Fronts". NASA CR 3002 (1978)
5. M. E. Teske, W. S. Lewellen: "The Prediction of Turbulence and Wind Shear Associated with Thunderstorm Gust Fronts", in *Proceedings of Conference on Atmospheric Environment of Aerospace Systems and Applied Meteorology,* American Meteorological Society, New York, NY, Nov. 14–16 (1978)
6. J. S. Turner: *Buoyancy Effects in Fluids* (Cambridge University Press, Great Britain 1973)
7. W. S. Lewellen, M. E. Teske: "Turbulent Transport Model of Low-Level Winds in a Tornado", in *Proceedings of 10th Conference on Severe Local Storms,* American Meteorological Society, Omaha, NE, Oct. 18–21 (1977) pp. 291–298
8. W. S. Lewellen, M. E. Teske, Y. P. Sheng: "Wind and Pressure Distributions in a Tornado". *Fifth International Conference on Wind Engineering,* Colorado State University, July 8–13 (1979)
9. W. S. Lewellen, Y. P. Sheng: "Influence of Surface Conditions on Tornado Wind Distributions", in *Proceedings of the 11th Conference on Severe Local Storms,* American Meteorological Society, Kansas City, Missouri, Oct. 2–5 (1979)
10. N. R. Wilson, R. H. Shaw: A higher order closure model for canopy flow. J. Appl. Meteorol. **16,** 1197–1205 (1977)
11. R. H. Shaw, G. den Hartog, K. M. King, G. W. Thurtell: Measurements of mean wind flow and three-dimensional turbulence intensity within a mature corn canopy. Agric. Meteorol. **13,** 419–425 (1974)
12. R. H. Shaw, R. H. Silversides, G. W. Thurtell: Some observations of turbulence and turbulent transport within and above plant canopies. Boundary-Layer Meteorol. **5,** 429–449 (1974)
13. W. S. Lewellen, Y. P. Sheng: "Modeling of Dry Deposition of SO_2 and Sulfate Aerosols". A. R. A. P. Rpt. No. 410 (1979)
14. R. A. Brown: *Analytical Methods in Planetary Boundary-Layer Modeling* (Wiley and Sons, New York 1974) pp. 94–108
15. M. E. Teske, W. S. Lewellen: "Horizontal Roll Vortices in the Planetary Boundary Layer", in *Proceedings of Fourth Symposium on Turbulence, Diffusion, and Air Pollution,* American Meteorological Society, Reno, NE, Jan. 15–18 (1979) pp. 456–463.
16. R. A. LeMone: Modulation of turbulence energy by longitudinal rolls in an unstable planetary boundary layer. J. Atmos. Sci. **33,** 1308–1320 (1976)
17. J. W. Deardorff: Numerical investigation of neutral and unstable planetary boundary layers. J. Atmos. Sci. **29,** 91–115 (1972)
18. J. C. Kaimal: Horizontal velocity spectra in an unstable surface layer. J. Atmos. Sci. **35,** 18–24 (1978)
19. G. Sandri: "Recent Results Obtained in the Modeling of Turbulent Flows by Second-Order Closure". A.R.A.P. Rpt. No. 332 (1978)

The Core-Bulk Structure Associated with Diffusing Clouds

P. C. Chatwin

Department of Applied Mathematics and Theoretical Physics
University of Liverpool, Liverpool, U.K.

P. J. Sullivan

Department of Applied Mathematics
University of Western Ontario, London, Canada

Abstract

It has recently been proposed by the authors [5] that certain statistical properties associated with clouds of contaminant dispersing in some turbulent flows have a core-bulk spatial structure. This structure is briefly summarized and illustrated by some specific examples. The basis of the earlier work was that ensemble means were considered only after studying the dispersion of the cloud during each realization. Little use was made, therefore, of the equations governing the evolution of the statistical properties. Hence the principal aim of the present paper is to examine whether the proposed structure is consistent with these equations. It is shown that it is, and detailed discussion of the matching procedure across the core-bulk interface is given for some special cases. The paper concludes with a discussion of the general case and the effects of molecular diffusion.

Nomenclature

C, c Ensemble mean, and fluctuation, of Γ

$L = L(t)$ Measure of cloud diameter, defined in Eq. (6)

L_0 Measure of initial cloud diameter, defined in Eq. (7)

$L_1 = L_1(t)$ Measure of core thickness

$L_2 = L_2(t), L_3 = L_3(t)$ Scales used in analysis

p, q, m Constant powers used in analysis

Q Total quantity of contaminant

$R = x/L$

t Time elapsed since release of cloud

U, u Ensemble mean, and fluctuation, of Υ

x Position vector relative to (instantaneous) centre of mass of cloud

$\Gamma = \Gamma(x, t)$ Distribution of concentration in one realization

ϵ Ensemble mean rate of dissipation of kinetic energy per unit mass

κ Molecular diffusivity

λ_c Conduction cut-off length, defined in Eq. (8)

ν Kinematic viscosity

$\rho = x/L_1$

$\Upsilon = \Upsilon(x, t)$ Velocity field in one realization

Introduction

The paper is concerned with the following situation, the basic one of turbulent diffusion. At time $t = 0$, a finite quantity Q of contaminant is released in a specified way in an incompressible turbulent flow. Let the distribution of contaminant subsequently be $\Gamma(x, t)$. It is important here to define $\boldsymbol{x} = \mathbf{0}$, the origin of coordinates. For definiteness this paper will be concerned entirely with *relative* turbulent diffusion which requires that, *throughout* each realization of the dispersion, $\boldsymbol{x} = \mathbf{0}$ is the centre of mass of the cloud of contaminant. More explanation of this choice of axes is given in [9]. In a laboratory frame, therefore, the origin used in this paper is moving in a random manner so that, for all t,

$$\int \boldsymbol{x}\Gamma(\boldsymbol{x}, t)dV(\boldsymbol{x}) = \mathbf{0}, \tag{1}$$

where the integral is over all space, as will be all integrals in this paper. The practical and theoretical advantages of relative diffusion are summarized in [1, 5, 6, 10]. Despite the restriction to relative diffusion, much of the work in this paper applies, with obvious changes in definitions, to absolute diffusion (when the origin is fixed, or moving in a predetermined way).

The equation governing $\Gamma(\boldsymbol{x}, t)$ is

$$\frac{\partial\Gamma}{\partial t} + \boldsymbol{\Upsilon}.\nabla\Gamma = \kappa\nabla^2\Gamma, \tag{2}$$

where κ is the molecular diffusivity, and $\boldsymbol{\Upsilon} = \boldsymbol{\Upsilon}(\boldsymbol{x}, t)$ is the turbulent velocity field in the frame defined above. [Note that (2) is a statement of conservation of mass and therefore does not depend on Newton's Second Law; hence it is irrelevant that the axes are not inertial.] Since the flow is incompressible,

$$\nabla.\boldsymbol{\Upsilon} = 0. \tag{3}$$

Further conditions that Γ will be required to satisfy are that $\Gamma(\boldsymbol{x}, 0)$ is specified, and that the total mass of contaminant is conserved. Hence, for all t,

$$\int \Gamma(\boldsymbol{x}, t)dV(\boldsymbol{x}) = Q, \tag{4}$$

where Q is the constant quantity of contaminant; this is ensured provided $\boldsymbol{n}.\nabla\Gamma = 0$ on all (stationary) boundaries, where $\boldsymbol{n}$ is the normal to the boundary, and provided $\Gamma \to 0$ as $|\boldsymbol{x}| \to \infty$, if the flow region is unbounded.

Denote ensemble means by overbars and write

$$\overline{\Gamma} = C, \quad \Gamma = C + c; \quad \overline{\boldsymbol{\Upsilon}} = \boldsymbol{U}, \quad \boldsymbol{\Upsilon} = \boldsymbol{U} + \boldsymbol{u}. \tag{5}$$

Note that ensemble means cannot be found by integration, either over space or time, since all statistical properties of Γ, like $C(\boldsymbol{x}, t)$ and $\overline{c^2}(\boldsymbol{x}, t)$ are inherently inhomogeneous in $\boldsymbol{x}$ (because the total quantity of contaminant is finite) and non-stationary in t (because the cloud spreads as t increases). The work to be described in this paper is about the way these statistical properties vary with $\boldsymbol{x}$ and t.

Several length scales will be prominent in the discussion. Define $L = L(t)$ by

$$QL^2(t) = \int |\boldsymbol{x}|^2 C(\boldsymbol{x}, t)dV(\boldsymbol{x}), \tag{6}$$

so that L is a length typical of the mean cloud diameter at time t. It is known that for the very high Reynolds numbers common in nature, L^2 is eventually approximately proportional to ϵt^3, where ϵ is the ensemble mean rate of dissipation of kinetic energy per unit mass. This result was first enunciated partially by *Richardson* [12], and a summary of later work is given by *Csanady* [10]. Secondly, define the constant length L_0 by

$$Q^2 L_0^{-3} = \int \Gamma^2(\boldsymbol{x}, 0) dV(\boldsymbol{x}). \tag{7}$$

L_0 is a length typical of the initial cloud size with L_0^3 being equal to the initial volume of the cloud when $\Gamma(\boldsymbol{x}, 0)$ is spatially uniform. Finally denote the conduction cut-off length by λ_c, so that

$$\lambda_c = (\nu\kappa^2/\epsilon)^{1/4}, \tag{8}$$

where ν is the kinematic viscosity.

The Dispersion of a Cloud of Marked Fluid Particles

Summary of Previous Work

When $L_0 \gg \lambda_c$ the cloud will be unaffected by molecular diffusion for a long period following release. In view of the fact that λ_c is of the order of fractions of a millimetre in many natural flows [6, 11], this condition is usually satisfied in practice. It is also sensible to begin a theoretical study by neglecting κ [1], since there is no hope of understanding the real problem, in which κ ultimately plays a crucial role, if the simpler problem is too difficult. When $\kappa = 0$, the quantity of contaminant within any fluid particle remains constant, and many general conservation relations follow. Apart from (4), the most important is that [5]

$$\int \overline{\Gamma^2}(\boldsymbol{x}, t) dV(\boldsymbol{x}) = \int C^2(\boldsymbol{x}, t) dV(\boldsymbol{x}) + \int \overline{c^2}(\boldsymbol{x}, t) dV(\boldsymbol{x}) = \frac{Q^2}{L_0^3}, \tag{9}$$

where (5) has been used to express $\overline{\Gamma^2}$ in terms of C^2 and $\overline{c^2}$, and (7) has been used to obtain the value of the integral.

Suppose now that $C(\boldsymbol{x}, t)$ and $\overline{c^2}(\boldsymbol{x}, t)$ have the same order of magnitude over the whole cloud. It follows then from mass conservation (4), and the definition of $L(t)$ in (6) that

$$C(\boldsymbol{x}, t) = O(QL^{-3}). \tag{10}$$

Initially Γ is specified so that, at $t = 0$, $\overline{c^2}$ is zero and only the integral of C^2 contributes in (9). But when (10) holds, the integral of C^2 is of order $Q^2 L^{-3}$. Since $L/L_0 \to \infty$ as $t \to \infty$, it follows that, as $t \to \infty$, the integral of $\overline{c^2}$ in (9) tends to $Q^2 L_0^{-3}$. Subject to the assumption made above that $\overline{c^2}$ has the same order of magnitude for all $\boldsymbol{x}$ it now follows, with an accuracy increasing as $t \to \infty$ [5], that

$$\overline{c^2}(\boldsymbol{x}, t) = O(Q^2 L^{-3} L_0^{-3}). \tag{11}$$

It is valuable to derive (10) and (11) by a simple probability argument. When $\kappa = 0$ the volume occupied by marked fluid is constant for all t and of order L_0^3 by (7). But the shape of this volume is random. What can be said is that this volume of order L_0^3 is, whatever its shape, located in a volume of order Γ^3 surrounding its centre of mass. Thus the probability of any point in the volume of order L^3 lying in marked fluid is of order $(L_0/L)^3$, if this probability has the same order of magnitude for all points. [The proviso is important.] Then the ensemble mean of any function of Γ, $f(\Gamma)$, has order of magnitude $f(QL_0^{-3})(L_0/L)^3$. The estimate (10) is recovered by taking $f(\Gamma) = \Gamma$, and the estimate (11) by taking $f(\Gamma) = \Gamma^2$, and using $L/L_0 \gg 1$. But return to the proviso made above. Suppose the cloud is such that, at $t = 0$, its centre of mass is embedded within it – normally this will be true (but not, for example, if the initial cloud is shaped like a torus). By continuity it follows that, unlike most points, the centre of mass is certain to lie in marked fluid for some time following release; also by continuity the same statement is true for points sufficiently near the centre of mass, i.e. sufficiently near $\boldsymbol{x} = \boldsymbol{0}$. For such points the ensemble mean of $f(\Gamma)$ is order $f(QL_0^{-3})$ (and equal to $f(QL_0^{-3})$ when the initial marking is spatially uniform); thus (10) and (11) are replaced for sufficiently small $|\boldsymbol{x}|$ by

$$C = O(QL_0^{-3}), \quad \overline{c^2} = O(Q^2L_0^{-6}). \tag{12}$$

The period of time following release, and the size of the region surrounding $\boldsymbol{x} = \boldsymbol{0}$, for which (12), not (10) and (11), are the correct estimates will depend both on the precise details of the initial cloud configuration and concentration, and on the statistical properties of the turbulent velocity field. In the only case when exact solutions are available, which is that when the velocity field is a uniform strain converting initial spheres into ellipsoids whose axes are oriented randomly in space, the estimates (12) hold provided $|\boldsymbol{x}|$ is less than $L_1(t)$, where $L_1(t)$ is the smallest semi-axis of the ellipsoid, while the estimates (10) and (11) hold for $|\boldsymbol{x}|$ much greater than L_1. Since $L_1 \to 0$ as $t \to \infty$, it follows that the distributions of $C(\boldsymbol{x}, t)$ and $\overline{c^2}(\boldsymbol{x}, t)$ have a core-bulk spatial structure. Within the core, defined by $|\boldsymbol{x}| \leqslant L_1$, the magnitudes of C and $\overline{c^2}$ are given by (12), and are therefore much greater than in the bulk, defined by $|\boldsymbol{x}| \gg L_1$, where the magnitudes of C and $\overline{c^2}$ are given by (10) and (11). Reference should be made to [5] for further discussion of the exact solution. Figure 1 shows the core-bulk structure of C in the case when the straining motion is such that spheres are converted into flat ellipsoids of revolution ("discuses"). In Fig. 1a the initial distribution of Γ is uniform, and in Fig. 1b it is a Gaussian function of $|\boldsymbol{x}|$. The important points in connection with these figures are that the magnitude of C at $\boldsymbol{x} = \boldsymbol{0}$ is constant for all t, and that the graphs for times later than that shown would have a wider spread but a narrower core.

Although the velocity field giving the exact solution is special, it is known [3, 13, 14] that it is realistic for clouds whose dimensions are no greater than $5(\nu^3/\epsilon)^{1/4}$. Because of continuity, slightly larger clouds and slightly different velocity fields will give a similar core-bulk structure. What is not clear is that much larger clouds and vastly different velocity fields will also give rise to this structure. But if they do not, there are several questions that need answering, as is clear from later work in this paper.

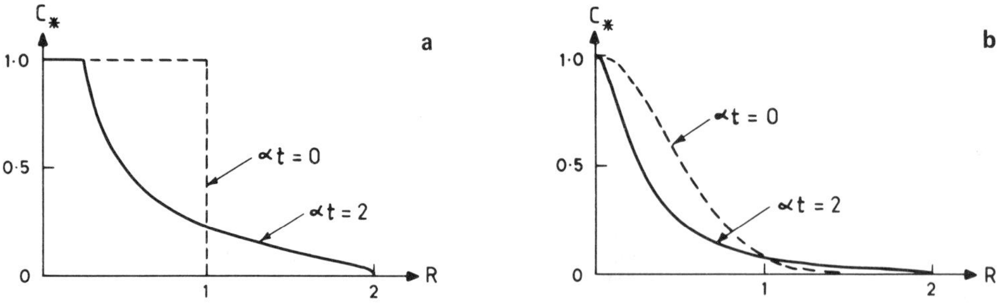

Fig. 1a, b. Two examples of the core-bulk structure for the uniform strain motion converting spheres into discuses. In both **(a)** and **(b)**, α is the maximum rate of strain, R is $|\boldsymbol{x}|$ divided by $5\,L\,(0)/3$, and $C_* = C(\boldsymbol{x}, t)/C(\boldsymbol{0}, 0)$

The Equations Governing the Evolution of C and $\overline{c^2}$

The equations governing $C(x, t)$ and $\overline{c^2}(x, t)$ are obtained from (2), (3) and (5). When $\kappa = 0$ they reduce to [10]:

$$\frac{\partial C}{\partial t} + \nabla\cdot(UC) + \nabla\cdot(\overline{uc}) = 0; \tag{13}$$

$$\frac{\partial \overline{c^2}}{\partial t} + \nabla\cdot(U\overline{c^2}) + \nabla\cdot(\overline{uc^2}) + 2\,\overline{uc}\cdot\nabla C = 0. \tag{14}$$

The main aim of this paper is to examine how far the core-bulk structure proposed for certain flows is consistent with these equations.

In the bulk of the cloud the orders of magnitude of C and $\overline{c^2}$ are QL^{-3} and $Q^2L^{-3}L_0^{-3}$, given in (10) and (11). Suppose that, in (13), the term involving $\overline{uc}$ is at least of the same order of magnitude as that involving UC. This is normally true and is certainly so if, for example, $U = \mathbf{0}$. It then follows from (13) that $\overline{uc}$ has order of magnitude $Q\dot{L}L^{-3}$, where $\dot{L}$ is dL/dt. Then, in (14), the production term $2\,\overline{uc}\cdot\nabla C$ has order of magnitude $Q^2\dot{L}L^{-7}$. But this is less, by orders of magnitude, than $\partial\overline{c^2}/\partial t$ which, using (11), has order of magnitude $Q^2\dot{L}L^{-4}L_0^{-3}$. Thus, to a very good approximation, no *production* of $\overline{c^2}$ takes place in those parts of the cloud – the bulk – where C and $\overline{c^2}$ have orders of magnitude QL^{-3} and $Q^2L^{-3}L_0^{-3}$ respectively.

However some production of $\overline{c^2}$ does take place during the dispersion. Initially $\overline{c^2}$ is everywhere zero and ultimately, by (9), the integral of $\overline{c^2}$ over all space is $Q^2L_0^{-3}$. There is therefore an important problem to investigate, namely when and/or where does the production of $\overline{c^2}$ take place? As shown in [5], the answer to this question for the case of a uniform strain is that it all takes place essentially in the core region of decreasing thickness surrounding $x = \mathbf{0}$. It will be shown below that this answer appears to be consistent with (13) and (14) for a wide class of turbulent velocity fields, which includes the uniform strain as a special case.

At first sight, of course, the existence of the proposed core-bulk spatial structure seems contrary to intuition (although it has to be stressed that such a structure is proposed only for the *statistical properties* of Γ and not for what is *observed* in an experiment, namely Γ itself). It is therefore worth pausing to see whether any other possibilities exist. One that has been suggested to the authors by an anonymous referee is that essentially all the production of $\overline{c^2}$ takes place at a very early stage in the dispersion, so early in fact that the estimates (10) and (11) do not then hold. This possibility is not in any way consistent with what happens in a uniform strain and, more seriously for general velocity fields, assumes, in effect, that the magnitudes of C and $\overline{c^2}$ decay with time at a rate which is of the same order of magnitude over the whole cloud. For a core would develop if the decay rate were not spatially uniform. There is no justification for this assumption of spatial uniformity; on the other hand it could be correct in some flows. Further investigation is needed on this point, which also arises later.

In the remainder of this paper, however, it will be assumed that a core-bulk structure exists. More specifically it will be assumed, in line with observations of plumes [6], that both the core structure and the bulk structure are self-similar. This means, according to (10) and (11), that in the bulk there are functions F and G such that

$$C(x, t) = QL^{-3}F(R), \quad \overline{c^2}(x, t) = Q^2L^{-3}L_0^{-3}G(R); \quad R = |x|/L. \tag{15}$$

One further assumption that will be made is that $\boldsymbol{U} = \mathbf{0}$ since this makes the mathematics more tractable but does not introduce any critical simplification of the physics. When (15) is substituted into (13) with $\boldsymbol{U} = \mathbf{0}$ it follows that

$$\overline{\boldsymbol{u}c} = Q\dot{L}L^{-3}\boldsymbol{R}F, \tag{16}$$

which is everywhere outwards since $\dot{L}$ and F are positive. As explained above, the production term in the equation for $\overline{c^2}$ is negligible, and substitution of (15) into (14) with $\boldsymbol{U} = \mathbf{0}$ then gives

$$\overline{\boldsymbol{u}c^2} = Q^2\dot{L}L^{-3}L_0^{-3}\boldsymbol{R}G, \tag{17}$$

which is also everywhere outwards. An important conclusion from (16) and (17) is therefore that changes in C and $\overline{c^2}$ at a fixed place in the bulk occur entirely as the result of transfer *outwards* from the central regions, which is, of course, consistent with production of $\overline{c^2}$ being confined to a central core region.

As explained above the structure in the core will be assumed to be self-similar; otherwise it could not match on to the assumed, and observed, self-similar structure in the bulk. Suppose therefore that for $|\boldsymbol{x}| \leqslant L_1(t)$, where L_1 is the core thickness to be determined later, there are functions f and g such that

$$C(\boldsymbol{x}, t) = QL_2^{-3}f(\rho), \quad \overline{c^2}(\boldsymbol{x}, t) = Q^2L_3^{-6}g(\rho); \quad \rho = |\boldsymbol{x}|/L_1. \tag{18}$$

Here $L_2 = L_2(t)$ and $L_3 = L_3(t)$ are length scales to be determined later. The scalings in (18) are more general than those given in (12) which were obtained on the assumption that the probability of the centre of mass lying in marked fluid was of order unity. Physically, therefore, (18) allows the possibility of this probability being different both from order unity and from the bulk value which is of order $(L_0/L)^3$.

The first of (18) gives

$$\frac{\partial C}{\partial t} = -3\,Q\dot{L}_2L_2^{-4}f - Q\dot{L}_1L_1^{-1}L_2^{-3}\rho f', \tag{19}$$

so that, from (13) with $\boldsymbol{U} = \mathbf{0}$,

$$\frac{1}{\rho^2L_1}\frac{\partial}{\partial\rho}(\rho^2\overline{u_rc}) = 3\,Q\dot{L}_2L_2^{-4}f + Q\dot{L}_1L_1^{-1}L_2^{-3}\rho f', \tag{20}$$

where u_r is the radial component of $\boldsymbol{u}$. Hence, on rearranging (20),

$$\frac{\partial}{\partial\rho}(\rho^2\overline{u_rc}) = \frac{\partial}{\partial\rho}(Q\dot{L}_1L_2^{-3}\rho^3f) - 3\,QL_2^{-4}(L_2\dot{L}_1 - L_1\dot{L}_2)\rho^2f; \tag{21}$$

since $\overline{\boldsymbol{u}c} = \mathbf{0}$ at $|\boldsymbol{x}| = 0$, (21) can be integrated to give

$$\overline{\boldsymbol{u}c} = Q\dot{L}_1L_2^{-3}\boldsymbol{\rho}f - 3\,QL_2^{-4}(L_2\dot{L}_1 - L_1\dot{L}_2)\,\boldsymbol{\rho}\rho^{-3}\int_0^\rho z^2f(z)dz. \tag{22}$$

This form of $\overline{uc}$ has to evolve smoothly as ρ increases into the form given by (16). Postulate the existence of an *overlap* region satisfying

$$L_1 \ll |\boldsymbol{x}| \ll L, \tag{23}$$

within which *both* (16) and (19) are valid, and examine whether this is possible. To proceed further requires knowledge of the behaviour of either $F(R)$ for small R or $f(\rho)$ for large ρ. Here a fairly general possibility is considered, namely an arbitrary power-law dependence. As will be seen later the uniform strain velocity fields are included as special cases. Thus suppose

$$F(R) \approx AR^{-p} \qquad \text{for } R \ll 1, \tag{24}$$

where A and p are non-dimensional constants. Since $LR = |\boldsymbol{x}| = L_1\rho$, (24) can be written

$$F(R) \approx A\rho^{-p}(L/L_1)^p. \tag{25}$$

Hence the expressions for $C(\boldsymbol{x}, t)$ in (15) and (18) are the same in the overlap region satisfying (23) provided

$$f(\rho) \approx A\rho^{-p} \quad \text{for} \quad \rho \gg 1, (L_2/L) = (L_1/L)^{(1/3)p}. \tag{26}$$

Furthermore, when (26) holds, the expression for $\overline{uc}$ in (22) becomes, for $\rho \gg 1$,

$$\overline{uc} \approx AQ\rho\rho^{-p}\left[\dot{L}_1L^{p-3}L_1^{-p}\left(1 - \frac{3}{3-p} + \frac{p}{3-p}\right) + \dot{L}L^{p-4}L_1^{1-p}\right]. \tag{27}$$

Hence

$$\overline{uc} \approx AQ\rho\rho^{-p}\dot{L}L^{p-4}L_1^{1-p} = AQ\dot{L}L^{-3}RR^{-p}, \tag{28}$$

which is the same as the expression for $\overline{uc}$ in (16) for $R \ll 1$. Therefore (26) ensures that *all* terms in the equation for $C(\boldsymbol{x}, t)$ in the bulk match on to the corresponding terms in the equation for $C(\boldsymbol{x}, t)$ in the core.

Consider now the equation for $\overline{c^2}(\boldsymbol{x}, t)$ and suppose, consistent with (24), that the power-law dependence of $\overline{c^2}$ is such that $G(R)$ in (15) satisfies

$$G(R) \approx BR^{-q} \quad \text{for} \quad R \ll 1. \tag{29}$$

The expressions for $\overline{c^2}(\boldsymbol{x}, t)$ in (15) and (18) are then the same in the overlap region satisfying (23) provided

$$g(\rho) \approx B\rho^{-q} \quad \text{for} \quad \rho \gg 1, \quad (L_3^2/LL_0) = (L_1/L)^{(1/3)q}. \tag{30}$$

The reason for postulating a core requires that the production term in the equation for $\overline{c^2}$ has the same order of magnitude in the core as the other terms. From (18) and (30) it follows that, in the core,

$$\frac{\partial\overline{c^2}}{\partial t} \approx B(q-3)Q^2\rho^{-q}\dot{L}L_0^{-3}L^{q-4}L_1^{-q}, \tag{31}$$

and from (18), (26) and (28) that, also in the core,

$$2\,\overline{uc}.\nabla C \approx -2pA^2Q^2\rho^{-2p}\dot{L}L^{2p-7}L_1^{-2p}. \tag{32}$$

These two terms have, by hypothesis, the same order of magnitude when ρ is of order unity. Hence

$$(L_1/L) = (L_0/L)^{3/(2p-q)}, \quad (L_2/L) = (L_3/L) = (L_0/L)^{p/(2p-q)}, \tag{33}$$

where the results for (L_2/L) and (L_3/L) follow from (26) and (30).

Comments, and Interpretation of the Analysis

There are certain obvious physical restrictions on the constants p and q appearing in the preceding analysis. First of all, C and $\overline{c^2}$ must decrease as ρ increases, and C cannot be anywhere greater than its initial value which is of order QL_0^{-3}. Hence, by (18), (26), (30) and (33),

$$p > 0, \quad q > 0, \quad p \geqslant q. \tag{34}$$

Note that it follows from (33) and (34) that the core thickness L_1 is much less than the mean cloud diameter L. But a more severe restriction is that the core thickness must be much less than the initial cloud diameter L_0. From (33) this requires

$$2p - q < 3. \tag{35}$$

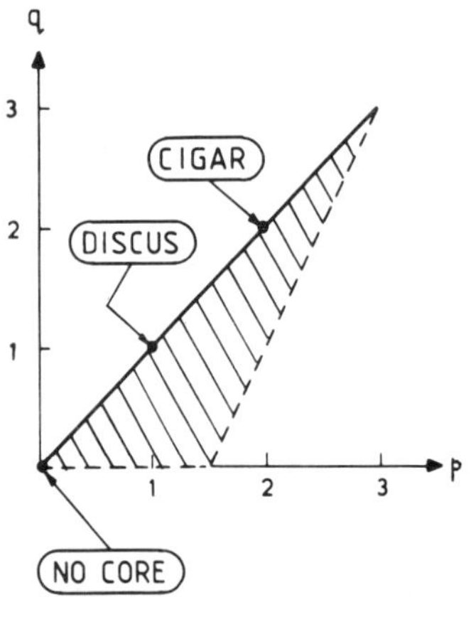

Fig. 2. Sketch illustrating permissible values of p and q

These restrictions are illustrated in Fig. 2. Values of p and q corresponding to points in the hatched region are consistent with (34) and (35). The two dashed boundaries are not included in the region but the solid boundary is. In fact $L_2 = L_0$ on the solid boundary, and this gives the core scalings for C and $\overline{c^2}$ in (12). A uniform strain velocity field is one of those flows giving these scalings, and marked on the solid boundary in Fig. 2 are two points corresponding to special cases of this flow. The point $(p, q) = (1, 1)$ occurs when the strain converts spheres into flat ellipsoids of revolution ("discuses"), and the point $(p, q) = (2, 2)$ when spheres are converted into thin ellipsoids of revolution ("cigars"). Details of these special cases are given in [5].

In view of the earlier discussion it is interesting to note that the point $(p,q) = (0, 0)$ is, strictly speaking, not in the region of permissible points. But it is on the solid boundary and corresponds, physically, to the case when there is *no* core, i.e. when C and $\overline{c^2}$ have the bulk estimates (15) throughout the cloud. In this case, however, it is still necessary to explain when the production takes place; this is not possible with the present analysis in view of the obvious singularity when $(p, q) = (0, 0)$.

One consequence of the present analysis is interesting. The equation for $\overline{c^2}$, namely (14) with $\boldsymbol{U} = \boldsymbol{0}$ can be written in terms of $\boldsymbol{\rho} = \boldsymbol{x}/L_1$ as

$$\nabla . (\overline{uc^2}) = -L_1 \frac{\partial \overline{c^2}}{\partial t} - 2\,\overline{uc}.\nabla C, \tag{36}$$

where ∇ is taken with respect to $\boldsymbol{\rho}$. Use of the results above shows that at the edge of the core this becomes

$$\nabla . (\overline{uc^2}) = \frac{Q^2 \dot{L}}{L_0^3 L^3} \left(\frac{L_0}{L} \right)^m [B(3-q)\rho^{-q} + 2A^2 p\rho^{-2p}], \quad m = \frac{3(1-q)}{(2p-q)}. \tag{37}$$

The terms in B and A come from the terms in (36) involving $\partial \overline{c^2}/\partial t$ and $2\,\overline{uc}.\nabla C$ respectively. It is therefore self-evident that the production term has the same order of magnitude as $\partial \overline{c^2}/\partial t$, since B and A are of order unity, which merely confirms the correctness of the analysis leading to (33). But since $2p > q$ from (34), it also follows from (37) that the production term in the equation for $\overline{c^2}$ is, at the core-bulk interface, falling to zero more rapidly than $\partial \overline{c^2}/\partial t$. But this result was not used in the above analysis. Hence it is evidence in support of the analysis, since the production term is known to be negligible compared with $\partial \overline{c^2}/\partial t$ in the bulk of the cloud.

An analysis analogous to that given in the present paper can be carried out for the case of a statistically steady plume of contaminant formed by its release at a constant rate from a fixed source. For, as shown in [6], the arguments about the production of $\overline{c^2}$ that suggest the existence of a core-bulk structure in a cloud also apply to a plume.

The Effects of Molecular Diffusion

Physically, the core thickness $L_1(t)$ is an average value of the minimum dimensions of the cloud, e.g. the diameters of the threads observed in the later stages of the dispersion of all clouds of contaminant. If there were no molecular diffusion, advection would cause L_1 to tend to zero as t tended to infinity. [Note that this is consistent with (33) and (35) above.] Hence the gradients of Γ would increase indefinitely. In reality, therefore, the flux of contaminant due to molecular diffusion increases as t increases. There is eventually a dynamic balance set up in which the tendency for L_1 to decrease indefinitely is offset by the spreading due to molecular diffusion. As a result L_1 does not, in reality, shrink to zero but remains constant when it reaches a value of order λ_c, the conduction cut-off length, defined in (8). Essentially, the physical arguments leading to this conclusion are given by *Batchelor* [2, 3].

It was argued by the authors [5, 7, 8] that for a real cloud with $L_0 \gg \lambda_c$ the core-bulk structure develops independently of κ (in the way summarized in earlier sections of this paper) until the core thickness is of order λ_c. But after that, molecular diffusion profoundly changes the description given above in many respects. In particular the total amount of $\overline{c^2}$ decays to zero. Because points in the core are much more likely to be in marked fluid than points in the bulk, it was argued in [5] that the direct effects of κ are, essentially confined to the core. They affect the bulk only because they are then transferred outwards.

When $\kappa \neq 0$ the equations governing $C(\boldsymbol{x}, t)$ and $\overline{c^2}(\boldsymbol{x}, t)$ which replace (13) and (14) are [10]:

$$\frac{\partial C}{\partial t} + \nabla \cdot (\boldsymbol{U}C) + \nabla \cdot (\overline{\boldsymbol{u}c}) = \kappa \nabla^2 C; \tag{38}$$

$$\frac{\partial \overline{c^2}}{\partial t} + \nabla \cdot (\boldsymbol{U}\overline{c^2}) + \nabla \cdot (\overline{\boldsymbol{u}c^2}) + 2\,\overline{\boldsymbol{u}c} \cdot \nabla C = \kappa \nabla^2 \overline{c^2} - 2\kappa \overline{(\nabla c)^2}. \tag{39}$$

Because of ensemble mean mass conservation the magnitude of C in the bulk must be of order QL^{-3} as it is when $\kappa = 0$ – see (10). The term involving κ in (38) is then negligible in the bulk compared with the other terms. But in the core the magnitude of C is changed by molecular diffusion once L_1 is of order λ_c, and the term involving κ becomes comparable with the other terms. The problem of matching across the core-bulk interface than develops a singular perturbation character. While this could perhaps be handled as far as the equation for C is concerned, there is no obvious way of proceeding with the term $\overline{(\nabla c)^2}$ in equation (39) for $\overline{c^2}$, so the analysis of matching would be thwarted by the normal closure problem. Furthermore, the inherent inhomogeneity and non-stationarity of the situation considered in this paper prevents the use of many of the approximate techniques that have been developed to resolve the closure problem. In particular, second-order closure methods cannot be used because, as Professor B. E. Launder pointed out to the authors, the spectrum of Γ is not "filled" since the distribution in wave-number space is always changing with time.

However the case when the velocity field is a uniform strain can be solved exactly even when $\kappa \neq 0$. It is shown in [5] that the *core* magnitudes of C and $\overline{c^2}$, as well as the *bulk* magnitude of $\overline{c^2}$, depend on the precise details of the uniform strain. As an example, the velocity field which converts spheres into discuses (giving rise to the graphs in Fig. 1 when $\kappa = 0$) gives *core* magnitudes of C and $\overline{c^2}$ as

$$C = O(QL^{-2}\lambda_c^{-1}), \quad \overline{c^2} = O(Q^2L^{-4}\lambda_c^{-2}), \tag{40}$$

while the *bulk* magnitude of $\overline{c^2}$ is of order $Q^2L^{-5}\lambda_c^{-1}$. It is also interesting that the *shapes* of the distributions of both C and $\overline{c^2}$ in the core, and in the bulk, are unaffected by κ in the exact solution. It is possible that this may be true for more general velocity fields, in which case matching may be technically feasible.

Concluding Remarks

The arguments suggesting a core-bulk structure are independent of whether the mean velocity field $\boldsymbol{U}$ is sheared or not. But the detailed shapes of the distributions of $C(\boldsymbol{x}, t)$, $\overline{c^2}(\boldsymbol{x}, t)$ and other statistical properties of $\Gamma(\boldsymbol{x}, t)$ will be affected by the presence of shear when L is comparable with the length over which $\boldsymbol{U}$ changes appreciably. A general discussion of this problem, particularly in the context of estuaries, has been given by the authors in [4]. It is evident, for example, that in such circumstances self-similarity could not always be expected so that the detailed matching would be more complicated than presented here. However there is no reason to suppose that shear will change the *magnitudes* of the statistical properties in any fundamental way except in so far as it changes L. It is also worth stressing that in large scale natural flows, the Reynolds numbers are frequently so high that *relative* (but not absolute) diffusion is largely independent of direct shear effects, since L lies within the range of eddy sizes described by Kolmogoroff's inertial subrange.

Much work on turbulent diffusion has concentrated on cases where Γ is statistically homogeneous in x and statistically stationary in t and, in particular, has used techniques, like spectral analysis, which were developed earlier for describing turbulent velocity fields. Unfortunately most real turbulent diffusion problems do not have these simplifying features, and the authors hope that their work will stimulate others to consider the more realistic problem of a finite cloud considered in this paper. However such research should be based on the special features of a finite cloud [for example, Eq. (4) above, which cannot be written down when Γ is statistically homogeneous in x]. Spectral analysis therefore seems inappropriate and it would seem more promising to try to use generalized Hermite polynomials, as discussed in [8] and [9].

Acknowledgement. Paul J. Sullivan acknowledges the financial support of the National Research Council of Canada during the period when this work was done.

References

1. G. K. Batchelor, "Diffusion in a field of homogeneous turbulence. II. The relative motion of particles". Proc. Cambridge Philos. Soc. **48**, 345–362 (1952)
2. G. K. Batchelor: "The effect of homogeneous turbulence on material lines and surfaces". Proc. R. Soc. London Ser: A, **213**, 349–366 (1952)
3. G. K. Batchelor: Small-scale variation of convected quantities like temperature in turbulent fluid. Part 1. General discussion and the case of small conductivity. J. Fluid Mech. **5**, 113–133 (1959)
4. P. C. Chatwin, P. J. Sullivan: "How Some New Fundamental Results on Relative Turbulent Diffusion can be Relevant in Estuaries and Other Natural Flows", in *Hydrodynamics of Estuaries and Fjords,* ed. by J. C. J. Nihoul (Elsevier, Amsterdam, 1978)
5. P. C. Chatwin, P. J. Sullivan: The relative diffusion of a cloud of passive contaminant in incompressible turbulent flow. J. Fluid Mech. **91**, 337–356 (1979)
6. P. C. Chatwin, P. J. Sullivan: Measurements of concentration fluctuations in relative turbulent diffusion. J. Fluid Mech. **94**, 83–102 (1979)
7. P. C. Chatwin, P. J. Sullivan: "Experiments on the Complete Mixing of a Passive Contaminant in Turbulent Diffusion", in *Proceedings of the Fifth Biennial Symposium on Turbulence,* Rolla, Missouri 1977, ed. by G. K. Patterson, J. L. Zakin (Science Press, Princeton 1979)
8. P. C. Chatwin, P. J. Sullivan: "The Basic Structure of Clouds of Diffusing Contaminant", in *Mathematical Modelling of Turbulent Diffusion in the Environment,* ed. by C. J. Harris (Academic Press, London 1979)
9. P. C. Chatwin, P. J. Sullivan: Some turbulent diffusion invariants. J. Fluid Mech. **97**, 405–416 (1980)
10. G. T. Csanady: *Turbulent Diffusion in the Environment* (Reidel, Dordrecht 1973)
11. A. S. Monin, A. M. Yaglom: *Statistical Fluid Mechanics: Mechanics of Turbulence,* Vol 2, ed. by J. L. Lumley (The M.I.T. Press 1975)
12. L. F. Richardson: Atmospheric diffusion shown on a distance-neighbour graph. Proc. R. Soc. London Ser; A, **110**, 709–737 (1926)
13. P. G. Saffman: On the fine-scale structure of vector fields convected by a turbulent fluid. J. Fluid Mech. **16**, 545–572 (1963)
14. A. A. Townsend: The diffusion of heat spots in isotropic turbulence. Proc. R. Soc. London Ser: A, **209**, 418–430 (1951)

Index of Contributors

Turbulent Shear Flows I

Selected Papers from the First International Symposium on Turbulent Shear Flows. The Pennsylvania State University, University Park, Pennsylvania, USA, April 18–20, 1977

Editors: F. Durst, B. E. Launder, F. W. Schmidt, J. H. Whitelaw

1979. 256 figures, 4 tables. VI, 415 pages
ISBN 3-540-09041-X

Contents:
Free Flows. – Wall Flows. – Recirculating Flows. – Developments in Reynolds Stress Closures. – New Directions in Modelling.

Here is a survey of the latest developments in the calculation of turbulent shear flows, emphasizing their flow and heat transfer properties. The improvement of physical understanding and related measurements is considered essential throughout. Skillfully edited and carefully selected, a third of the papers presented at the symposium have been included in this volume.

There are five distinct themes:
- Free shear flows
- Wall flows
- recirculating flows
- Stress transport models
- Modelling developments

Each section is preceded by an introductory review article contributed by the editors so that the volume provides a major and unified statement on the field. Researchers in most fields of turbulence will find the book a valuable aid in assessing likely trends of future work. It is a volume destined to have a lasting impact on the field.

Springer-Verlag
Berlin
Heidelberg
New York